C0-AKK-483

Ecological Studies, Vol. 84

Analysis and Synthesis

Ecological Studies

Volume 66
Forest Hydrology and Ecology at Coweeta (1987)
Edited by W.T. Swank and D.A. Crossley, Jr.

Volume 67
Concepts of Ecosystem Ecology: A Comparative View (1988)
Edited by L.R. Pomeroy and J.J. Alberts

Volume 68
Stable Isotopes in Ecological Research (1989)
Edited by P.W. Rundel, J.R. Ehleringer, and K.A. Nagy

Volume 69
Vertebrates in Complex Tropical Systems (1989)
Edited by M.L. Harmelin-Vivien and F. Bourliere

Volume 70
The Northern Forest Border in Canada and Alaska (1989)
By J.A. Larsen

Volume 71
Tidal Flat Estuaries: Simulation and Analysis of the Ems Estuary (1988)
Edited by J. Baretta and P. Ruardij

Volume 72
Acidic Deposition and Forest Soils (1989)
By D. Binkley, C.T. Driscoll, H.L. Allen, P. Schoeneberger, and D. McAvoy

Volume 73
Toxic Organic Chemicals in Porous Media (1989)
Edited by Z. Gerstl, Y. Chen, U. Mingelgrin, and B. Yaron

Volume 74
Inorganic Contaminants in the Vadose Zone (1989)
Edited by B. Bar-Yosef, N.J. Barnow, and J. Goldshmid

Volume 75
The Grazing Land Ecosystems of the African Sahel (1989)
By H.N. Le Houérou

Volume 76
Vascular Plants as Epiphytes: Evolution and Ecophysiology (1989)
Edited by U. Lüttge

Volume 77
Air Pollution and Forest Decline: A Study of Spruce *(Picea abies)* on Acid Soils (1989)
Edited by E.-D. Schulze, O.L. Lange, and R. Oren

Volume 78
Agroecology: Researching the Ecological Basis for Sustainable Agriculture (1990)
Edited by S.R. Gliessman

Volume 79
Remote Sensing of Biosphere Functioning (1990)
Edited by R. J. Hobbs and H.A. Mooney

Volume 80
Plant Biology of the Basin and Range (1990)
Edited by B. Osmond, G.M. Hidy, and L. Pitelka

Volume 81
Nitrogen in Terrestrial Ecosystem: Questions of Productivity, Vegetational Changes, and Ecosystem Stability (1990)
By C.O. Tamm

Volume 82
Quantitative Methods in Landscape Ecology: The Analysis and Interpretation of Landscape Heterogeneity (1990)
Edited by M.G. Turner and R.H. Gardner

Volume 83
The Rivers of Florida (1990)
Edited by R. J. Livingston

J.G. Goldammer (Ed.)

Fire in the Tropical Biota

Ecosystem Processes and Global Challenges

With 116 Figures

Springer-Verlag Berlin Heidelberg New York
London Paris Tokyo Hong Kong Barcelona

Dr. Johann Georg Goldammer
Fire Ecology and Fire Management Research Unit
Institute of Forest Zoology
University of Freiburg
Bertoldstraße 17
7800 Freiburg, FRG

ISBN 3-540-52115-1 Springer-Verlag Berlin Heidelberg New York
ISBN 0-387-52115-1 Springer-Verlag New York Berlin Heidelberg

Library of Congress Cataloging-in-Publication Data. Fire in the tropical biota : ecosystem processes and global challenges / J.G. Goldammer (ed.). p.cm. -- (Ecological studies ; vol. 84) Papers from the Third Symposium on Fire Ecology held at Freiburg University in May 1989 and sponsored by the Volkswagen Foundation. Includes bibliographical references and index. ISBN 0-387-52115-1 (acidfree paper) 1. Fire ecology–Tropics–Congresses. 2. Wildfire–Tropics–Congresses. 3. Biotic communities–Tropics–Congresses. 4. Botany–Tropics–Congresses. 5. Fires–Tropics–Congresses. I. Goldammer, J.G. (Johann Georg), 1949– . II. Symposium on Fire Ecology (3rd : 1989: Freiburg University) III. Volkswagen Foundation. IV. Series: Ecological Studies ; v. 84. QH545.F5F575 1990 574.5′2623--dc20 90–10007

Printed in Germany

Typesetting: International Typesetters Inc., Makati, Philippines
2131/3145(3011)-543210 – Printed on acid-free paper

Preface

In 1977, the Volkswagen Foundation sponsored the first of a series of International Symposia on Fire Ecology at Freiburg University, Federal Republic of Germany. The scope of the congresses was to create a platform for researchers at a time when the science of fire ecology was not yet recognized and established outside of North America and Australia. Whereas comprehensive information on the fire ecology of the northern boreal, the temperate, and the mediterranean biotas is meanwhile available, it was recognized that considerable gaps in information exist on the role of fire in tropical und subtropical ecosystems. Thus it seemed timely to meet the growing scientific interest and public demand for reliable and updated information and to synthesize the available knowledge of tropical fire ecology and the impact of tropical biomass burning on global ecosystem processes.

The Third Symposium on Fire Ecology, again sponsored by the Volkswagen Foundation and held at Freiburg University in May 1989, was convened to prepare this first pantropical and multidisciplinary monograph on fire ecology[1]. The book, in which 46 scientists cooperated, analyzes those fire-related ecosystem processes which have not yet been described in a synoptic way. Following the editor's concept, duplication at previous efforts in describing tropical vegetation patterns and dynamics was avoided. Extensive bibliographical sources are given in the reference lists of the chapters.

The contributions in this book cover tropical and subtropical terrestrial fire ecology, fire history, climatology, atmospheric chemistry, and remote sensing. The fire-soil and fire-wildlife interactions will be covered in a separate volume. The main emphasis in this book is laid on the analysis of the dual role of fire in tropical vegetation. It is recognized that natural and anthropogenic fire regimes in the tropics are undergoing dramatic changes. The increase of fire frequency, together with the enormous demand for fuelwood and

---

[1]The contributions on fire ecology in Mediterranean and northern biotas given in the first part of the symposium are published separately (J.G. Goldammer and M.J. Jenkins [eds.] 1990. Fire in Ecosystem Dynamics Mediterranean and Northern Perspectives. SPB Academic Publ., The Hague).

for agricultural and grazing lands, have led to severe degradation processes in forests and other vegetation. Vast amounts of aerosol and gaseous emissions from savanna fires and forest conversion have considerable impact on the chemistry of the atmosphere. The regional contributions in the first part of the book cover the terrestrial fire ecology in the main vegetation types of tropical Asia, America, and Australia. Information on advanced fire management in the savanna and grassland biomes comes from Southern Africa. Detailed information is also given on the use and the effects of prescribed fire in the management of industrial plantations with introduced pine species.

The prehistorical and historical views on wildland fire intend to clarify the role of man in the development of tropical fire regimes. Three chapters on remote sensing of tropical fires provide the most recent knowledge on the state of the art and highlight the perspectives and needs of monitoring tropical fires from space. In three contributions substantial emphasis is placed on the emissions from tropical biomass burning, the injection of trace gases into the atmosphere, and on the formation of photochemical smog. The final chapter deals with scenarios of vegetation and wildfire response to a warming global climate expected to develop due to technological emissions and due to conversion and burning of tropical vegetation.

In the *Freiburg Declaration on Tropical Fires* the scientists participating at the Symposium underscored the need for immediate action to reduce tropical forest conversion and vegetation degradation by fire. In the declaration it was also pointed out that the International Geosphere-Biosphere Program (IGBP) should serve as a vehicle for international multidisciplinary research in global fire ecology. The initiative is taken by this book.

J.G. GOLDAMMER

Contents

1 Fire in Tropical Ecosystems and Global Environmental Change: An Introduction
D. MUELLER-DOMBOIS and J.G. GOLDAMMER (With 1 Figure) . . . 1

1.1 Introduction . . . 1
1.2 Deforestation in the Tropics . . . 2
1.3 Patterns of Succession . . . 3
1.4 Changes in Fire Regimes and Biodiversity . . . 4
1.5 The Future of Tropical Forests and Forestry . . . 5
1.6 Global Impacts on Climate and Soil . . . 7
1.7 Conclusions . . . 9
References . . . 9

2 The Impact of Droughts and Forest Fires on Tropical Lowland Rain Forest of East Kalimantan
J.G. GOLDAMMER and B. SEIBERT (With 6 Figures) . 11

2.1 Introduction . . . 11
2.2 Climatic Variability and Fire Regimes . . . 12
2.3 The 1982-83 ENSO, its Predecessors and the Wildfires . . . 16
2.3.1 The 1982-83 ENSO . . . 16
2.3.2 Predecessors . . . 19
2.3.3 The Wildfire Scenario in 1982-83 . . . 21
2.4 Forest Regeneration after the 1982-83 Fires . . . 23
2.4.1 The Regeneration Process . . . 25
2.4.1.1 Primary Forest . . . 25
2.4.1.2 Logged-Over Forest . . . 26
2.5 Conclusions . . . 27
References . . . 28

3 The Role of Fire in the Tropical Lowland Deciduous Forests of Asia
P.A. STOTT, J.G. GOLDAMMER, and W.L. WERNER (With 5 Figures) . . . 32

3.1 Fire: An Alien Ecological Pressure? . . . 32

3.2 Fire Patterns in Time and Space 34
3.2.1 Timing and Origins 34
3.2.2 Fire and Fuel 36
3.3 Fire Management 41
References 43

4 Fire in the Pine-Grassland Biomes of Tropical and Subtropical Asia
J.G. GOLDAMMER and S.R. PEÑAFIEL
(With 6 Figures) 45

4.1 Introduction 45
4.2 Adaptive Traits of Tropical Pines to Fire 46
4.2.1 Character of Bark 47
4.2.2 Rooting Habit 48
4.2.3 Basal Sprouting 50
4.2.4 Site and Fuel Characteristics 50
4.3 Origin and Extent of Fires 50
4.4 Management Considerations 52
4.4.1 Distribution of the Pine-Grassland Fire Climax in Luzón 52
4.4.2 Main Ecological Challenges 55
4.4.2.1 Grass Species Composition and Site Degradation . . 55
4.4.2.2 Fire-Host Tree-Insect Interactions 58
4.5 Conclusions and Outlook 59
References 60

5 Fire in Some Tropical and Subtropical South American Vegetation Types: An Overview
R.V. SOARES (With 6 Figures) 63

5.1 Introduction 63
5.2 Tropical Rain Forest 64
5.3 Trade Wind Forest of Venezuela and Columbia . . 65
5.4 "Babaçu" Palm Forest 66
5.5 Steppe 67
5.6 Savanna or "Cerrado" 67
5.7 Coastal Rain Forest 67
5.8 Subtropical Forest 68
5.9 Brazilian Pine Forest 69
5.10 Steppic Savanna or "Chaco" 74
5.11 Subtropical and Araucaria Forests of Chile and Argentina 75
5.12 Nonforested Areas 76
5.13 Exotic Planted Forests 77
References 80

6 Fire in the Ecology of the Brazilian Cerrado
L.M. COUTINHO (With 8 Figures) 82

6.1 Introduction 82
6.2 Regimes and Causes of Fire in the Cerrado 86
6.3 The Abiotic Effects of Fire 88
6.3.1 Air Temperature 88
6.3.2 Soil Temperatures 88
6.3.3 Cycling of Mineral Nutrients 90
6.4 The Biotic Effects of Fire 96
6.4.1 Resistance to Fire 96
6.4.2 Primary Productivity 97
6.4.3 Stability of the Vegetation 98
6.4.4 Flowering 99
6.4.5 Dispersion of Seeds 100
6.4.6 Germination of Seeds 100
6.4.7 Fire and Fauna 100
6.5 Management by Fire 101
References 103

7 Fire in the Tropical Rain Forest of the Amazon Basin
P.M. FEARNSIDE (With 1 Figure) 106

7.1 Ancient and "Natural" Fires 106
7.2 Deforestation and Burning in Amazonia Today . . . 107
7.3 Types and Qualities of Burning 109
7.4 Impacts of Burning on Amazonian Vegetation . . . 112
7.5 Indirect Effects of Burning 114
References 115

8 Interactions of Anthropogenic Activities, Fire, and Rain Forests in the Amazon Basin
J.B. KAUFFMAN and C. UHL (With 1 Figure) 117

8.1 Introduction 117
8.2 The Fire Environment 118
8.2.1 Fire History of Tropical Rain Forests 118
8.2.2 Fuel Biomass and Arrangement 119
8.2.2.1 Defining Tropical Rain Forest Fuels 119
8.2.2.2 Variability in Fuel Loads 119
8.2.2.3 Effects on Disturbance on Fuel Biomass 120
8.2.3 Microclimates and Fire in Tropical Rain Forests . . 121
8.2.4 Susceptibility of Tropical Rain Forest Ecosystems to Fire 123
8.2.5 Fire Behavior and Biomass Consumption in Tropical Rain Forests 125
8.3 Vegetation Adaptations and Responses to Fire . . . 125

8.3.1 Bark Properties 126
8.3.2 Anomalous Arrangement of Stem Tissues 127
8.3.3 Vegetative Sprouting 127
8.3.4 Seedbanks 129
8.3.5 Dispersal Mechanisms as Adaptations for Fire Survival 130
8.3.6 Fire-Enhanced Flowering 131
8.4 The Winners and the Losers 131
References 133

9 Social and Ecological Aspects of Fire in Central America
A.L. KOONCE and A. GONZÁLES-CABÁN
(With 5 Figures) 135

9.1 Introduction 135
9.2 Socio-Economic Factors 137
9.3 Tropical Forest Resources 138
9.4 Fire in the Tropical Forests of Central America . . . 141
9.4.1 Fire Effects on Soils 143
9.4.2 Fire Effects on Pine Forests 146
9.4.3 Fire Effects on Dry Forests 150
9.4.4 Fire Effects on the Aripo Savannas 154
9.4.5 Fire Effects on Montane Forests 154
9.5 Closing Remarks 155
References 156

10 Fires and Their Effects in the Wet-Dry Tropics of Australia
A.M. GILL, J.R.L. HOARE, and N.P. CHENEY
(With 4 Figures) 159

10.1 Introduction 159
10.2 Location and Landscape 160
10.3 Proneness to Fires and the Fire's Characteristics . . 161
10.3.1 Fuels 161
10.3.2 Fire Climate 162
10.3.3 Ignition Sources and Fire Frequencies 162
10.3.4 Fire Characteristics 164
10.4 Fire's Impact on Plants: Demographic Aspects . . . 165
10.5 Fire and Communities 168
10.5.1 Eucalypt Forests and Woodlands 168
10.5.2 "Fire-Sensitive" Communities 170
10.6 Fire Management 172
10.6.1 National Parks 172
10.6.2 Cattle Raising 173
10.6.3 Invasive Plants 174

10.6.4 Emissions 175
10.7 Conclusions 175
References 176

11 Fire Management in Southern Africa: Some Examples of Current Objectives, Practices, and Problems
B.W. van WILGEN, C.S. EVERSON, and W.S.W. TROLLOPE (With 13 Figures) 179

11.1 Introduction 179
11.2 Major Vegetation Types of Southern Africa 179
11.3 Management of Southern African Areas Using Fire 181
11.4 Fynbos Catchments in the Western Cape Province . 182
11.4.1 Aims of Management 182
11.4.2 Fire Frequency 185
11.4.3 Fire Season 189
11.4.4 The Control of Alien Woody Weeds in Fynbos Catchments 190
11.4.5 Wildfires as a Complicating Factor in Prescribed Burning 191
11.5 Grassland Catchments in the Natal Drakensberg . . 192
11.5.1 Aims of Management 192
11.5.2 Fire Regime 192
11.5.3 Fire Frequency 193
11.5.4 Fire Season 195
11.5.5 Prescribed Burning 196
11.6 Fire in Savannas: Basic Principles 198
11.6.1 Natural and Modified Fire Regimes 198
11.6.2 Grass/Bush Dynamics 200
11.6.3 The Importance of Fire Intensity 200
11.7 Agricultural Areas in the Eastern Cape 202
11.7.1 Aims of Management 202
11.7.2 Research Background 202
11.7.3 Current Management 202
11.8 The Hluhluwe/Umfolozi Game Reserves Complex 203
11.8.1 Aims of Management 203
11.8.2 Background 204
11.8.3 Current Fire Management 205
11.9 The Pilanesberg National Park 205
11.9.1 Aims of Management 205
11.9.2 Research Background 206
11.9.3 Current Management 206
11.10 The Kruger National Park 207

11.10.1 Aims of Management 207
11.10.2 Research Background 207
11.10.3 Management 208
11.11 The Etosha National Park 210
11.11.1 Aims of Management 210
11.11.2 Fire Management 210
11.12 Conclusions 211
References 212

12 Prescribed Fire in Industrial Pine Plantations
C. de RONDE, J.G. GOLDAMMER, D.D. WADE, and R.V. SOARES (With 9 Figures) 216

12.1 Introduction 216
12.2 Prescribed Burning Objectives 218
12.2.1 Wildfire Hazard Reduction 218
12.2.2 Prepare Sites for Planting 220
12.2.3 Other Objectives 220
12.3 Fuel Appraisal 221
12.3.1 Natural Vegetation 222
12.3.2 Available Fuel 223
12.3.3 Fuel Moisture 225
12.3.4 Evaluating Fuel Inputs 226
12.4 Weather and Topographic Considerations 227
12.4.1 Wind 228
12.4.2 Relative Humidity 230
12.4.3 Temperature 231
12.4.4 Precipitation and Soil Moisture 232
12.4.5 Slope 233
12.5 Fire Behavior Prediction 233
12.5.1 Descriptors 234
12.5.2 Fire Behavior Models 236
12.5.3 Predicting Crown Scorch Height 236
12.6 Prescribed Burning Techniques 237
12.6.1 Backing Fire 238
12.6.2 Strip-Heading Fire 240
12.6.3 Point Source (Grid) Ignition 240
12.6.4 Edge Burning 242
12.6.5 Center and Circular (Ring) Firing 243
12.6.6 Pile and Windrow Burning 243
12.7 Prescribed Burning Plans 244
12.7.1 The Written Plan 245
12.7.2 Preparing for the Burn 250
12.7.3 Executing the Burn 250
12.7.4 Evaluating the Burn 251

12.8 Fire Effects 252
12.8.1 Effects on Trees 252
12.8.2 Effects on Woody and Herbaceous Understory Vegetation 257
12.8.3 Effects on Forest Floor Dynamics 258
12.8.4 Effects on Soil 261
12.9 Conclusions 265
References 265

13 Landscapes and Climate in Prehistory: Interactions of Wildlife, Man, and Fire
W. SCHÜLE (With 2 Figures) 273

13.1 Introduction 273
13.2 Natural Fires 275
13.2.1 Vegetation, Herbivores, and Fire 278
13.2.2 Fire and Evolution 278
13.2.3 Fire and Mammals 280
13.2.4 Fire and Hominoidea 281
13.2.4.1 Neogene Africa 282
13.2.4.2 End-Tertiary African Primates 285
13.2.4.3 Plio-Pleistocene Hominidae and Megaherbivores . . 287
13.3 *Homo* sp. and Fire 288
13.3.1 Hominid Use of Fire in Africa 288
13.3.2 Fire and the Evolution of the Brain 290
13.3.3 Increased Fire Frequency and Climate 290
13.4 Anthropogenic Fire in Eurasia 291
13.4.1 Tropical and Subtropical Eurasia 291
13.4.2 Old World Temperate Zones and the Eutrophic Line 292
13.4.3 Man and Landscapes in the Hinterland 293
13.4.3.1 The Pre-Agricultural Mediterranean Region 294
13.4.4 Life and Fire Outside the Eutrophic Line 295
13.5 The Conquest of the Forbidden Countries 295
13.5.1 Celebes-Sulawesi, the Philippines, and the Wallacean Islands 296
13.5.2 Australia, New Guinea, and Tasmania ("Sahul") 296
13.5.3 America 298
13.5.4 Madagascar 301
13.5.5 New Zealand 303
13.5.6 Small Off-Shore Islands 304
13.6 Agriculture and Domestic Ungulates 306
13.7 Conclusions 310
References 315

14 Fire Conservancy: The Origins of Wildland Fire Protection in British India, America, and Australia
S.J. PYNE 319

14.1 Introduction 319
14.2 Home Fires: A Synoptic Fire History of Britain . . . 320
14.3 An Empire Strategy: Fire Protection in British India 323
14.4 An American Strategy: Systematic Fire Protection . 326
14.5 An Australian Strategy: Bringing System to Burning Off 330
14.6 Stirring the Ashes: Concluding Thoughts 333
References 335

15 The Contribution of Remote Sensing to the Global Monitoring of Fires in Tropical and Subtropical Ecosystems
J.-P. MALINGREAU (With 9 Figures) 337

15.1 Introduction 337
15.2 Satellite Monitoring of Vegetation Dynamics 338
15.3 Fires in Vegetation – Data Needs 341
15.4 Fire Detection Using the AVHRR Instrument . . . 343
15.5 Fires and Environmental Conditions 345
15.6 Post-Fire Landscapes 350
15.7 Fire in Tropical Ecosystems 351
15.7.1 The Amazon Basin 351
15.7.2 South East Asia 356
15.7.3 Africa 358
15.7.4 West Africa 359
15.7.5 Central Africa 362
15.7.5.1 Transition Rain Forest – Seasonal Forest-Woodland Savanna 363
15.7.5.2 The Central Congo Basin 365
15.8 A Global Fire Monitoring System: Conclusions . . . 366
References 368

16 Remote Sensing of Biomass Burning in the Tropics
Y.J. KAUFMAN, A. SETZER, C. JUSTICE, C.J. TUCKER, M.G. PEREIRA, and I. FUNG (With 12 Figures) . . . 371

16.1 Introduction 371
16.2 The NOAA-AVHRR Series 378
16.3 Remote Sensing of Fires, Smoke, and Trace Gases . 378
16.4 Remote Sensing of Aerosol Characteristics 382
16.5 Satellite Estimation of Gaseous Emission from Biomass Burning 383

16.5.1 Estimate Based on the Average Emission of Particulates per Fire 383
16.5.1.1 Basic Assumptions 384
16.5.1.2 Estimation of the Emission Rates per Fire 384
16.5.1.3 Remote Sensing of Fires and Total Emitted Mass . . 386
16.5.1.4 Accuracy Estimates 386
16.5.1.5 Application of the Techniques 387
16.5.2 Estimate Based on Average Biomass Burned per Fire 391
16.6 Discussion 394
16.7 Conclusions 396
References 397

17 NOAA-AVHRR and GIS-Based Monitoring of Fire Activity in Senegal – a Provosional Methodology and Potential Applications
P. FREDERIKSEN, S. LANGAAS, and M. MBAYE (With 7 Figures) 400

17.1 Introduction 400
17.2 Methodology 401
17.2.1 Definition of a Scene Model 402
17.2.2 Field Radiometric Measurements 403
17.2.3 Integrated Camera and Radiometer Measurements . 404
17.2.4 AVHRR Image Processing and Field Verification 405
17.2.5 GIS Manipulation 406
17.3 Results and Discussion 406
17.3.1 The Spectral Evolution of a Burned Area 406
17.3.2 Fractional Cover Burned 409
17.3.3 Preliminary Bushfire Statistics 410
17.4 Conclusions and Further Work 415
References 416

18 Factors Influencing the Emissions of Gases and Particulate Matter from Biomass Burning
D.E. WARD (With 9 Figures) 418

18.1 Introduction 418
18.2 Forest Fuels Chemistry 419
18.3 Combustion Processes 421
18.4 Smoke Production 422
18.4.1 Release of Carbon 422
18.4.2 Formation of Particles 423
18.4.3 Fuel Chemistry Effects on Particle Formation . . . 424
18.4.3.1 Particle Number and Volume Distribution 425

18.4.3.2 Emission Factors for Particulate Matter 426
18.4.4 Emissions of Gases 430
18.4.4.1 Nitrogen Gases 430
18.4.4.2 Sulfur Emissions (Carbonyl Sulfide) 432
18.4.4.3 Methyl Chloride 432
18.4.4.4 Carbone Monoxide 433
18.4.4.5 Methane and Nonmethane Hydrocarbons 433
18.5 Summary 434
References 435

19 Ozone Production from Biomass Burning in Tropical Africa. Results from DECAFE-88
M.O. Andreae, G. Helas, J. Rudolph, B. Cros, R. Delmas, D. Nganga, and J. Fontan (With 1 Figure) 437
References 439

20 Estimates of Annual and Regional Releases of CO_2 and Other Trace Gases to the Atmosphere from Fires in the Tropics, Based on the FAO Statistics for the Period 1975-1980
Wei Min Hao, Mei-Huey Liu, and P.J. Crutzen (With 1 Figure) 440
20.1 Introduction 440
20.2 Computational Approach 442
20.3 Results 445
20.3.1 Closed Nonfallow Forests 445
20.3.2 Closed Fallow Forests 445
20.3.3 Open Nonfallow Forests 448
20.3.4 Open Fallow Forests 450
20.3.5 Grass Layer of Open Forests (Humid Savannas) 450
20.3.6 Distribution of CO_2 Emissions in 5° x 5° Grid Cells 452
20.4 Discussion 456
20.5 Conclusions 459
References 460

21 Global Change: Effects on Forest Ecosystems and Wildfire Severity
M.A. Fosberg, J.G. Goldammer, D. Rind, and C. Price (With 11 Figures) 463
21.1 Introduction 463
21.2 Scientific Bases for the Greenhouse Effect 463

21.3 Modeling the Atmospheric Response to the Greenhouse Effect 466
21.4 Coupling the Biosphere to the Geosphere 467
21.5 Sensitivity of the Ecosystem Forecasts to Uncertainties in the GCM's 468
21.6 Ecosystem Stresses Which Have Not Been Included 472
21.7 Ecosystems and Potential Fire Behavior 474
21.8 The Effect of Climate Change on Lightning 476
21.9 Regional Predictions: Impact of Climate Change on Distribution of Forest Biomes and Other Effects 478
21.9.1 North America 478
21.9.2 The Tropics 482
21.10 The Future 482
References 484

Appendix: The Freiburg Declaration on Tropical Fires 487

Subject Index 491

List of Contributors

You will find the addresses at the beginning of the respective contribution

Andreae, M.O. 437
Cheney, N.P. 159
Coutinho, L.M. 82
Cros, B. 82,437
Crutzen, P.J. 440
Delmas, R. 437
de Ronde, C. 216
Everson, C.S. 179
Fearnside, P.M. 106
Fontan, J. 437
Fosberg, M.A. 463
Frederiksen, P. 400
Fung, I. 371
Gill, A.M. 159
Goldammer, J.G. 1,11,32, 45,216,463
Gonzáles-Cabán, A. 135
Helas, G. 437
Hoare, J.R.L. 159
Justice, C. 371
Kauffman, J.B. 117
Kaufman, Y.J. 371
Koonce, A.L. 135
Langaas, S. 400
Malingreau, J.-P. 337
Mbaye, M. 400
Mei-Huey Liu 440
Mueller-Dombois, D. 1
Nganga, D. 437
Peñafiel, S.R. 45
Pereira, M.C. 371
Price, C. 463
Pyne, S.J. 319
Rind, D. 463
Rudolph, J. 437
Schüle, W. 273
Seibert, B. 11
Setzer, A. 371
Soares, R.V. 63,216
Stott, P.A. 32
Trollope, W.S.W. 179
Tucker, C.J. 371
Uhl, C. 117
van Wilgen, B.W. 179
Wade, D.D. 216
Ward, D.E. 418
Wei Min Hao 440
Werner, W.W. 32

1 Fire in Tropical Ecosystems and Global Environmental Change: An Introduction

D. MUELLER-DOMBOIS[1] and J.G. GOLDAMMER[2]

1.1 Introduction

Fire has always been an important ecological stress factor in the seasonal tropics. The dry deciduous forest and the savanna grasslands, in general, have evolved with fire. The plant life forms of these tropical biomes can cope with fire through various adaptive traits (Budowski 1966).

No such categorical statement can be made about the humid or wet tropics. Under the present climatic conditions, fire plays a less important role in the year-round wet biome of the undisturbed equatorial rain forest. The permanently warm temperature, in combination with the continuously high atmospheric and soil moisture levels, has created a natural greenhouse climate which encourages high rates of decomposition. Sudden accumulations of organic waste on the forest floor can occur, but they are usually spotty, noncontinuous, and transient. Such accumulations are brought about by certain rain forest trees that drop their foliage during flowering and fruiting or through trees that break down and create tree-fall gaps. For this reason, fire, which could start naturally from lightning, has only a small chance of gaining access into primary rain forest.

There are, however, exceptions related to climatic oscillations and to larger-scale disturbances. Recent investigations have shown that climatic changes between the Pleistocene and today have created drought conditions that have favored the spread of forest fires in long-return intervals (Sanford et al. 1985; Uhl et al. 1988; Goldammer and Seibert 1989 this Vol.).

Certain rain forests, particularly in the subequatorial tropics (from 10 to 23° latitude), are frequently subject to hurricane damage. Such broader-area damage promotes the invasion of vines, which can contribute to foliar biomass accumulation on the soil surface in forest openings, particularly during occasional dry spells. Also some extreme soils, those with imbalanced nutrients and toxic elements, can slow or stall decomposer activity and thereby permit accumulation of spatially continuous fuel loadings that can lead to fires in tropical rain forests.

The spread of frequent fires into the wet tropics is a more recent development, attributable largely to human population pressures. During the last two decades, the use of fire has dramatically increased with the rapid

[1] Department of Botany, University of Hawaii at Manoa, Honolulu, Hawaii 96822, USA
[2] Department of Forestry, University of Freiburg, 7800 Freiburg, FRG

clearing of tropical rain forests for economic development. This in turn has promoted the geographic expansion of the pyrophytic grass life form (Mueller-Dombois 1981; Uhl et al. 1982; Goldammer and Peñafiel this Vol.).

Fire has recently assumed a prominent role in the wet tropics with ecological consequences that will impoverish biodiversity and the floristic potential of the humid tropics, decrease its future forest and agricultural potential, promote weed-grass invasion, negatively impact the soil resource, and modify the earth's atmosphere.

Although qualitatively a predictable degradation, there are at present many uncertainties about this new quantitative impact of fire in tropical and subtropical ecosystems.

As a general introduction, we may focus briefly on five topics that appear to be of particular relevance in this context: (1) deforestation in the tropics, (2) patterns of succession, (3) changes in fire regime and biodiversity, (4) the future of tropical forests and forestry, and (5) global impacts on climate and soil.

It is an objective of this book to review current knowledge and to develop hypotheses that should be addressed by interdisciplinary research under the International Geosphere-Biosphere Program (IGBP). The aim is to deepen our understanding of the ecology of fire and to promote future management policies that are based on sound ecological principles.

1.2 Deforestation in the Tropics

Many estimates exist as to how rapidly the tropical forest is disappearing (Myers 1980; FAO 1982). Woodwell (1984) presented a list that shows the rate of forest clearing to vary widely from 33,000 to 201,000 km^2 per year. A symposium held in Woods Hole in 1978 had the principal objective of clarifying the role of terrestrial vegetation in the global carbon cycle. Short of resolving the problem, the symposium began to look in more detail at the use of satellite imagery for determining the rate of forest loss. Many complications were recognized and an improved methodology suggested.

More recently, Lugo (1988) decried the great variation of these estimates and their lack of proper substantiation. Myers (1988) presented statistics which reasonably match earlier average values and were said to be based on improved estimates.

Out of an area of potentially 15 to 17 million km^2 tropical moist forest, which approximates 7% of the world's forests, 9.5 million km^2 were still forested in the early 1970's. The average annual rate of loss, according to Myers, was estimated as 95,000 km^2 yr^{-1}. This amounts to 1% annual loss of forest. However, great variations exist between different areas in the tropics. Several are recognized as "hotspots" of destruction (Table 1).

The second aspect of concern is the loss of species. According to Raven (1988), a world estimate is 250,000 terrestrial plant species. Of these, two thirds

Table 1. Examples of "hotspots" of forest destruction in tropical Africa and Asia. (Myers 1988)

	Area of Original primary forest area (km²)	Area left in 1988 (%)	Species Total number of original plant species	Estimated as being eliminated (%)
Madagascar	62,000	16	6,000	82
Brazil, NE Coast	1,000,000	2	10,000	50
West Ecuador	27,000	9	10,000	25

are said to occur in the tropics, and 25,000 are expected to disappear by the year 2000. This amounts to a loss rate of 2000 to 2100 plant species per year or approximately five species per day.

1.3 Patterns of Succession

In most temperate forests throughout North America and Northern Europe, fire is recognized as one of the most important factors initiating a new ecological succession (Kozlowski and Ahlgren 1974; Mooney et al. 1981; Goldammer and Jenkins 1990). Thus fire successions are well known as natural rejuvenation processes in temperate forests. In the humid tropics, forest rejuvenation is now widely seen as resulting from "patch dynamics" (Pickett and White 1985); that is, from small area succession patterns created in the absence of fire by an occasional tree breakdown in the otherwise green matrix of primary forest. While this may apply to many multi-species tropical lowland forests in the equatorial belt, tropical rain forests in the subequatorial belt (between 10 and 23° latitude) are also subjected to larger area destructions by hurricanes, for example the Australian rain forest in North Queensland.

In these more broadly disturbed areas, successional patterns include the spread of vines, often as a dominant life form. With variations in weather patterns, the vine life form can become a significant source of fuel loading for natural fires.

In terms of vegetation structure, successional patterns in forests are seen generally as starting with a sudden reduction in standing biomass. This is then followed by a rebuilding of biomass over time.

A natural secondary succession can be described (after Woodwell 1984) as following four general sequential patterns. Starting with denudation (non-forest) the succession proceeds through a herb stage into a shrub stage and back to forest as follows:

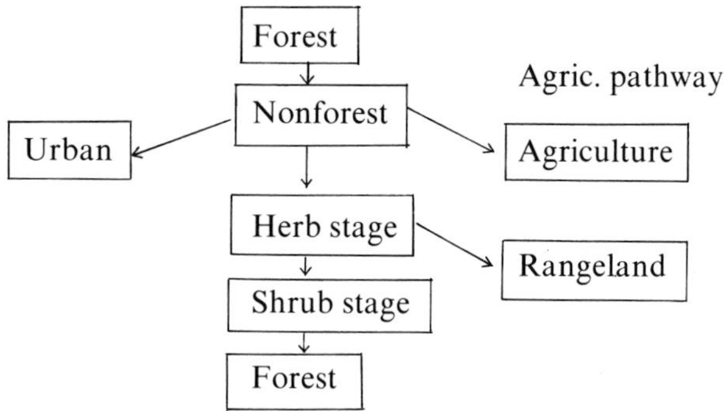

With interference through human activity, however, a few more patterns become evident. The most important diversion is through an agricultural pathway, whereby a rebuilding of biomass will be restricted in various ways.

The time in each stage leads to an important question for research in landscape ecology. In most situations, human influences create great departures from the idealized natural patterns of "patch dynamics".

The spatial factor is extremely important here and has rarely been given adequate attention. In general, the larger the disturbance, the longer the successional stage, and very large area disturbances can reach a point of no return (Mueller-Dombois and Spatz 1975; Mueller-Dombois 1981).

1.4 Changes in Fire Regime and Biodiversity

When fire is used for slash-burning in tropical rain forests after logging, herbaceous weeds will soon become dominant. Among these, pyrophytic grasses assume a prominent role. These are caespitose (bunch-forming) and/or stoloniferous (mat-forming) grasses. Some are annuals, such as *Imperata cylindrica* in tropical Asia or *I. brasiliensis* in tropical America. Because of their annual life cycle, these grasses dry up after flowering and seeding and then form an acute fire hazard by means of their dead shoot material. They re-invite fire. Similarly, perennial bunch- and mat-forming grasses, typified, for example, by *Hyparrhenia rufa* and *Melinis minutiflora*, which originated in the seasonal tropics, often maintain their evolutionarily induced phenological dormancy by drying up, at least partially, during the time when the dry season occurs in the neighboring seasonal tropics. Moreover, as short-lived perennials they accumulate dry shoot material with age, and finally lose vigor. At that time they cause an increased fire hazard in their senescing life-phase. This life-phase is also favorable for the re-invasion of woody plants, because competition by grasses is then at a minimum. A shift to woody vegetation at this time depends on the availability of disseminules of fast-growing woody perennials. This is a precarious stage, because fire can easily recur and destroy young woody plants. Fire can also rejuvenate senescing grasses by removal of their dead shoot material without destroying their roots. In this way "savannization" of tropical rain forest can become a self-perpetuating process.

A further retrogression to permanent savanna or a return to forest depends on two factors: (1) the presence/absence of sources of ignition, and (2) the availability of suitable woody pioneers to continue the process of succession to forest.

The availability of suitable woody colonizers becomes questionable with an increasing size of disturbed area and the concomitant loss of woody species adapted for colonizing grass-invaded areas. Two tropical areas can serve as opposite examples.

In the Hawaiian Islands, the progression of introduced pyrophytic grasses into native woody vegetation is currently proceeding even without the aid of fire (Smith and Tunison 1989); but when fire follows, there is a further explosive invasion and densification of such grasses. The natural lack of native woody colonizers able to cope with grass-invaded areas is a major cause of this "savannazation" process. In contrast, the Australian rain forest in North Queensland, which has been widely retrenched by aboriginal use of fire, is rapidly regaining territory, replacing sclerophyll (mixed eucalypt and acacia) forests, where these are protected from fire (Ellis 1985). Again, the availability of suitable woody colonizers is the key to the reconversion of savanna woodland to rain forest.

Thus, impoverishment of biodiversity, in the case of woody species capable of invading grasslands and those others completing the successional return to forest, can become the limiting factor in the recovery process.

Figure 1 summarizes the ecological and anthropogenic parameters influencing the various tropical and subtropical fire regimes (Goldammer 1986).

1.5 The Future of Tropical Forests and Forestry

Where large-area forest destruction has led to loss of the natural recovery process as, for example, in the Talasiga (the "sunburned land") of Fiji, the only course to forest reestablishment is plantation forestry. Here, the "Fiji Pine Commission" has launched a great effort to re-forest the vast Talasiga grasslands of western Viti and Vanua Levu with *Pinus caribaea*. This operation has been successful, at least in part. However, the pines do not reduce the constantly looming fire hazard, and many expensively established plantations go up in flames every year.

Cattle are being used to keep the mission grass (*Pennisetum polystachyon*) down; but this grass is palatable only in its young stage and after fire. It continues to grow among the pine trees and, together with the accumulating pine needles, combines into a continuous fuel complex. Clearly, suppression of this grass is not enough. As pine stands mature, their needle litter accumulates together with branches from self-thinning, and thus the fire hazard increases with stand age. Thus, a high fire hazard remains a constant threat to the reforestation effort.

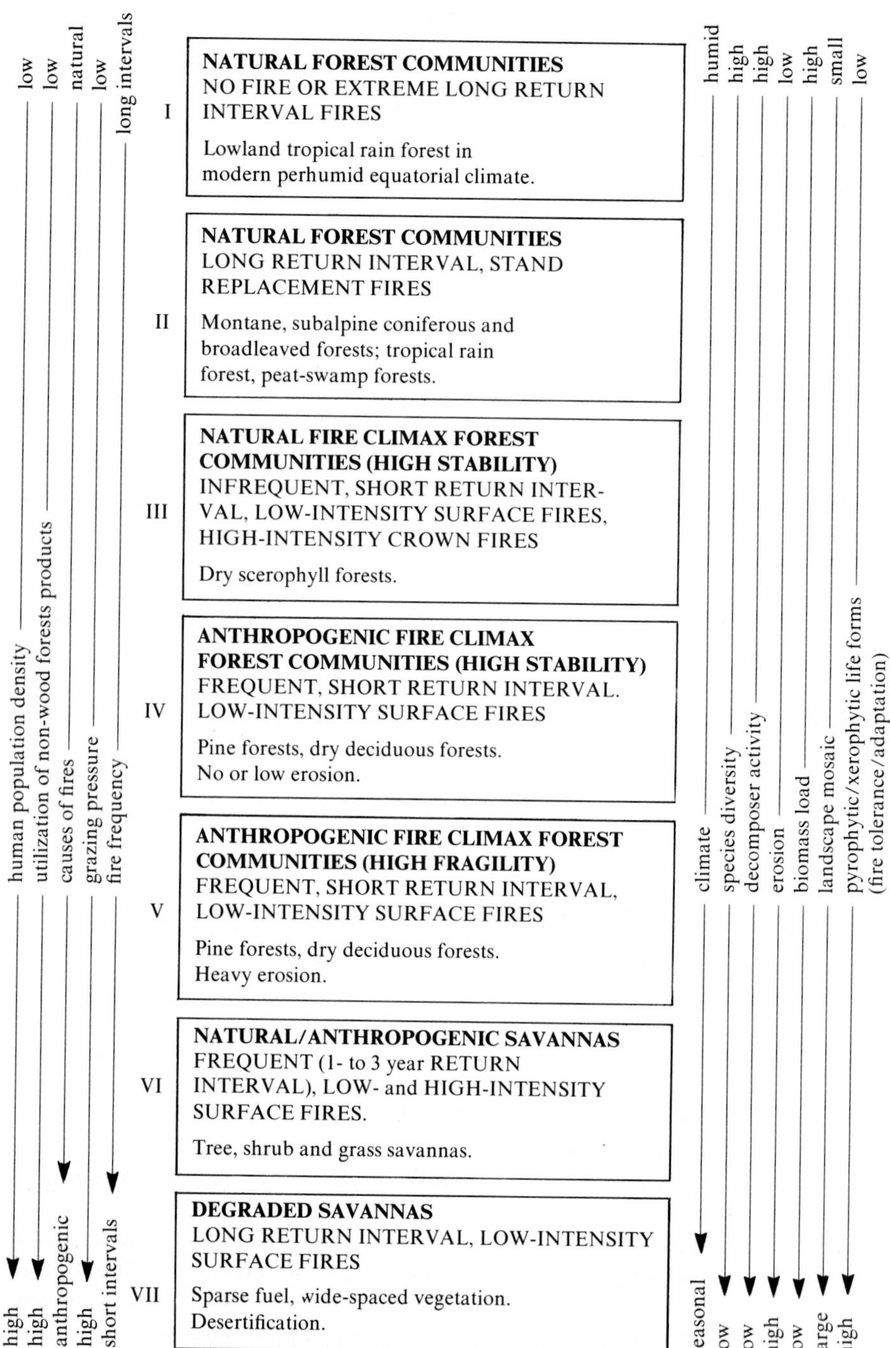

Fig. 1. Types of tropical/subtropical fire regimes as related to ecological and anthropogenic gradients. Exemptions from this generalized scheme such as higher species diversity in certain fire-climax communities must be noted. (After Goldammer 1986)

This is not an isolated case. The creation of large plantations is commonly done with species that produce poorly decomposable litter. Eucalypts are another example of preferred plantation trees in the subtropics that create their own fire regime in areas which had no fire hazard under natural conditions.

The new forest demography promoted through industrial forestry operations becomes a consequence of large-area forest destruction. Forests similar to those favored by industrial forestry have developed naturally in biogeographically isolated islands and mountains (Sprugel 1976; Wardle and Allen 1983; Kohyama 1988; Itow and Mueller-Dombois 1988). Here colonizing species evolved that form large cohort stands following catastrophic stand breakdowns. These biogeographically impoverished island and mountain forests are characterized by low canopy species diversity and a simplified forest structure made up of mosaics of even-aged stands. These are easily destabilized as they contain the ingredients of stand-level breakdown.

One of these destabilizing factors relates to the demographic instability which increases when stands have passed their most vigorous life stage. The other relates to the effects of monoculture, which promotes a one-sided exploitation of soil nutrients and the accumulation of metabolic waste products. These can result in allelopathy and eventually autotoxicity, a process recognized long ago in agriculture. Here crop rotation and periods of fallow have been practiced as needed for sustainable production.

Crop rotation is certainly an agrobiological principle based on the application of biodiversity as functionally important in a chrono-sequential sense. Shifting agriculture in the tropics has traditionally served the same purpose. It is well to consider these proven practices in future forestry operations by managing for a natural succession combined with tree crop rotation to restore soil fertility in tree plantations.

The third instability ingredient is the already discussed accumulation of undecomposable litter in so many plantation stands (see de Ronde et al. this Vol.). Clearly, attention has to be given also to undergrowth management of plantation forests for keeping the decomposer function intact and thereby reducing future fire hazards at their root causes.

1.6 Global Impacts on Climate and Soil

The conversion of so much of the world's forests to fire-maintained grasslands, savannas, and smaller areas of fire-prone plantation forests, will most likely have global consequences that can be perceived in three testable hypotheses requiring interdisciplinary work under the IGBP:

1. Large-area forest conversion with an enlarged fire regime will change the soil hydrology and atmospheric moisture at least on regional scales.
2. Large-area forest conversion with an enlarged fire regime will decrease biophilic nutrient budgets and lower the sustainability of the soil resource for organic production.

3. Large-area forest conversion with an enlarged fire regime will alter the atmospheric chemistry on our globe and contribute to changes in the world's climate.

1. The soil hydrology hypothesis. The inevitable penetration of the pyrophytic grass life-form and its shift into a more dominant role in formerly tree-covered areas is by itself a global environment change. It promotes the spatial enlargement of the fire regime into the world's humid tropical biomes, which under natural conditions rarely included fire as an ecological force factor.

When large areas in the humid tropical zone lose their forest cover, hydrological changes are inevitable (Salati and Vose 1984). The moist ground is then no longer affected by the "tall-wick" effect of trees, but only by the "short-wick" effect of grasses. This difference, further aggravated by the differences in root systems, may substantially reduce the amount of moisture being re-circulated to the atmosphere. On an area basis the differences can be very large. For example, an *Andropogon virginicus* grass cover in a moist rain forest area in Hawaii transpired approximately 50 mm/month, while an *Eugenia cumini* tree cover recirculated approximately 150 mm/month under the same environmental conditions (Mueller-Dombois 1973). The difference is 100,000 liters per hectare.

While loss of forest may not decrease the rainfall over the deforested area, it will most likely decrease it in an area lying somewhere in the wind direction of the denuded area. Drying effects may thus be noticeable in neighboring areas (see Fosberg et al. this Vol.).

2. The soil nutrient hypothesis. A lowering of the soil nutrient potential is tied to the absence of a protective tree cover especially when the onslaught of incoming moisture from rain storms causes accelerated erosion and loss of soil nutrients. Repeated burning, as is well known, destroys organic matter in the upper few centimeters of the mineral soil, and nitrogen is removed in gaseous form. In the humid tropics, organic matter is often the only source of soil nutrients because of advanced laterization. Thus, at the very least, the cycles of cations, sulfur, phosphorus, and nitrogen will be much reduced. While grasslands can have initially the same rates of production as forests, repeated grass fires will eventually result in skeletal soils highly impoverished of nutrients (see Goldammer and Peñafiel this Vol.).

3. The atmospheric chemistry hypothesis. Destruction of tropical rain forest and subsequent burning of organic matter emits CO_2. This has been estimated to be an important contributory source of the steadily increasing CO_2 level in the global atmosphere. Earlier estimates (Woodwell et al. 1978) varied from zero net emission to 18×10^{15} g C of CO_2 per year. More conservative estimates (Bolin 1977; Wong 1978; Seiler and Crutzen 1980) have varied between zero to 2×10^{15} g. The estimate published in this Volume (Hao et al. this Vol.) is in the range of 0.9 to 2.5×10^{15} g C of CO_2 net release from fires in the tropics in addition to the ca. 5.0×10^{15} g $\times$ yr^{-1} C of CO_2 contributed by industrial emissions.

In this sense, fire in the tropics can be said to contribute annually between 15 and 33% (0.9/5.9 and 2.5/7.5 ratio of biomass vs. fossil fuel burning) of the noncycling and accumulating CO_2 in the global atmosphere. This estimate will require further scrutiny, but it indicates the important role of fire in presentday atmospheric chemistry. This effect is now believed by many atmospheric chemists and climatologists to cause a human-induced trend of global warming and associated climatic disturbances with worldwide impacts.

1.7 Conclusions

The notion of a global deterioration due to the enlarged role of fire needs testing. This requires the acquisition of new data which are not often considered part of standard ecological studies. Examples relate to the:

1. Effects of fire on loss of biodiversity in terms of colonizers, successional and climax species. Disruptions in the "self-repairing" process of vegetation may be the inability of forest to re-invade a burned area, or an "arrested succession", when scrub vegetation is no longer replaced by forest.
2. Effects of tree monocultures on suppressing decomposer fauna and flora, on demineralization of soils and the development of autotoxicity.
3. Effects of climatic perturbations on forest stability. Global warming is expected to aggravate climatic extremes. In stands of trees this may be reflected by more frequent and vigorous physiological shocks and generally in a more violent global forest dynamics. Permanent plot studies are of greatest value for the acquisition of base-line data.

Information from such research is expected to help formulate appropriate policy and revised management practices to restore forest in denuded landscapes, to recoup soil hydrological and soil nutrient and organic decomposition optima, to reduce CO_2 emissions from fire by balancing these with patterns of plantations and semi-natural forests that serve as CO_2 sinks.

Fire protection will undoubtedly continue to remain a most important forest management objective; but fire hazards in the future should be reduced by a more vigorous application of biological principles.

References

Bolin B (1977) Changes of land biota and their importance to the carbon cycle. Science 196:613–615

Budowski G (1966) Fire in tropical American lowland areas. In: Proc Ann Tall Timbers Fire Ecol Conf 5:5–22, Tallahassee, Florida

Ellis RC (1985) The relationship among eucalypt forest, grassland and rainforest in a highland area in north-eastern Tasmania. Austr J Ecol 10:297–314

FAO (1982) Tropical forest resources. FAO Forest Paper 30, Rome, 106 pp

Goldammer JG (1986) Feuer und Waldentwicklung in den Tropen und Subtropen. Freiburger Waldschutzabhandl 6:43–57, Inst For Zool, Univ Freiburg

Goldammer JG, Seibert B (1989) Natural rain forest fires in Eastern Borneo during the Pleistocene and Holocene. Naturwissenschaften 76:518–520

Goldammer JG, Jenkins MJ (eds) (1990) Fire in ecosystem dynamics. Mediterranean and northern perspectives. Proc 3rd Int Symp Fire Ecology (Part I), Univ Freiburg, 16–20 May 1989, SPB Academic Publishing, The Hague (in press)

Itow S, Mueller-Dombois D (1988) Population structure, stand-level dieback and recovery of *Scalesia pedunculata* forest in the Galápagos Islands. Ecol Res 3:333–339

Kohyama T (1988) Etiology of "Shimagare" dieback and regeneration in subalpine *Abies* forest of Japan. Geo J 17(2):201–208

Kozlowski TT, Ahlgren CE (eds) (1974) Fire and ecosystems. Academic Press, New York, 542 pp

Lugo AE (1988) Estimating reductions in the diversity of tropical forest species. In: Wilson EO, Peter FM (eds) Biodiversity. National Acad Press, Washington, pp 58–70

Mooney HA, Bonicksen TM, Christensen NL, Lotan JE, Reiners WA (eds) (1981) Fire regimes and ecosystem properties. USDA For Serv Gen Tech Report WO-26, 594 pp

Mueller-Dombois D (1973) A non-adapted vegetation interferes with water removal in a tropical rain forest area in Hawaii. Trop Ecol 14(1):1–18

Mueller-Dombois D (1981) Fire in tropical ecosystems. In: Mooney HA, Bonnicksen TM, Christensen NL, Lotan JE, Reiners WA (eds) Fire regimes and ecosystem properties. USDA For Serv Gen Tech Report WO-26, pp 137–176

Mueller-Dombois D, Spatz G (1975) The influence of feral goats on the lowland vegetation in Hawaii Volcanoes National Park. Phytocoenologia 3:1–29

Myers N (1980) Conversion of tropical moist forests. Nat Acad Sci, Washington, DC, 205 pp

Myers N (1988) Tropical forests and the botanist's community. In: Greuter W, Zimmer B (eds) Proc XIV Int Bot Congr. Koeltz, Königstein/Taunus, pp 291–300

Pickett STA, White PS (eds) (1985) The ecology of natural disturbance and patch dynamics. Academic Press, San Diego, 472 pp

Raven PH (1988) Our deminishing tropical forests. In: Wilson EO, Peter FM (eds) Biodiversity. National Acad Press, Washington DC, pp 119–122

Salati E, Vose PB (1984) Amazon Basin: a system in equilibrium. Science 225:129–138

Sanford RL, Saldarriaga J, Clark KA, Uhl C, Herrera R (1985) Amazon rain-forest fires. Science 227:53–55

Seiler W, Crutzen PJ (1980) Estimates of gross and net fluxes of carbon between the biosphere and the atmosphere from biomass burning. Climatic Change 2:207–247

Smith CW, Tunison JT (1989) The modification of Hawaiian fire regimes by alien plants: research and management implications. In: Stone CP, Smith CW, Tunison JT (eds) Alien plant invasions in Hawaii. Univ Hawaii Press (in press)

Sprugel DG (1976) Dynamic structure of wave-regenerated *Abies balsamea* forests in the northeastern United States. J Ecol 64:889–911

Uhl C, Clark H, Clark K (1982) Successional pattern associated with slash-and-burn agriculture in the upper Rio Negro Region of the Amazon Basin. Biotropica 14:249–254

Uhl C, Kauffman JB, Cummings DL (1988) Fire in the Venezuelan Amazon 2: Environmental conditions necessary for forest fires in the evergreen rain forest of Venezuela. Oikos 53:176–184

Wardle JA, Allen RB (1983) Dieback in New Zealand *Nothofagus* forests. Pac Sci 37:397–404

Wong CS (1978) Atmospheric input of carbon dioxide from burning wood. Science 200:197–200

Woodwell GM (ed) (1984) The role of terrestrial vegetation in the global carbon cycle. Scope 23. Wiley, New York 247 pp

Woodwell GM, Wittaker RH, Reiners WA, Likens GE, Delwiche CC, Botkin DB (1978) The biota and the world carbon budget. Science 199:141–146

2 The Impact of Droughts and Forest Fires on Tropical Lowland Rain Forest of East Kalimantan*

J.G. GOLDAMMER[1] and B. SEIBERT[2]

2.1 Introduction

Lowland tropical rain forests have generally been regarded as ecosystems in which natural fire was excluded by fuel characteristics and the prevailing moist environment (Richards 1966; Mutch 1970; Mueller-Dombois 1981). However, recent findings demonstrate that climatic conditions since the late Pleistocene have favored the occurrence of natural and anthropogenic fires in the Amazon Basin and in East Kalimantan (Sanford et al. 1985; Saldarriaga and West 1986; Goldammer and Seibert 1989). It has also been demonstrated that the fuel characteristics, and the influence of drought on the microclimate and flammability of rain forest, may create conditions suitable for the occurrence and spread of long-return interval wildfires in today's primary rain forests (Uhl and Kauffmann this Vol.). Modern human impact on tropical forest lands is rapidly increasing, causing overall degradation, and conversion of rain forest vegetation to pyrophytic life forms with increased flammability and fire frequency (Mueller-Dombois and Goldammer this Vol.; Goldammer 1991).

The 1982–1983 wildfires in Borneo left behind more than 5×10^6 ha of burned primary and secondary rain forest. These fires were the result of extensive land-clearing activities, and an extreme drought which was attributed to the El Niño-Southern Oscillation complex. In the aftermath of the fires, which the media largely regarded as an "ecological disaster" (Seibert 1988), a strong and growing scientific interest developed, focusing on the impact of fire on rain forest vegetation.

This chapter will provide a synthesis of various studies intended to clarify the role and the impact of fire in the dynamics of lowland tropical dipterocarp rain forest in East Kalimantan. In order to understand past and current fire regimes, a brief description of the influence of climatic change and climatic oscillation on rain forest flammability and fire occurrence follows.

*Kalimantan is the Indonesian part of the Island of Borneo with four provinces in its center, west, south; and east. Sarawak and Sabah in its northwest and north are states of Malaysia; Brunei is on the northwest coast of Borneo.

[1] Department of Forestry, University of Freiburg, 7800 Freiburg, FRG
[2] Faculty of Forestry, University of Mulawarman, Samarinda 75001, East Kalimantan, Indonesia

2.2 Climatic Variability and Fire Regimes

It is generally recognized that during the last Ice Age the transfer of water from the oceans to continental ice caps lowered the sea level of the earth by at least 85 m (McIntyre et al. 1976).

Besides exposing land, especially on the Sunda Shelf, the drop in ocean water levels caused the development of an overall arid climate at that time. In the highlands of Malesia, reliable palynological and radiometric information has clarified the climatic and vegetational history since the last glaciation, as well as human impact (Flenley 1979b; Morley 1982; Maloney 1985; Flenley 1988; Newsome 1988; Newsome and Flenley 1988). In his holistic appraisal of the geologic and biogeographic history of the equatorial rain forest, Flenley (1979a) suggested that the most acute differences in the Quaternary climate of equatorial Indo-Malesia, compared to present conditions, occurred during the period ca. 18,000 to ca. 15,000 B.P. At that time, for example, the upper forest limit in the New Guinea highlands was 1700 m below its present value, while the mean temperature was 8 to 12°C lower.

Although palynological evidence from the tropical lowlands is still very scarce (Flenley 1982), it must be assumed that lowland vegetation was generally that of areas with a more pronounced dry season. Lowland pollen analyses from West Malesia are, or appear to be, of Holocene age (Maloney 1985), and no data from East Kalimantan are available.

However, the only study available from South Kalimantan may serve as an auxiliary argument for the climatic change which occurred during the Holocene. Morley (1981) suggested that the ombrogenous peat development of a peat swamp in the Sebangau river region had been initiated by a change from a more continental to a less seasonal climate during the mid-Holocene. This implies that the lowland climate of East Kalimantan, which today is still slightly seasonal (Whitmore 1984), must have been considerably drier within the period between the last glaciation and the development of today's rain forest climate. At that time, fuel characteristics and flammability of the prevailing vegetation must have created conditions suitable for the occurrence of wildfires, as was assumed, although never proved, for north Sumatra at ca. 17,800 B.P. (Maloney 1985).

The first evidence of ancient wildfires in the eastern part of Borneo was found by Goldammer and Seibert (1989). The radiometric age determination of charcoal recovered along an east-west transect between Sangkulirang at the Strait of Makassar, and about 75 km inland, showed that fires had occurred between ca. 17,510 and ca. 350 B.P. These events must have occurred in situ, as upper-slope hill terrain was selected for sampling, to avoid data-sampling from charcoal dislocated by sedimentation, deposits of which were found in the lower areas of Kutai National Park (Shimokawa 1988) and dated ca. 1040 B.P. (Goldammer and Seibert 1989). Charcoal residues suggesting ancient forest fires were also recently found in several places in Sabah (Marsh personal commun.).

The fire dates of 350 to 1280 B.P., as presented in the study, reveal that wildfires occurred not only during the dry Pleistocene, but also after the present wet, rain forest climate stabilized, at about 10,000 to 7000 B.P. These fires can be explained by periodic droughts such as those caused by the El Niño-Southern Oscillation complex (ENSO).

The ENSO phenomenon, which has been comprehensively described (Troup 1965; Julian and Chervin 1978; Philander 1983a; Mack 1989), is regarded as one of the most striking examples of inter-annual climate variability on a global scale. It is caused by complicated atmospheric-oceanic coupling which is not yet entirely understood (Behrend 1987). The event is initiated by the Southern Oscillation, which is the variation of pressure difference between the Indonesia low and the South Pacific tropical high. During a low pressure gradient, the westward trade winds are weakened, resulting in the development of positive sea surface temperature anomalies along the coast of Peru and most of the tropical Pacific Ocean. The inter-tropical convergence zone and the South Pacific convergence zone then merge in the vicinity of the dateline, causing the Indonesian low to shift its position into that area. Subsequently, during a typical ENSO event, the higher pressure over Malesia leads to a decrease of rainfall, the severity of the dry spells depending on the amplitude and persistence of the climatic oscillations.

In the rain forest biome these prolonged droughts drastically change the fuel complex and the flammability of the vegetation. Once the precipitation falls below 100 mm per month, and periods of 2 or more weeks without rain occur, the forest vegetation sheds its leaves progressively with increasing drought stress. In addition, the moisture content of the surface fuels is lowered, while the downed woody material and loosely packed leaf-litter layer contribute to the build-up and spread of surface fires. Aerial fuels such as desiccated climbers and lianas become fire ladders leading to crowning fires, or the "torching" of single trees.

Peat swamp forests found in the lowlands of Borneo represent another fuel type. With increasing precipitation deficit and the dropping of the water table in the peat swamp biome, the organic layers progressively dry out. During the 1982–1983 ENSO, various observations in East Kalimantan confirmed a desiccation of more than 1 to 2 m (Johnson 1984). While the spread of surface and ground fires in this type of organic terrain is not severe, deep burning of organic matter leads to the toppling of trees and a complete removal of standing biomass. It is further assumed that smoldering organic fires may persist throughout the subsequent rainfall period, to be reactivated as an ignition source in the next dry spell (Goldammer and Seibert 1989). This re-ignition is similar to the patterns of fire behavior in the organic soils of northern boreal and circumpolar biotas.

The climatic variability during the past 18 millennia, with long-term changes and the short-term oscillations, may give sufficient explanation for environmental prerequisites for wildfire occurrence. However, the origins of the fires are not clear and cannot be interpreted through the ^{14}C-data of charcoal. Under the drier and more seasonal climate of the last glaciation, early anthropogenic fires and frequent lightning fires may have played a role similar to the conditions in today's deciduous savanna forests of continental South Asia (see

Stott et al. this Vol.). Volcanism as another natural fire source may have influenced vegetation development on Southeast Asian islands with high volcanic activities, e.g., the highlands of Sumatra and Java.

Long-lasting fires in coal seams extending to, or near, the surface (Fig. 1), are found in various rain forest sites in East Kalimantan and are another important natural fire source there (Goldammer and Seibert 1989). It was assumed that all of the 29 coal fires known to be burning at present had been ignited by the 1982–1983 wildfires. This is questioned by Goldammer and Seibert (1989), since there are numerous oral reports about burning coal seams made before the 1982–1983 drought.

Goldammer and Seibert (1989) also investigated the "baking" effects of subsurface fires on sediment or soil layers on top of the coal seams. The effects of ancient, no longer burning coal seam fires can still be seen today. The material, locally called "baked mudstone", is utilized at present for road construction purposes. Thermoluminescence analysis of burnt clay, collected on

Fig. 1. Subsurface coal fire under primary dipterocarp rain forest near Muara Lembak, East Kalimantan (January 1989)

top of an extinguished coal seam in the vicinity of active coal fires, proved a fire event 13,200 to 15,300 years B.P. It must be assumed that both old, and most ongoing, coal fires have been ignited by lightning.

The edges of the burning coal seams progress slowly through the ground of the rain forest and cannot be extinguished by water. Even a water body cascading over the edge of a burning coal seam cannot affect the combustion process, as observed by Goldammer and Seibert (1989).

During the 1987 ENSO, the authors witnessed the ignition of a forest fire by a burning coal seam and its spread from there into the Bukit Soeharto forest reserve (Fig. 2). Figure 3 shows how trees, and other forest debris which falls over the progressing fire edge, may carry the fire into the surrounding forest land.

These observations, together with the data on ancient fires and the longevity of coal fire occurrence, suggest that burning coal seams represent a permanent fire source from which wildfires spread whenever a drought occurs and the fuel conditions are suitable for carrying a fire. This interaction between climatic variability, fire sources, and wildfires seems to be unique. However, the phenomenon may well help to clarify the role and impact of long-return interval disturbances, like fire, in the evolutionary process of the rain forest biome which, until now, has generally been considered to be infinitely stable.

Fig. 2. Surface fire originated at a burning coal seam edge in Bukit Soeharto National Park during the drought of 1987 (September 1987)

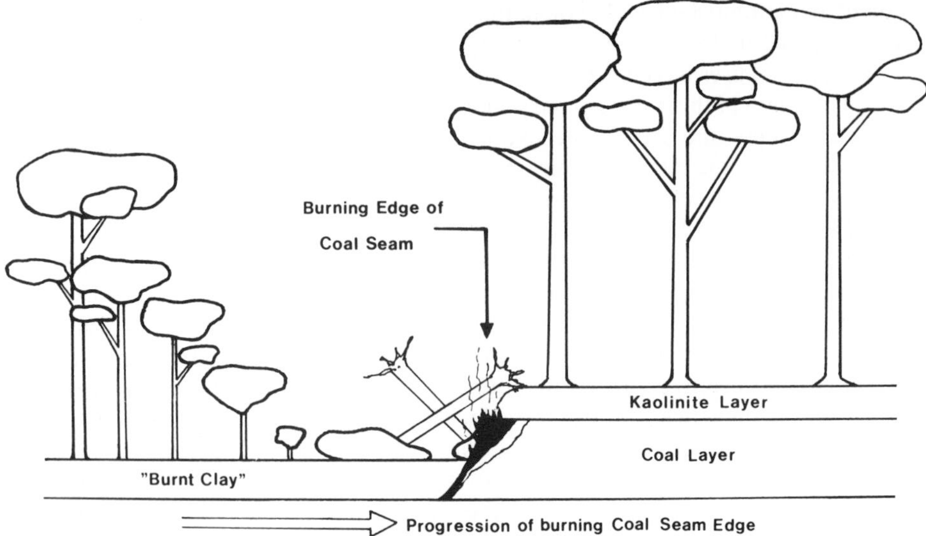

Fig. 3. Schematic progression model of a burning subsurface coal seam edge. (Goldammer and Seibert 1989)

2.3 The 1982-83 ENSO, its Predecessors and the Wildfires

2.3.1 The 1982-83 ENSO

The 1982–1983 drought in Malesia was the result of an extreme ENSO of exceptionally large amplitude and persistence (Philander 1983b). In north and east Borneo, the decrease of rainfall began in July 1982, and lasted until April 1983, interrupted only by a short rainy period in December 1982. Monthly precipitation dropped below critical values along the coast and up to around 200 km inland. In Samarinda, near the east coast of East Kalimantan, the rainfall between July 1982 and April 1983 was only 35% of the mean annual precipitation (Departemen Perhubungan, undated; Fig. 4a). Further inland, rainfall recordings from Kota Bangun (100 km from the coast; Fig. 4b) and Melak (150 km from the coast; Fig. 4c) still show critical deficits. The precipitation did not fall below the critical margin of 100 mm in Long Sungai Barang, Bulungan (300 km inland; Fig. 4d). These recordings support Brünig's (1969) observation from Sarawak, that drought stress occurs more frequently in coastal areas than in the hinterland.

Rainfall conditions in northern Borneo during the 1982–1983 drought were similar. Five stations in Sabah recorded an average precipitation decrease of 60% (Woods 1987). No significant drought and no fires were observed in Sarawak at that time (Marsh personal commun.).

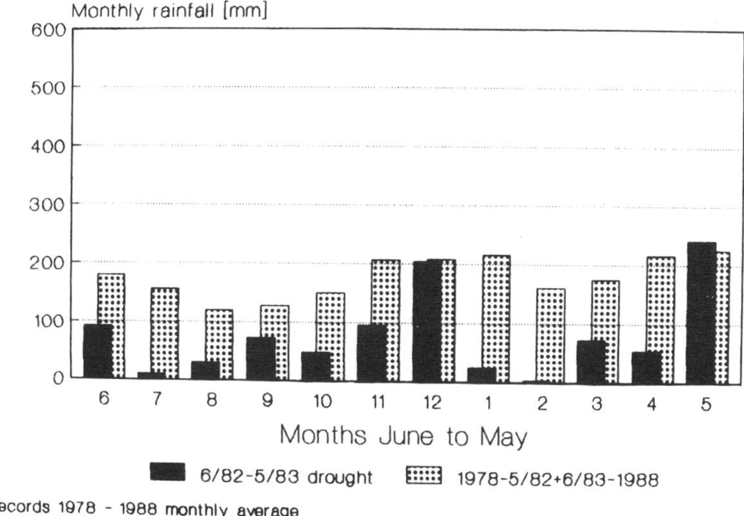

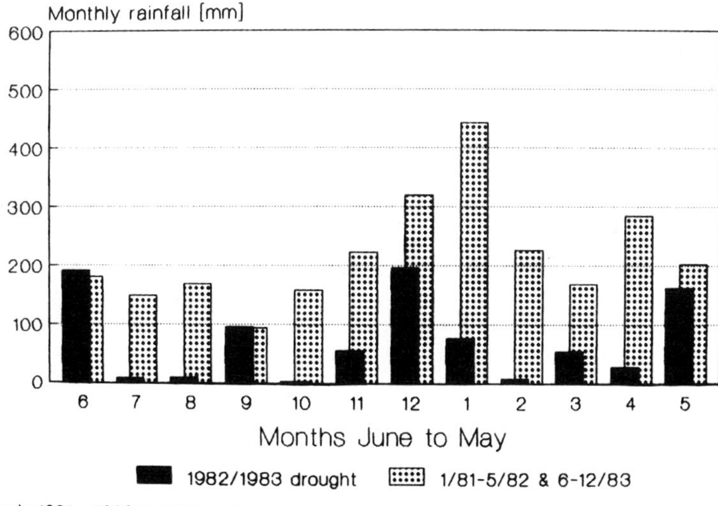

Fig. 4a-d. Monthly rainfall in East Kalimantan during the 1982–83 drought, compared with average years (Fig. 4c,d see page 18)

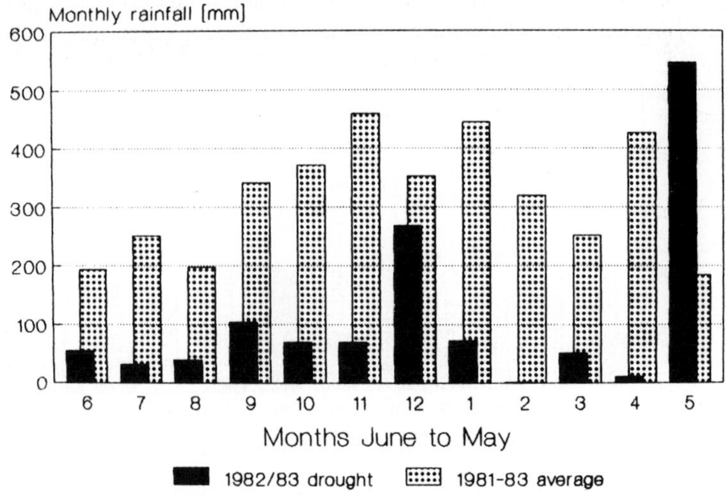
Station Melak, T.A.D.
Monthly rainfall [mm]
600
500
400
300
200
100
0
6 7 8 9 10 11 12 1 2 3 4 5
Months June to May
1982/83 drought
1981-83 average
Records 1981 - 1983 monthly average
6/1982 - 5/1983 monthly readings
c

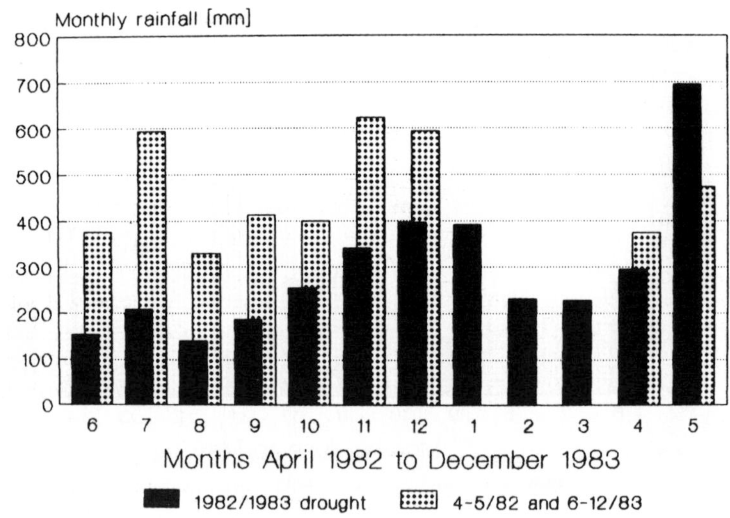
Station Long Sungai Barang, Bulungan
Monthly rainfall [mm]
800
700
600
500
400
300
200
100
0
6 7 8 9 10 11 12 1 2 3 4 5
Months April 1982 to December 1983
1982/1983 drought
4-5/82 and 6-12/83
Records 4/1982 - 12/1983 monthly reading
d

Berlage (1957) found that between 1830 and 1953, about 93% of all droughts in Indonesia occurred during an ENSO event. In a systematic evaluation of precipitation data since 1940 (Leighton 1984), most of the 11 droughts recorded in 1941/42, 1951, 1957, 1961, 1963, 1969, 1972, 1976, 1979/80, 1982/83, and 1987 accompanied an ENSO event. The worst droughts during that period were in 1941/1942, 1972, and 1982–1983, while the 1961 drought occurred independently of an ENSO event, and the 1965 ENSO did not cause drought in Indonesia.

2.3.2 Predecessors

The first written information about the impact of an extreme drought in East Kalimantan is given by Bock (1881). This Danish zoologist traveled through the lowlands of the Kutai district of East Kalimantan in 1878 and reported drought and famine which had occurred in the year before his visit. He recorded that about one third of the tree population in the forests around Muara Kaman in the Middle Mahakam area had died due to the drought. More recent observations in various peat swamp forests of the Middle Mahakam Area of East Kalimantan confirm that significant disturbances of this ecosystem must have occurred around 80–100 years B.P. (Weinland 1983). The rainfall records of Jakarta (Java) 1877/78 explain these observations: between May 1877 and February 1878, rainfall in Jakarta was reduced by two thirds; a second severe precipitation deficit followed in the period July to December 1878 (Kiladis and Diaz 1986).

Bock did not report any forest fires. Nevertheless, the authors looked for evidence of forest fires in East Kalimantan and were informed by Amansyah (personal commun.) that, according to his grandmother, fires had occurred in the area of Muara Lawa on the Kedang Pahu river during the period of investigation. Also Grabowsky (1890) mentions a forest fire which had occurred some years before his visit in 1881–1884, on two mountains, Batu Sawar and Batu Puno, in the central part of South Kalimantan, about 70 km inland. These two mountains, according to Grabowsky, had been totally deforested by the fire, and the approximate date of the fire coincides with Carl Bock's remarks on the severe drought in East Kalimantan.

In 1914–15, forest fires were again reported from Borneo. Published records were found for Sabah, where an area of 80,000 ha of rain forest and its superficial peat soil layer were destroyed by fire after an exceptionally dry period (Cockburn 1974). This area now forms the Sook Plain grassland of Sabah. Amansyah (personal commun.) also reported fires, during the same period, in the Muara Lawa area. Endert (1927) confirms these reports in his reference to fires which had occurred about 10 years before his visit to East Kalimantan in 1925. According to the farmer Rajab (personal commun.) of Modang (Pasir District of East Kalimantan) serious fires swept through his farmland from the coast, in the same year, before proceeding inland. Rainfall records from Balikpapan (Fig. 5a), and Samarinda (Fig. 5b) close to the east coast of East

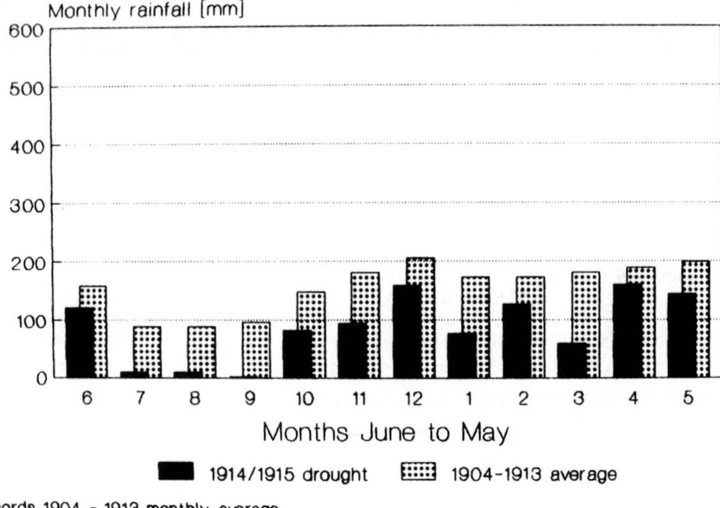

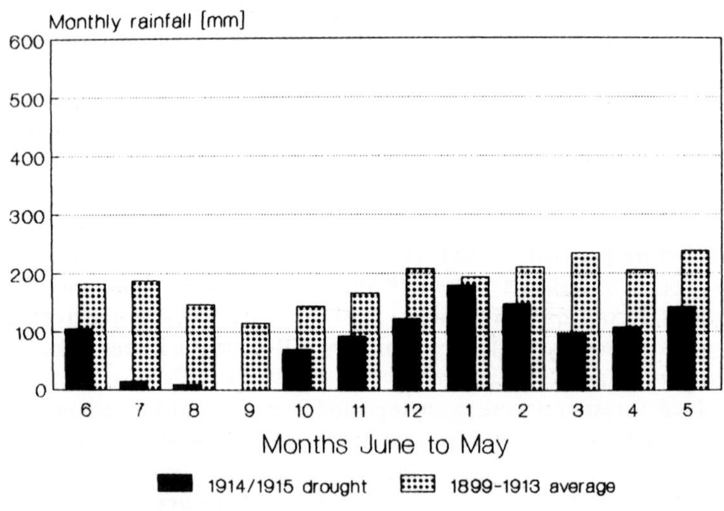

Fig. 5. Monthly rainfall in East Kalimantan during the 1914–15 drought, compared with average years

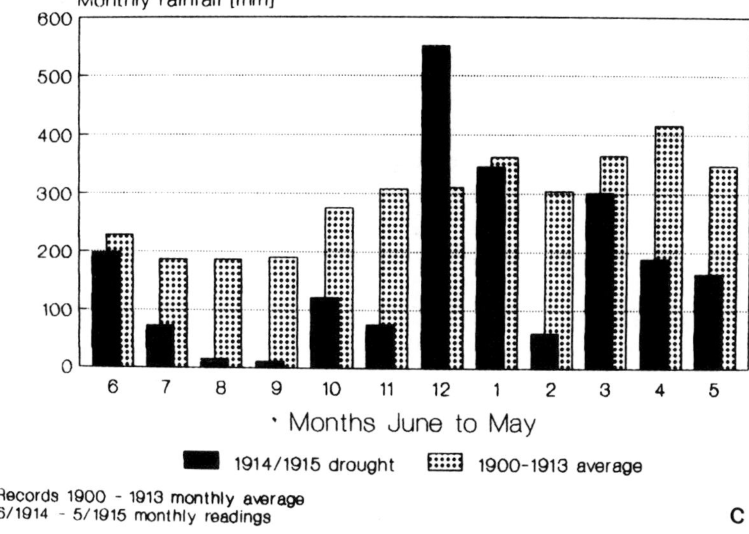

Kalimantan, and from Long Iram, about 180 km inland (Fig. 5c), prove a severe drought in 1914–1915.

Severe forest fires in Brunei following a drought of 6 weeks in 1958 were observed by Brünig (1971). Smaller fires in lowland dipterocarp forests and in *Dacrydium elatum* forests were recorded for 1969 and 1970 in Sabah and Brunei by Fox (1976, cited by Woods 1987). Brünig (1971) has also described an exceptional drought at this time, but no fires.

2.3.3 The Wildfire Scenario in 1982-83

The wildfire scenario in Borneo in 1982/83 was set by the extensive drought and by numerous slash-and-burn land-clearing activities from which the fires ran out of control. The extent and the immediately visible impact of these fires have been described by several authors and teams (Wirawan and Hadiyono 1983; Johnson 1984; Leighton 1984; Lennertz and Panzer 1984; Malingreau et al. 1985; Woods 1987).

It is assumed that the overall land area of Borneo affected by fires exceeded 5×10^6 ha. In East Kalimantan alone, a total of 3.5×10^6 ha were affected by drought and fire. Of the total area, 0.8×10^6 ha was primary rain forest, 1.4×10^6 ha logged-over forest, 0.75×10^6 ha secondary forest (mainly in the vicinity of settlement areas), and 0.55×10^6 ha peat swamp biome (Lennertz and Panzer 1984).

One of the first aerial and ground surveys of the fire damage was carried out in a totally burned area in the Kutai National Park, west of heavily logged and farmed areas (Leighton 1984). It was found that fire damage was higher in the secondary forest than in primary forest, although the degree of damage varied greatly. The fires had swept twice through the ITO timber concession southwest of the Kutai National Park, the first causing defoliation of many trees and lianas; the second completely burning this accumulated litter. No surviving trees were observed in areas which had burned twice. It can be assumed that this also occurred in the Porodisa timber concession, north of the Park, and in the Sylva Duta timber concession, west of the Park.

In his 1983 ground survey of the northern part of the National Park near the Mentoko research station, Leighton (1984) found that the primary forest had been badly damaged. He was unable to report any unburned primary forest on hills, ridges, or slopes which could have served as a control plot to distinguish damage by drought or fire. He inferred that the drier soils of the hillside and hilltop areas, and also their shallowness (as argued by Whitmore 1984), could be an important factor determining the water deficit during prolonged drought seasons.

Narrow, 5–20-m-wide belts of unburned primary forest flanking streams were also observed, but these account for only 5–10% of the total area. In the burned areas, 99% of the trees below 4 cm DBH had died, although about 10% were resprouting from the ground. In the diameter 20–25 cm DBH class, 50% had died, with 20–35% of trees above cm DBH. Among the larger trees which had died, only Bornean ironwood (*Eusideroxylon zwageri* T & B) was observed resprouting from the ground. Lianas and strangling figs had been virtually eliminated from burned parts of the forest, apparently being particularly sensitive to drought.

Wirawan (1983) made a ground survey in a less damaged area in the southern part of the National Park. Fire extended there about 30–40 km from the coast, but was able to sweep further inland wherever previous logging activities had produced suitable fuels, particularly in the surroundings of logging roads. The healthiest areas were in the southwest of the Park further inland. Dead stems of emergent trees sticking out of the canopy have apparently died from drought, not from fire. Canopy and subcanopy had regreened in these areas, while in burned forests, the subcanopy was usually defoliated due to heat rising from the fire.

Wirawan's (1983) unpublished report shows that in unlogged, unburned dipterocarp forest the effect of the long drought period was more severe on larger trees than on smaller trees: 70% of the trees above 60 cm DBH were dead, while only 40% in the DBH class of 30 to 60 cm, and 20–25% in the class below 30% respectively. Drought has produced a diameter distribution similar to that in an unburned logged-over dipterocarp forest in that area.

Only 15% of the trees in all diameter classes had died in unlogged, unburned ironwood forest. Dead individuals were mainly *Shorea* species and others, but ironwood was able to survive even on ridges. In unlogged, but burned ironwood forest, damage occurred particularly on smaller individuals: 75% of the trees below 5 cm DBH and 50% of the trees from 5 to 10 cm DBH were dead, compared with only 8–15% of the trees above 10 cm DBH. Many of the latter, however, sustained bark damage and may die in the near future.

Fire intensity in previously logged areas was directly related to the intensity of logging. The fires were severe but did not completely destroy moderately logged stands where, after the fire, a few trees with green foliage could still be observed, although spaced and scattered.

In heavily logged forest areas, where remaining trees are widely spaced, shrub had formed a thick ground cover, providing an excellent biomass source for the fires after the extensive drought. Here the fuel consumption was more complete.

Lennertz and Panzer (1984) made several ground surveys in seven timber concessions throughout the burned area, confirming the findings of Wirawan (1983) and Leighton (1984) on a larger-area scale. The damage was generally heavier in logged-over than in primary forests. On an average, diameter classes between 35 and 65 cm DBH were least damaged, while smaller trees were severely affected. Damage to trees above 65 cm DBH could also be due to drought.

Areas attributed to three damage classes:

1. Up to 10% of the tree crowns dry, indicating drought but no fire.
2. 10–50% of canopy and undergrowth trees dead.
3. Over 50% dead trees, indicating severe fires in logged-over forests.

About 75% of the total area affected by fires had to be classified as severely damaged (class 3). The damage degree generally decreased towards the west of the area.

Damage was total in most secondary forest previously used for shifting cultivation, and was also severe – with the area looking like a mikado game from the air – in the burned peat swamp forests. Only areas on shallow peat layers were less affected.

2.4 Forest Regeneration After the 1982–83 Fires

The regeneration of damaged forests after the fires was observed by several research teams between 1983 and 1987, but to date there has been no conclusive study which includes all occurring forest types, damages related to site, and previous utilization of the respective forests. For this preliminary evaluation, we distinguish the following forest types:

- Primary forest: Mainly in the Kutai National Park around 100 km north of Samarinda, and in forest reserves between Samarinda and Balikpapan (Fig. 6). Primary forest was mainly affected in the driest parts of the Province along the coastline.

Fig. 6. Recovery process of primary dipterocarp lowland rain forest in Bukit Soeharto National Park, 5 years after being affected by drought and fire in 1982

- Logged-over forest: Besides the distance from the coast, the intensity of previous logging and the time since logging are mainly decisive in the degree of damage.
- Secondary forest (after disturbance by man, mainly through shifting cultivation): These areas were, in most cases, totally deprived of any remaining vegetation and had begun their regeneration process from zero.
- Swamp and peat swamp forests: Wherever this type was affected by fires, the damage was total, because in most cases the structure of the organic underground broke down.

The two latter forest types are not considered in this overview. Damage by fires in secondary forests is similar to the development of areas after shifting cultivation, and relevant information can be found in the respective literature (Riswan and Kartawinata 1989, Goldammer 1991, see also synoptic overview in Peters and Neuenschwander 1988). Swamp and peat swamp forests were destroyed to such a degree that the area is lost for forestry for an unforeseeable time, and recovery processes leading to forest could not be observed.

2.4.1 The Regeneration Process

Three to 4 months after the fire, a carpet of seedlings was already covering a large part of the forest floor, dominated by pioneers, particularly *Macaranga* species and grasses. Seeds, especially those of *Macaranga* species, as naturally occurring pioneers, are assumed to originate from the site's seed bank and not from outside dispersal (Leighton 1984). Dipterocarp seedlings were still absent in August 1983 (Suyono 1984), but grew after a considerable flowering of many dipterocarp species, in early 1984.

2.4.1.1 Primary Forest

A direct observation of succession after fire was carried out in a protected forest near Samarinda. Sixty percent of the area of a primary forest plot originally established for the UNESCO-MAB project and measured for the first time in 1976 (Riswan 1976, 1982) had burned, in 1983. The area was remeasured in 1983 (Riswan and Yusuf 1986). In 1983, a burnt part (0.6 ha) of this primary forest plot was measured by Suyono (1984); this plot was again measured by Boer et al. (1988a,b) in 1987. Although not consistent in size, the plots are compared in Table 1. Results of the 1988 study indicate fast recovery of slightly damaged primary forests 4.5 years after the fire. The recovery process has already increased both the number of trees > 10 cm DBH, and the basal area, by approximately 10% – halfway back to the situation before the fire. Dipterocarpaceae and Euphorbiaceae are the leading families in the diameter classes above 10 cm DBH, while *Eusideroxylon zwageri* is the single most important species.

Some of the primary forest in the Kutai National Park, particularly in areas within 40 km of the coastline, was far more badly damaged than observed in

Table 1. Damage and recovery after fire of primary forest in Lempake, Samarinda. Sources: Riswan and Yusuf (1986; data of 1976, 3. and 12.1983); Suyono (1984; data of 8.1983); and Boer et al. (1988a,b; data of 11.1987)

Date	Remarks	No. of sp.	No. of trees/ha	Basal area/ha (m^2)
12.1976		209	445	33.7
3.1983	Natural death		63	
3.1983	New in class > 10 cm DBH	42	32	
12.1983	Burnt trees			
	(whole plots)		79	
	(weighed[a])		132	
12.1983	1.6-ha plot	199	335	26.5
8.1983	0.6-ha plot	112	403	26.5
11.1987	0.6-ha plot	116	430	29.7

[a] Assuming that the whole plots had burned; in fact only 60% of the primary forest were damaged by fire.

areas around, and south of Samarinda. The area of the National Park and north of it receives precipitation, even in normal dry seasons, at the lower limit of the 100 mm/month margin determining the occurrence of tropical rain forest. Wherever fires were strong enough to erase the upper canopy, both number and composition of species changed drastically, and no considerable recovery of primary forest communities could be seen 4.5 years later: Miyagi et al. (1988) and Tagawa et al. (1988), both based on field research in 1986–87, distinguish several shrub types in burnt primary and logged-over forests, all but one with only few dipterocarps, but dominated by site-specific pioneers. The scarcity of regenerating dipterocarps as the leading tree family in some undisturbed lowland forest communities in East Kalimantan raises questions as to what type of forest will develop in large heavily damaged areas in this region.

2.4.1.2 Logged-Over Forest

About 1.5 years after the fire, several species, including dipterocarps, had flowered, possibly also induced by stress due to the opening of the canopy and the restricted crowns of the partly damaged trees.

Consequently, in slightly logged and burned forest areas where dipterocarps still occur in the upper canopy, regeneration was often dominated by dipterocarp species (Noor 1985) able to establish easily under the moderate shelter of the upper canopy which usually covers more than 75% of the surface area. In these stands, the dipterocarps faced less competition from vigorous pioneer species. Among both saplings and seedlings, dipterocarps are considerably represented with 21 and 42% respectively (Hatami 1987), indicating that this habitat may recover fairly rapidly. Among the seedlings observed in 1986, pioneer species play only a minor role (7% of individuals).

In moderately damaged areas, where some dipterocarp mother trees had survived, dipterocarp seedlings of several species were observed in considerable number (Boer 1984). In these areas, dipterocarps still hold the same rank among the trees, 5 years after the fires (Boer and Matius 1988), but the regeneration of dipterocarps as the leading commercial species group is marginal, indicating that the first regeneration after the fires did not survive. In some areas, a fire-resistant palm, *Borassodendron borneensis*, is the leading element of the pole class, while saplings and seedlings are dominated by *Macaranga* species and other pioneer and shrub species of minor value. Crown closure is still in the range of only 50%.

Heavily damaged areas without surviving seed trees of primary forest species undergo the same recovery processes as the areas left behind by shifting cultivators. In areas close to settlements, where abandoned land is a persistent *Imperata cylindrica* seed source, many damaged areas were invaded by this weed and are practically lost for forestry (see also Seavoy 1975).

It is still impossible to quantify the surface attributed to the referred classes of damage and the related recovery processes. A more detailed study is presently being carried out by the International Tropical Timber Organization (ITTO).

Based on ground survey inventories and followed by the interpretation of recent aerial imagery, this study should provide answers to many currently open questions.

2.5 Conclusions

What was new about the 1982–83 drought and fires in Borneo was not the fact of their occurrence, but the extent. It is the large size of contiguous burnt areas and the high quantities of available fuel, caused by human interference with the tropical rain forest that has made the damage so bad. As demonstrated above, and supported by numerous observations in selectively logged forests, dipterocarps are able to germinate and to grow in disturbed sites. Their frequent occurrence in groups indicates the gap-opportunism of many dipterocarp species. But dipterocarp seeds, although winged, do not fly farther than about 100 m, will not keep longer than about 20 days, and symbiotic ecto-mycorrhizal fungi are no longer present once the dipterocarps have vanished from an area. If this area exceeds a certain size, it may take centuries for dipterocarps to migrate back to a former habitat.

A surprising response to fire is shown by Bornean ironwood (*Eusideroxylon zwageri*). In all surveys, ironwood trees play an important role in species composition after fire. The resprouting capability of old trees and stumps after fire is high, as is fruiting of survivors. Producing a timber of extreme density and a specific weight above that of water, ironwood is supposed to grow relatively slowly, also because it is usually a tree of the second canopy layer. Several authors refer to ironwood as the most drought- and fire-resistant species in the lowland forests of Borneo (Wirawan 1983; Leighton 1984).

Why, then, does ironwood occur so frequently and so often dominate the lowland forests of East Kalimantan? Although it is a species of relatively slow growth in the understory of the forest, ironwood has been successfully cultivated in pure stands in South Kalimantan. Could ironwood serve as an indicator of a long history of episodic drought and fires in Kalimantan? If this is so, why are dipterocarps still a leading group of species in the lowlands although they are so susceptible to drought and fire? Must they always be attributed to the "climax" phases of a forest, or are there species acting as primers, as gap and even damage opportunists? Several dipterocarp species, especially *Shorea lamellata*, were often observed growing wild on roadsides on barren soil, where no other trees had yet been able to establish. Does the clumped occurrence of many dipterocarps further indicate their opportunistic regeneration after disturbances?

There are more questions. The recent discovery of prehistoric fires in today's humid tropics of Amazonia and Borneo highlights the need to reconsider the theory of a never-disturbed and ageless climax of the tropical rain forest. In its evolution the rain forest has obviously survived long-term and short-term climatic variability, fires, storms, and flooding (van der Hammen 1974; Absy 1982; Colinvaux 1989). The role of disturbance in rain forest evolution, however, is not yet clear. One of the recent hypotheses, the controversially debated forest refuges theory, postulates that evolutionary diversification took place in isolated areas. Among those who support this theory there is still debate as to whether, during the last glaciation, the wetter uplands had become refuges for isolated evolution of rain forest species, or rain forest survived and diversified in those lowlands which were warmer but nonarid (Haffer 1969; Beven et al. 1984; Lewin 1984; Whitmore 1984; Kam-biu Liu and Colinvaux 1985; Connor 1986; Salo 1987; Colinvaux 1989; Goldammer 1991).

The mosaic of recent findings on climatic variability and forest fires in East. Kalimantan does raise more questions but at least may provide some answers. The highest impact of drought and fires during the 1982–1983 ENSO was largely restricted to the eastern lowlands and east of today's 3–3500 mm isohyet. This land area, at the steep eastern edge of the Sunda shelf, had already been a coastal ecotone when the Sunda shelf became dry during the last glaciation. Borneo's inland mountains at that time may have been more continental but still received more rainfall than the seasonal lowlands. A pronounced moisture gradient between these ecotones must have resulted in the development of a wet rain forest refuge in the highlands, eventually somewhere west of today's 3–3500 mm isohyet. The vegetation surviving on the lowland at that time had to cope with frequent droughts and fires.

Most of this former lowland vegetation of the Sunda shelf was then inundated and lost during the Pleistocene-Holocene transition. The remaining lowlands are presently covered by a highly diverse moist rain forest biome, although repeatedly disturbed by droughts and fire.

Is the lowland rain forest the result of a rapid expansion from the wet highland habitat? Or can diversity of the lowland rain forest be explained by frequent disturbances (Connell 1978)? Forest die-back due to drought stress and fire provides great evolutionary possibilities because it creates gaps and speeds up regeneration dynamics. It also prevents take-over by a few dominant species which would lead to diversity-poor forest communities.

The basis for this evolutionary hypothesis is still fragmented and inconclusive. However, as has now been demonstrated, lowland dipterocarp rain forest may recover from drought stress and fire if these events are not excessive and not associated with human interference. Further, its corollary highlights the opportunity selective silviculture methods may offer towards sustained ecosystem utilization and management.

Acknowledgments. This research was sponsored by the Volkswagen Foundation and supported by the German Agency for Technical Cooperation (GTZ).

References

Absy ML (1982) Quaternary palynoligical studies in the Amazon Basin. In: Prance GT (ed) Biological diversification in the tropics. Columbia University Press, New York, pp 67–73

Amansyah (personal communication) trader from Muara Lawa, East Kalimantan

Behrend H (1987) Teleconnections of rainfall anomalies and of the Southern Oscillation over the entire tropics and their seasonal dependence. Tellus 39 A:138–151

Berlage HP (1957) Fluctuations in the general atmospheric circulation of more than one year, their nature and prognostic value. Roy Neth Met Inst, Mededlingen an Verhandelingen 69

Beven S, Connor EF, Beven K (1984) Avian biogeography in the Amazon basin and the biological model of diversification. J Biogeogr 11:383–399

Bock C (1881, reprinted 1985) The Head-Hunters of Borneo: a narrative of travel up the Mahakam and down the Barito; also, journeyings in Sumatra. Singapore Oxford Univ Press, Oxford, New York, 1985. XIII + 344 pp. First published by Sampson Low, Marston, Searle and Rivington, London, 1881 and under the title: Unter den Kannibalen auf Borneo. Eine Reise auf diese Insel und auf Sumatra. Jena: Hermann Costenoble. (1882) XX + 407 pp

Boer Ch (1984) Studi tentang derajat kerusakan akibat kebakaran hutan di Long Nah, S Kelinjau, dan S Separi Kalimantan Timur (Study on the degree of damage due to forest fire in Long Nah, S Kelinjau and S Separi, East Kalimantan) S1 Thesis IPB Bogor, 194 pp (unpublished)

Boer Ch, Matius P (1988) Suksesi setelah tebangan dan kebakaran hutan di PT. Kayu Lapis Indonesia, Unit Separi, Kalimantan Timur (Succession after logging and forest fire in PT. Kayu Lapis Indonesia, Unit Separi, East Kalimantan). Report, Fakultas Kehutanan Unmul – Deutsche Gesellschaft für Technische Zusammenarbeit (GTZ), 38 pp

Boer Ch, Matius P, Sutisna M (1988a) Suksesi setelah kebakaran hutan primer di hutan pendidikan Lempake, Samarinda, Kalimantan Timur. Report, Fakultas Kehutanan UNMUL – Deutsche Gesellschaft für Technische Zusammenarbeit (GTZ), 41 pp

Boer Ch, Matius P, Sutisna M (1988b) Succession after fire in the Lempake primary forest near Samarinda, East Kalimantan. GFG-Report 11:27–34

Brünig EF (1969) On the seasonality of droughts in the lowlands of Sarawak (Borneo). Erdkunde 23(2):127–133

Brünig EF (1971) On the ecological significance of drought in the equatorial wet evergreen (rain) forest of Sarawak (Borneo). Transact. First Aberdeen-Hull Symp on Malesian Ecology, 1970. Univ Hull, Dept Geogr, Misc Ser 11:66–97

Cockburn PF (1974) The origin of the Sook Plain. Malayan For 37:61–63

Colinvaux PA (1989) The past and future Amazon. Sci Am 260(5):102–108

Connell JH (1978) Diversity in tropical rain forests and coral reefs. Science 199:1302–1310

Connor EF (1986) The role of Pleistocene forest refugia in the evolution and biogeography of tropical biotas. Tree 1:165–168

Departement Perhubungan (Undated) Badan Meteorologi dan Geofisika, Jakarta (Rainfall data for Indonesia)

Endert FH (1927) Floristisch Verslag. Chapter III (pp 200–291) In: Indisch Comite voor Wetenschappelijke Onderzoekingen (ed) Midden-Oost-Borneo Expeditie 1925. 423 pp

Flenley JR (1979a) The equatorial rain forest: a geological history. Butterworth, London, 162 pp

Flenley JR (1979b) The late Quaternary vegetational history of equatorial mountains. Prog Phys Geogr 3:488–509

Flenley JR (1982) The evidence for ecological change in the tropics. Geogr J 148:11–15

Flenley JR (1988) Palynological evidence for land use changes in South-East Asia. J Biogeogr 15:185–197

Fox JED (1976) Environmental constraints on the possibility of natural regeneration after logging in tropical moist forest. J For Ecol Man 1:512–536

Goldammer JG (1991) Feuer in Waldökosystemen der Tropen und Subtropen. Birkhäuser, Basel (in preparation)

Goldammer JG, Seibert B (1989) Natural rain forest fires in Eastern Borneo during the Pleistocene and Holocene. Naturwissenschaften 76:518–520

Grabowsky F (1890) Streifzüge durch die malayischen Distrikte Südost-Borneos, II. Globus 57:219–222

Haffer J (1969) Speciation in Amazonian forest birds. Science 165:131–137

van der Hammen T (1974) The Pleistocene changes of vegetation and climate in tropical South America. J Biogeogr 1:3–26

Hatami M (1987) Analisis permudaan alam pada hutan bekas kebakaran di Bukit Soeharto (Analysis of natural regeneration in a forest after fire in Bukit Soeharto) S1 Thesis For Fac, Univ Mulawarman, Samarinda, 52 pp (unpublished)

Johnson B (1984) The great fire of Borneo. Report of a visit to Kalimantan Timur a year later. WWF Report, 22 pp

Julian PR, Chervin RM (1978) A study of the southern oscillation and Walker Circulation phenomenon. Mon Wea Rev 106:1438–1451

Kam-biu Liu, Colinvaux PA (1985) Forest changes in the Amazon basin during the last glacial maximum. Nature (Lond) 318:556–557

Kiladis KN, Diaz HF (1986) An analysis of the 1877–1879 ENSO episode and comparison with 1982–1983. Mon Wea Rev 114:1035–1047

Leighton M (1984) The El Niño-Southern Oscillation event in Southeast Asia: Effects of drought and Fire in tropical forest in eastern Borneo. WWF Report, 31 pp

Lennertz R, Panzer KF (1984) Preliminary assessment of the drought and forest fire damage in Kalimantan Timur. Report by DFS German Forest Inventory Service Ltd. for Deutsche Gesellschaft für Technische Zusammenarbeit (GTZ) 45 pp + annexes

Lewin R (1984) Fragile forests implied by Pleistocene pollen. Science 226:36–37

Mack F (1989) Das El Niño-Southern Oscillation-Phänomen (ENSO) und die Auswirkungen dadurch bedingter Klimaschwankungen auf die Dipterocarpaceenwälder von Ost-Kalimantan, Borneo (The El Niño-Southern Oscillation (ENSO) phenomenon and the effects of climatic oscillations on the dipterocarp forests of East Kalimantan, Borneo). Diploma thesis, Dep For, Univ Freiburg, FRG, 164 pp

Malingreau JP, Stephens G, Fellows L (1985) Remote sensing of forest fires: Kalimantan and North Borneo in 1982–83. Ambio 14(6):314–321

Malingreau JP, Tucken CJ, Laporte N (1989) AVHRR for monitoring global tropical deforestation. Int J Rem Sens 10:855–867

Maloney BK (1985) Man's impact on the rainforests of West Malesia: The palynological record. J Biogeogr 12:537–558

Marsh C (personal communication). P.O. Box 11623, 88817 Kota Kinabalu, Sabah, Malaysia

MCDW (Monthly Climate Data of the World). NOAA, Nat Climatic Data Center, Asheville, NC

McIntyre et al. (CLIMAP Project Members) (1976) The surface of Ice-Age earth. Science 191:1131–1137

Miyagi Y, Tagawa H, Suzuki E, Wirawan N, Oka N (1988) Phytosociological study on the vegetation of Kutai National Park, East Kalimantan, Indonesia. L.c. Tagawa H, Wirawan N (eds) pp 51–62

Morley RJ (1981) Development and vegetation dynamics of a lowland ombrogenous peat swamp in Kalimantan Tengah, Indonesia. J Biogeogr 8:383–404

Morley RJ (1982) A paleoecological interpretation of 10,000 year pollen record from Danau Padang, Central Sumatra, Indonesia. J Biogeogr 9:151–190

Mueller-Dombois D (1981) Fire in tropical ecosystems. In: Mooney HA et al. (eds) Fire regimes and ecosystem properties. USDA For Ser Gen Tech. Rep. WO-26, pp 137–176

Mutch RW (1970) Wildland fires and ecosystems – a hypothesis. Ecology 51:1046–1051

Newsome J (1988) Late Quarternary vegetational history of the Central Highlands of Sumatra. I. Present vegetation and modern pollen rain. J Biogeogr 15:363–386

Newsome J, Flenley JR (1988) Late Quaternary vegetational history of the Central Highlands of Sumatra, II. Pelaeopalynology and Vegetational History. J Biogeogr 15:555–578

Noor M (1985) Pengamatan permudaan alam jenis Dipterocarpaceae pada hutan primer setelah kebakaran hutan di Bukit Soeharto Kalimantan Timur (An Evaluation of natural regeneration of Dipterocarpaceae in a primary forest after fire in Bukit Soeharto, East Kalimantan) S1 Thesis For Fac, Univ Mulawarman Samarinda, 126 pp (unpublished)

Peters WJ, Neuenschwander LF (1988) Slash and burn. Farming in the Third World forest. University of Idaho Press, Moscow, Idaho, 113 pp

Philander SGH (1983a) El Niño Southern Oscillation Phenomena. Nature (Lond) 302:295–301

Philander SGH (1983b) Anomalous El Niño of 1982–83. Nature (Lond) 305:16

Rajab (personal communication). A farmer in Modang, Pasir District of East Kalimantan

Richards PW (1966) The tropical rain forest. Cambridge Univ Press, 450 pp

Riswan S (1976) Penelitian suksesi hutan tanah rendah tropika di Gunung Kapur, Lempake, Samarinda (Succession study on lowland tropical forest at Gunung Kapur, Lempake, Samarinda). Frontir, Univ Mulawarman 6:25–27

Riswan S (1982) Ecological studies on primary, secondary and experimentally cleared mixed diopterocarp forest and kerangas forest in East Kalimantan, Indonesia. Ph.D. Diss. Univ Aberdeen, UK

Riswan S, Yusuf R (1986) Effects of forest fires on trees in the lowland dipterocarp forest of East Kalimantan, Indonesia. In: BIOTROP Special Publ No 25 (1986). Proc Symp Forest Regeneration in Southeast Asia. Bogor, Indonesia, 9–11 May 1984. SEAMEO-BIOTROP Bogor, Indonesia, pp 155–163

Riswan S, Kartawinata K (1989) Regeneration after disturbance in a lowland mixed dipterocarp forest in East Kalimantan, Indonesia. Ekologi Indonesia 1:9–28

Saldarriaga JG, West DC (1986) Holocene fires in the northern Amazon basin. Quat Res 26:358–366

Salo J (1987) Pleistocene forest refuges in the Amazon: evaluations of the biostratigraphical, lithostratigraphical and geomorphological data. Ann Zool Fenn 24:203–211

Sanford RL, Saldarriaga J, Clark KE, Uhl C, Herrera R (1985) Amazon rain-forest fires. Science 227:53–55

Seavoy RE (1975) The origin of tropical grasslands in Kalimantan, Indonesia. J Trop Geogr 40:48–52

Seibert B (1988) Forest fires in East Kalimantan 1982–83 and 1987: The press coverage. Fac For, Univ Mulawarman, Samarinda, East Kalimantan (Mimeograph)

Shimokawa E (1988) Effect of a fire of tropical rain forest on soil erosion. L.c. Tagawa H, Wirawan N (eds) pp 2–11

Suyono H (1984) Kerusakan tegakan hutan akibat kebakaran pada hutan pendidikan Universitas Mulawarman di Lempake, Samarinda (Damage of a forest stand caused by fire in the educational forest of Mulawarman University in Lempake, Samarinda). S1 Thesis For Fac, Univ Mulawarman, Samarinda, 59 pp (unpublished)

Tagawa H, Wirawan N (eds) (1988) A research on the process of earlier recovery of tropical rain forest after a large scale fire in Kalimantan Timur, Indonesia. Univ Kagoshima, Res Cent South Pacific, Occasional Papers 14 (1988); 136 pp

Tagawa H, Suzuki E, Wirawan N, Miyagi Y, Oka NP (1988) Change of vegetation in Kutai National Park, East Kalimantan. L.c. Tagawa H, Wirawan N (eds) pp 12–50

Troup AJ (1965) The Southern Oscillation. JR Meteorol Soc 91:490–505

Weinland G (1983) Unpublished internal report. Deutsche Gesellschaft für Technische Zusammenarbeit, GTZ (unpublished)

Whitmore TC (1984) Tropical rain forests of the Far East. Clarendon, Oxford, 352 pp

Wirawan N (1983) Progress in the management of protected areas in Kalimantan and consequences of recent forest fires. (WWF Project 1687) (unpublished)

Wirawan N, Hadiyono (1983) Survey to the eastern part of the proposed Kutai National Park June 17–27, 1983 (Mimeograph)

Woods P (1987) Drought and fire in tropical forests in Sabah. An analysis of rainfall pattern and some ecological effects. In: Kostermans AJGH (ed) Proc Third Round Table Conference on Dipterocarps. Samarinda, East Kalimantan, 16–20 April 1985. Unesco, Regional Office for Science and Technology for Southeast Asia, Jakarta, Indonesia, pp 367–387

3 The Role of Fire in the Tropical Lowland Deciduous Forests of Asia

P.A. STOTT[1], J.G. GOLDAMMER[2], and W.L. WERNER[3]

Each dry season from January onwards the litter on the jungle floor ignites spontaneously and creeping ground fires clear away old matted vegetation, making way for the new season's growth. Some of these fires reach considerable size and provide one of the most impressive of jungle sights when viewed from a distance on a dark night. Long, jagged fronts of flame advance across the black backdrop of hillsides, exploding into pyrotechnics as clumps of dry bamboo are engulfed by the everchanging scarlet patterns of destruction . . . .

A teak-wallah's description of forest fire in Northern Thailand (H.N. Marshall, Elephant Kingdom, London: The Travel Book Club, 1959, p. 133).

3.1 Fire: An Alien Ecological Pressure?

Throughout the lowlands of tropical Asia, in regions experiencing either a monsoon forest (Köppen's Am) or a savanna forest (Aw) climate, both natural and human-induced annual or biennial fires have long been a common ecological phenomenon during the dry season, which may extend from three to seven months. Such burns occurred quite naturally through a number of causes before the appearance, around 12,000 years ago, of agriculture in the region, but, with the advent of Neolithic communities, swiddening, and stubble burning, dry-season fires inevitably became much more frequent and widespread. In modern times, this fire pressure has grown even more intense, and the ecological role of fire is now a matter for serious debate, as it is seen to affect diminishing forest resources in closer and closer proximity to human settlements (Goldammer 1988). In Burma, for example, the total annually burnt area of forested land has been estimated at 3.5 to 6.5×10^6 ha yr^{-1} (Goldammer 1986). For Thailand, data collected from 1984 to 1986 show that fire affected 20.92% of the forested area or 3.1×10^6 ha yr^{-1} (Royal Forest Department 1988), while worldwide, for all ecosystems, a figure of 630–690 $\times 10^6$ ha yr^{-1} has been suggested by Seiler and Crutzen (1980), 98% of which occurs in the tropics and subtropics. In response to this, the clear temptation is for governments

[1]Department of Geography, School of Oriental and African Studies, University of London, London WC1H OXG, United Kingdom

[2]Department of Forestry, University of Freiburg, 7800 Freiburg, FRG

[3]Department of Geography, South Asia Institute, University of Heidelberg, 6900 Heidelberg, FRG

throughout the Asian tropics to try to outlaw dry-season burning completely, particularly in protected habitats, like national parks, forest reserves, and wildlife sanctuaries. Unfortunately, this policy often proves impossible to enforce effectively, and the result is an unpredicted wildfire, feeding on the abundant fuels that have been allowed to accumulate under the regime of fire protection. The cure can all too easily become worse than the disease. There is thus an increasing need to understand fully the ecological role of fire in the tropical lowland deciduous forests of Asia, so that we can manage fire safely as a tool, making humans the masters, not the fire (Mather 1978).

The ecological impact of fire, however, varies markedly from vegetation formation to formation, and it is quite wrong to assume that a program of fire management which is suitable for one habitat can be automatically transferred to another. Even within the same habitat, different management practices will be required to achieve different ecological and productive goals. Unlike lowland tropical rain forest, most of the vegetation formations characterizing the seasonal tropics of Asia, which range from the celebrated monsoon forests with teak (*Tectona grandis* L. f.), through the drier sāl (*Shorea robusta* Gaertn. f.) forests of the Indian subcontinent and the dry deciduous dipterocarp forests (savanna forests) of mainland South East Asia, to the more localized open grassland savannas and thorn forests, are all naturally preadapted, in differing degrees, to the ecological stress of fire. Indeed, in certain areas, fire will be the stabilizing factor, preventing the formation from progressing to a more evergreen alternative. The dry deciduous dipterocarp forests of Manipur State, Burma, Thailand, Laos, Kampuchea (Cambodia), and Viet-Nam, for example, are now known to possess ancient core communities, which were largely determined by topographic and edaphic factors, but from which the formation has since been spread by fire and the axe into more fire-sensitive formations, such as lowland tropical semi-evergreen rain forest, where savanna forests are now maintained by dry-season burning (Barrington 1931; Stott 1976, 1984, 1988a). In the core areas, fires also occur, but they are clearly not the key ecological factor, and fire protection will not normally cause one formation to succeed another, although it may alter the structural balance between trees, shrubs, and grasses. In contrast, outside these core areas, fire protection will trigger the succession to a more evergreen type. It is thus vital to decide what is required, and to exclude or to use fire accordingly.

In perhumid equatorial Asia, annual fire is largely an alien ecological force, although there is increasing evidence that even here fire may have affected vegetation patterns on a long-term cycle. On the other hand, throughout the seasonal tropics of Asia, where the dry season is prolonged, the total annual rainfall is usually much less than 2000 mm, yet the mean temperature of the coldest month rarely less than 20° C, fire is an integral ecological element, year in, year out, one which needs, however, to be managed correctly to produce the required results. In some circumstances, the exclusion of fire may, in fact, be itself alien to the environment; in others, as on steep slopes, fire may lead to severe biological disruption and to considerable geomorphological problems, such as increased soil erosion. The right application of fire to limited fuels at the appropriate time under sensible controls in the correct location is the key objective. It is the purpose of this chapter to explore the implications of these ecological subtleties for the drier seasonal forests of lowland tropical Asia.

3.2 Fire Patterns in Time and Space

Throughout the seasonal tropics of Asia, dry season fires vary markedly in character according to a whole range of variables, but above all in relation to timing, fuels, and topography, the first two of these, however, being themselves largely determined by the overarching variable of climate. Understanding the timing, origins, and fuel characteristics of the fires is especially vital if we are to learn how to manage fire safely and effectively.

3.2.1 Timing and Origins

The end of the great South West Monsoon is constant in neither space nor time, so that the potential for ignition can be delayed from early November to late January in any given year. In mainland South East Asia, the peak period for forest fires normally lies between late December and early March, by which time most sites will have been fired, although some burning may continue even into May. This peak period correlates strongly with the optimal provision of ignitable fuels, in that some 90% of leaf-fall will be generally completed by the end of December (Nalamphun et al. 1969), while the standing crop of grasses and pygmy bamboos of the genus *Arundinaria* will have dried and be ready for sustained burning (Chanatip 1973; Ruangpanit and Pongumphai 1983). However, enormous regional variations in timing are apparent, savanna forests around Chiang Mai in Northern Thailand, for example, tending to be fired early in December and January, whereas those near Korat in North East Thailand, by contrast, usually burn much later in February and March, following the late January rains that can come up suddenly from the Gulf of Thailand. The fiercest burns seem to be mid-season fires, each regionally defined, which tend to be "hot" fires, with high scorch heights. In contrast, early- and late-season burns are less dramatic, so that early-season burning in particular, defined in relation to the potential for fire ignition and sustainability in any given year and location, is often the best management option to follow to prevent uncontrolled and damaging mid-season conflagrations. Such a pattern closely parallels the ancient aboriginal burning practices recorded for the eucalypt savanna woodlands of Northern Australia (Braithwaite and Estbergs 1985; Bowman 1987; Stott 1989). Although floristically distinct, these eucalypt savanna woodlands are strikingly similar to the savanna forests of tropical Asia in terms of their structure and physiognomy, while both formations share clear resemblances to the miombo (*Brachystegia-Julbernardia* woodlands) of Central and Southern Africa (Chidumayo 1988; Guy 1989). All three are surprisingly homogenous in structure when compared with savanna and savanna forest formations in other parts of Africa and Latin America.

The percentage of burns that are natural is difficult to assess, although it is not uncommon to encounter fires far from any human settlement. One possibly classic example is reported to occur annually in the west of Thailand, in Kanchanaburi province, where the sun's rays are thought to be deflected into the forest from a crystallized quartzite cliff [Saranarat (Ôy) Kanjanavanit personal commun. 1988]. A more common cause is lightning strike. Lightning scars can be observed on many trees, such as the resinous *Pinus merkusii* Jungh. and De Vriese, and, although an immediate outbreak of fire is often prevented by the accompanying rain, the tree can continue to smolder within, leading to fire after the storm has passed. Not all lightning strikes, however, will lead to widespread burns, for many will simply result in the local "torching" of a tree, although this is less likely in the dry savanna forests. Other causes may include friction, particularly of dry bamboo stalks in the mixed deciduous forests or of pygmy bamboos (*Arundinaria* spp.) in the savanna forests. It is clear, however, that there are sufficient natural causes of fire to put at risk even the best protected of seasonal forests, which must, therefore, have a proper system of fire breaks and observation posts if the risk of widespread wildfire is to be minimized.

Today, however, the main causes of the dry-season burns are undoubtedly human, and it is especially difficult to protect dry tropical forests from the pyrogenic activities of humans in the densely peopled lands of Asia (Goldammer 1987). Many deliberate fires are actually set within the forest itself, to produce new grazing or thatching material (such as the large leaves developed by *Dipterocarpus tuberculatus* Roxb.), wood-oil from the dipterocarps, edible shoots of plants like *Melientha suavis* Pierre or tasty fungi on the forest floor, paths easy of access, a "drive" in hunting, or nutrients which are later rain-washed down to cultivated land with the yearly return of the South West Monsoon (Stott 1988b). In contrast, other deliberate fires are set outside the forest, but then get out of control, later spreading or "jumping" into nearby forest vegetation. The clearing of roadside vegetation and shifting cultivation (swiddening or slash-and-burn agriculture) are clearly the main culprits in this category of fire causes. On a world scale, Seiler and Crutzen (1980) have estimated that some 41×10^6 ha yr^{-1} of forest land are burnt or cleared through shifting cultivation alone. Deliberate fires are also employed for more specialized reasons, such as the stimulation of better growth in "tendu" leaves (*Diospyros melanoxylon*), which are used for cigarette (bidi) wrapping in Maharashtra, India.

Many fires, however, are not deliberate at all, but arise accidentally from a wide range of causes. The sheer number and variety of people crossing the forests of Asia, such as graziers, collectors of one sort or another, migrant hill peoples, prospectors, monks, guerillas, not to mention whole armies, in training or live combat, clearly provoke a high probability for accidental wildfires. One of us (Stott) has witnessed evidence of the potential for fire from army training camps in Thailand, while Goldammer (1986) shows that, in Burma, the predilection of the rural population for smoking "cheroot" cigars when working in or walking through forest lands may be a serious hazard. These cheroots normally comprise a mixture of tobacco leaves and stem particles, with the latter glowing for a considerable time and acting like sparks. In Thailand, one estimate suggests that no less than 66% of all forest fires are caused by people traversing the forests (Royal Forest Department 1988). Finally, fire may be set purely for the sake of fire itself, the phenomenon having a mesmeric effect on both children and adults alike.

The regularity of burning throughout the seasonal tropics of Asia is therefore apparent, although the exact pattern varies from formation to formation. One of the present authors (Werner), using data from 118 experimental plots in Thailand, has shown that for plots studied during January to March 1988, 62% of the pine-dipterocarp forest plots had been burnt, but only 14% of the plots in pine-oak forest. For plots studied during May-June of the previous year, the figures were 69 and 50% respectively, although, in this case, there was no way of checking whether the fires had all occurred in 1987. In January-March 1988, 35% of plots in the pine-savannas had been burnt, despite the fact that the majority lay within a national park, where complete fire prevention was being attempted, yet more evidence that a policy of total fire exclusion is often impossible to enforce.

3.2.2 Fire and Fuel

The nature, amount, and spatial distribution of ignitable fuel will largely govern the character of the fire in any forest location. In the tropical lowland forests of Asia, fires may be low-level litter fires, groundcover fires of various intensities and heights, or mixtures of the two, where there is a complex or mosaic of rock outcrops, litter, and groundcover (Stott 1986, 1988a,b). Each of these fire types, however, will in turn also vary in character with prevailing climatic conditions, such as humidity and wind, whether a backfire or headfire, and in relation to topography. It is important to recognize, therefore, a whole range of fire types, each of which has different direct and indirect effects on the vegetation formation involved, and which pose varying levels of threat from the human point of view. The extremes are slow-moving litter backfires, with a maximum flame height of 0.5 m and a fireline intensity of around 230 kW m^{-1}, which contrast with fast-moving, jumping, and "spotting" groundcover headfires, with mean flame heights of well over 2 m and a fireline intensity of anything from 2000 to 5000 kW m^{-1}, which is beyond the limits of human control, even with heavy mechanical equipment (Stott 1986, 1988b). Where the fuel load is kept down by regular burning on an annual or biennial basis, the latter dangerous type of wildfire is unlikely to occur, and the fires will tend to be moderate fires, which are largely within human control and which cause little *direct* ecological damage (Fig. 1).

The ecological significance of fire is therefore complex and depends above all on the nature and the distribution of the fuels available for burning. Because the majority of regular fires in the lowland deciduous forests of Asia tend to be

Fig. 1a-e. A series of photographs showing **a** fuel type, **b** low-intensity fire behavior, and **c** fire effects in an indaing forest dominated by *Dipterocarpus tuberculatus* Roxb., Yezin, Burma. **d** A high-intensity fire at the peak of the fire season, and **e** its resultant effects on the southern tropical dry deciduous forest biome of Central India (Maharashtra). Over a long period, such fires, with high flame lengths, will cause a significant deterioration in the tree layer, and give a clear advantage to pyrophytic life-forms, particularly the grasses. (Photos: J.G. Goldammer)

small-scale surface fires feeding on limited fuels, it can be argued that they pose little direct threat to the main forest types, which are clearly adapted to moderate levels of fire. However, in taking this view, it is important to remember that different formations will respond in contrasting ways, and, above all, that these low-level fires will still affect the balance between trees, shrubs, and grasses in a formation, controlling, in particular, the rate of tree seedling survival and persistence. The synergism of fire and forest will therefore be immensely subtle, and will be perceived very differently by foresters, scientific ecologists, botanists, zoologists, landscape planners, politicians, peasants, and the general public. Hence the management dilemma. Foresters will largely want to exclude fire to improve tree stocks, but this policy may prove disastrous if it results in the build-up of immense fuel banks, which will eventually undoubtedly feed severe wildfires. Whatever the policy adopted, it must include some element for the control of fuel build-up, whether litter or ground cover. If the controlled application of prescribed fire is not appropriate, then it must be replaced by systems of grazing, cutting, trampling, or other forms of management. The policing alone of fire high risk areas will never prove sufficient in crowded Asia. Fire education and the development of distinct policies for different forest areas and vegetation units will also be required. In many zones, the application of early-season prescribed burning on limited fuels will still be the best option.

The work of one of us (Werner in Northern Thailand) has clearly demonstrated how different fuels and their attendant fires can markedly affect patterns of seedling and forest survival (see also Santisuk 1988). In pine-dipterocarp savannas, pine savannas, pine plantations, and pine-montane oak woods, young seedlings of *Pinus kesiya* Royle ex Gord. and *P. merkusii* clearly go through their most difficult phase in the first year of growth when they are especially susceptible to fire damage. The chances of survival, however, appear to be much higher in a pure litter burn than in a groundcover burn. The seedlings of *Pinus kesiya* have short needles between 22 and 33 cm in length. The lower of these tend to wilt during the dry season, while the younger leaves around the growing point are covered by a layer of blue-green wax. Low flames of the type characteristic in litter fires tend to consume the needles, but they do not damage the stem. In contrast, severe groundcover fires will kill the whole plant. *Pinus merkusii* possesses the additional protection of a "grass stage" (Fig. 2), although this naturally delays progression to the tree form. In Northern Thailand, the young pines can persist in this stage for up to 7 years. During the grass stage the plant has a strong rootstock, and it is remarkably resistant to all but the severest of groundcover fires. Unfortunately, the grass stage means that *P. merkusii* is not favored by foresters, who tend to want quick results, although the forms of *P. merkusii* found in North Eastern Thailand around Ubon Ratchathani and Sisaket tend to have a shorter grass stage of 1–3 years, while plants exhibiting a remarkable biogeographical disjunct distribution in Sumatra have no grass stage at all.

The effects of dry-season fire on the regeneration systems of the different formations and taxa is thus of great significance, and will vary with both the frequency and the intensity of the fires being encountered. For example, it is

Fig. 2. Grass stage in *Pinus merkusii* Jungh. and De Vriese, Ban Wachan, Chiang Mai, Northern Thailand. (Photo: W.L. Werner)

quite clear from studies that the recovery rates of *Shorea obtusa* Wall. ex Blume seedlings after low-intensity fires is good (Stott 1986), whereas high-intensity fires can kill 95% of *S. obtusa* saplings that are less than 1 m high, although a mere 7% of taller saplings will be lost (Khemnark 1979). In some systems, therefore, short-term fire protection until saplings can withstand fire may prove an important management option, whereas long-term fire protection would trigger a change to an unwanted or less deciduous formation. Moreover, in some instances, burned stands will exhibit greater species diversity than unburned. Sukwong and Dhamanitayakul (1977) recorded 223 individuals of 30 identified species (excluding grasses) in a 320-m^2 stand of savanna forest at Sakaerat, North East Thailand, but only 181 individuals of 25 species in a neighboring unburnt plot of the same area. The number of seedlings of some important constituent tree species, e.g., *Shorea obtusa* and *Xylia kerrii* Craib and Hutch., was virtually the same in both, suggesting that the germination of dry deciduous dipterocarp tree species is little hampered by fire. The relationships between fire and regeneration are therefore far from simple, and much more detailed work is required in nearly all the formations of seasonal Asia.

Throughout the lowland deciduous forests of Asia, there is thus a complex spectrum of adaptation to different levels of fuel and fire. The most obviously adapted is the dry deciduous dipterocarp forest of mainland South East Asia, where the dominant deciduous dipterocarps, such as *Dipterocarpus intricatus* Dyer, *Dipterocarpus obtusifolius* Teijsm., *D. tuberculatus*, *Shorea obtusa*, and

Fig. 3. *Dipterocarpus tuberculatus* Roxb. has a remarkable ability to recover from fire damage. This particular specimen is aged about 85 years, having been established in 1901 and cut in 1986. It was injured by a high-intensity fire in 1967, which created the fire scar. After 1967, the tree was affected by several low-intensity surface fires, which did not, however, delay the healing process. (Photo: J.G. Goldammer)

Shorea siamensis Miq., all possess a wide range of physiognomic, physiological, and phenological defences against both drought and fire (Stott 1986, 1988a,b; Fig. 3), and where fires rarely exceed moderate levels, unless fuel build-up is permitted to take place without any checks. Similarly, the terai forests of North India, which comprise both pure and mixed stands of sāl (*Shorea robusta*) are likewise fire-adapted (Goldammer 1987), old sāl trees having a thick, insulating bark, and the rootstock remaining normally unaffected by the low-intensity fires. Vigorous shoots soon appear after the fire has passed through, and these provide good grazing for cattle, so that local people are easily tempted to set fires deliberately in these formations. However, the frequent fires clearly affect the age-class distribution of the trees by widening the gap between the mature overstorey and the regeneration processes (Goldammer 1988; Fig. 4).

Mixed deciduous forests with teak (*Tectona grandis*) combined with a complex of other species, such as *Pterocarpus macrocarpus* Kurz and *Terminalia alata* Heyne ex Roth., also exhibit a range of adaptations to fire, but here tall bamboos are common and provide fuels for hot and damaging fires that can carry the flames high into the canopy. Indeed, in all formations, intense local burns can torch individual trees, particularly where the fire takes hold of roots, so that a localized build-up of groundcover and litter can have a damaging effect within an otherwise low-level fire. The type, the mosaic, and the loading of fuel

Fig. 4. Short-return interval fires frequently eradicate natural regeneration under old, fire-resistant sāl trees (*Shorea robusta* Gaertn. f.). The sāl stands may become moribund if not protected from fire and grazing during the critical phase of regeneration establishment. Terai forest, southern Nepal. (Photo: J.G. Goldammer)

in relation to the adaptive levels of the vegetation formation involved will therefore govern whether or not a fire has serious ecological and human consequences; in the lowland deciduous forests of Asia, the correlations are remarkably complex, and show that there can be no one solution for the management of fire in the region.

3.3 Fire Management

In the lowland deciduous forests of tropical Asia, there is therefore a vital need for a whole range of fire management systems, rather than a simple policy of "fire out" or fire control (Goldammer 1988; Rodgers 1986). Unfortunately, the latter approach is still adopted by many government organizations throughout the region (Royal Forest Department of Thailand 1988), partly through ignorance, but also because fire control divisions find it easier to battle for money and resources if they are seen as national fire fighters. The real subtleties of fire ecology are not, unfortunately, quite so easy to sell politically.

If we are really to progress, however, each situation, habitat, landscape unit, or fire problem must be approached in its own right, and the correct fire regime established, bearing in mind the main management purpose or aim at that particular location. It cannot be stressed too fully that we should choose the fire management system which is the most suited to achieving a given goal. For example:

– The general management of a tract of savanna forest in a protected wildlife area will probably demand a program of prescribed burning carried out early in the dry season on a regular rotation, so that new grazing is provided for wildlife and fuel is prevented from accumulating to the level where it would feed a severe groundcover fire that could have serious ecological consequences. A program of fire education in prescribed burning could be established for local villagers and visitors.

– If the aim is to combine high quality timber production with grazing, then a combined system of prescribed burning and prescribed grazing will be necessary (e.g., Liacos 1986; Goldammer 1988). Such systems may involve the protection of young saplings from the stresses of both fire and grazing on a short-term basis, but both grazing and fire will later be used to reduce the risk of severe fires and to maintain the basic ecosystem.

– If the aim is to control severe soil erosion on watersheds, steep slopes, or sensitive areas, then a much stricter system of fire control may be required, combined with other controls on the build-up of fuel loads. One of us (Werner) has studied such systems in *Pinus kesiya* plantations on watersheds in Northern Thailand, where fire protection tends to change the plantation into a pine-oak forest, which gives good soil protection, as well as a valuable timber and fuelwood resource. In this case, the pines are simply acting as pioneers. In other, more lowland, regions, however, the forest would remain fire-prone, with a dangerous build-up of fuel. In such areas, prescribed burning, very early in the dry season, but at longer intervals, combined with grazing, cutting, trampling, and fire protection during the intervening years, may be the necessary option. In contrast to such a regime, the regular late-season burning of teak forest near Haldwani (Uttar Pradesh, India) leads to severe soil erosion when the monsoon rains hit the newly exposed soil, one stand having lost 2000 m^3 of topsoil since its establishment 30 years ago (Goldammer 1987, 171; Fig. 5).

– In mixed deciduous forest with teak, light prescribed burning will have an important role to play in the silvicultural practices for the encouragement of teak (Kutintara 1970; Khemnark 1979, 38–39).

These, then, are just a few of the options for the management of fire in different socio-economic and ecological situations in the seasonal tropics of Asia. Fire is a tool which has long been used by humans (Sauer 1962). In many circumstances, even in crowded tropical Asia, any attempt to exclude fire completely will be bound to failure, and may, in fact, lead to disaster. However, wildfire cannot be allowed to have the mastery to itself. In the lowland deciduous forests of tropical Asia, there is a desperate need to manage fire carefully and correctly in order to produce the agroforestry systems which are required to protect the landscape, increase its productive value, and maintain ancient ecosystems, long adapted to moderate levels of burning. We cannot achieve this, however, without a full understanding of the basic ecology of forest fire in the

Fig. 5. Severe erosion under a 30-year-old teak (*Tectona grandis* L. f.) plantation caused by the impact of heavy monsoon rains on the soil exposed by annual fires (see text). Uttar Pradesh, India. (Photo: J.G. Goldammer)

region, a subject about which we know far too little. Research must come first, then a program of education, then the safe use of this ancient force in the monsoon lands of "fire and flood".

Acknowledgments. We are most grateful to Dr. Thawatchai Santisuk, Bangkok, and to Dr. Dietrich Schmidt-Vogt and Prof. Dr. U. Schweinfurth, Heidelberg, for helpful comments on an early draft of this paper.

References

Barrington AHM (1931) Forest soil and vegetation in the Hlaing Forest Circle, Burma. Burma For Bull 25, Ecol Ser, Rangoon: Govt Printing and Stationery

Bowman DMJS (1987) Stability amid turmoil?: towards an ecology of north Australian eucalypt forests. Proc Ecol Soc Aust 15:149–158

Braithwaite RW, Estbergs JA (1985) Fire patterns and woody vegetation trends in the Alligator Rivers region of northern Australia. In: Tothill JC, Mott JJ (eds) Ecology and management of the world's savannas. Aust Acad Sci, Canberra, pp 359–364

Chanatip K (1973) Seasonal variation in biomass and primary productivity of yaa phet (*Arundinaria pusilla*) in dry dipterocarp at Sakaerat Experimental Station. MSc thesis, Bangkok, Univ Kasetsart, Fac For (Thai: English abstract) (unpublished)

Chidumayo EN (1988) A re-assessment of effects of fire on miombo regeneration in the Zambian Copperbelt. J Trop Ecol 4:361–72

Goldammer JG (1986) Burma: forest fire management. (Report prepared for the Government of Burma). UNDP, FAO, Rome

Goldammer JG (1987) Wildfires and forest development in tropical and subtropical Asia: outlook for the year 2000. In: Proc Symp Wildfire 2000. Berkeley, California: Pacific Southwest Forest and Range Experiment Station, Forest Service, US Dep Agric, Gen Tech Rep PSW-101, pp 164–176

Goldammer JG (1988) Rural land-use and wildfires in the tropics. Agrofor Syst 6:235–252

Guy PR (1989) The influence of elephants and fire on a *Brachystegia-Julbernardia* woodland in Zimbabwe. J Trop Ecol 5:215–226

Khemnark C (1979) Natural regeneration of the deciduous forests in Thailand. Trop Agr Res Ser 12:31–43, Tropic Agric Res Cent, Minist Agric For Fish, Jpn

Kutintara U (1970) Regeneration of teak in Thailand. Msc Report. Colorado State Univ, Fort Collins, Colorado

Liacos L (1986) Ecological and management aspects of livestock animal grazing and prescribed burning in Mediterranean warm coniferous forests. In: ECE/FAO/ILO Seminar on Methods and equipment for the prevention of forest fires. Valencia, Spain, Icona, Madrid, pp 168–175

Marshall HN (1959) Elephant Kingdom. Travel Book Club, London

Mather AD (1978) Prescribed burning. Tech Note 17. Chiang Mai: FAO/Royal Forest Department, Thailand (unpublished)

Nalamphun A, Santisuk T, Smitinand T (1969) The defoliation of teng (*Shorea obtusa* Wall.) and rang (*Pentacme suavis* A. DC.) at ASRCT Sakaerat Experiment Station (Amphoe Pak Thong Chai, Changwat Nakhon Ratchasima). Report 27/8, 1 ASRCT, Bangkok

Rogers WA (1986) The role of fire in the management of wildlife habitats: a review. Indian For 112:846–857

Royal Forest Department (1988) Forest fire control in Thailand. Royal Forest Department, Forest Fire Control Sub-Division, Bangkok

Ruangpanit N, Pongumphai S (1983) Accumulation of energy and nutrient content of *Arundinaria pusilla* in the dry dipterocarp forest. Univ Kasetsart, Fac For, Bangkok (Thai: English abstract) (unpublished)

Santisuk T (1988) An account of the vegetation of Northern Thailand. Geoecol Res 5. Steiner, Stuttgart

Sauer CO (1962) The agency of man on the Earth. In: Wagner PL, Mikesell MW (eds) Readings in Cultural Geography. Univ Press, Chicago, pp 539–557

Seiler W, Crutzen PJ (1980) Estimates of gross and net fluxes of carbon between the biosphere and the atmosphere from biomass burning. Climatic Change 2:207–247

Stott P (1976) Recent trends in the classification and mapping of dry deciduous dipterocarp forest in Thailand. In: Ashton P, Ashton M (eds) The classification and mapping of Southeast Asian ecosystems. Trans IV Aberdeen-Hull Symp Malesian Ecol Hull: Univ Hull, Dep Geogr, pp 22–56

Stott P (1984) The savanna forests of mainland South East Asia: an ecological survey. Prog Phys Geogr 8:315–335

Stott P (1986) The spatial pattern of dry season fires in the savanna forests of Thailand. J Biogeogr 13:105–113

Stott P (1988a) The forest as Phoenix: towards a biogeography of fire in mainland South East Asia. Geogr J 154:337–350

Stott P (1988b) Savanna forest and seasonal fire in South East Asia. Plants Today 1:196–200

Stott P (1989) Lessons from an ancient land: Kakadu National Park, Northern Territory, Australia. Plants Today 2:121–125

Sukwong S, Dhamanityakul P (1977) Some effects of fire on dry dipterocarp community. In: Proc Symp Management of Forest production in Southeast Asia. Bangkok, pp 380–390

4 Fire in the Pine-Grassland Biomes of Tropical and Subtropical Asia

J.G. GOLDAMMER[1] and S.R. PEÑAFIEL[2]

4.1 Introduction

The genus *Pinus* is one of the most widely distributed genera of trees in the extra-tropical northern hemisphere (Critchfield and Little 1966; Mirov 1967). In the evolutionary history of its approximately 105 species two main centers of speciation are recognized. These are in southeastern Eurasia and southern North America, from where *Pinus* extends into the tropics. The range of extension of the pines in tropical South Asia has been described by Critchfield and Little (1966), Kowal (1966), Cooling (1968), and Stein (1978). Schweinfurth (1988) has compiled additional information and references on the distribution of pines in South East Asia provided by vegetation maps of the Himalayas, China, Cambodia, Viet Nam, Sumatra, and Thailand.

In tropical South Asia the biogeographical range of the genus *Pinus* is confined to the zone of lower montane rain forest and dry sites with a slight to distinct seasonal climate. In the perhumid equatorial rain forest biome of the lowlands of South East Asia the pines do not occur naturally. However, palynological data give evidence of *Pinus* occurrence in Northern Borneo up till the Pliocene (Muller 1972), thus reflecting the different climatic conditions at that time.

In a brief and general survey on the occurrence of conifers in tropical forests of Asia, Whitmore (1984) stated that the range of pines has been extended by anthropogenic and natural disturbances. The pines are pioneers (seral species) that easily colonize landslide sites and abandoned cultivation lands. Furthermore, pines are strongly adapted to and favored by fire. Both the flammability of the pine fuels and the various adaptive traits of pines to fire characterize most pine forest communities as a fire climax.

In tropical and subtropical Asia, most of the lands bearing pine forests are under increasing human pressure. Slash-and-burn cultivation extending into higher altitudes and steeper slopes, excessive fuelwood cutting, and grazing practices have brought an increasing frequency of man-caused fires into the mid-elevation forests. In many places the "fire-hardened" pine forests become more and more degraded because of shorter fire-return intervals and secondary

[1]Department of Forestry, University of Freiburg, 7800 Freiburg, FRG
[2]Protected Areas of Wildlife Bureau, Department of Environment and Natural Resources, Quezon City, The Philippines

Fig. 1. The regular influence of fire over centuries has created open, park-like stands of *Pinus kesiya* in the Central Cordillera of Luzón. The young stand in the background is reforested and protected from fire

effects of fire. Nowhere else can the ambiguous role of fire be better recognized than in the tropical pine forest biomes.

This chapter aims to highlight this dual role of fire in ecosystem processes of pine forests and associated vegetation in South Asia. Main emphasis is laid on the pine forests and grasslands of Northern Luzón, The Philippines (Fig. 1). Additional information on the fire ecology of pine forest communities in mainland South Asia and in Central America is found in this Volume (Stott et al.; Koonce and González-Cabán). The use and the impact of fire in pine plantations is covered by the contribution of De Ronde et al. (this Vol.).

4.2 Adaptive Traits of Tropical Pines to Fire

Forests and other wildlands, e.g., the various types of tree, brush, and grass savannas, which are regularly influenced by natural or anthropogenic short- to medium-return interval fires can be considered as a fire climax. In a fire climax the dominating tree species and associated vegetation are characterized by either fire tolerance or fire dependence (effects of fire selection). Furthermore, fire-dependent plant communities burn more readily than nonfire-dependent communities because natural selection has favored development characteristics that make them more flammable (Mutch 1970).

The elimination of regular fire influence from these plant communities would allow the gradual development of seral stages finally leading to a nonfire climax. In the course of a seral development toward a nonfire climax, the distribution pattern and biomass load of fire-tolerant/dependent plants vs. fire-susceptible plants and the flammability of the plant community change considerably.

The main pine species forming fire climax communities in the tropical and subtropical climate zone of Asia and within its immediate area of influence (e.g., along the southern slopes of the Himalayas) are *Pinus kesiya* Royle ex Gordon, *Pinus merkusii* Jungh. and de Vriese, and *Pinus roxburghii* Sarg. These species have developed adaptive traits to fire which are described below.

4.2.1 Character of Bark

Of all the protective mechanisms of the tree, the bark is the most important (Martin 1963; Brown and Davis 1973). The heat-insulating capacity of the bark layer depends on those characteristics which have influence on heat conduction (structure, density, moisture content, thickness). With most of the pine species, bark thickness is strongly related to age. Old trees of the Asian pines *P. kesiya*, *P. merkusii*, and *P. roxburghii* generally develop a thick bark in the lower part of the bole, thus making the tree resistant to low- to medium-intensity surface fires. Kowal (1966) found that the survival of *P. kesiya* in the Central Cordillera of Luzón (ca. 2200 m) after wildfire was practically assured when DBH was 6 cm (at the age of ca. 11 yrs) and more.

An altitudinal gradient of bark thickness and fire resistance of *P. roxburghii* (and *Shorea robusta* as well) in the Central Himalaya is described by Singh and Singh (1984, 1987). The findings show that relative density of thick-barked trees is highest in elevations up to ca. 1800 m where fire is frequent, and declines rapidly above ca. 1800 m, where the influence of frequent burning is negligible. In the elevations of ca. 2500 m and above, forests consist primarily of fire-susceptible evergreen broadleaf species among which *Pinus wallichiana* is mixed.

The ability of healing physical damage by fire (bole injury, fire scar) is another characteristic feature of tropical Asian pines. Surface fires driven by wind or running uphill create hotter and more persistent flames in the lee (or uphill) side of tree stems. If the living tissue of the downwind parts of the stem is killed, the affected bark sloughs off. The surviving cambium forms callus tissue at the edge of the killed area, and the lateral wood rings eventually meet and cover the dead scar. This phenomenon, which was first described in *P. roxburghii* in India (Champion 1919), is an excellent source of dating historical fires and reestablishing historical fire regimes of tropical and subtropical pine forests (Fig. 2).

Fig. 2. Fire scars reveal the fire history of *Pinus kesiya* forests. This 72-year-old tree was recovered near Bobok, Benguet Province, Luzón

4.2.2 Rooting Habit

Roots generally have thin cortical covering and, if stretching near the surface, are easily affected by fire. The deep vertical rooting habit of the tropical Asian pines makes these species less susceptible to fire damage unless they are exposed by erosion. The seedling or "grass" stage is the most critical phase of development and growth of pines in a fire environment. Both *P. roxburghii* and *P. merkusii* have developed remarkable adaptive traits to this fire environment by formation of an extremely thick bark at the basal part of the seedling and by resprouting (see below). Figure 3 shows a *P. roxburghii* seedling which had repeatedly been burned over and affected by cattle trampling and browsing. The cross-sections show stimulated cortex formation at the upper part of the rootstock which was exposed at the soil surface, thus minimizing the impact of the frequent needle litter fires on the plant.

Fig. 3a-d. Repeatedly burned *Pinus roxburghii* seedling with carrot-like root, recovered near Nainital, Uttar Pradesh, India (**a**). The cross-section in the upper part of the root (in the area of fire impact) shows stimulated cortex formation (**c**). Cortex formation ca. 5 cm below soil surface is considerably less (**b**, same scale as **c**). Dormant buds are found in the thickened cortex of the stem base (**d**)

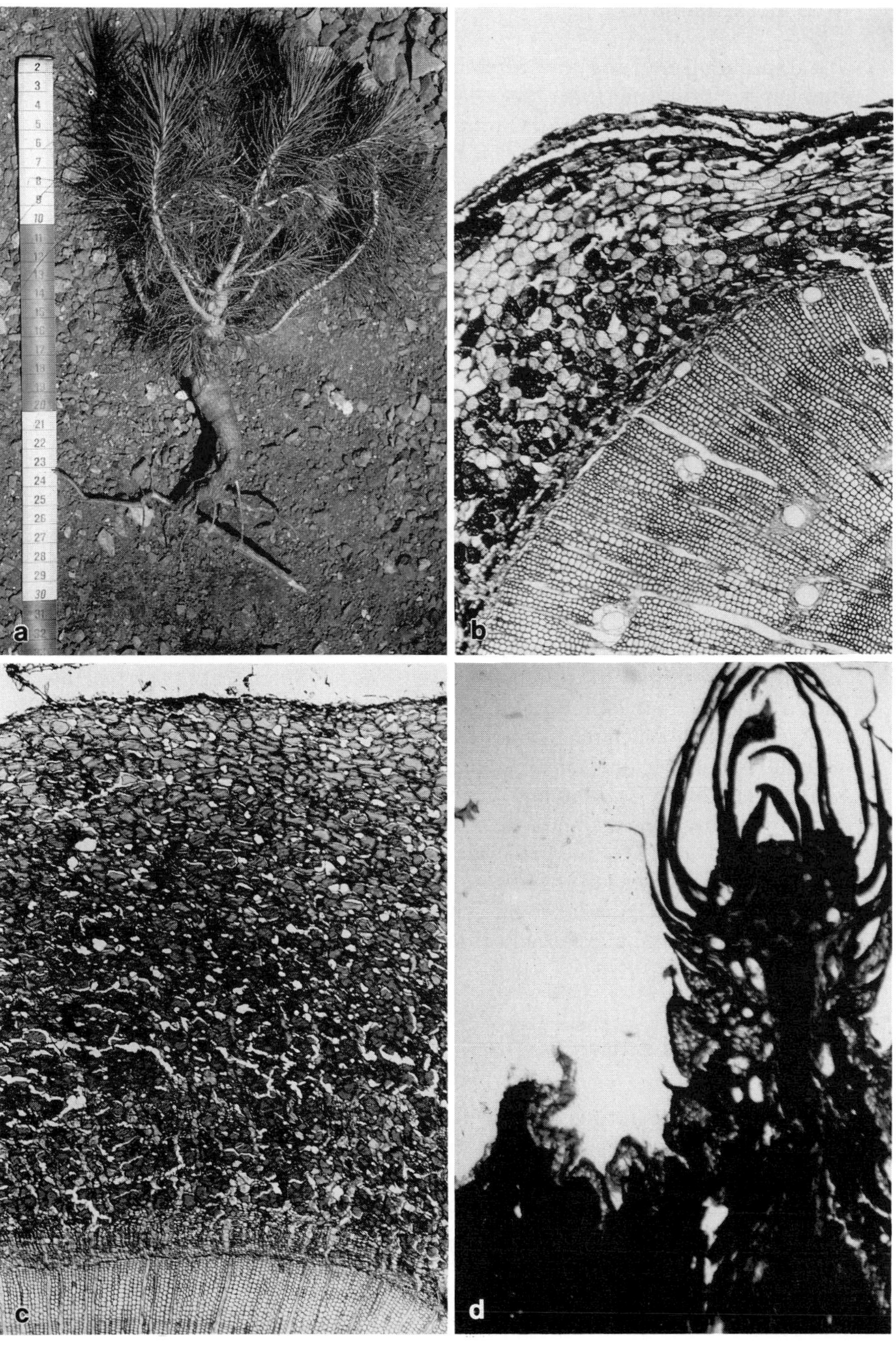
a
b
c
d

4.2.3 Basal Sprouting

Basal sprouting of young pines after fire injury has been described for a limited number of North American and Central American pines (*P. echinata*, *P. taeda*, *P. rigida*, *P. oocarpa*) (Stone and Stone 1954; Little and Somes 1960; Phares and Crosby 1962; Venator 1977). The resprouting capability is a prerequisite adaptive trait of *P. merkusii* to survive in a grass stage as described by Stott et al. (this Vol.). The ability of *P. roxburghii* seedlings to coppice after fire injury or browsing was first reported by Troup (1916). Our investigations show that dormant buds of *P. roxburghii* are embedded in the thickened cortex of the stem base (Fig. 3d). Both the heat insulating and the sprouting capability of the cortex characterize this plant as highly specialized to cope with a fire environment.

4.2.4 Site and Fuel Characteristics

Needle litter fall and fuel accumulation has been described for *P. roxburghii* (Mehra et al. 1985; Chaturvedi 1983) and for various subtropical and tropical pine plantations (Goldammer 1983; De Ronde et al. this Vol.). The accumulation of needle litter layer in *P. roxburghii* forests is explained by the high C:N ratio found in the litter and soils of the pine forest sites (Singh and Singh 1984). Furthermore, the decomposers on litter with high C:N ratio immobilize available N from the soil solution. It was suggested by Singh et al. (1984) that this fact, in addition to the fire-promoting character of the needle layer, is the main strategy through which pine invades a disturbed oak area and is partially able to hold the site against possible reinvasion by N-demanding broadleaved species. The heavy lopping and fuelwood cutting of oaks in the Himalayan mixed oak-pine forests reduces the N return through leaf litter fall and thus gives additional advantage to the low-N-demanding pine. Altogether these site and fuel characteristics lead to an increase of fire occurrence and gradual expansion of pure fire climax pine lands into sites formerly occupied by mixed broadleaf-pine forests and pure broadleaf forests.

4.3 Origin and Extent of Fires

In the seasonal pine forests and mixed pine-broadleaf forest biomes of tropical Asia lightning-struck trees and lightning fires have been reported occasionally (e.g., Osmaston 1920). Under today's man-caused fire pressure, however, lightning as a natural fire source becomes relatively less important compared to the extent and frequency of anthropogenic fires. The main causative agencies of tropical wildfires which are related to the social and cultural environment are escaped shifting cultivation fires, fires set by graziers, hunters, and collectors of nonwood forest products and fires originated at the wildland/residential interface (Goldammer 1988); in addition, numerous fires are started by carelessness (cigarettes, torches, playing children).

Although a variety of local and traditional slash-and-burn cultivation methods exist, the fires are generally set at a similar time of the dry season and time of day when weather conditions are suitable for successful burning. Many of the land-clearing fires then escape into the drought-stressed surrounding forest lands because wildfire risk and fire behavior are poorly understood and preventive measures to confine the fires are hardly taken. Fires started by forest travelers and collectors of nonwood-forest products are very common throughout the seasonal mixed pine-broadleaf forests of tropical Asia. Fires set by graziers and hunters aimed to stimulate growth of palatable grasses are encountered in many rural cultures of the world (see Bartlett 1955, 1957, 1961). Occasionally, local particularities evolve, such as the habit of graziers to burn on the steep slopes of the Indian Himalaya. The needle litter layer of *P. roxburghii* occupying these slopes is extremely slippery and dangerous for cattle and therefore is burnt annually (Goldammer 1988).

The extent of wildfire occurrence and the area of seasonal forests annually affected by fires are not known because of the lack of systematic monitoring of the fire scene (see Malingreau this Vol.). Most fire statistics based on ground observations greatly underestimate the fire occurrence. This is not only due to the general difficult accessibility of the terrain. Fire reporting in many cases is restricted to plantation-type forests and not related to other forest land. Fire figures in many places are kept low in order to prevent sanctions against responsible fire control officers.

It is interesting to note that two independent assessments on annual fire occurrence in Burma (Goldammer 1986) and Thailand (Royal Forest Department 1988) show a similar dimension. In Burma the appraisal was based on extensive communication with various authorities of the Forest Department and yielded a total estimate of up to 6.5×10^6 ha $\times$ yr^{-1} of forest fires (equaling ca. 14% of the total forest cover of the country). This number comprises the pine forests, which covered a total area of ca. 300,000 ha in the early 1980's (FAO/UNEP 1981), and which are predominantly burned in short-return intervals.

The Thailand figures reveal annual fires on 3.1×10^6 ha equaling ca. 21% of the forested area. The source does not provide information about the extent of fires in the pure or mixed pine forests.

Little exact information about the area affected by fires in the *P. kesiya* forests in the Philippines is available. Most of the ca. 238,000 ha of closed and open pine forests is found in the Central Cordillera of Luzón (DENR 1988). The small reported fraction of severe wildfires does not represent the total amount of pine forest-grassland biomes regularly burned over (see Goldammer 1985, 1987a). Similar uncertainties about fire occurrence and size in the pine forest biomes exist in all other South Asian countries, especially in India, Kampuchea, Lao, Nepal, Pakistan, and Viet Nam (Goldammer 1987b).

4.4 Management Considerations

This section aims to elaborate on a case study by demonstrating the complexity of the *Pinus kesiya* fire climax forest in Northern Luzón, The Philippines. As was stressed in the introductory remarks, fire plays an extremely ambivalent role in the development and future of the tropical pine forest ecosystems. On the one hand, the steadily increasing impact of swidden agriculture (kaingin) and carelessly escaping kaingin fires has resulted in extended deterioration of the steep highlands of Northern Luzón. On the other hand, the influence of kaingin and grazing fires has created a forest environment capable of providing sustained yield of valuable timber and other forest resources, ensuring at the same time the landscape potential and the survival of the indigenous mountain population. Both the ecological and managerial challenges are highlighted in the following.

4.4.1 Distribution of the Pine-Grassland Fire Climax in Luzón

Pinus kesiya Royle ex Gordon (formerly *Pinus insularis* Endl.) occurs mainly in the Central Cordillera of Northern Luzón at 120° W and between 15°30′N and 18° 15′N within an altidudinal range of 750 to 2450 m. The pine, locally called Benguet Pine, is also found between 600 and 1400 m on the spurs and ridges of the Caraballo mountains, and a small provenance occurs in the Zambales mountains at altitudes from 450 to 1400 m (Armitage and Burley 1980). Recently, plantations have been established below these boundaries.

As was suggested by Kowal (1966), the original vegetation of the Central Cordillera, before being disturbed by kaingin and grazing fires, most likely was a broadleaf forest (lowland rain forest at the lower elevations, grading into tropical montane forest and "mossy" forest in the higher elevations). Kowal (1966) stressed that the pines, being essentially pioneers, expanded into gaps of broadleaf forest associations which were created by human disturbance. Kowal's impressions of the history of fire-influenced development of the mountain forest vegetation and the potential future development are shown in his slightly modified model (Fig. 4). This model shows an increase of formerly patchy pine occurrence toward a broad submontane fire climax pine belt. At the interface between pine and broadleaf forest two main processes are recognized:

- Pine forest advancing. When fires are fairly frequent, each fire injures or kills plants at the edge of the fire-susceptible broadleaf forest. As a result, the pine forest advances slowly into the broadleaf forest area.
- Broadleaf forest advancing. When fires are less frequent, of lesser intensity, or even suppressed, the broadleaf forest reoccupies the fire climax pine land.

Within the pine belt it can be observed that the broadleaved species largely survived along water bodies, mainly in wet gullies and narrow valleys.

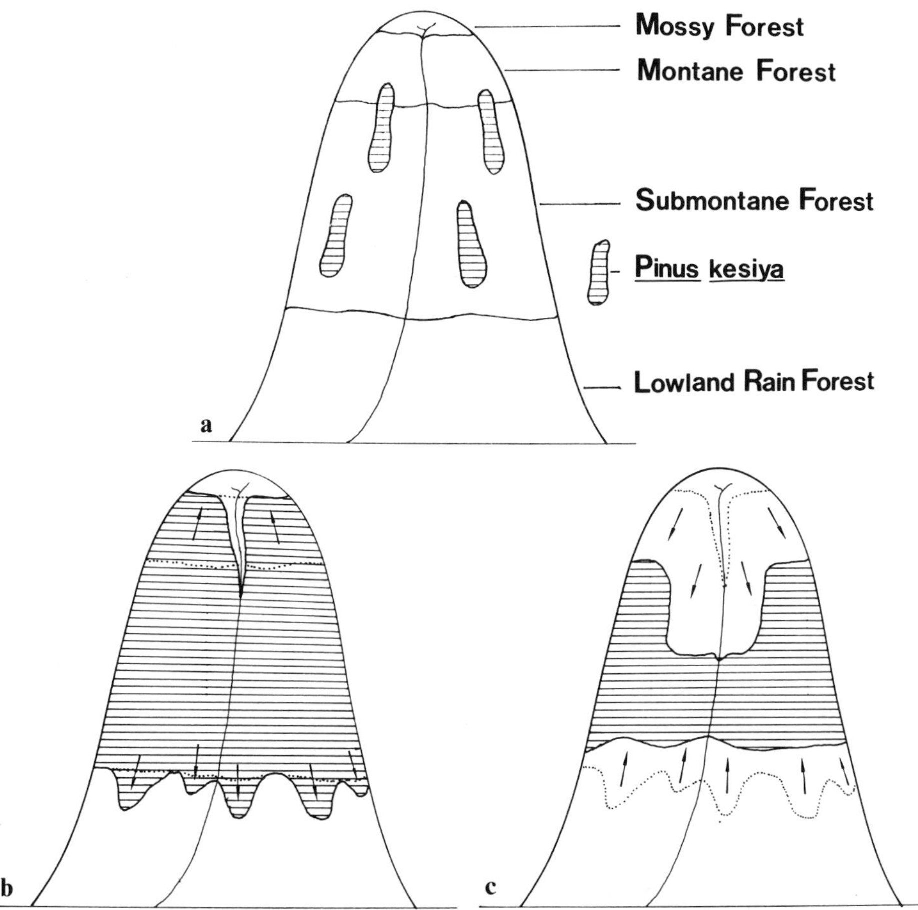

Fig. 4a-c. Probable changes of forest types in the Central Cordillera of Luzón (Philippines) during the past, characterized by scattered occurrence of *Pinus kesiya* in the submontane/montane broadleaf dipterocarp forest (**a**), present expansion of the fire climax pine belt (**b**), and probable future retreat of pines if forests are protected from fire (**c**). (After Kowal 1966)

P. kesiya forms extensive, more or less even-aged stands which, at higher elevations above 1500 m, may be densely stocked but which become more open at lower altitudes. Most of the forests consist of two strata: the pine canopy and the herbaceous layer dominated by grasses such as *Themeda triandra*, *Imperata cylindrica*, *Eulalia trispicata*, *Eulalia quadrinervis*, *Miscanthus sinensis*, and bracken fern (*Pteridium aquilinum*). During the dry season the cured grasses and the highly flammable pine litter favor the spread of surface fires which tend to kill pine seedlings and other upcoming fire-sensitive vegetation. Those pine stands that are regularly affected by short-return interval fires (1- to 3-year intervals) show hardly any successful establishment of pine regeneration (Goldammer 1985).

Fire intensity and fire impact vary according to the return interval. Annual fires usually consume the grass layer and the annual shed of needles and other dead organic matter, not exceeding a total of more than 3 to $5 t \times ha^{-1}$. After long-lasting fire exclusion wildfires tend to be of extreme intensity due to high fuel accumulation and its spatial distribution (surface fuels, understory, draped fuels); the fuel load varies considerably, depending on the understory vegetation. Since the annual rainfall in the Central Cordillera (above 2500 mm) is concentrated in the rainy season between May and October, most fires take place between the middle and the end of the dry season (end of January until the break of the rains in May).

The other anthropogenic fire climax is found in the lower elevations, mainly in the foothills of the Cordillera at the interface of agricultural and forest lands. These are grass savannas with scattered brushes and trees. Among the grasses the most important species are *Imperata cylindrica*, *Themeda triandra*, *Chrysopogon accilculatus*, and *Cappilipodium parviflorum*. The associated trees are *Piliostigma malabricum*, *Antidesma frutescens*, *Syzygium cumini*, and *Albizzia procera*. Thick bark, hard seat coats, and resprouting organs characterize these species as xerophytic and pyrophytic remnants of formerly closed broadleaved forests being capable of coping with a fire environment (Fig. 5). The annual grass fires which originate in the pasture lands generally advance upslope to the edge of the dipterocarp forest and gradually increase the grasslands.

Fig. 5. As a typical xerophytic and pyrophytic remnant of the former lowland dipterocarp rain forest, *Sycygium cumini* shows the formation of thick bark enabling this species to survive in a fire-degraded environment

The shrinking area of the low-elevation dipterocarp forest is not only due to the progressing ecotones induced by fires from above (pine belt) and from below (foothill grass-tree savannas). With increasing population pressure the formerly observed fallow periods of the kaingin agriculture cycles have resulted in the expansion of grassland in the remaining broadleaved forest belt.

4.4.2 Main Ecological Challenges

The ambivalence of fire in ecological processes in the seral pine forests and grasslands can be demonstrated by some selected examples of species composition, site degradation, and insect infestation.

4.4.2.1 Grass Species Composition and Site Degradation

In unburned *Imperata cylindrica*-dominated grasslands thick swards of undecomposed grasses cover the mineral soil. The slow decomposition process and dense growth of *Imperata* shoots has been used as explanation for the lack of other short grasses and legumes (Peñafiel 1980). In *Themeda triandra* grasslands, the net primary production of 90 days on unburned sites was ca 30 $g \times m^{-2}$ vs. ca. 135 $g \times m^{-2}$ on burned sites (Peñafiel 1980). On the other hand, burning enhances the tillering capacity of *Imperata* (Sajise and Codera 1980). With the increase of fire frequency and dominance of *Imperata*, the amount of nutrients removed by volatilization and surface runoff gradually leads to soil impoverishment and site deterioration (see Soepardi 1980; Tjitrosemito 1986; Dela Cruz 1986).

The effects of soil depletion by regular burning on surface runoff and erosion has been demonstrated in the pine-grassland biomes of the Central Cordillera. Table 1 shows the significantly higher surface runoff in a freshly

Table 1. Comparison of monthly total surface runoff[a] collected in burned and unburned plots. (Costales 1980)

Month	Total Rainfall (mm)	Total surface runoff		Difference (mm)
		Burned	Unburned	
June	57.06	7.45	1.67	5.78*
July	558.53	57.78	32.01	25.77**
August	921.95	113.28	72.82	40.46**
September	242.60	39.65	23.04	16.61**
October	105.00	25.21	17.58	7.63*
November	53.51	2.19	2.24	–0.05 NS
Total	1,938.65	245.56	149.36	96.20

[a] Mean of ten surface runoff plots in the burned and unburned areas.
NS Not significant.
* Significant at 5% level.
** Significant at 1% level.

underburned pine forest at the beginning of the rainy season. At the end of the rainy season these differences between burned and unburned sites disappear. The same trend is observed in sediment yields (Table 2). Figure 6 shows the typical erosion patterns in the pine-grassland biomes of tropical South Asia.

Comparing the annual sediment yields of an annually burned grassland, a dipterocarp forest, and a reforested grassland, Dumlao (1987) reported that the

Table 2. Comparison of monthly total sediment yield[a] collected in burned and unburned plots. (Costales 1980)

Month	Total rainfall (mm)	Total sediment yield Burned	Unburned	Tons/hectare Difference (mm)
June	57.06	0.42	0.10	0.32*
July	586.83	6.51	2.27	4.24**
August	921.95	21.28	13.52	7.76**
September	242.60	4.08	1.34	2.74**
October	105.00	3.18	1.04	2.14**
November	53.51	0.07	0.16	-0.09 NS
Total	1938.65	35.54	18.43	17.11

[a] Mean of ten surface runoff plots in the burned and unburned areas.
NS - Not significant.
* - Significant at 5% level.
** - Significant at 1% level.

Fig. 6a-c. Increasing fire and grazing pressure in the submontane fire climax pine forest lead to severe erosion processes. The photographs show the degrading *Pinus kesiya* forests in Benguet Province, Luzón, Philippines (**a**), the *P. merkusii* remnants around Lake Toba, Sumatra (**b**), and the impact of cattle trampling, browsing and fire in *P. roxburghii* in Uttar Pradesh, India (**c**)

yearly burned grassland catchment had a mean annual sediment load of ca. 3.1 $t \times ha^{-1}$ compared to ca. 1.2 $t \times ha^{-1}$ and ca. 0.8 $t \times ha^{-1}$ for a secondary dipterocarp forest and reforestation catchment, respectively.

4.4.2.2 Fire-Host Tree-Insect Interactions

The occurrence of bark beetle species (Coleoptera: Scolytidae) is a common phenomenon throughout the pine forest biomes of the northern hemisphere. In the ecology of the patchy occurrence of *P. kesiya* within the undisturbed dipterocarp (- oak) forests, scolytids played a minor role. The proliferation of the pines by fire and the formation of an almost pure pine belt, however, have created host type stands suitable for the build-up and heavy outbreaks of bark beetle populations subsequently introduced from Central America.

With the introduction of the *Ips interstitialis* (Eichhoff) [probably *Ips calligraphus* (Germar)] from North America, most likely between 1935 and 1945 (Marchant and Borden 1976; Lapis 1985), the dynamic equilibrium between fire and pine forest gradually became destabilized. Still in the early 1960's only small bark beetle populations were recorded in the Central Cordillera (Caleda and Veracion 1963). In the early 1980's the beetle had dramatically increased its range and the severity of impact on the pine biome, and is now regarded as the major threat to the mountain forests of Luzón (Grijpma 1982; Lapis 1985).

Widely accepted observations suggest that, besides drought stress (e.g., the impact of the ENSO event on Asia in 1982–83, see Goldammer and Seibert this Vol.), the increasing fire occurrence and fire-induced damages were responsible for this recent development.

In a series of investigations employing monitor traps since 1985, Goldammer and Peñafiel (unpublished data) found that the occurrence of *I. interstitialis* in freshly burned-over forests was higher than in unburned control sites. The immediate increase of the number of bark beetles caught in pheromone traps up to three times during the first 24 h after burning and the subsequent decrease to the pre-burn population density level was observed in various experiments. The success of infestation, however, was generally nil if no crowning fire had occurred in the pine stands. Partially scorched trees without short-term and long-term (1 year) change in the regime of oleoresin exudation pressure (see Vité 1961; Goldammer 1983) provided resistant to successful infestation.

The area attractiveness of freshly burned stands to increased aggregation of bark beetles is explained by the release of monoterpenes through the heating process in resinous fuels (due to low-intensity fire and partial combustion of fuel). Field tests revealed that a high dosage of (−)-α-pinene, the major monoterpene in *P. kesiya* resin, provides in fact considerable area attractiveness (Table 3).

The ongoing research shows the complexity of the interactions between fire (man-induced)-insect (introduced by man)-host tree (fire-induced) in an en-

Table 3. Field response of *Ips interstitialis* to flight barriers baited with ipsdienol, (s)-cis-verbenol and (−)-α-pinene (1:1:18) on fresh mini clearcuts and simulated clearcuts with different treatments within pure *Pinus kesiya* stands in the Central Cordillera of Luzón, The Philippines

Series A[a]		Series B[a]	
Treatment of mini clearcuts (50 m^2)	Response of *I. interstitialis* $\bar{x}^b$ ± S.E.	Treatment of simulated clearcuts (50 m^2)	Response of *I. interstitialis* $\bar{x}^b$ ± S.E.
Fire Surface fire consuming needle litter and other ground fuels, heating/scorching of stumps	1.8a,b ± 0.60	Fire Fire in container with needle litter enriched with resin	4.5a ± 1.76
α-Pinene dispenser (5 ml)	5.2b,c ± 1.60	α-Pinene dispenser (5 ml)	20.0b ± 3.54
Resin Additional resin of *P. kesiya* in open Petri dishes	8.5c ± 2.19	Resin Resin of *P. kesiya* in open Petri dishes	4.3a ± .1.44
Control Freshly cut resinous stumps	0.3a ± 0.25	Control Paper dishes, no treatment	02.0a ± 1.00

[a] Series A: Bobok, 17–18 February 1987; Series B: Loakan, Forest Research Institute site, 21–22 February 1987. Experiment with four replications, twice interchanged. The low number of beetles is due to overall population decline in that area since 1985–86.

[b] Column means followed by different letters are significantly different: Kruskal-Wallis H-Test (Series A: F = 0.0199; Series B: F = 0.0044) and two-tailed Mann-Whitney U-Test ($p < 0.05$).

vironment which has undergone very fundamental ecological changes through a complicated pathway from stability to instability.

4.5 Conclusions and Outlook

In forest management and land-use planning, decision makers face a dilemma: Obviously, fire plays a key rôle in maintaining the tropical submontane pine forests, which at the same time offer adequate habitability conditions for man. If used properly in time and space, fire creates a highly productive conifer forest, granting landscape stability and sustained supply of timber, fuelwood, resin, and grazing land. If fire runs out of control, as it obviously does in many places of the Asian pine forest lands, the losses are great: increased runoff (floods), erosion (siltation), soil denudation (landslides) and forest degradation (destruction of regeneration, insect pests).

The pros and cons of fire protection and "let burn" in tropical seasonal broadleaf and coniferous forests have been discussed extensively since fire protection was introduced to India by Brandis in 1863 (Shebbeare 1928; see also Pyne this Vol.). Soon it became obvious that fire exclusion would bring tremendous problems in the regeneration of teak (*Tectona grandis*) and sal (*Shorea robusta*) stands (see Stott et al. this Vol.) and increase the wildfire hazard in the *Pinus roxburghii* forests. "Controlled early burning" was introduced in 1877 (Shebbeare 1928) and was one of the focal points of discussion in India's forest management in the following half century.

However, throughout the pine forests of tropical Asia no real progress has been made since then. The forest managers and fire control officers are still running behind the wildfire problem. They are not adequately prepared for an approach toward integrated fire management in which fire prescribed in time and space would allow them to profit from the benefit of fire and to exclude the negative impacts of uncontrolled wildfire.

In the Philippines, Peñafiel (1982) showed that the loss of forests by wildfires was greater than the reforestation capabilities. The call for an integrated fire management approach was mandatory. A program initiated by FAO and accompanied by a research component tried to base its concept on the historical and social fire ecology of the environment of the Central Cordillera (Goldammer 1985, 1987a). The implementation of strategies elaborated, however, stagnates due to economic constraints.

Will there be a realistic chance to pursue the ideas of integrated fire management within the tropics? The answer is yes, because the pines are one of the few genera able to cope with an environment stressed by multiple disturbance and degradation processes due to the increasing pressure of human population, fire, and grazing. The ecological plasticity of pine-grassland agroforestry systems stabilized by prescribed burning and prescribed grazing (Goldammer 1988) may even cope with the yet unpredictable climate change and physiological transition stresses. Regardless of the combustion and fermentation processes involved, these agroforestry systems may contribute to halt the net carbon flux from tropical land use into the global carbon cycle.

Acknowledgments. The authors wish to acknowledge the field support by M. Pogeyed and his fire crew, and the administrative support by R. Goze and J. Gumayagay, Department of Environment and Natural Resources, Cordillera Administrative Region, Republic of the Philippines. Cross sections of *Pinus roxburghii* were prepared by S. Fink, University of Tübingen, Federal Republic of Germany. The research was sponsored by FAO and the Volkswagen Foundation.

References

Armitage FB, Burley J (eds) (1980) *Pinus kesiya* Royle ex Gordon (Syn. *P. khasya* Royle; *P. insularis* Endlicher). Com For Inst, Trop For Pap 9, Oxford

Bartlett HH (1955) Fire relation to primitive agriculture and grazing in the tropics: annotated bibliography, Vol. 1; Vol. 2 (1957); Vol. 3 (1961). Mimeo Publ Univ Mich Bot Garden, Ann Arbor

Brown AA, Davis KP (1973) Forest fire. Control and use. McGraw-Hill, New York, 686 pp

Caleda AA, Veracion VP (1963) Destructive insect pests of Benguet pine (*Pinus insularis* Endl.). For Leaves 14 (1):19–20

Champion HG (1919) Observations on some effects of fires in the Chir (*Pinus longifolia*) forests of the West Almora Division. Ind For 55:353–364

Chaturvedi OP (1983) Biomass structure, productivity and nutrient cycling in *Pinus roxburgii* forest. Ph.D. thesis, Kumaun Univ, Nainital, India

Cooling ENG (1968) *Pinus merkusii*. Fast-growing timber trees of the lowland tropics. No 4 Com For Inst, Oxford

Costales EG Jr (1980) Some effects of forest fire on the hydrologic and soil properties of Benguet Pine watershed. MS thesis, Univ Philippines, Los Baños College, Laguna

Critchfield WB, Little EL (1966) Geographic distribution of the pines in the world. USDA For Serv Misc Publ 991, Washington, 97 pp

Cruz dela RE (1986) Constrains and strategies for the regeneration of Imperata grasslands. BIOTROP Spec Publ 25:23–34, Bogor, Indonesia

DENR (Forest Management Bureau, The Philippines) (1988) Natural forest resources of the Philippines. Philippine-German Forest Resources Inventory Project, Summary Report, Manila

Dumlao TP (1987) Soil erosion and surface run-off in secondary dipterocarp forest, annually burned and reforested grassland catchment. MS thesis, Univ Philippines, Los Baños College, Laguna

FAO/UNEP (1981) Tropical forest resources assessment project. Forest resources of tropical Asia. FAO (UN 32/6.1301–78–04, Tech Rep 3), Rome, 475 pp

Goldammer JG (1983) Sicherung des südbrasilianischen Kiefernanbaues durch kontrolliertes Brennen. Hochschul-Verlag, Freiburg i Br, 183 pp

Goldammer JG (1985) Multiple-use forest management, The Philippines. Fire management FAO (FO:DP/PHI/77/011, Working Paper 17), Rome, 65 pp

Goldammer JG (1986) Forestry and forest industries technical and vocational training, Burma. Forest fire management. FAO (FO:DP/BUR/81/001, Tech Note No 5), Rome, 60 pp

Goldammer JG (1987a) TCP assistance in forest fire management, The Philippines. FAO (FO:TCP/PHI/66053 T, Working Paper No 1), Rome, 38 pp

Goldammer JG (1987b) Wildfires and forest development in tropical and subtropical Asia: Outlook for the year 2000. In: Proc Wildland Fire 2000. USDA For Serv Gen Tech Rep PSW- 101, pp 164–176

Goldammer JG (1988) Rural land-use and wildland fires in the tropics. Agrofor Syst 6:235–252

Grijpma P (1982) Multiple-use forest management, The Philippines. Possibilities for the control of *Ips calligraphus* (Germar) in the Philippines. FAO (FO:DP/PHI/77/011, Working Paper No 7, Vol I + II), Rome, 70 + 26 pp

Kowal NE (1966) Shifting cultivation, fire, and pine forest in the Cordillera Central, Luzon, Philippines. Ecol Monogr 36:389–419

Lapis E (1985) Some ecological studies on the six spined engraver beetle, *Ips calligraphus* (Germar), infesting *Pinus kesiya* (Royle ex Gordon) in the Philippines. MS thesis, Univ Philippines, Los Baños College, Laguna

Little S, Somes HA (1960) Sprouting of loblolly pines. J For 58:195–197

Marchant MR, Borden JH (1976) Worldwide introduction and establishment of bark and timber beetles (Coleoptera: Scolytidae and Platypodidae). Simon Fraser Univ Pest Manage Pap 6

Martin RE (1963) Thermal and other properties of bark related to fire injury of tree stems. Ph.D. Diss., Univ Michigan, Ann Arbor

Mehra MS, Pathak U, Singh JS (1985) Nutrient movement in litter fall and precipitation components for Central Himalayan forests. Ann Bot 55:153–170

Mirov NT (1967) The genus *Pinus*. Ronald, New York

Muller J (1972) Palynological evidence for change in geomorphology, climate and vegetation in the Mio-Pliocene of Malesia. In: Ashton P, Ashton M (eds) The Quaternary era in Malesia. Univ Hull, Dept Geography

Mutch RE (1970) Wildland fires and ecosystems – a hypothesis. Ecology 51:1046–1051

Osmaston AE (1920) Observations on some effects of fires and on lightning-struck trees in the Chir forests of the North Garhwal Division. Ind For 56:125–131

Peñafiel SR (1980) Early post-fire effects in a *Themeda* grassland. MS thesis, Univ Philippines, Los Baños College, Laguna

Peñafiel SR (1982) Forest fire costs and losses in the Upper Agno River Basin over a four-year period. Philippine Lumberman 28(2):20–21

Phares RE, Crosby JS (1962) Basal sprouting of fire-injured shortleaf pine trees. J For 60:204–205

Royal Forest Department Thailand (1988) Forest fire control in Thailand. Bangkok, 14 pp

Sajise PE, Codera VT (1980) Effects of fire on grassland vegetation. Upland Hydroecology Program, Ann Rep Univ Philippines, Los Baños College, Laguna

Shebbeare EO (1928) Fire protection and fire control in India. Third British Empire Forestry Conference. Govt Printing Office, Canberra

Schweinfurth U (1988) *Pinus* in Southeast Asia. Beitr Biol Pflanzen 63:253–269

Singh JS, Singh SP (1984) An integrated study of eastern Kumaun Himalaya, with emphasis on natural resources. Vol 1. Studies with regional perspectives. Final Report (HCS/DST/187/76). Kumaun Univ, Nainital

Singh JS, Singh SP (1987) Forest vegetation of the Himalaya. Bot Rev 53:80–192

Singh JS, Rawat YS, Chaturverdi OP (1984) Replacement of oak forest with pine in the Himalaya affects the nitrogen cycle. Nature (Lond) 311:54–56

Soepardi G (1980) Alang-alang [*Imperata cylindrica* (L.) Beauv.] and soil fertility. BIOTROP Spec Publ 5:57–69, Bogor, Indonesia

Stein N (1978) Coniferen im westlichen malayischen Archipel. Biogeographica XI, Junk, The Hague

Stone EL, Stone MH (1954) Root collar sprouts in pine. J For 52:487–491

Tjitrosemito S (1986) The use of herbicides as an aid for reforestation in *Imperata* grassland. BIOTROP Spec Publ 25:201–211, Bogor, Indonesia

Troup RS (1916) *Pinus longifolia* Roxb.; a sylvicultural study. Ind For Mem, Sylviculture Ser 1:1

Venator CR (1977) Formation of root storage organs and sprouts in *Pinus oocarpa* seedlings. Turrialba 27:41–46

Vité JP (1961) The influence of water supply on oleoresin exudation pressure and resistance to bark beetle attack in *Pinus ponderosa*. Contr Boyce Thompson Inst 21:37–66

Whitmore TC (1984) Tropical rain forests of the Far East. Clarendon, Oxford, 352 pp

5 Fire in Some Tropical and Subtropical South American Vegetation Types: An Overview

R.V. SOARES[1]

5.1 Introduction

The great majority of the forests of the world have been burned over in various return intervals for many thousands of years. Even the perhumid tropical rain forest, where there is limited evidence of recent fires, was certainly affected by fire in the past. Sanford et al. (1985), studying charcoal fragments found in a tropical rain forest soil near San Carlos de Rio Negro, Venezuela, found radiometric ages which ranged from 250 to 6260 years B.P., indicating that fires occurred in that period (see also Saldarriaga and West 1986). Most likely these fires were associated with extremely dry periods or human disturbances.

In the past, natural fires influenced forest dynamics. Old timber stands were replaced by young trees; sometimes one type of forest was replaced by another (USDA Forest Service 1972). There is ample evidence that fire is nature's way of eliminating old stands and replacing them with new ones (Canadian Forest Service 1975).

Fire has long been considered an important element in some ecosystems. It plays a fundamental role in the succession of certain communities. Pine forests and grassland are examples of plant communities which foster high flammability and frequent fires. According to Mutch (1970), they are, by definition, fire-dependent communities.

General observations have often indicated that fire has been selective in the species it favors. The idea that fire selects individual species is derived from the view that it has always been present in certain ecosystems. Myers (1936), for instance, said he had never seen a South American savanna, however small or distant from settlement, which did not show signs of more or less frequent burning.

The effect of fire in ecosystems can vary considerably, depending on its successional stage, condition of the vegetation and fuels, and the nature of fires. According to Martin et al. (1976), vegetative succession moves plant communities from pioneer, through seral, to the climax stage, which remains on the site unless a disturbance occurs. When a severe disturbance, such as wildfire, is introduced, succession is usually retarded or moved back to an earlier stage of succession. However, moderate disturbance such as light fires may advance or set back succession, or merely hold it at a given stage. Therefore, repeated light

[1] School of Forestry, Federal University of Paraná, 80.001 Curitiba, Paraná, Brazil

fires can maintain succession in a seral stage, as occurs in the savannas of central Brazil, for instance.

It should be emphasized that the effect of fire on vegetation depends primarily upon fire intensity. A wildfire occurring in areas where exclusion has allowed heavy fuel accumulation would cause damaging effects even to fire-resistant species. On the other hand, a light fire may have minimal or beneficial effects. Natural differences in fire resistance between species and individual trees becomes of practical importance in appraising the effects of less intense or prescribed fires.

The objective of this chapter is to present a general review of the effects and influences of fire on the dominant vegetation types of South America. This contribution is amended by the chapter on the fire ecology of the "cerrado" (Coutinho this Vol.) and the lowland tropical rain forest (Kauffman and Uhl this Vol.; Fearnside this Vol.).

For the purpose of this presentation, the South American continent was divided into 11 main or dominant vegetation types (Fig. 1) as follows:

1. Tropical rain forest
2. Trade wind forest of Venezuela and Columbia
3. "Babaçu" palm forest
4. Steppe
5. Savanna ("cerrado")
6. Coastal rain forest
7. Subtropical forest
8. Brazilian pine forest
9. Steppic savanna ("chaco")
10. Subtropical and araucaria forests of Chile and Argentina
11. Nonforest areas. *a* High savannas of Guyana; *b* Orenoco plains; *c* High elevations of the Andes; *d* Pacific coast deserts; *e* Prairies of Uruguay and Argentina; *f* Xeromorphic shrublands; *g* Patagonia steppes and deserts

A brief description of every vegetation type listed above will be presented when discussing fire effects and influences, separately, on each ecosystem.

Besides these indigenous vegetation types, it is important to note the increasing importance of exotic forests planted on the continent, especially *Eucalyptus* spp and *Pinus* spp plantations in Brazil, and *Pinus radiata* in Chile. These species have been planted mainly in the ecosystems dominated by tropical rain forests, savanna, coastal rain forest, subtropical forest, Brazilian pine forest, and subtropical forest of Chile. A general discussion on the influences of fire on these planted forests will also be presented.

5.2 Tropical Rain Forest

The fire ecology of South America's tropical rain forest is covered by the contributions of Kauffman and Uhl (this Vol.) and by Fearnside (this Vol.).

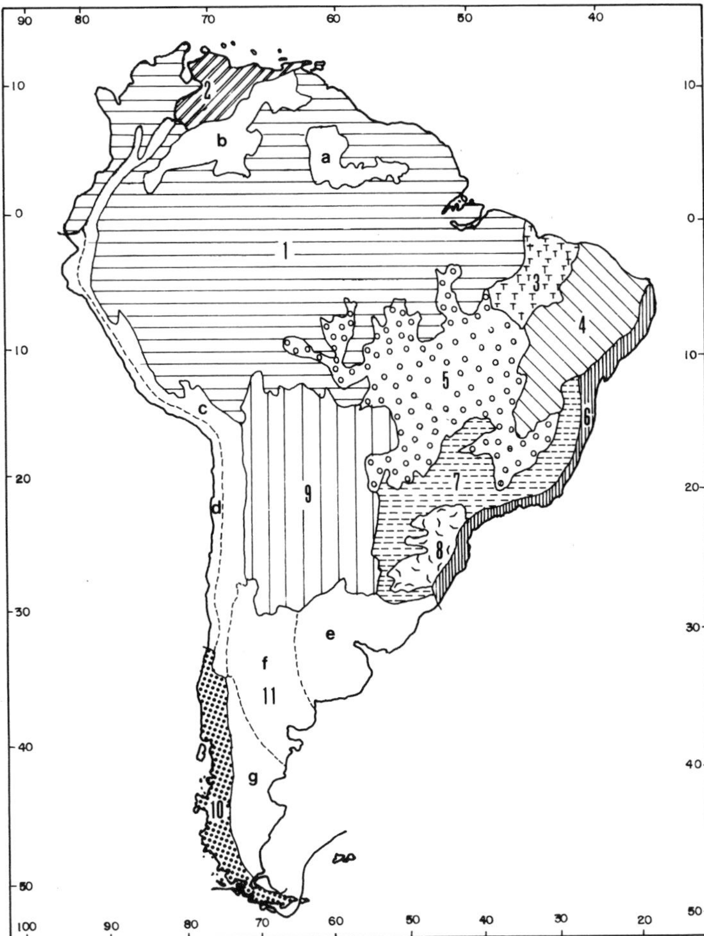

Fig. 1. Dominant vegetation types of South America. (After Hueck 1978)

5.3 Trade Wind Forest of Venezuela and Colombia

The trade wind forest, located at the extreme north of the South American continent, is a region between the nonforested plains of the Orinoco basin and the tropical rain forests from the Pacific and Caribbean.

This forest type is highly influenced by the trade winds, that with variable seasonal intensity significantly affect the landscape, climate, and vegetation of the region. The trade winds cause a notable variation in the rainfall regime of the region, with moist periods from May to December, and a dry season during the remaining months. The annual total precipitation ranges from 1200 to 2000 m, while the annual mean temperatures oscillate between 26 and 28 °C, with small variations in the monthly averages (Hueck 1978).

while the annual mean temperatures oscillate between 26 and 28 °C, with small variations in the monthly averages (Hueck 1978).

The dominant type of vegetation is deciduous trees. The forest is about 30 m high, and presents variable density. In this forest grow the economically most important trees of Venezuela. The main species found in this forest type are: *Cedrela mexicana*, *C. odorata*, *Switenia macrophylla*, *Anacardium rhinocarpus*, *Spondias mombin*, *Bombacopsis sepium*, *Ceiba pentandra*, *Centrolobium paraense*, *Hura crepitans*, *Cordia alliadora*, *Tabebuia* spp., and others (Hueck 1978).

The risk of fire is very high in the trade wind forest. Fires from the adjacent savannas or from the burning of agricultural debris move into the forest. The large amount of fuel, the dry season, and the strong winds create conditions favorable for the occurrence and propagation of fires, which can reach catastrophic dimensions.

Due to the continued felling of trees for agricultural purposes, the posterior abandonment of the land, and extensive grazing and burning practices, large areas of this forest type have been converted to savannas.

5.4 "Babaçu" Palm Forest

In northeastern Brazil, between the steppe (east side), and the Amazonian tropical rain forest (west side) is a large area covered with "babaçu" palm (*Orbignya martiana*), that completely dominates the ecosystems.

The "babaçu" palm region has a total precipitation between 1500 and 2200 mm a year. The wettest months are March and April, with over 400 mm per month. There is a remarkable dry season, from August to December. The mean annual temperature is around 26 °C, with very small monthly variation.

The "babaçu" palm forest is very homogeneous, and few species of secondary importance, like *Copernicia cerifera*, *Marutitia vinifera*, *Euterpe* spp., and species of the genera *Aspidosperma*, *Jacaranda*, *Tecoma*, *Caesalpinia*, and *Piptadenia* can be found in the ecosystem (Hueck 1978; Ferri 1980).

The seeds of "babaçu" palm forest have a proportion of 60 to 65% of fat matter. The very hard seed shell produces a high quality charcoal, with a heat content of approximately 8000 kcal/kg. The weight of the seeds is about 10% of the total weight of the fruit (Hueck 1978).

The "babaçu" palm forest is a secondary formation, resulting from felling and burning of the pre-existent forests, for agricultural purposes. The former dense forests were limited to a few palms, since palm seedlings are not able to grow in a shaded environment; but once the trees were felled and the site burned, the "babaçu" sprouted very intensely, and the palm, a pioneer species, fully dominated the site (Aubréville 1961; Pires 1964; Rizzini 1979). The periodical burning of the site has maintained this aggressive pioneer as the dominant species in the ecosystem. The continued use of fire in the region tends to maintain the dominance of the "babaçu" palm vegetation type.

5.5 Steppe

The Brazilian steppe or "caatinga" is typical of most of northeastern Brazil, covering an area of approximately 800,000 km^2. The main characteristic of this environment is the severe drought. In most of the "caatinga" region, the total annual precipitation is less than 700 mm, and in the center of the area, the "drought polygon", the rainfall ranges from 250 to 500 mm a year, or less. The relative humidity is very low, varying annually from 50 to 60%, or even less in some areas. The mean annual temperature is high, over 26°C, with small monthly variation (Ferri 1980; Hueck 1978).

The Brazilian steppe is very heterogeneous in physiognomy and structure. However, its composition is quite uniform, with a predominance of trees and scrub species, and several Cactaceae scattered throughout the area. The vegetation is mainly deciduous, thorn, and xerophytic, composed of woody species, and quite rich in rigid Cactaceae and Bromeliaceae.

The main and typical arboraceous species of this ecosystem are *Zizyphus joazeiro*, *Cavallinesea arborea*, *Chorisia crispiflora*, *Hymenea* spp., *Melanoxylum brauna*, *Anacardium occidentale* (cashew), *Dalbergia* spp., *Caesalpinia* spp., *Caryocar glabrum*, *Astronium urundeuva*, *Amburana cearensis*, *Aspidosperma pyrifolium*, and others less common (Rizzini 1979; Hueck 1978).

The creeping vegetation, herbs, and grasses of this environment sprout only in the wet season, and for this reason they are not found during most of the year. As a consequence, fire is not used in the "caatinga" for pasture improvement. The use of fire is limited to some agricultural areas, to clean the land after cutting the existing vegetation. Climate rather than fire is responsible for this typical environment of northeastern Brazil.

5.6 Savanna or "Cerrado"

The fire ecology of the "cerrado" savannas is covered by the contribution of Coutinho (this Vol.).

5.7 Coastal Rain Forest

The coastal rain or Atlantic forest follows almost all of the eastern coast of Brazil, from the State of Rio Grande do Sul in the south, to the State of Rio Grande do Norte in the northeast. The structure of this forest is similar to the Amazonian forest, with several species common to both.

The Atlantic forest was the first vegetation type to be exploited and destroyed by the Brazilian colonizers. The first forest species to be exploited by Portuguese and French colonizers, beginning in the year 1500, was the "pau brasil" (*Caesalpinia echinata*). Due to intensive exploitation, it is now practically extinct in its natural environment. Several valuable broadleaf species occur in the Atlantic forest of Brazil. These include *Dalbergia nigra* ("jacarandá"), probably the most valuable wood in all South America, *Bowdichia virgiloides, Aspidosperma peroba, Cedrela fissilis, Cabralea cangerana, Myrocarpus fastigiatus, M. erythroxylon, Machaerium lanatum, M, firmum, Andira anthelmintica, Nectandra* spp., *Piptadenia comunis*, and several others (Hueck 1978; Rizzini 1979).

The total annual precipitation in the Atlantic forest ecosystem varies from 1000 to a little over 2000 mm. In northeastern Brazil, 1000 mm is about the minimum required to maintain the rain forest. The annual mean temperature varies from about 16°C in the southern limit of the forest up to 25°C in the northern limit. Ecologically, there is no dry season in this ecosystem, although in abnormally dry years, some dryer months, especially in the winter, may occur.

The Atlantic forest has been nearly destroyed by exploitation, and land occupancy for agricultural and pastural purposes. Only in the southern Brazilian states of Sao Paulo, Parana, and Santa Catarina there are significant areas of this forest type remaining.

In the center of the ecosystem, in the states of Bahia and Espirito Santo, where the most valuable species of this forest type occurred, the devastation was very intense. The forest was clearcut, the most valuable species sold to sawmills, others used for charcoal production, and the residues broadcast burned for preparing the land for agriculture and pasture. Presently, large areas are being planted with *Eucalyptus* spp in this part of the former Atlantic forest in areas abandoned by agriculture, in order to supply wood for two big pulmills established nearby.

Forest fires have occurred periodically in the Atlantic forest, from fires used for burning agricultural residues, pasture improvement, and cleaning sugar plantations before harvesting. However, there is no evidence of fire occurrence in this forest type in the past. The environment was quite wet, and probably fire was not part of the natural ecosystem.

5.8 Subtropical Forest

On the western side of the central and southern coastal rain forest emerges an extensive forest region that presents characteristics different from that of the rain forest. Most of the trees are evergreen, with a certain proportion of deciduous species.

The development of this type of forest is related to the existence of two distinct climatic seasons, one wet and the other dry, or an accentuated thermal variation, i.e., one warm the other cold. These climates determine a foliar seasonality of the dominant arboraceous elements, which present an adaptation to the hydric deficit, or to low temperatures in the coldest months. In the tropical areas of this ecosystem there are different periods, one wet and the other dry, with an annual mean temperature around 22°C. In the subtropical portion, the annual mean temperature is approximately 18°C, with a cold period with temperatures in the vicinity of 15°C or less, but without a marked dry season, excepting for short periods in some dryer years (Veloso and Góes-Filho 1982).

Economically, this is also a very important region for the lumber industry. Several valuable species are present, including *Machaerium* spp., *Myrocarpus frondosus*, *Aspidosperma peroba*, *Cedrela fissilis*, *Cassia ferruginea*, *Enterolobium contorsiliquum*, *Schizolobium excelsum*, *Apuleia ferrea*, *Balfourodendron riedlianum*, *Phoebe porosa*, *Nectandra* spp., *Luehea divaricata*, *Cordia trichotoma*, *Tabebuia* spp., *Euterpe edulis*, and many others (Hueck 1978; Rizzini 1979).

Extensive areas of this forest type were destroyed from the beginning of this century until the 1950's for the establishment of the large coffee plantations in the southern Brazilian states São Paulo and Paraná. Presently, vast soybean and sugar cane plantations have been established in the area formerly occupied by the subtropical forest. The most fertile Brazilian soils are located in this ecosystem, and for this reason, the expansion of agriculture drastically reduced this type of forest, which at the present occupies a very small area.

Land occupation for agricultural purposes in this area was similar to other regions of South America: clearcutting the forest, utilization of the valuable species by the lumber industry, and drying and burning the residues. Fire is still used in the region for debris burning and pasture improvement, sometimes causing severe wildfires, because the farmers burn in the winter or early spring, when fire danger is high. In part of the area affected, a big fire occurred in the State of Paraná in 1963, burning about 2×10^6 ha in the subtropical forest ecosystem, although most of the fire burned in the Brazilian pine forest.

5.9 Brazilian Pine Forest

Brazilian pine or paraná pine (*Araucaria angustifolia*) is the most important indigenous species in the forest economy of South America and the only valuable conifer to occur in Brazil. Besides the natural stands still existing in the species distribution area, approximately 12% of the new forests planted in the state of Paraná are Brazilian pine trees (Soares 1977).

The natural stands of araucaria originally occupied an area of approximately 197,000 km² along the southern states of Brazil and a small part of Argentina at elevations of 500 m above sea level in the southern limit (Hueck 1978). Presently, most of this valuable forest is gone, and the species itself is threatened by extinction. Exploitation is the main reason for this threat, but another important factor is often ignored: the lack of knowledge on the behavior and requirements of the species, especially regarding its natural regeneration. Several old and decadent Brazilian pine stands are progressing toward a hardwood community and there is no natural regeneration of the species beneath those stands.

The araucaria forest is, in fact, a mixed conifer-hardwood forest. The pine stands out because mature individuals tower above the other species and because its characteristic umbrella shape distinguishes it from other trees.

Important hardwood species are associated with the araucaria, like *Phoebe porosa*, *Cedrella fissilis*, *Ilex paraguariensis*, *Luehea divaricata*, *Nectandra* spp., and other less important ones. There are two other conifers associated with the araucaria forests, *Podocarpus lambertii*, a smaller tree, and *Podocarpus sellowii*, both of little economic value.

Although a dominant tree, the araucaria is a seral species, and if succession progresses toward advanced stages, the hardwoods would dominate the forest. The Brazilian pine is not a climax species and there is no natural regeneration in the understory of a dense native stand. For this reason the species remains on the site only if a disturbance impedes succession.

The Brazilian pine does not exhibit the basic characteristics of a pioneer species either, e.g., high mobility, resistance to harsh environments, and the need for full sunlight to regenerate. Araucaria is not mobile because its seeds are big (4 to 7 cm long), heavy, wingless, and lose their germination capacity very quickly. The seedlings need some protection against adverse climatic conditions even in the natural range of the species. Rogers (1953) related that poles up to 3 m high can be killed by severe frosts in their natural environment. Araucaria, when a young plant, grows better in a partially shaded environment. Inoue and Torres (1980) concluded that light intensities of about 25% produced taller young plants. Maack (1968) related, based on experimental plantations, that better tree growth was obtained when araucaria was planted under the protection of crown of pioneer species. In commercial plantations of the species it was also observed (not experimentally) that the seedlings grew better when corn was cultivated between the araucaria rows.

Based on this evidence, it seems that araucaria could be more properly defined as an intermediate or seral species. This hypothesis would explain the lack of natural regeneration of the species in old natural stands. Araucaria needs some kind of disturbance, that partially opens the canopy and eliminates the competition of creeping species to regenerate. This fact can be observed in native araucaria forests where the leaves and twigs of *Ilex paraguariensis* ("mate") are harvested. In order to create conditions for working in the forest and pruning the "mate" trees, it is necessary to partially clean the understory, and as a consequence araucaria regeneration becomes abundant. In this case, a disturbance was produced, favoring natural regeneration of the species.

It is probable that in the past periodic fires occurred that produced favorable conditions for the species to regenerate and kept succession in seral stages, dominated by pine trees.

In addition to the use of fire by Indians that lived in the region, the association between lightning-caused fires and araucaria forests has also been demonstrated. Soares and Cordeiro (1974) studied forest fire occurrence in the central part of Paraná for 10 years and concluded that lightning was responsible for 20% of the fires in that period. When the native araucaria forests started to be harvested and replaced by *Eucalyptus* spp, the number of lightning-caused fires decreased to the present level of 1% in the same region. Therefore, a close relationship between araucaria forests and natural fires seems to have existed. The local thunderstorms are always accompanied by rain, and only highly flammable vegetation can support a fire under those conditions. Araucaria has a high oleoresin content in its leaves and can keep a low intensity fire burning even under high humidity conditions.

Considering these facts, and according to the theory of Mutch (1970), araucaria could be considered a fire-dependent species, and it exhibits important characteristics of the fire-dependent species. It is very flammable and resistant to fire, and at maturity possesses thick bark (5 to 10 cm), a branchless bole, and a high crown (Figs. 2, 3). Therefore, low intensity fires kill the understory and competitive broadleaves, but do not harm the araucaria, which regenerates maintaining the seral stage. The fire favors the conifers over the broadleaves for the occupancy of the site. This fact may be responsible for the existence of the Brazilian pine forests, that, without disturbance, could not have competed with the more aggressive and specialized broadleaves. For this reason, low intensity prescribed fires could be a good alternative in the management of araucaria forests, mainly for inducing natural regeneration.

Fires in the araucaria forest region are more frequent between July and September, a period of lower precipitation, and also when fire is intensively used

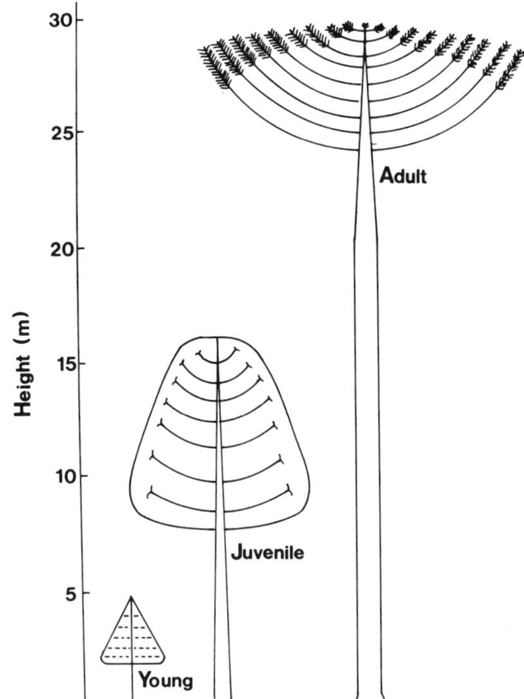

Fig. 2. Characteristic shape of the crown of Brazilian pine (*Araucaria angustifolia*) at different stages of life. (Soares 1977)

Fig. 3. Characteristic shape of crowns of *Arancaria augustifolia* in Paraná, Brazil. (Pholo by J.G. Goldammer)

for agricultural land cleaning. During these months temperature and relative humidity are lower. Therefore, the fire season occurs in the winter, and forest fires are related more to relative humidity than air temperature (Soares 1972).

In the range of natural distribution of the araucaria forest, total rainfall is close to 1400 mm a year, exceptionally surpassing 2000 mm, but never less than 1000 mm (Hueck 1978). The mean annual temperatures oscillate between 13 and 19°C, seldom more or less. A typical example of the araucaria forest climate is that from Curitiba region, in Paraná State (Table 1).

Dry winters have been observed at 3- to 4-year intervals in the state of Paraná during the last three decades. In these periods, relative humidity can reach values as low as 15% or less, as in August 1963. Due to these adverse climatic conditions, the largest forest fire ever to occur in Brazil burned an area of approximately 2×10^6 ha, mainly in the araucaria forest region (Fig. 4). About 500,000 ha of native forests (Soares 1972), principally natural Brazilian pine stands, were affected by the fire, and the mature trees of this species were the most resistant to the fire due to the thick bark and high crowns. However, despite its resistance to fire, the araucaria will probably not outlive the present intensive exploitation, and the extinction of this highly valuable forest type can be predicted.

Fire is also an important factor in the regeneration and management of the "bracatinga" (*Mimosa scabrella*), a native pioneer species of the araucaria forest ecosystem. The "bracatinga" exhibits a fast initial growth, and is used by small farmers to produce firewood and to protect the soil. The clearcutting of

Table 1. Monthly mean temperature and total precipitation in the region of Curitiba, Paraná State, Brazil, from 1931 to 1960. (Ministry of Agriculture, Brazil)

Month	Mean temperature (°C)	Total precipitation (mm)
January	20.1	198.5
February	20.1	173.2
March	19.2	123.7
April	16.8	78.4
May	14.5	85.0
June	13.2	87.5
July	12.5	81.2
August	14.0	82.7
September	14.8	119.4
October	16.2	130.3
November	17.4	105.4
December	18.9	147.4
Average/total	16.5	1412.7

Fig. 4. In August 1963 an area totaling ca. 2×10^6 ha was burned in the State of Paraná. Most affected were native araucaria forests and afforestations of *Pinus* spp., *Eucalyptus* spp., and *Araucaria angustifolia*. The photo shows the wildfire effects at the interface between native araucaria forest and pine plantation, Klabin do Paraná, September 1981. (Photo by J.G. Goldammer)

"bracatinga" stands is done when they reach 6 to 8 years of age. After the clearcut, the land is burned because the heat provided by the fire is necessary to break the dormancy of seeds and to enhance abundant germination. After burning, the farmers seed corn and beans, thinning the "bracatinga" seedlings to provide more space for the crops. The soil, fertilized by the ash, favors good grain production, and after harvesting the crops, only the "bracatinga" remains on the site until the next clearcut. In this way, the soil will be able to recuperate from the burning, and after the next "bracatinga" clearcut, will be in a condition to afford another good crop of grains, without using any fertilizer.

5.10 Steppic Savanna or "Chaco"

The extensive region of the "chaco" covers an area of more than 800,000 km^2 in the central part of South America, mainly Paraguay, Argentina, and Bolivia, and a very small part of Brazil. The designation steppic savanna (Veloso and Góes-Filho 1982) was adopted because it presents characteristics of both environments: arboreous and shrub strata similar to the steppes, and a creeping grass vegetation like the savannas.

The arboreous stratum is dominated by small trees, semi-deciduous, and many Cactaceae and Bromeliaceae. The main arboreous species are *Schinopis quebracho-colorado*, *S. balansae*, *S. haenkeana*, *Aspidosperma quebracho-beanco*, *Prosopis alba*, *P. nigra*, *Caesalpinia paraguariensis*, *Zizyphus mistol*, *Tipuana tipu*, *Piptadenia macrocarpa*, *Jacaranda mimosifolia*, and others (Hueck 1978; Rizzini 1979). The "chaco" is the most important forest formation of the central region of South America, and sustains a quite important tanin industry.

In the central zone of the "chaco" the annual precipitation is around 500 mm, gradually increasing toward the periphery of the area to 800 to 1000 mm. There is a clearly defined dry season between March and October. The annual mean temperature ranges from 18 to 25°C, with a thermal amplitude (difference between the hottest and coldest months) of 6 to 14°C (Hueck 1978).

When the white man arrived in the region, the "chaco" was covered by a dense forest. The steppic savanna or "chaco" is, therefore, an anthropogenic formation (Veloso and Góes-Filho 1982). The Indians that lived in the region also had some influence on the change in vegetation. When burning the area for agriculture or grazing, they opened clearings in the forest and started modifying the structure of the vegetation (Hueck 1978). However, the colonizers of the region really produced the greatest change in the vegetation, through intensive deforestation, the exploitation of the "quebracho" for the tannic acid industry, and the use of fire for clearing the land for agriculture and grazing. The utilization of wood for firewood and charcoal has also been very intense in the region.

5.11 Subtropical and Araucaria Forests of Chile and Argentina

The subtropical and araucaria forest of Chile and Argentina can be divided into three main types: the Valdivian rain forest, the *Nothofagus* forest, and the *Araucaria araucana* forest.

The Valdivian rain forest is located in a high precipitation region, usually over 2000 mm a year, sometimes reaching 25000 mm, with 160 to 207 days of rain per year. The abundant precipitation, uniform temperature, ranging from 10 to 12°C as annual average, and the high air moisture content allow the development of a luxuriant forest 40 to 50 m high, with a predominance of evergreen species. The main species of this forest type are the conifers *Fitzroya cupressoides*, *Pilgerodendron uviferum*, and *Saxegothaea conspicus*, and the broadleaves *Laurelia* spp., *Myrcengenella apiculata*, *Aetoxicum punctatum*, and *Persea lingue*. The most important species in the ecosystem, *Fitzroya cupressoides*, is threatened by extinction, due to intensive exploitation. For this reason, the governments of Chile and Argentina have taken protective action to preserve the species (Hueck 1978; Corporacion Nacional Forestal 1987). Due to high humidity, fire is not an important factor in this forest type. References to fire occurrence in this ecosystem were not found.

The *Nothofagus* forest grows in an area with precipitation ranging between 1000 and 2000 mm a year, or up to 3000 mm in the Andes region. The temperature records indicate a warm summer, with a mean temperature of 14°C, while in the winter the average remains lower than 10°C for several months. The most important component of this ecosystem is the "roble" (*Nothofagus obliqua*) forest. In the southern limit, the dominant component changes to the "rauli" (*Nothofagus procera*) forest. *Nothofagus* spp. are deciduous species that can reach 40 m in height, and 2 m in diameter. They always grow in deep and fertile soils. The "rauli" is probably the most valuable tree in Chile (Hueck 1978; Corporacion Nacional Forestal 1987). Little information about fire occurrence could be found on this forest type.

The *Araucaria araucana* forests are economically very important in Chile and Argentina. The untouched araucaria forests normally present all age classes, from abundant regeneration to very old trees, with estimated ages of more than 1000 years. The climatic requirements of the species are different, depending upon the region. In the Argentine precipitation is lower, from 600 to 1200 mm a year, while in Chile there is more rain. The winter temperatures are quite low, down to -10°C on the Chilean side and -20°C in the Argentinian side. Along with the araucaria forest, there is another important species for the forest economy of Chile and Argentina, the "cypres" (*Libocedrus chilensis*), a conifer that can reach 30 m in height and 1 m in diameter.

One of the main problems in the araucaria forests has been fire. Fighting wildfires in these forests is very hard, due to the local conditions, especially the difficulty of access to the forests in hilly areas (Corporacion Nacional Forestal 1986).

Because of the seriousness of the fire problem, the Chilean forest fire control service is the most organized and efficient in South America. Even so, fire has burned large areas in Chile, in both planted and native forests. According to the Corporacion Nacional Forestal (1986), from 1972 to 1986 around 587,661 ha of land were burned in Chile, comprising 69,997 ha of planted forests and 517,664 ha of native vegetation. From the total native vegetation, 116,478 ha consisted of forested areas.

5.12 Nonforested Areas

The nonforested or poorly forested areas of South America occur in several places on the continent. From the seven areas listed in Fig. 1, only three will be referred to here: the high savannas of Guyana, the plains of Orenoco, and the prairies of Uruguay and Argentina.

The high savannas of Guyana, that advance in the direction of Venezuela and Brazil, are similar to those of central Brazil, and seem to originate without human interference. Part of this savanna resembles a steppe with grasses, without any woody species (Hueck 1978). The effect of fire in this ecosystem is similar to that presented in the Brazilian savanna ("cerrado") section.

The Orinoco plains are dominated by grass savannas, with a high dominance of graminaceous species, and are dryer than the Brazilian "cerrado" (Rizzini 1979). According to Hueck (1978) there is evidence that the present vegetation of this ecosystem is not natural, but a consequence of the repeated use of fire. The occurrence of some species indicates that in the past the region may have been covered by a vegetation similar to that of the trade wind forest. Fire and grazing probably changed the forest into a grassland environment. Rizzini (1976) noted that local populations used to set fires on the plains to kill snakes and ticks. Evidence that this ecosystem is anthropogenic is derived from the fact that it tends toward natural forest succession when protected from fires. According to Rizzini (1976), the trees start to invade the savanna from the periphery. Blydenstein (1962), Vareschi (1962), and Foldats and Rutkis (1965) studied the role of fire in the Orinoco plains, including the use of controlled fires, and also concluded that fire is a basic ecological factor in maintaining this grass savanna ecosystem.

The prairies of Uruguay and Argentina are an extensive treeless area. The region exhibits characteristics of a grass steppe. It appears that these prairies are a natural or autochthonous vegetal formation. No native forest species of the region are known (Hueck 1978). The first colonizers that arrived in this region found a treeless grassland. Presently, most of this ecosystem is occupied with agriculture and pasture, and fire seems not to be ecologically important in the region.

5.13 Exotic Planted Forests

When studying the effects and influences of fire in South America, it is not possible to disregard the exotic planted forests. In Brazil, there is an area of approximately 60,000 km^2 planted with exotic species, mainly *Eucalyptus* spp (ca. 60%) and *Pinus* spp (ca. 40%), and a few other species, such as *Cunninghamia lanceolata*, *Cryptomeria japonica*, and *Gmelina arborea*, covering small areas. In Chile, from 1929 to 1985, a total of 11,970 km^2 were planted, especially with *Pinus radiata* and *Eucalyptus* spp (Corporacion Nacional Forestal 1987; Kuschel and Ruiz-Tagle 1979).

The exotic species have been planted in other regions of the continent, and as result of the changes they produce in the ecosystems, are always threatened by fire. The forests planted in the savanna, for example, produce a change in the type and an increase in the amount of available fuel in the ecosystem (Soares 1979; Goldammer 1983; De Ronde et al. this Volume). The Brazilian savannas are a typical fire-climax ecosystem, and the periodical fires that occur in the region are surface fires of low or medium intensity. When a pine or eucalyptus forest is established in the ecosystem, the amount of available fuel and the density of the stands are higher than the native vegetation. Therefore, the fires are more intense, sometimes crown fires develop, severely damaging the forests (Fig. 5).

When exotic forests are planted in replacement of the moist rain forests, like the tropical or coastal forests, they create a dryer environment, because the understory strata of the native forests are not present. For this reason, although fire is not a component of the original ecosystem, the environmental changes produced by the planted forests create favorable conditions for the occurrence of large and intense fires.

The leading cause of fire in the planted forests in Brazil, and probably all South America, is debris burning, mainly for agricultural and grazing purposes. Other important causes are incendiary and recreation fires. Soares (1989), analyzed 1754 reported fires in Brazil from 1983 to 1987 and concluded that only 2.1% of the fires were natural, i.e., lightning-caused (Table 2). In Chile, forest and agricultural activities were responsible for 26.1% of the total number of fires occurring from 1977 to 1986 (Corporacion Nacional Forestal 1986).

Climatic conditions coincide with the best season to burn for clearing the land and make forest fire occurrence in Brazil much more frequent in the last 6 months of the year, with a critical period between August and November, as shown in Table 3.

According to Soares (1989), between 1983 and 1987, fires destroyed or damaged approximately 50,523 ha of planted forests in Brazil. *Eucalyptus* plantations were the most affected forest type, as it is the leading planted species in the country. Considering only the planted forests, 77.2% of the reported fires in the country from 1983 to 1987 affected eucalypt forests, while 18.4% occurred in pine forests, and 4.4% in other planted forests. Regarding the burned areas, approximately 89.0% corresponded to eucalypt plantations, 10.6% to pine plantations, and only 0.4% to other planted forests. This high percentage of

Fig. 5. A fire-killed *Pinus caribaea* stand established in the Brazilian savanna. (Photo by R.V. Soares)

Table 2. Number of fires and burned area by group of cause, occurred and reported in Brazilian protected lands between 1983 and 1987. (Soares 1989)

Cause group	Reported fires		Burned area	
	Number	%	ha	%
Lightning	27	2.1	175.42	0.2
Debris burning	435	33.6	59,064.10	63.7
Smoking	104	8.0	2,648.76	2.9
Incendiary	386	29.8	13,589.06	14.7
Railroad	12	0.9	496.18	0.5
Recreation	141	10.9	10,778.23	11.6
Forest operations	87	6.7	1,875.23	2.0
Miscellaneous	104	8.0	4,052.03	4.4
Subtotal	1296	100.0	92,679.01	100.0
Not determined	458	–	41,428.30	–
Total	1754	–	134,107.31	–

Table 3. Reported fires and burned area distribution by month in Brazil from 1983 to 1987. (Soares 1989)

Month	Reported fires		Burned area	
	Number	%	ha	%
January	62	3.5	327.05	0.3
February	62	3.5	1,612.12	1.2
March	55	3.1	312.13	0.2
April	35	2.0	427.79	0.3
May	30	1.7	258.34	0.2
June	60	3.4	659.08	0.5
July	134	7.7	22,733.06	17.0
August	339	19.3	26,416.53	19.7
September	308	17.6	31,309.18	23.3
October	341	19.5	27,393.56	20.4
November	226	12.9	13,732.84	10.3
December	102	5.8	8,888.63	6.6
Total	1754	100.0	134,107.31	100.0

Fig. 6. Wildfire burning in a *Pinus radiata* plantation in Chile. (Photo by R.V. Soares)

burned area in eucalypt plantations is in part due to the fact that most stands of this species were established in the savanna regions, that have a 3- to 4-month-long dry season, and fire, as part of the ecosystem, occurs periodically.

In Chile the forest fire season lasts from November to April, with a peak between December and March (Corporacion Nacional Forestal 1985). In this country, from 1972 to 1986, 51,520 forest fires were reported, burning a total of 627,700 ha, including 69,997 ha of planted forests, 116,478 ha of native forests, 174,467 ha of scrub vegetation, and 226,719 ha of grasslands. Considering only the planted forests, the fires affected 57,054 ha (81.5%) of *Pinus radiata* plantations, 11,165 ha (16.0%) of eucalypts, and 1778 ha (2.5%) of other planted species (Corporacion Nacional Forestal 1986). Fire is also used in Chile in the management of forests. In the 1984/1985 fire season 64,028 ha of forest land was burned by 1327 prescribed fires (Corporacion Nacional Forestal 1985).

References

Aubréville A (1961) Étude écologique des principales formations vegetales du Brésil et contribution à la connaissance des forêts de l'amazonie brasilienne. Cent Tech For Tropic, Paris, 265 pp

Blydenstein J (1962) La sabana de Trachypogon de alto llano. Bol Socedad Venezolana de Ciencias Naturales 22:132–206

Canadian Forest Service (1975) Forest fire control in Canada. Ottawa, Information Canada, Cat No F 061-3, 32 p

Corporación Nacional Forestal (1985) Informativo programa manejo del fuego. Santiago, Vol 2 (2), 11 p

Corporación Nacional Forestal (1986) Estadisticas de ocurrencia y daño de incendios forestales – temporadas de 1964 a 1986. Santiago, Informe Estadistico No 20, 36 pp

Corporación Nacional Forestal (1987) Información escolar. Santiago, Without typographic notes

Ferri MG (1980) Vegetacao brasileira. Univ, São Paulo, São Paulo, 157 pp

Foldats E, Rutkis E (1965) Influência mecânica del suelo sobre la fisionomia de algunas sabanas del llano venezolano. Bol. Sociedad Venezolana de Ciencias Naturales 25 (108):335–392

Galvão F, Kuniyoshi YS, Roderjan CV (1990) Caracterização das difernetes unidades florísticas da Floresta Nacional de Irati. Floresta 19 (2) (in press)

Goldammer JG (1983) Sicherung des südbrasilianischen Kiefernanbaus durch kontrolliertes Brennen. Hochschul-Verlag, Freiburg, 183 pp

Hueck K (1978) Los bosques de Sudamérica – ecologia, composición e importancia económica. GTZ, Eschborn, 476 pp

Inoue MT, Torres DV (1980) Comportamento do crescimento de mudas de *Araucaria angustifolia* em dependência da intensidade luminosa. Floresta 11 (1):07–11

Kuschel AH, Ruiz-Tagle EV (1979) Estudio del comportamiento del fuego en los incendios forestales de la V Region. Univ Chile, Santiago, 78 pp

Maack R (1968) Geografia física do Estado do Paraná. BADEP, Curitiba, 350 pp

Martin RE, Dell JD, Juhl TC (1976) Preliminary prescribed burning guidelines for eastern Oregon and Washington. Paper presented to the Eastside Prescribed Fire Workshop, US Forest Service, Region 6, Bend, Oregon, 54 pp

Mutch RV (1970) Wildland fires and ecosystems – a hypothesis. Ecology 51:1046–1051

Myers JG (1936) Savanna and forest vegetation of the interior Guiana Plateau. J Ecol 24:162–184

Pires JM (1964) Sobre o conceito "zona dos cocais" de Sampaio. Anais XIII Congr da Sociedade Botânica do Brasil, pp 271–275

Rizzini CT (1976) Tratado de fitogeografia dc Brasil – aspectos ecologicos. Univ São Paulo, São Paulo, 327 pp

Rizzini CT (1979) Tratado de fitogeografia do Brasil – aspectos sociológicos e floristicos. Univ São Paulo, São Paulo, 374 pp

Rogers RL (1953) Problemas silviculturais da *Araucaria angustifolia*. Anuário Brasileiro de Economia Florestal 6 (6):308–359

Saldarriaga JG, West DC (1986) Holocene fires in the northern Amazon basin. Quat Res 26:358–366

Sanford JR, Saldarriaga J, Clark KA, Uhl C, Herrera R (1985) Amazon rain-forest fires. Science 227:53–55

Soares RV (1972) Determinação de um índice de perigo de incendio para a regiao centro-paranaense, Brasil. MSc thesis, IICA, Turrialba, 72 pp

Soares RV (1977) The use of prescribed fire in forest management in the state of Paraña, Brazil. Ph.D. Diss., Univ Washington, Seattle, 203 pp

Soares RV (1979) Determinacao da quantidade de material combustivel acumulado em plantios de *Pinus* spp. na regiao de Sacramento, MG Floresta 10 (1):48–62

Soares RV (1989) Forest fires in brazilian industrial plantations and protected public lands. Paper presented to the 3rd Int Symp on Fire Ecology, Freiburg, FRG

Soares RV, Cordeiro L (1974) Análise das causas e épocas de ocorrência de incêndios florestais na região centro-paranaense. Floresta 5 (1):46–49

USDA Forest Service (1972) The role of fire. Missoula, Northern Region, RI-72-033, 22 pp

Vareschi V (1962) La quema como factor ecologico en los llanos. Bol Sciedad Venezolana de Ciencias Naturales 23 (101):9–26

Veloso HP, Góes-Filho L (1982) Fitogegrafia brasileira-classificação fisionòmico-ecológica da vegetação neotropical. Salvador, RADAMBRASIL, Série Vegetação No 1, 85 pp

6 Fire in the Ecology of the Brazilian Cerrado

L.M. Coutinho[1]

6.1 Introduction

In the broad sense, the cerrado may be considered part of a large ecocline, occurring in Brazil under a single, tropical, seasonal climate (climate zone II of Walter 1970) and determined primarily by gradients of soil fertility and the incidence of fire. In fact, the cerrado represents the final portion of the transition from eutrophic tropical seasonal forest to a pedo- and peinobiome of tropical moist savanna and grassland (Fig. 1).

The cerrado is not uniform in physionomy. As part of a larger gradient, it includes the cerradão (a scleromorphic, dystrophic forest formation), the cerrado sensu stricto, the campo cerrado, the campo sujo (intermediate, moist dystrophic savanna formations), and the campo limpo (a dystrophic grassland formation). The savanna formations may be considered as wide ecotones between the cerradão and the campo limpo, since a great number of the species present are common to both these extreme formations. Their herbaceous/undershrub species are not shade-tolerant or shade-dependent, as might be expected from the undergrowth strata of integrated communities; they are heliophytic species, as are the trees and shrubs (Coutinho 1976, 1978). The number, and canopy and basal areas of the tree/shrub and herbaceous/undershrub species change along the gradient campo sujo/cerradão in an inverse ratio (Goodland 1969, 1971a), demonstrating a clearly negative relationship, suggesting the ecotonal nature of the savanna intermediary formations (Coutinho 1976, 1978a).

This physiognomic complex of oreadic vegetation [the term oreadic refers to a floristic province recognized by Martius (1840–1906) in his Flora Brasiliensis, named *Oreades*] represented by the cerrado has a core zone centered on the great plateau of Central Brazil, covering some 1,500,000 km^2. Additionally, there are islands or patches of cerrado enclosed within the Amazon region in the north, the Caatinga in the northeast, the tropical seasonal forests in the southeast and the Pantanal in the west. The total area of the cerrado may attain approximately 1,800,000 km^2 (Fig. 2).

Several of the characteristics of the cerrado may be encountered within a short distance on crossing cerrado stands. In fact, the enormous area covered with cerrado is a micromosaic of its different formation types. Such a dis-

[1] Department of Ecology, Institute of Biosciences, University of São Paulo, 05499 São Paulo, Brazil

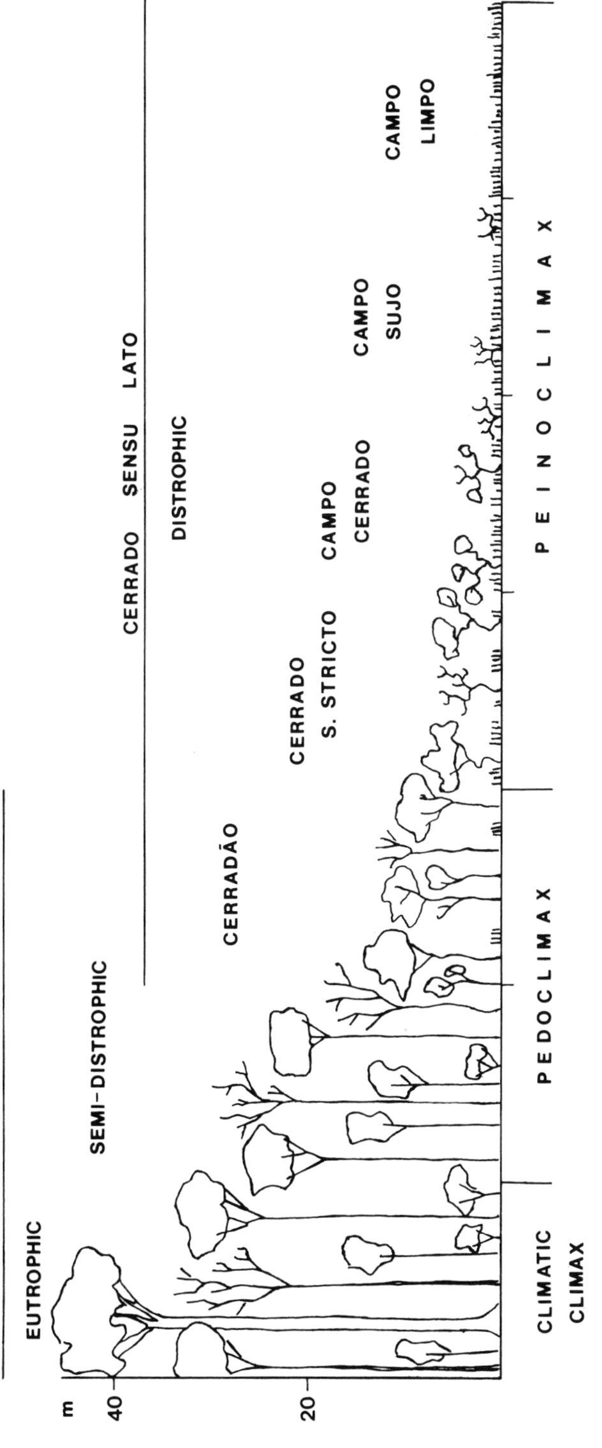

Fig. 1. The forest-grassland ecocline in central Brazil

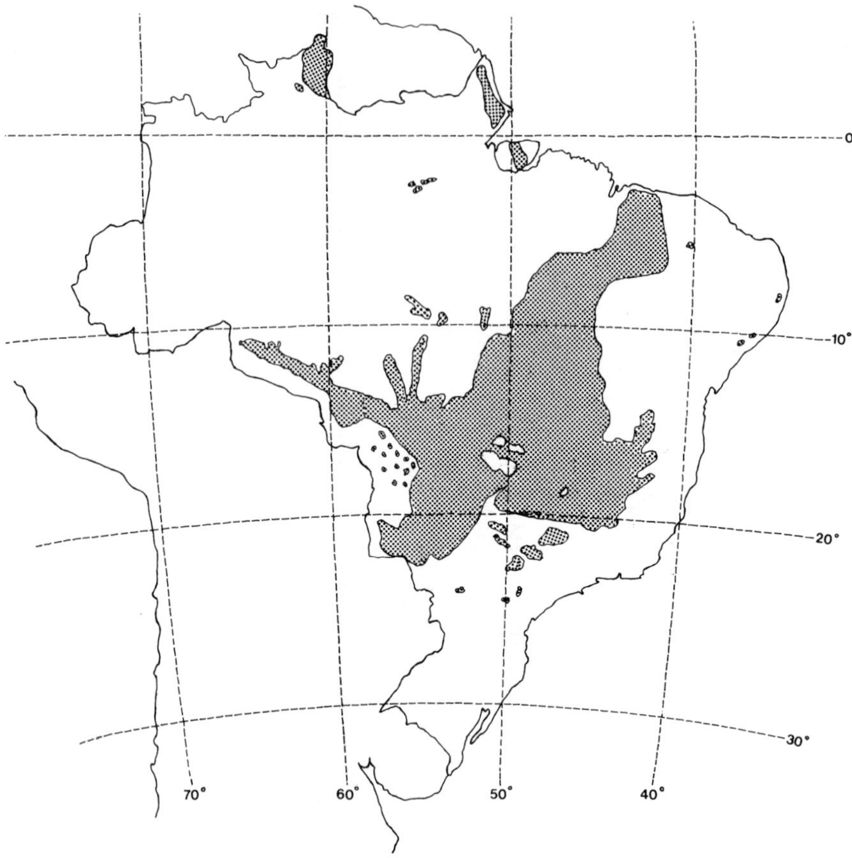

Fig. 2. Distribution of the cerrado in Brazil

tribution pattern is determined primarily by the mosaic pattern of soil form distribution and/or fire action.

The flora of the cerrado is not yet completely known, although a large number of species have already been described. Two fundamental types of flora may be distinguished in the cerrado: the oreadic grassland flora and the oreadic scleromorphic forest flora, although some species are more typical of the ecotonal formations. According to Heringer et al. (1977), 774 woody plant species are known. The herbaceous/undershrub flora may comprise more than double this number of species. An estimate of the total number of species in the cerrado would be around 2400.

According to Sick (1965), it is difficult to establish the concept of a typical cerrado fauna in the sense of an endemic fauna restricted to the cerrado formation. In Vanzolini's (1963) interpretation "three ecological factors are thought to be especially significant for the distribution of animals [in the cerrado]: intense radiation exchange at ground level, penetrability of the soil to deep levels, and presence of an endemic flora. Intense insolation during the day and irradiation at night, causing surface temperature oscillations of as much as 45°C in 24 h is a general characteristic of open formations. On the contrary, the large relative volume of roots and stems reaching levels as far as 20 m deep is peculiar to the cerrado. The joint action of these factors favors an "evasion" of the fauna to nocturnal and subterranean life. An endemic flora must condition an endemic fauna of phytophagous invertebrates and biologically associated forms. This possibility has not been exploited". For Vanzolini, a vertebrate fauna peculiar to the cerrado apparently does not exist.

Since the first half of the last century, the origin and principal determinant factors of the cerrado vegetation have been discussed (Saint-Hilaire 1824; Warming 1892). In his classic work *Lagoa Santa*, Warming considered the cerrados to be a climatic climax determined by the dry conditions predominant in the period from May to September, coinciding with winter. Rawitscher (1942b), Rawitscher et al. (1943), Ferri (1944), and Rachid-Edwards (1956) gave greater relevance to the factor fire, although restricted to the more southern cerrados. Alvim and Araujo (1952), Arens (1958, 1963), Beiguelman (1962), and Goodland (1969, 1971a,b) have emphasized the role of soil dystrophy and aluminum toxicity in the determination of the presence of cerrados. However, these authors, among various others interested in the problem of cerrado ecology, have taken a reductionist view of the problem, finding in one single environmental factor an explanation for the existence of a complex of physiognomic vegetation forms of subcontinental extension such as the cerrado is. In our opinion, all these factors play a role in the determination of the cerrados, each on a different scale. Thus, the limits of distribution of the core area of the cerrado may be understood from the great coincidence of the seasonal tropical climate predominating there. However, climate alone does not permit an explanation of the occurrence within the core area of seasonal tropical forest communities beside the cerrado communities, with frequently abrupt transitions. In these cases to a lesser degree, at a level closer to that of the community, the mosaic distribution of soil patches and the irregularity of the areas affected by recurrent fires allows an understanding of the distribution of the vegetation. Within this concept, the cerrados cannot be considered a climatic climax but rather a pedoclimax or a pyroclimax. The climatic climax of the whole Central Brazilian region, dominated by the seasonal tropical climate (climatic zone II of Walter 1970) would be the eutrophic seasonal tropical forest.

Although the factor fire has been discussed since the classic works of Saint-Hilaire (1824, 1847), Warming (1892), and Löfgren (1898), experimental research on the ecological effects of fire on the cerrados has been relatively scarce and fragmentary, a fact noted by Labouriau (1963, 1966) in reviews of the physio-ecological problems of the cerrado and related literature. Since 1969, our interests and those of our students have been concerned with fire ecology in cerrado ecosystems. However, with the expansion of agricultural borders within areas of cerrado, notably those located on the southern limits, the allocation of medium- to long-term experiments in these ecosystems is becoming increasingly difficult. Recently, research groups at the University of Brasilia (UnB), Instituto

Brasileiro de Geografia e Estatistica (IBGE) and Empresa Brasileira de Pesquisas Agropecuárias (EMBRAPA) have initiated interesting projects on burn-offs in areas located within the Federal District in the core area of the cerrados, with which I have had the satisfaction of collaborating.

6.2 Regimes and Causes of Fire in the Cerrado

Fire in the cerrado has not been attributed to natural causes in the scientific literature to date. However, this represents the consequence of an absolute lack of research on the problem rather than a result of systematic observation. Farmers and inhabitants of the cerrado cattle ranches report that lightning occasionally causes burning of the vegetation; there is, in fact little reason for this not to occur, especially in areas unaffected by fire for several years. Although scientific data are not presently available, as investigations proceed in our cerrado parks and reserves, proof of this natural cause of fire should become established.

There is absolutely no doubt that man has been the principal cause of fire in the cerrado from the earliest times. Recent anthropological research has shown that man has inhabited central Brazil for more than 32,000 years (Guidon and Delibrias 1986), increasing previous estimates threefold. This primitive man already made use of fire, according to the anthropological data. At the time of the colonization of Brazil, Indians were using fire for many purposes. Several descriptions of their habits mention the common practice of burning vegetation for hunting and tribal wars. According to Anderson and Posey (1987), the Kayapó Indians still use fire as a form of management of areas of the cerrado to limit the development of certain undesirable species and to stimulate the production of certain native fruit-bearing trees. Thus, either naturally, or through the influence of primitive human populations, there is no doubt that fire was present in central Brazil long before the arrival of the Portuguese colonialists. Evidence for this can be found in charcoal fragments recovered from cerrado soil near Brasilia (DF), which C^{14} dating reveals to be 1600 years old (Berger and Libby 1966). Coutinho (1981) has reported charcoal pieces similarly dated at 8600 years from a campo cerrado soil horizon lying at 2 m depth near Pirassununga (SP). Probably new findings of charcoal in the cerrado soils will extend these dates of ancient fires in areas currently occupied by the cerrado vegetation. The tolerance of and dependence on fire of a great number of species of the rich cerrado flora represent further evidence that fire is an old and major ecological factor in this ecosystem.

The more open forms of the cerrado are frequently used by cattle ranchers as natural pastures. However, the grazing capacity of such pastures is extremely low, forcing the use of extensive areas. In the dry season, which lasts some 3 to 4 months in central Brazil, the cattle suffer from the lack of palatable, green feed; weight loss is accentuated and milk production diminishes. Firing the cerrado in the second half of the dry season (August-September) constitutes the cheapest management practice undertaken by the cattle ranchers; since a few days or weeks after a burn-off the vegetation sprouts, thus furnishing the cattle with palatable, green feed, rich in protein, cellulose, and salts. This is the principal cause of fires in the cerrado regions.

The demand for foodstuffs has grown as a result of the population increase in Brazil and of the increase in exports; since the cerrados represent low-cost land and provide favorable conditions for intensive cultivation of cereals, the greatest expansion of the agricultural frontiers has been exactly in this phytogeographic region. The participation of the cerrados in the national production of rice, beans, corn, soybean, and wheat reached 30% in 1988 with productivities attaining 3.1, 2.0, 7.6, 4.0, and 2.5 tons/hectare, respectively, values which may be doubled by irrigation techniques (Goedert in press). By irrigating in autumn/winter, it is possible to obtain two harvests per year. Thus, great areas of the cerrado are cleared and burned at the end of the dry season (August-September) to bring in these new agricultural areas. This seems to be the second great cause of burn-offs in the cerrado at present.

There are still some burn-offs which may result from various other causes, such as the control of shrubs in pastures, pest control, carelessness in fire management in intentionally burned areas (such as during the cutting and burning of vegetation while cleaning the edges of highways and railroads), the falling of balloons with the wicks still alight during the June religious festivals. According to B.F.S. Dias (personal commun.) from the University of Brasilia (DF), carelessness with cigarettes does not seem to be a relevant cause of fire in the cerrado. Experiments in which a large number of lit cigarettes were thrown into the vegetation under meteorological conditions of high fire risk showed no evidence of combustion. These data thus appear to corroborate the opinion of Vareschi (1962) regarding the same problem in the Venezuelan llanos.

Fires in the cerrado region usually begin in May, coinciding with the beginning of the dry season; incidence increases during June and July, attaining a maximum in August. In September-October with the arrival of the rains the occurrence of fires drops markedly. During the wet season (October-March), which coincides with spring and summer, burn-offs are not usually made, although the vegetation is susceptible to burning, particularly in areas where there have been no burn-offs for several years, and after a sequence of hot days in the absence of rain (veranico). From August to the beginning of September is the period of greatest risk, when meteorological conditions are particularly favorable for the propagation of fire. Relative air humidity during the hottest hours of the day (25–30°C) can reach below 20% at this time of the year. The risk of fire increases greatly in years of frosts, which provoke the death of a great part of the epigeous phytomass in the herbaceous/undershrub stratum, and the concomitant fall of leaves from many trees and shrubs, which accumulate on the soil as an easily dehydratable and highly combustible material.

When used as a form of pasture management, burn-offs are usually effected towards the end of the afternoon to reduce the risks of uncontrolled propagation. Such burn-offs are made by the local inhabitants of the region at mean intervals of some 3 years. With the transformation of the subsistence economy into a market economy brought about by the implantation of more ambitious agricultural/farming projects, and the transference of populations from other regions of the country, an increase in the frequency of pasture burn-offs has occurred, the interval dropping to 2 years or even 1 year, with serious risks to the ecological equilibrium of the whole region. In Mato Grosso and Mato Grosso do Sul States, most of which are covered by cerrado vegetation, atmospheric pollution in the form of smoke during certain days of the months of August and September actually impedes landing and take-off by commercial airliners.

Accidental fires when they occur during the day are usually difficult to control, propagate more easily, and cause more intense combustion of the vegetation.

6.3 The Abiotic Effects of Fire

6.3.1 Air Temperature

One of the more immediate consequences of a burn-off is the elevation of local temperature, particularly where the plant material is in full combustion. The quantity of thermal energy released by a fire depends on a series of factors such as the mass of combustible material, degree of desiccation, aeration, etc. Since the phytomass of the cerrado is not uniform but varies from one physiognomic form to the next, a certain variation in the maximum temperatures attained can be expected, depending on the type of cerrado being burned.

There is very little information available on air temperature during burning of the cerrado. Cesar (1980) measured a temperature of 800 °C in the flames of a campo sujo burn-off in Brasilia (DF), suggesting that flame temperature may reach as high as 1000 °C. Cesar also showed that the higher the surrounding air temperature and the drier the air, the more rapid is the burn-off and the less elevated the flame temperature. In a slow-burning plot, flame temperature was almost double that observed in another plot which burned two and a half times more quickly.

6.3.2 Soil Temperature

The degree of heating of the soil during a fire is a function of various factors such as combustible phytomass per unit area, phytomass humidity, and humidity of the soil. Evidently, the greater the quantity of material to be burned, the higher the temperatures attained and the greater the quantity of ashes and coals deposited on the soil surface, resulting thus in a greater temperature increase. It is for this reason that forest fires, for example, induce a much greater increase in soil temperature than do field fires. The humidity of the phytomass is important because the drier the phytomass, the more rapid the burn-off, resulting in a diminished effect on soil temperature. The humidity of the soil is of extreme relevance, since humid soils heat up much less during fires as a result of their greater specific heat and better thermal conduction; furthermore, the evaporation of water helps to reduce the heating effect.

There have been only few measurements made of soil temperature during fires in open formations. Coutinho (1976, 1978b) has made some measurements on Brazilian cerrados during experimental burn-offs, the highest value encountered being 74°C at the surface (Fig. 3). At 1, 2, and 5 cm depth, the maximum values were 47, 33, and 25°C, respectively. Cesar (1980) has noted widely differing values for experimental fires at a campo sujo in Brasilia (DF), registering 280°C at 1 cm depth in one area and no elevation at all at the same depth in another. Temperatures as high as that registered by Cesar are probably not common; at 280°C, all the plant structures present would burn and those usually present on the soil surface after fires (stems, twigs, dry leaves) would no longer be visible; the buds of cryptophytic and hemicryptophytic species so common in the herbaceous/undershrub stratum would die and the vigorous sprouting observed only a few days or weeks after the fire thus could not occur.

In addition to the direct effect of fire on soil temperature, the indirect effects are also of interest in that the burn-off exposes the soil surface directly to the diurnal solar radiation and to nocturnal irradiation, thus increasing the amplitude of the daily thermal variation in the post-burning period. Since the burn-offs generally occur in seasons when nocturnal thermal irradiation is aided by the lack of cloud cover, the daily temperature fluctuations at the soil level, the importance of which have been pointed out by Vanzolini (1963), are still more intense in recently burned areas.

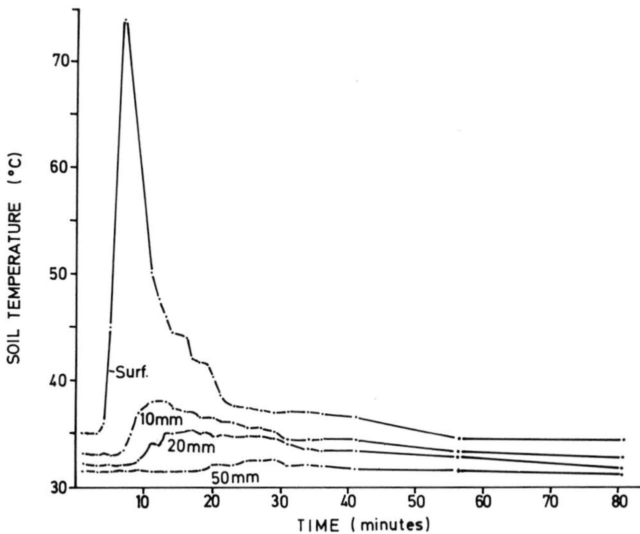

Fig. 3. Soil temperatures at depths of 0, 1, 2, and 5 cm during the burning of a campo cerrado at Emas, Pirassununga, São Paulo State. (Coutinho 1976, 1978b)

6.3.3 Cycling of Mineral Nutrients

Mineral nutrient cycling is an aspect of enormous relevance when dealing with fires in the vegetation. The habit of firing the covering vegetation has been condemned by most people not only because of its destructive effects but also because of the undesirable effects on soil characteristics, specifically chemical properties.

A more complete analysis of this problem must take into account aspects such as the type of vegetation burned, its degree of adaptation to fire, the resilience of the system, the number and frequency of burn-offs, the short-, medium-, and long-term effects of the fire, the soil characteristics, etc.

In the case of the cerrado, given the great variation in physiognomy as well as floristic composition, phytomass, susceptibility, and resistance to fire, etc., a certain variation may be expected in results when a particular experiment is repeated in areas of different physiognomy. It would be unlikely, in fact, that the same results be obtained after the firing of a cerradão and a campo limpo. Generally, the soils underlying the cerrado are rather poor in mineral nutrients and are distinctly acidic. They exhibit a low cationic exchange capacity, low saturation by bases, and a high saturation by aluminum (Alvim and Araujo 1952; Arens 1958, 1963; Ranzani 1963; Goodland 1971b; Lopes 1975; Lopes and Cox 1977). Thus, until a few decades ago, the cerrado soils had little commercial value in spite of exhibiting physical properties and a relief favorable for agriculture. The pronounced scleromorphism of the woody plants of the cerrado has been attributed to the oligotrophism of its soils (Arens 1958, 1963; Beiguelman 1962), a hypothesis criticized by Labouriau (1963), but supported and extended by Goodland (1969, 1971b) with a consideration of aluminum toxicity.

Fire is intimately related to these nutritional problems since it is involved with the cycling of mineral nutrients. It is through the action of fire that most of the epigeous biomass is rapidly mineralized, resulting in a temporary increase in the level of mineral nutrients in the surface layers of the soil as a function of ash deposit. The cycling of mineral nutrients is thus accelerated. The total quantity of nutrients and other elements of special ecological interest found per hectare in the aerial parts of the herbaceous/undershrub layer of a campo cerrado stand with an annual plant biomass production of 5 to 7 ton/ha (Coutinho et al. 1982) is approximately (in kg) N 70, P 3.5, K 18, Ca 4, Mg 2, S 2, Al 4, and Si 120 (André-Alvarez 1979). Batmanian (1983) studying a campo sujo in Brasilia (DF), found an annual phytomass production after burning-off of only 2.5 ton/ha; the nutrient content of this phytomass (in kg/ha) was N 21, P 2.8, K 8.8, Ca 2.6, Mg 2.0, and Al 3.3.

Cavalcanti (1978) has performed analyses of the soil at different depths after a campo cerrado burn-off to monitor the vertical movement of ash nutrients into the soil by rain action (Fig. 4). A large increment in certain nutrients was noted at the soil surface (0–5 cm) immediately after burning, diminishing over the following 3 months. At deeper levels there was no obvious increase in nutrient concentration that might indicate greater lixiviation and, conversely, the data

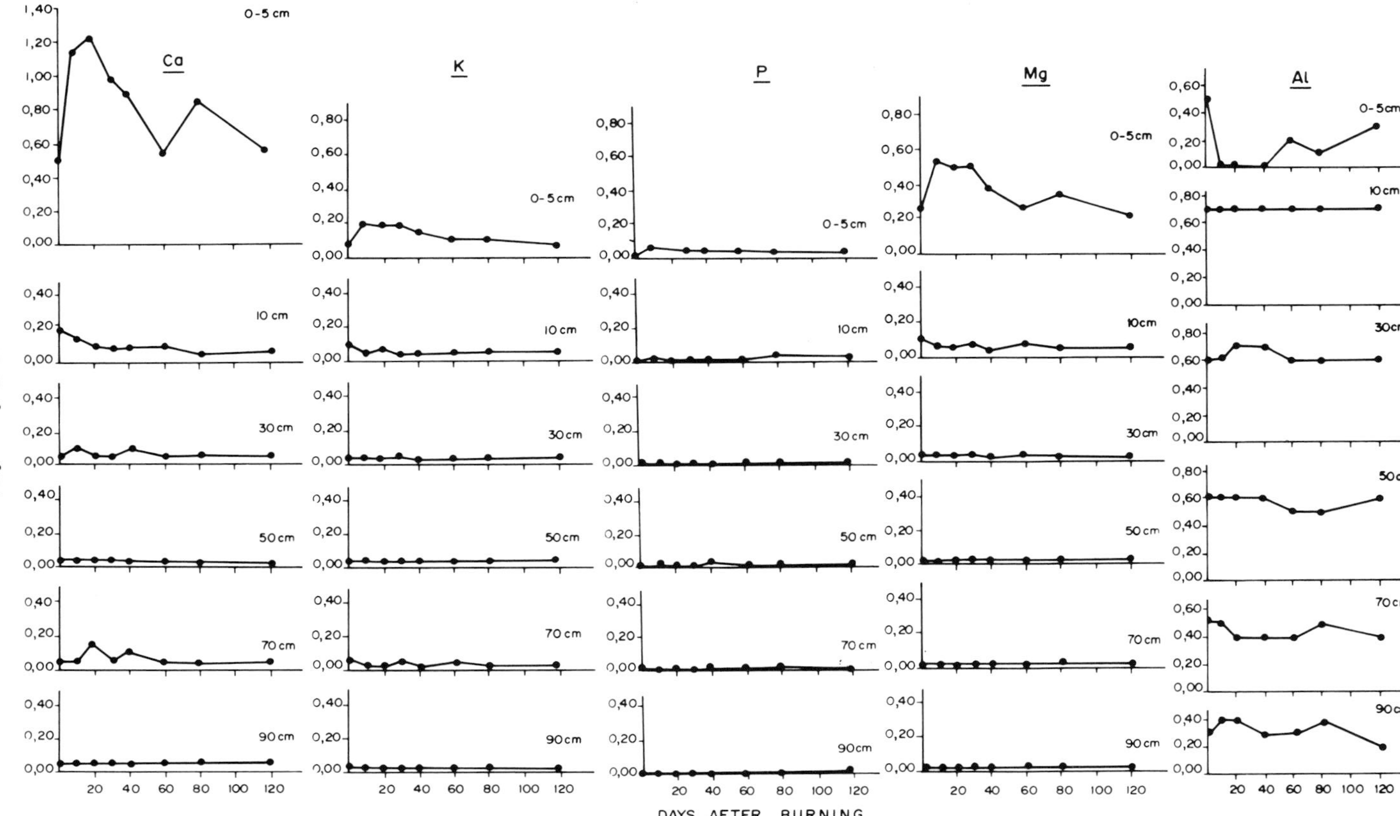

Fig. 4. Concentrations of some mineral nutrients at different soil depths and at different times after the burning of a campo cerrado at Emas, Pirassununga (SP). (After Cavalcanti 1978)

suggest that most of the mineral nutrients deposited with the ash on the soil surface are reabsorbed by the superficial root systems of the herbaceous plants. An opposing phenomenon was seen regarding aluminum: at the soil surface, the concentration/time curve is notably different from other elements like Ca, K, and Mg. Over a 40-day period after burning, aluminum concentration remained at zero and at deeper levels no significant changes were noted. Such results indicate that the ash produced by campo cerrado fires is highly beneficial to herbaceous/undershrub species with superficial root systems, since these are furnished with mineral nutrients and the aluminum toxicity of the upper soil layer is temporarily eliminated. Cavalcanti's data also show that fire, while beneficial to the herbaceous cerrado plants, is detrimental to the tree and shrub species which frequently possess deep root systems (Rawitscher 1942a,b; Rawitscher et al. 1943); i.e., burning of the leaves, flowers, fruits, and branches of the tree and shrub species causes nutrients to be transferred to the herbaceous and undershrub plants.

Several species in the herbaceous/undershrub layer exhibit special woody underground organs denominated lignotubers or xylopodia, the morphological nature of which has not been well studied. These organs are thick and lignified and present many dormant buds in their upper region at ground level or some centimeters below. The root systems of many species such as *Lantana montevidensis*, *Isostigma peucedanifolium*, *Macrosiphonia martii*, and *Stylosanthes capitata* are superficial. Some authors have assumed the physiological function of the xylopodia to be water storage. Owing to our interest in nutrient cycling, and given Cavalcanti's (1978) results, we measured the mineral nutrient concentration of the first two species above over a 1-year period before and after burning (Coutinho et al. 1978); these species were selected owing to their very typical xylopodia. Our data for *Lantana montevidensis* and *Isostigma peucedanifolium* demonstrate that the xylopodia are mineral nutrient storage organs rather than water reservoir organs. Water content was low and very constant throughout the year. In contrast, ash content was high and exhibited a pronounced increase in August 1 month after burning of the experimental area, the value being double those for February, May, and November (Table 1). The P, K, Ca, and Na contents of the xylopodia were also exceptionally high in August, suggesting intense and efficient absorption of the ash nutrients returned to the soil after burning, confirming the interpretation of Cavalcanti's (1978) data (Fig. 5). Although these species possess only few, short, and little-ramified roots, their absorption efficiency appears to be great. A possible explanation for this efficiency may be the presence in many roots of vesicular-arbuscular mycorhyzae (Thomasini 1972; Teixeira 1986).

Table 1. Ash content (% dry matter) of xylopodia from *Lantana montevidensis* and *Isostigma peucedanifolium*. (Coutinho et al. 1978)

Species	February	May	August	November
Lantana montevidensis	10.5	11.6	23.2	11.3
Isostigma peucedanifolium	11.3	11.6	22.7	11.6

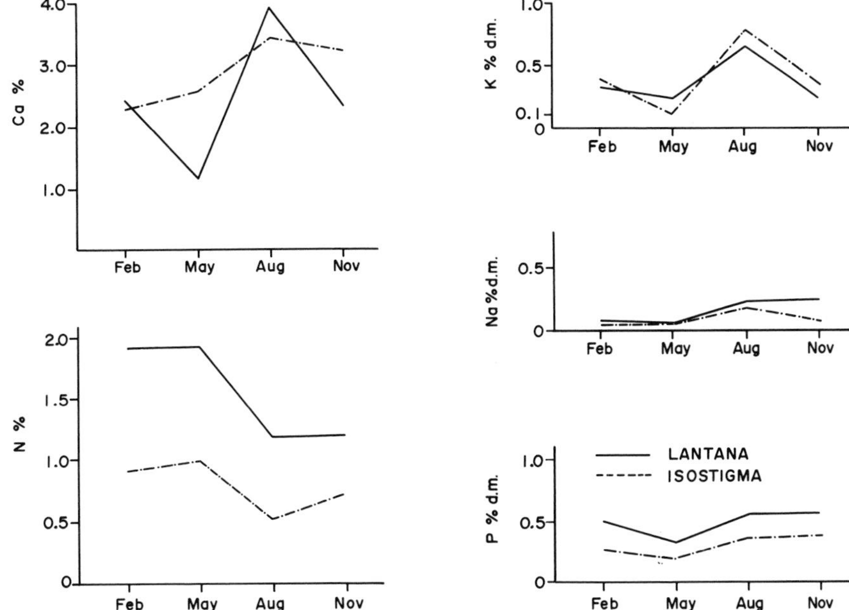

Fig. 5. Mineral nutrient content (% dry matter) in xylopodia at different times of the year before and after a July burn-off. Data are for *Lantana montevidensis* and *Isostigma peucedanifolium* from a campo cerrado at Emas, Pirassununga (SP). (Coutinho et al. 1978)

Batmanian (1983) has observed that during 3 months following a burn-off in a campo sujo there is an increase in the availability of K, Ca, and Mg in the soil to a depth of 60 cm with a concomitant decrease in Al concentration, concluding that the herbaceous/undershrub layer does not cycle most of the nutrients, resulting in their lixiviation. These data clearly demonstrate the importance of the herbaceous/undershrub biomass in the cycling efficiency of ash nutrients. Cavalcanti's (1978) results for the campo cerrado at Pirassununga (SP) show a production of 8 ton ha^{-1} $year^{-1}$ compared to only 2.4 ton ha^{-1} $year^{-1}$ for the campo sujo in Brasilia.

Soares and Souza (personal commun.), taking advantage of an accidental fire in the cerradão at São Carlos (SP), analyzed variation in soil nutrient levels to a depth of 1 m during the subsequent 6 months. pH was slightly increased only in the first few centimeters of soil; Al declined in the first two or three months after the fire; over the profile examined, Mg, P, Ca, K, and N levels were altered after the fire, suggesting a certain lixiviation of these elements. These results once again show the importance of the herbaceous/undershrub stratum in the rapid and efficient cycling of mineral nutrients. In the cerradões, this stratum is very poorly developed.

The transference of nutrients from the tree/shrub to the herbaceous/undershrub layer, possibly brought about by fire action, appears to be antagonized by the leaf-cutter ants (various species of the genus *Atta*) which are frequently encountered in the cerrado at a density of two to three nests/ha or more. Through their foraging activity such ants transfer nutrients to their fungal chambers and eventually to their refuse chambers located in the soil up to 6 or 7 m depth. Here the detritus decomposes, releasing the mineral nutrients to the soil (Fig. 6). Although this process occurs in a very localized way, i.e., only where the ants have their nests, over the long term, nutrients will be transferred to the tree/shrub stratum which possesses deep roots, to the detriment of the herbaceous/undershrub layer (Coutinho 1984).

An extremely important aspect to be considered is the great loss of nutrients from the burned ecosystem in the form of smoke released with the flames. Nitrogen and S are easily volatilized in a fire and at temperatures above 600°C even P can be thus lost. The alkaline elements like Ca, K, and Mg, less volatile than the former elements, are lost in particulate form. While burned areas export this material to the atmosphere, other areas import it through the precipitation of the particles or their dissolution in rain water. We have measured nutrient transfer from the atmosphere to the cerrado at Emas, Pirassununga (SP), through monthly collections with ten pluviometers over a 1-year period (Coutinho 1979). Nutrient input (in kg ha^{-1} $year^{-1}$) was: K 2.5, Na 3.4, Ca 5.6, Mg 0.9, and PO_4 2.8 (Fig. 7). Schiavini (1984), working in Brasilia

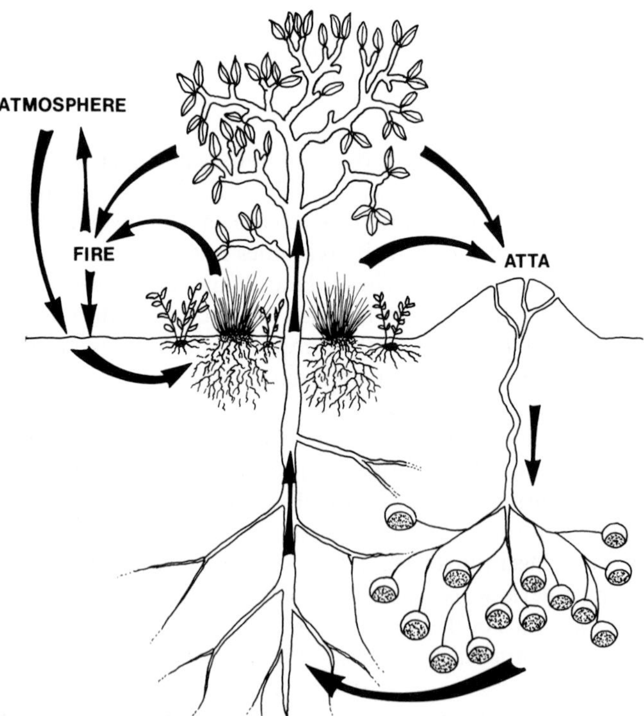

Fig. 6. Possible antagonistic effects between fire and *Atta* ants on mineral nutrient cycling in a campo cerrado. (Coutinho 1984)

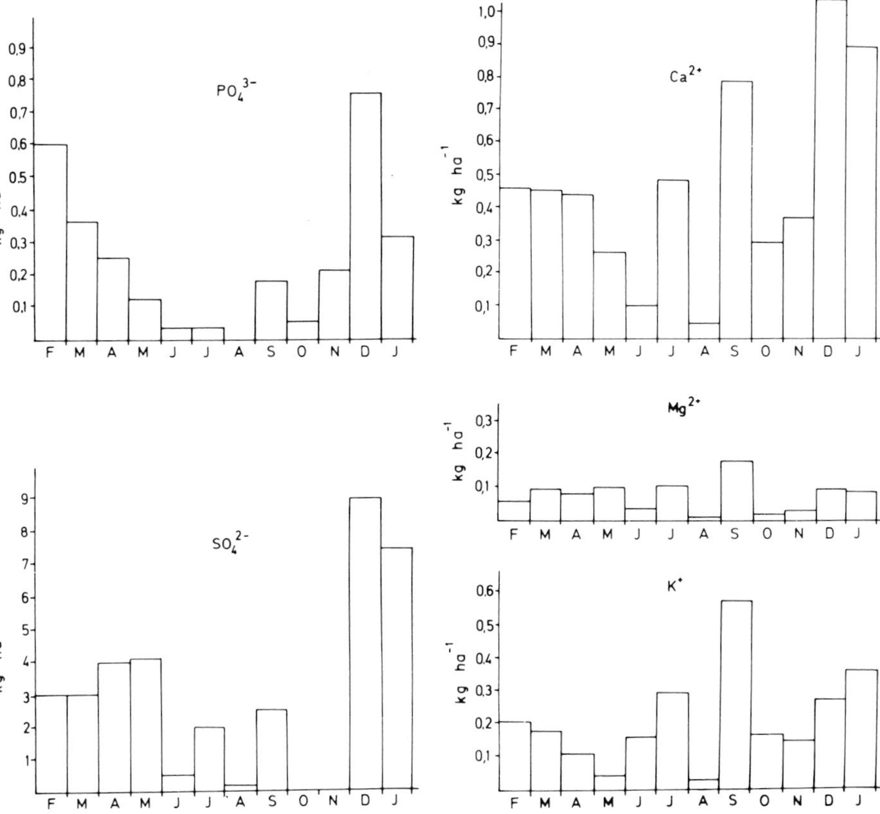

Fig. 7. Monthly mineral nutrient precipitation in a campo cerrado area at Emas, Pirassununga (SP). (Coutinho 1979)

(DF), encountered higher values for K and Mg (5.6 and 1.5, respectively) and lower values for Ca and Na (1.5 and 2.9, respectively).

Pivello-Pompéia (1985) has evaluated the transference of nutrients to the atmosphere during six burn-offs in Pirassununga (SP), obtaining the following mean values (in kg/ha): K 7.1, Ca 12.1, Mg 3.0, P 1.6, and S 3.2. When expressed as a percentage of their phytomass content these values represent 95% N, 51% P, 44% K, 52% Ca, 42% Mg, and 59% S. A negative correlation was found between Ca export and the degree of combustion of the vegetation. This may be explained by the negative correlation existing between the degree of combustion and the percentage humidity of the phytomass and the positive correlation between the latter and Ca concentration. These results demonstrate that the quantity of mineral nutrients lost in a fire can represent up to three times the amount returning per year. Effectively, if the burn-offs in such an area were made with a frequency of one every 3 years, then a relative equilibrium might be established between input and output. In this case, the fires will not substantially affect the nutritional qualities of the soil (Pivello and Coutinho in press). However, these results should not be extrapolated to other Brazilian cerrados, since the values obtained may vary widely from one region to another.

6.4 The Biotic Effects of Fire

6.4.1 Resistance to Fire

The flora of the open cerrado forms, and specifically the flora of the herbaceous/undershrub stratum, is typically pyrophytic. Throughout the evolution of this stratum, fire, in spite of its sporadic nature, although at the cost of a violent selective force, has selected characteristics that have allowed the species to survive this drastic condition inherent to the cerrado ecosystems. The predominant adaptations appear to be those enabling the species to avoid the dangerously high temperatures, different mechanisms having evolved in herbs and undershrubs than in trees and shrubs. The presence of a large number of such adaptive characters in the cerrado flora conveys the idea that fire is an ancient and natural component of this type of ecosystem. Man with his burn-offs has only accentuated the importance of this factor in the cerrado. Fortunately, a well-adapted flora is already in place.

Among the pyrophytic characteristics of the cerrado species is the strong suberization of the trunk and branches of the trees, permitting thermal isolation of the living internal tissues of these organs and thus survival at the high flame temperatures. Even so, the lateral, horizontal branches located at some 2 to 3 m height frequently show fire damage to their under surfaces. This is probably due to their exposure to the tops of the flames, where temperatures can attain 800 to 1000 °C. Occasionally, on this lower surface the bark is totally destroyed, leaving the wood exposed to subsequent fires; when this occurs the branch is generally eliminated. Among some of the woody plant species, this particular adaptive strategy is lacking and others have evolved such as, for example, the capacity to produce vigorous sprouts from the gemmiparous, subterranean roots despite the totally carbonized aerial branches. Even the seedlings of certain tree species may present this type of adaptation (Dionello 1978).

The dormant apical buds of some trees escape the destructive action of fire through the protective effect of dense, hairy cataphylls such as seen in *Aspidosperma tomentosum*. However, there are many species in which the dormant apical buds are very exposed and frequently die owing to the high flame temperatures. In these cases, adventitious buds may sprout from the branches some days later, resulting in sympodial growth of the stems and imparting a characteristic feature to the trees of the cerrado, i.e., marked tortuosity of the trunks and branches. This characteristic may not result from fire alone; insects may also destroy the apical buds and promote lateral budding. Other tree species possess apical or terminal inflorescences and, consequently, normal sympodial growth. Late fires may also be responsible for the destruction of entire growing branches and the final tortuosity of the trees.

It is among the herbaceous/undershrub flora that the majority of species highly resistant to fire can be found. Some are annuals, growing and developing only in the rainy season, thus escaping the dangers of fire in the form of seeds. Many perennial species exhibit subterranean organs such as bulbs, underground shoots, rhizomes, and xylopodia which, owing to the isolation provided by the superficial soil layer, also escape the destructive action of fire. Some days or weeks after a fire these organs sprout with full vigor. The buds of tunic-graminoids are protected inside the densely imbricated sheaths of their leaves, where combustion is limited due to inadequate aeration (Rachid-Edwards 1956)

There are species which, although woody, belong to the herbaceous/undershrub stratum. Such species develop their entire system of trunks and woody branches subterraneously, only the small vegetative branches or yearly reproductive sprouts protruding above the soil; these are truly subterranean trees such as, for example, *Anacardium pumilum* and *Andira humilis*. Similar examples of cryptophytism also may be found among the palms such as *Acanthococos emensis* and *Attalea exigua* (Rawitscher et al. 1943; Rawitscher and Rachid 1946; Lopez-Naranjo 1975).

6.4.2 Primary Productivity

The net primary production of the herbaceous/undershrub stratum of the cerrados seems to vary as a function of the form of cerrado under consideration, the yearly climatic conditions, the textural type of soil, and the incidence of fires. After the burn-off of a campo cerrado in São Carlos (SP), Souza (1977) measured an annual productivity of 5.6 ton/ha. In Pirassununga (SP), Cavalcanti (1978) obtained values above 8 ton ha^{-1} year^{-1} also from an area of campo cerrado. Cesar (1980) measured a productivity of 3.7 ton/ha in a campo sujo in Brasilia (DF) while Batmanian (1983) found values of only 2.5 ton ha^{-1} year^{-1}. Coutinho et al. (1982) have analyzed the effect of burn-off season on annual phytomass production, obtaining some 5.5 ton ha^{-1} year^{-1} after a campo cerrado burn-off in July. In another burn-off, in January, a slightly higher value of 7.0 ton ha^{-1} year^{-1} was obtained. Cavalcanti (1978) has observed that with the removal of the ashes, phytomass production in the subsequent year was significantly less, giving a difference of nearly 3.3 ton ha^{-1} year^{-1} (Fig. 8), demonstrating that the mineral nutrients present in the ash are important for subsequent production. Batmanian's (1983) data show that the maximum phytomass increase in a control area was approximately 1 ton ha^{-1} year^{-1} while in the burned area this increase was 2.4 ton ha^{-1} year^{-1}. This lesser increment in the control area may have resulted from the immobilization of nutrients in the dead biomass which were partially available in the burned area.

In ecosystems like the cerrado, fire may be considered a component of great importance in promoting primary productivity, given that it accelerates mineral

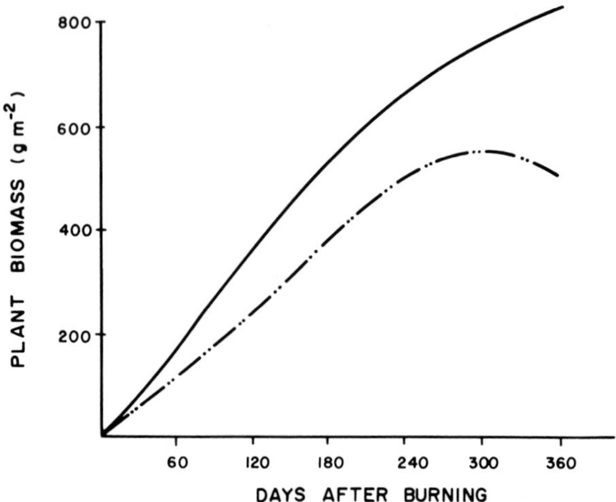

Fig. 8. Plant biomass production in two plots of campo cerrado at Emas, Pirassununga (SP) after a July burn-off. -.-.-, plot from which ash was removed; -----, control plot. (After Cavalcanti 1978)

nutrient cycling. Considering that production is equal to accumulation plus decomposition, any long-lasting accumulation in dystrophic ecosystems will be detrimental to future production, since many of the requisite nutrients are immobilized. Thus, the mineral nutrients immobilized in the inert dry phytomass become re-available to the roots after burning.

6.4.3 Stability of the Vegetation

Experiments begun in 1946 on a campo sujo in Pirassununga (SP) have shown over the subsequent 43 years that protection against fire in that area has brought about a gradual and progressive change in the flora and in the entire vegetation. The original campo sujo gradually became denser, transforming into a cerradão. In better protected areas the herbaceous/undershrub stratum disappeared completely. Similar phenomena have been observed in other cerrados and South American savannas such as those at Mogi-Guacu (SP), Paraopeba (MG), and the llano at Calabozo in Venezuela. At all these sites, one of the first symptoms of alteration to the flora and vegetation is the invasion of the area by exotic grasses such as *Melinis minutiflora* and *Hyparrhenia rufa*. According to B.F.S. Dias (personal commun.), certain stretches of the IBGE Reserve in Brasilia (DF), protected against fire for 15 years, are also becoming transformed into cerradão, although there has been no invasion by exotic grasses as yet.

It seems therefore, that fire is an essential condition for the floristic and physiognomic equilibrium of at least some stands of open cerrado forms. However, these observations should not be extrapolated to other areas of cerrado, since certain open forms may be determined in fact not by the incidence of fire but by other factors such as poor nutritional quality of the soil, reduced soil depth, presence of lateritic hard-pans, anthropic activity, etc.

6.4.4 Flowering

Some days or weeks after a fire, many species of the herbaceous/undershrub stratum in the cerrado begin to sprout and flower with intense vigor. In fact, many produce buds and flowers even before producing their vegetative organs. Such vigorous flowering after a fire is an interesting adaptive strategy since the flowers, fruits, seeds, and seedlings thus escape the destructive action of fire. Furthermore, the synchronization of flowering among most of the individuals of a given species favors cross-pollinization, thus playing an important role in the genetics of these populations. Coutinho (1976) has demonstrated and analyzed this effect under field conditions, defining the following behavioral patterns: (1) species that depend qualitatively or quantitatively on fire to flower, responding with intense flowering to burn-offs occurring during any season of the year; (2) species that depend qualitatively or quantitatively on fire to flower, flowering, however, only should the fire occur during the dry season, i.e., period of short daylengths; (3) species that are qualitatively or quantitatively independent of fire to flower, flowering in the dry season, i.e., during the period of short daylengths; (4) species that are qualitatively independent of fire to flower but which are quantitatively harmed by a burn-off, flowering generally in the rainy season, i.e., during the period of long daylengths; (5) species with a pluriannual flowering cycle apparently not stimulated by fire. The majority of species observed fall within pattern (1).

Laboratory experiments with four species from the herbaceous/undershrub stratum have shown that the effect of fire on the induction of flowering is not the result of thermal action or fertilization by the ashes or the gases emanating from combustion. Burning, cutting the plants close to the soil, or exposing them to a period of hydric stress causing the death of their epigeous parts, all resulted in a high percentage of flowering (see Table 2). The important factor appears to be the elimination of the epigeous parts in which an inhibitor of flowering may be produced. Anatomical studies of the buds made

Table 2. Percentage flowering recorded following the removal of epigeous organs by different methods. (Coutinho 1976)

Species	Burnt	Pruned	Dried	Control 1	Control 2
Lantana montevidensis	83	80	86	0	18
Stylosanthes capitata	86	76	76	0	20
Vernonia grandiflora	90	80	86	0	26
Wedelia glauca	83	86	80	0	20

1 – Sampling for 90 days after treatment.
2 – Sampling for 1 year after treatment.

prior to and after a burn-off demonstrate that the transformation of the vegetative into floral buds occurs only some days after the fire, indicating not a mere effect of distension of floral buds already present before the fire but rather a process of induction of flowering. These morphogenetic processes are thus designated pyroperiodism, i.e., induced by fire, and hydroperiodism, i.e., induced by dryness (Coutinho 1976, 1982).

6.4.5 Dispersion of Seeds

Many species in the herbaceous/undershrub stratum of the cerrado exhibit dehiscence of their fruits and seed dispersion shortly after a fire, e.g., *Anaemopaegma arvensis*, *Jacaranda decurrens*, *Gomphrena macrocephala*, and *Nautonia nummularia*. This fact suggests that fire may be beneficial to such species, since it promotes or facilitates the dispersion of their anemochoric seeds exactly when the soil surface is free of branches and dry grass, which might impede the dispersive action of the wind (Coutinho 1977).

6.4.6 Germination of Seeds

Coutinho and Jurkevics (1978) have conducted some experiments with the seeds of a cerrado undershrub species belonging to the genus *Mimosa*; such seeds possess an extremely hard integument which is impermeable to water, rendering germination difficult. When submitted to thermal shocks of between 70 and 100°C for 5, 10, and even 30 min duration such seeds show high indices of germination. At these temperatures, the seed coats appear to crack and permit water penetration and soaking of the tissues, conditions fundamental for germination. Fire apparently favors the reproduction of this species.

6.4.7 Fire and Fauna

One of the principal and as yet unresolved problems in the cerrado concerns the mortality of wild animals during fires. Certain environmentalists and conservationists question the use of fire as a management form for our open formations, alleging that fire irremediably destroys the fauna. Unfortunately, there are no data available with respect to the effects of fire on the population dynamics of the wildlife. It is undeniable that some larger animals such as the tamandua (*Myrmecophaga tridactyla*) may become surrounded by fire and succumb; this has been observed in accidentally burned national parks such as occurred recently at Emas National Park in southern Goias State. However, the death of these animals, while tragic, reveals nothing of their population dynamics; it is knowledge of exactly this factor that is necessary to evaluate the real effect of fire.

Although fire may cause death among wild animals, it is also necessary to consider possible beneficial effects on the population. It is well known that wild animals such as the campeiro deer (*Ozotocerus bezoarticus*) scour recently burned areas to lick the ashes which represent a source of elements and mineral salts. Further, they can graze on fresh, palatable forage rich in protein, in the form of tender shoots. With the sprouting that occurs within a few days or weeks, the period of food shortage after a fire is short. Food shortage also occurs during the dry season even in the absence of fires. The complete disappearance of the epigeous parts of many cryptophytic and hemicryptophytic species and the impalatability of the rough, dry leaves cause weight loss in the wild fauna and cattle, showing that such food shortage already exists. With the sprouting that occurs after a fire, this period of fasting may even be reduced. The intense production of flowers, fruits, and seeds after the fire is another effect that favors the nectarivorous, pollinivorous, frugivorous, and granivorous fauna represented principally by the insects. The great diversity of species sprouting after a fire, furnishing a host of alimentary niches favorable for herbivores, should also be considered. Furthermore, young leaves apparently contain lower concentrations of secondary metabolites repellent to phytophagous insects.

Adaptive strategies linked to the occurrence of fires can also be encountered amongst the wildlife. Many animals of the cerrado exhibit coat or plumage coloring facilitating camouflage in the black and gray of the burned areas. Since burning of the vegetation increases the visual horizon, exposing prey to predator more easily, such camouflage probably has its ecological significance. Examples of mimetism of the dark gray ground and burned trunks can be seen in the emu (*Rhea americana*), the tamandua (*Myrmecophaga tridactyla*), the peccary (*Taiassu pecari*), the southern lapwing (*Vanellus chilensis*), the buff-necked ibis (*Theristicus caudatus*), and even among invertebrates like cicadas, grasshoppers, etc.

To escape fire, smaller nonflying animals take shelter in holes and burrows in the soil made by armadillos, ants, etc. Larger animals can take cover in the less flammable, humid swamps and riparian forests. Fire probably causes greater harm to eggs, nestlings, and very young animals which cannot escape the flames. However, in the cerrados, the burning season does not coincide with the breeding season of most wildlife, which generally takes place in the spring and summer. Very old and sick animals may also be eliminated by fire.

Finally, populational studies of the wild animals in the cerrado are highly desirable to evaluate the real effects of cerrado fires with greater precision and rigor. Unfortunately, such studies are entirely lacking.

6.5 Management by Fire

The adequate management of an area depends on two basic premises: profound knowledge of what is to be managed, and the establishment of the objectives to be attained. Should these premises not be clearly defined, the risks of failure are great. Thus, as is the case in the management of areas of the cerrado for natural pastures or biological preservation, it is first necessary to invest in detailed knowledge of the area in question and delineate the desired objectives.

Accidental or criminal burning are harmful simply because they occur independently of the above premises, thus provoking undesirable alterations in the ecosystem affected.

Generally, management of open areas through burning should take burn-off frequency into account. This is of utmost importance since, as emphasized above, burn-off frequency can dangerously alter certain dynamic equilibria related to the transference of mineral nutrients to the atmosphere, for example. The empirical knowledge of the peasant farmer has led him to burn cerrado pastures at intervals of not less than 3 years; otherwise, he holds, the pasture weakens, diminishing production. This empirical knowledge, however, must be replaced with a solid technical base.

Other factors besides burn-off frequency are equally important in management by fire. The burning season, be it early or late, influences the results obtained. The time of day, with its associated meteorological characteristics of air temperature, relative humidity, and wind velocity, are factors to be considered. The intensity of the burn-off can vary considerably, depending on these atmospheric conditions, as can the risk of the fire getting out of control and escaping to surrounding areas. Popular knowledge teaches that the safest time for a burn-off is the late afternoon or early evening when the temperature clearly tends to decrease, relative humidity of the air is rising, and vision of the fire centers is facilitated.

The water content of the vegetation to be burned, whether green, dry, or very dry is another aspect that influences the type and intensity of the burn-off and must be taken into consideration.

Once these parameters have been established and the objectives defined, programmed burn-offs may become a useful tool to man in adequately managing the cerrado ecosystems. The lack of such management may allow the accumulation of dry grass over many years, resulting in accidental fires disastrous not only to the fauna but to the very vegetation. Such unexpected fires can spread to large areas including neighboring properties, becoming uncontrollable. Recent examples are the fires in Emas National Park, where cerrado vegetation predominates. Some 80,000 ha were burned between July 29 and August 5, 1988, i.e., over a 7 day- period. Through lack of adequate management, this park has suffered devastating and uncontrollable fires in cycles of approximately 3 to 4 years. This is the time necessary for the arrow grass (*Tristachia leyostachia*), the dominant grass in vast areas of the park, to flower and dangerously increase the combustible phytomass owing to its long and numerous inflorescences.

The necessity for an adequate fire management program in cerrado parks and reserves where preservation of the open formations is desired becomes more evident given the data available on the importance of fire in the maintenance of the stability of these physiognomic forms.

Special care should be taken concerning the invasion of such protected areas by exotic grasses such as *Mellinis minutiflora* and *Hyparrhenia rufa*, for example. Such invasion can spread throughout such disturbed areas as roads and even firebreaks. The prolonged absence of fires appears to reduce the vigor and competitive force of the native species of the herbaceous/undershrub stratum of the cerrado either by delaying mineral nutrient cycling or by a decrease in reproductive capacity. Thus, those exotic species which, when subjected to periodic burn-offs, are unable to propagate in the central areas of cerrado stands, normally being confined to their margins, now do so with ease, dominating the herbaceous/undershrub stratum and eliminating its native species.

References

Alvim PT, Araujo WA (1952) El suelo como factor ecologico en el desarrollo de la vegetación en el centro-oeste del Brasil. Turrialba 2:153–160

Anderson AB, Posey DA (1987) Reflorestamento indigena. Ciência Hoje 6:44–50

André-Alvarez M (1979) Teor de nutrientes minerais na fitomassa do estrato herbáceo subarbustivo do cerrado de Emas (Pirassununga, Est de São Paulo). MSc thesis, Univ São Paulo, São Paulo

Arens K (1958) O cerrado como vegetação oligotrófica. Univ São Paulo, Fac Filosofia, Ciências Letras, Bol 224, Bot 15:59–77

Arens K (1963) As plantas lenhosas dos campos cerrados como flora adaptada às deficiências minerais do solo. In: Symp sobre o Cerrado, EDUSP, São Paulo, pp 285–303

Batmanian GJ (1983) Efeitos do fogo sobre a produção primária e a acumulação de nutrientes no estrato rasteiro de um cerrado. MSc thesis, Univ Brasília, Brasília-DF

Beiguelman B (1962) Cerrado: vegetação oligotrófica. Ciência Cultura, São Paulo 14:99–107

Berger R, Libby WF (1966) UCLA Radiocarbon Dates V. Radiocarbon 8:467–497

Cavalcanti LH (1978) Efeito das cinzas resultantes da queimada sobre a produtividade do estrato herbáceo subarbustivo do cerrado de Emas. DSc thesis, Univ São Paulo, São Paulo

Cesar HL (1980) Efeitos da queima e corte sobre a vegetação de um campo sujo na Fazenda Água Limpa-DF. MSc thesis, Univ Brasilia, Brasilia-DF

Coutinho LM (1976) Contribuição ao conhecimento do papel ecológico das queimadas na floração de espécies do cerrado, Livre-Docente thesis, Univ São Paulo, São Paulo

Coutinho LM (1977) Aspectos ecológicos do fogo no cerrado. II As queimadas e a dispersão de sementes em algumas espécies anemocóricas do estrato herbáceo subarbustivo. Bol Bot, Univ São Paulo 5:57–64

Coutinho LM (1978a) O conceito de cerrado. Rev Brasil Bot 1:17–23

Coutinho LM (1978b) Aspectos ecológicos do fogo no cerrado. I A temperatura do solo durante as queimadas. Rev Brasil Bot 1:93–97

Coutinho LM (1979) Aspectos ecológicos do fogo no cerrado. III A precipitação atmosférica de nutrientes minerais. Rev Brasil Bot 2:97–101

Coutinho LM (1981) Aspectos ecológicos do fogo no cerrado. Nota sobre a ocorrência e datação de carvões vegetais encontrados no interior do solo, em Emas, Pirassununga, S.P. Rev Brasil Bot 4:115–117

Coutinho LM (1982) Ecological effects of fire in Brazilian Cerrado. In: Huntley BJ, Walker BH (eds) Ecology of Tropical Savannas. Ecol Stud 42:273–291

Coutinho LM (1984) Aspectos ecológicos do fogo no cerrado. A saúva, as queimadas e sua possível relação na ciclagem de nutrientes minerais. Bol Zool, Univ São Paulo 8:1–9

Coutinho LM, Jurkewics IR (1978) Aspectos ecológicos do fogo no cerrado. V-O efeito de altas temperaturas na germinação de uma espécie de *Mimosa*. Ciência Cultura, São Paulo 30 (Suppl), 420 pp

Coutinho LM, Pagano SN, Sartori AA (1978) Sobre o teor de água e nutrientes minerais em xilopódios de algumas espécies de cerrado. Ciência Cultura, São Paulo 30 (Suppl), 349 pp

Coutinho LM, De Vuono YS, Lousa JS (1982) Aspectos ecológicos do fogo no cerrado. A época da queimada e a produtividade primária líquida epigeia do estrato herbáceo subarbustivo. Rev Brasil Bot 5:37–41

Dionello SB (1978) Germinação de sementes e desenvolvimento de plântulas de *Kielmeyera coriacea* Mart. DSc thesis, Univ São Paulo, São Paulo

Ferri MG (1944) Transpiração de plantas permanentes dos "cerrados". Univ São Paulo, Fac Filosofia, Ciências Letras, Bol 41, Bot 4:155–224

Goedert WJ (in press) Culturas anuais: situação atual e perspectivas. In: VII Symp sobre o Cerrado. EMBRAPA, Brasilia-DF

Goodland RJA (1969) An ecological study of the cerrado vegetation of South central Brazil. DSc thesis, McGill Univ, Toronto

Goodland RJA (1971a) A physiognomic analysis of the "cerrado" vegetation of Central Brazil. J Ecol 59:411–419

Goodland RJA (1971b) Oligotrofismo e alumínio no cerrado. In: III Symp sobre o cerrado. EDUSP, São Paulo

Guidon N, Delibrias G (1986) Carbon-14 dates point to man in the Americas 32,000 years ago. Nature (Lond) 321:769–771

Heringer EP, Barroso GM, Rizzo JA, Rizzini CT (1977) A flora do cerrado. In: IV Symp sobre o cerrado. EDUSP, Belo Horizonte, pp 211–232

Labouriau LFG (1963) Problemas da fisiologia ecológica dos cerrados. In: Symp sobre o cerrado. EDUSP, São Paulo, pp 233–276

Labouriau LFG (1966) Revisão da situação da ecologia vegetal nos cerrados. In: II Symp sobre o cerrado. Anais Acad Brasil Ciências 38 (Suppl), pp 5–38

Löfgren A (1898) Ensaio para uma distribuição dos vegetais nos diversos grupos florísticos no Estado de São Paulo. Bol Comissão Geogr Geol São Paulo 11:1–50

Lopes AS (1975) A survey of the fertility status of soils under "cerrado" vegetation in Brazil. MSc thesis, North Carolina State Univ, Raleigh

Lopes AS, Cox FR (1977) Cerrado vegetation in Brazil: an edaphic gradient. Agron J 69:828–831

Lopes-Naranjo HJ (1975) Estrutura morfológica de *Anacardium pumilum* St. Hi. Anacardiaceae. MSc thesis, Univ São Paulo, São Paulo

Martius CF Ph von (1840–1906) Flora Brasiliensis. Tabulae physiognomicae, Vol 1, part 1, Lipsiae

Pivello-Pompéia VR (1985) Exportação de macronutrientes para a atmosfera durante queimadas realizadas no campo cerrado de Emas (Pirassununga – SP). MSc thesis, Univ São Paulo, São Paulo

Pivello VR, Coutinho LM (in press) Aspectos ecológicos do fogo no cerrado. Transferência de nutrientes minerais para a atmosfera durante queimadas em campo cerrado

Rachid-Edwards M (1956) Alguns dispositivos para proteção de plan tas contra a seca e o fogo. Univ São Paulo, Fac Filosofia, Ciências Letras, Bol 219, Bot 13:35–68

Ranzani G (1963) Solos do cerrado. In: Symp sobre o cerrado. EDUSP, São Paulo, pp 51–92

Rawitscher F (1942a) Algumas noções sobre a transpiração e o balanço d'água de plantas brasileiras. Anais Acad Brasil Ciências 14:7–36

Rawitscher F (1942b) Problemas de fitoecologia com considerações especiais sobre o Brasil meridional. Univ São Paulo, Fac Filosofia, Ciências Letras, Bol 28, Bot 3:3–111

Rawitscher F, Rachid M (1946) Troncos subterrâneos de plantas brasileiras. Anais Acad Brasil Ciências 18:261–280

Rawitscher F, Ferri MG, Rachid M (1943) Profundidade dos solos e vegetação em campos cerrados do Brasil meridional. Anais Acad Brasil Ciências 15:267–294

Saint-Hilaire A (1824) Histoire des plantes les plus remarcables du Brésil et du Paraguay, I. Belin Imprimeur-Libraire, Paris

Saint-Hilaire A (1847) Voyages dans l'interieur du Brésil, III partie – Voyages aux sources du Rio S. Francisco et dans la Province de Goyaz

Schiavini I (1984) Ciclagem de nutrientes no cerrado. I – Água da chuva. Ciência Cultura, São Paulo 36, (Suppl), 634 pp

Sick H (1965) A fauna do cerrado. Arquivos Zool, São Paulo 12:71–93

Souza MHAD (1977) Alguns aspectos ecológicos da vegetação perimetral da Represa do Lobo, SP. DSc thesis, Univ São Paulo, São Paulo

Teixeira S (1986) Análise ecológica da ocorrência de micorriza no cerrado: resultados preliminares. Anais I Reunião Brasil Sobre Micorrizas, Lavras, 191 pp

Thomasini LI (1972) Micorriza em plantas do cerrado. DSc thesis, Fac Filosofia, Ciências Letras, Rio Claro

Vanzolini, PE (1963) Problemas faunisticos do cerrado. In: Symp sobre o cerrado. EDUSP, São Paulo, pp 305–321

Vareschi V (1962) La quema como factor ecológico en los llanos. Bol Soc Venezolana Ciencias Naturales XXIII:9–26

Walter H (1970) Vegetation und Klimazonen – Grundriss der globalen Ökologie. Ulmer Stuttgart

Warming E (1892) Lagoa Santa – et Bidrad til den biologiske. Plantegeographi, Kjobenhavn

7 Fire in the Tropical Rain Forest of the Amazon Basin

P.M. Fearnside[1]

7.1 Ancient and "Natural" Fires

Fire has long played an important role in the formation of vegetation types in Amazonia. During the Pleistocene, a large part of Amazonia was covered by grassland, with forest confined to small refugia (the number, size, and evolutionary importance of which are the subject of controversy) (Prance 1982). This period coincided with the arrival of the first humans in the area. Fires started by precolumbian human groups would have slowed the progress of recolonization of the grasslands by forest (e.g., Budowski 1956). Human-initiated burning of grasslands could be expected to affect both the 165,000 km^2 of humid savannas of present-day Amazonia (such as those in Roraima in the Humaitá area of the state of Amazonas) and the limit between the forest and the cerrado or central Brazilian scrubland. Charcoal in the soil of the lavrados ("natural" grasslands) of Roraima indicate large-scale burning about 1000 years B.P. (Sternberg 1968).

Fire in clearings for shifting cultivation within the forested area also affects the composition of the vegetation, although not to the extent of preventing woody vegetation from returning. The density of human population was formerly much higher than it is today, as indicated by the large number of anthropogenic "black soil" sites in the region (Smith 1980). Charcoal has been found in soils under "virgin" Amazonian forest in Venezuela and Colombia (Sanford et al. 1985), in Altamira (Pará) (Fearnside 1978) and at Manaus (Amazonas) and Ouro Preto do Oeste (Rondônia). Recent recovery and radiometric age determination of charcoal in the lowland tropical rain forest of East Kalimantan (Indonesia) is described by Goldammer and Seibert in this Volume.

Fires can also be started by nonhuman agents such as lightning. Reasons for believing that the principal cause of precolumbian forest burning was human, however, include the very limited extent of lightning-caused fires in Amazonia today. Usually only a single tree or a very small patch is burned when lightning strikes. The forest normally has to be felled and the trees allowed to dry on the ground for a few weeks before they can be ignited.

[1] National Institute for Research in the Amazon (INPA), C.P. 478, 69.011 Manaus, Amazonas, Brazil

7.2 Deforestation and Burning in Amazonia Today

Burning in deforested areas in Amazonia today dwarfs the burning of past eras. By 1988 approximately 400,000 km^2, or 8% of Brazil's 5×10^6 km^2 Legal Amazon region had been cleared, and the cleared was increasing at about 35,000 km^2 annually (Fearnside in press). Of this cleared area, approximately 250,000 km^2 is in the 3.4×10^6 km^2 part of the Legal Amazon occupied by dense or "rain" forests (7% of the original area of these forests), while 170,000 km^2 is in the 1.3×10^6 km^2 portion occupied by cerrado (Mato Grosso and Tocantins, formerly northern Goiás) versus the remaining seven states, where dense forests predominate (Fearnside unpublished a). In addition to clearing, the cerrado vegetation is regularly burned without being deforested – burning is an event to which the fire-resistant species of this vegetation type are adapted.

The above estimates of areas deforested are based on linear projections to 1988 from the most recent 2 years of satellite data within each of the Brazilian Legal Amazon's nine states (Fearnside in press). These estimates are substantially lower than an estimate produced from AVHRR images of fires interpreted by the National Institute for Space Studies (INPE), indicating that in 1987 204,608 km^2 was burned, of which approximately 80,000 km^2 was considered to be deforestation in the dense forest area (Setzer et al. 1988, 28). Reasons for the much higher value for burning include the fact that burning is not the same thing as deforestation: pastures, secondary forests, and cerrado are burned without contributing to deforestation. The area covered by the AVHRR image included substantial areas of pasture and cerrado outside of the Legal Amazon, having included the full area of two states (Maranhão and Goiás) that are only partially within the Legal Amazon. The discrepancy between the 80,000 km^2 area considered new deforestation in the AVHRR estimate and the 35,000 km^2 figure derived from linear projections is due to saturation of the AVHRR sensor when recording the heat emanating directly from the fires. The AVHRR images are composed of picture elements or pixels of 120 ha each; only a small area burning within one of these pixels is sufficient to trigger the sensor to indicate the entire pixel as burning. As little as 30 m^2 on fire can saturate the sensor (Robinson in press). A constant correction factor of 70% was applied in the INPE study to adjust for partially burning pixels, but an adequate constant is difficult to derive for making this correction because of the great sensitivity to fire temperature, which varies depending on meteorological conditions, fuel moisture, time of day, etc. The explanation of the discrepancy in saturation of the AVHRR sensor is indicated by the results of an AVHRR image for Rondônia in the same year (1987) interpreted by Jean-Paul Malingreau (pers. commun. 1988). Using reflected light from deforested areas, this indicates 15.1% of Rondônia as deforested by that year, whereas the INPE study of burning indicated 18.7% as actually on fire. Since any given hectare in the deforested area can only be re-burned once every 2–3 years, the INPE result implies an area on the order of 40% deforested, which represents the same level of discrepancy as that between the 35,000 km^2 indicated as the rate of deforestation in the Legal

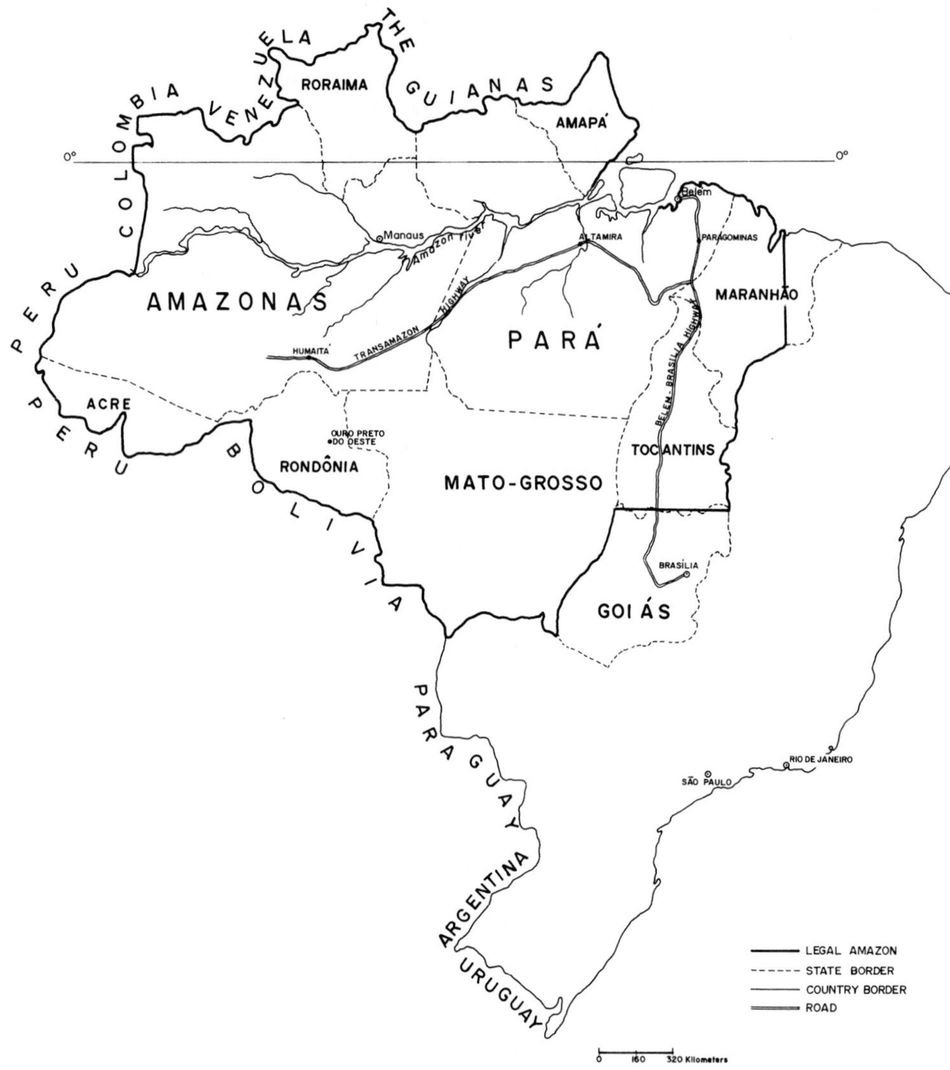

Fig. 1. Map of the Brazilian territory of the Amazon Basin, showing the boundaries of Legal Amazon, state boundaries, and main highways

Amazon by linear projections, versus the 80,000 km^2 indicated by AVHRR sensing of burning.

The above results from linear projections also differ from an estimate for 1988 deforestation made by a different sector of INPE using LANDSAT imagery (Brazil, INPE 1989). This study, released on 6 April 1989 in conjunction with President Sarney's Nossa Natureza (Our Nature) Program, calculates that a total of only 5% of the Legal Amazon had been cleared, and implied an annual rate of increase of only 16,674 km^2. However, the data for several states conflict with other LANDSAT studies carried out by the Brazilian Institute for Forestry Development (IBDF). The INPE results would imply decreases in deforested areas for Acre, Pará, and Mato Grosso based on previous LANDSAT studies, and in Rondônia based on AVHRR studies. In Roraima, although a decrease is not implied, the increase would be so slight as to be highly improbable. The discrepancy between the linear projection estimate of 8% and the Nossa Natureza Program estimate of 5% is not due to out-of-date data in the linear projection study: the states with the most out-of-date data sets (Amazonas and Amapá) do not result in positive discrepancies, whereas states such as Pará and Acre, where recent LANDSAT data are available, are precisely those where the greatest discrepancies with the Nossa Natureza study are found. Combining the two estimates to arrive at a "best estimate" for deforested area through 1988 in the Legal Amazon, a figure of 8.4% results if the linear projection estimates are used for all states for which the Nossa Natureza estimates conflict with other data, and 8.1% if Nossa Natureza data are used for Rondônia, where differences between the AVHRR and LANDSAT-TM sensors could provide an explanation for the discrepancies (Fearnside unpublished b). These figures tend to reinforce the 8% estimate as the most likely approximation of deforestation through 1988. Regardless of which estimates are correct, all show rapid increases in cleared areas and suggest that immediate measures must be taken to slow forest loss.

7.3 Types and Qualities of Burning

Burning in Amazonia has been classified into categories for virgin forest (forest not previously cleared by nonindigenous immigrants), secondary forest (in the case of slash-and-burn sites: abandoned areas at least 8 months uncultivated), weeds (areas 8 or fewer months uncultivated), and pasture. Small amounts of burning also take place in sugar cane, but this land use occupies only an insignificant portion of Amazonia.

Virgin forest is felled prior to burning, in such a way as to maximize the thoroughness of the burn. First the broca or underclearing is performed, cutting vines, understory plants, and saplings or poles that can be cut with a machete or brush hook. Valuable tree species may also be cut selectively prior to felling. The large trees are then cut in the derrubada (felling) phase. Most is done with chainsaws, although some small farmers use axes. Because of the onerous labor of cutting trees with axes, small farmers clearing their own land with axes are more likely to leave scattered trees standing in the cleared areas. Large ranchers, which contract third parties to clear the forest, generally have the least standing trees unfelled.

The time available during the dry season limits the area that small farmers can clear with the labor available to them (Fearnside 1980). If burning is delayed too long (usually with the intention of felling more area), then the always-unpredictable rainy season may begin, leaving the farmer with no land suitable for planting. In order to maximize the area cleared and the quality of the burn, clearing is often done in a circular pattern, leaving to be felled last an island of vegetation in the center of the field; the fire can then converge on the center of the clearing to consume this last-felled area despite its still being relatively green. The felled trees cannot be left too long before burning because resprouting green vegetation will shade the fuel bed and keep it moist and because the dry leaves will fall off the downed trees (making them more difficult to ignite). Variation in the dryness of the vegetation, nearness to the forest edge, topography, wind, and other conditions result in a large amount of variability in burn quality over short distances within the clearing. In one study of a 200-ha burn near Manaus (Amazonas), a visual assessment of 200 sampling stations classed burns into five categories: 22.0% of the points were classed as "excellent" (at least some trunks burned to ash), 45.5% were classed as "good" (burned vines and thick branches), 17.0% were "medium" (burned leaves and thin branches), 12.0% were "poor" (only leaves burned), and 3.5% were classed as "none" (not even dry leaves on the ground burned) (Fearnside et al. unpublished). This burn had consumed 29% of the pre-burn aboveground dry weight biomass, based on comparison of six pre-burn with ten post-burn destructive quadrats of 10 × 10 m each.

The percentage of biomass in the trunks, as opposed to vines and other finer material, appears to have a strong influence on the proportion of the biomass combusted. Finer material burns much more thoroughly: in the study near Manaus, 100% of the leaves burned, 75.8% of the vines, 49.1% of the branches, and 20.9% of the trunks. In a study of burns near Altamira (Pará), where a higher percentage of the forest biomass is composed of vines, a higher percentage of the biomass disappeared in the burn than was the case at Manaus.

Great variability in burn quality exists between farmers and between years. In a study among colonists on the Transamazon Highway near Altamira, burns were classified on a six-level scale: "none" (no burn attempted); "0" (burn attempted but did not burn); "1" (bad burn: only leaves and small twigs combusted; "2" (patchy burn: a mixture of class 1 and 3 burns); "3" (good burn: burned some wood as well as leaves and twigs); "4" (overburned: large logs burned completely to ashes) (Fearnside 1986a, 1989). Grouping qualities "0" and "1" as "bad" burns and qualities "2" and "3" as "good" burns, of 247 virgin burns, 76 (30.8%) were "bad", while 171 (69.2%) were "good." Burn quality is a critical factor affecting soil fertility and agricultural productivity on the Transamazon Highway, and thereby plays a key role in limiting the human carrying capacity of the area. A good burn is necessary to remove the physical encumbrance of the downed vegetation, to deposit nutrients in the form of ashes, to reduce competition from weeds, and to reduce the acidity of the soil. Burn quality is significantly associated with raising soil pH, lowering the concentration of toxic aluminum ions, and raising the levels of soil phosphorus (Fearnside 1986a, 188–191). A poor soil with a good burn often produces a greater agricultural yield than a good soil with a poor burn. The quality of the burn depends partly on luck in having little or no rain during the period when the cut vegetation is drying, and partly on the skill of the farmer in judging when to set the fire. The range of behavior is very great: of 138 virgin burns performed by Transamazon Highway colonists, an average of 44 days elapsed between felling and burning with a standard deviation of 65 (CV = 148%). Felling dates ranged from May through December, with the most popular months being September (49.5%), October (22.6%), November (10.2%), and August (8.5%). The quality of the burn could be correctly predicted in 74% of 247 cases from meteorological factors: rain, evaporation, and insolation between felling and burning and in the 15 days prior to burning (Fearnside 1986a, 1989). Very high variability in meteorological parameters in the area guarantees high variability in burn quality from year to year (Fearnside 1984).

Second growth burns can be predicted from meteorological factors in a way similar to that for virgin burns. Of 54 second growth burns surveyed in Altamira, 31 (57.4%) were "bad" (classes 0 and 1) and 23 (42.6%) were good (classes 2 and 3). Mean time elapsed between cutting and burning was 53 days (SD = 96, N = 79), with a coefficient of variation of 183%. The tremendous variability in clearing behavior explains some of the high variation in burn quality. The range of cutting dates is greater than for virgin burns, extending from June to January. It is possible to get away with cutting secondary forest closer to the beginning of the rainy season than is the case for primary forest. The most popular month for cutting secondary forest on the Transamazon Highway is September (21.6% of 111 cases), followed by October (21.6%), November (18.0%) and December (9.0%). The dates for burning in other parts of the region vary with the annual seasonal cycle. In Rondônia and Acre the most popular month is August, while in Amazonas it is September or October.

The fires are usually set at about 13:00 h, and burn throughout the afternoon and into the night. By the next morning one can normally walk through the burned areas, although large trunks may smolder for several days. The different species of trees burn with varying thoroughness. A few, such as samaúma (*Ceiba pentandra*) commonly burn completely to ashes.

Trunks of many species are reduced to ashes in the relatively rare cases where, under unusually favorable burning conditions, an "overburn" (class 4 burn) occurs. This "burns the earth," and results in stunted crop growth. Secondary succession is also altered: in the Altamira area overburned sites sometimes develop a cover of bracken ferns that delays entry of woody secondary forest species for several years.

Following either a virgin or a second growth burn the farmer may cut unburned branches and small trunks and pile them in mounds where they can be burned in a second burn prior to planting. The piling and re-burning operation, known as coivara, is usually done if upland rice is to be the crop planted; it is occasionally done for maize and never for pasture. The task is a laborious one, but varies greatly depending on the quality of the burn and the thoroughness with which the coivara is done. For virgin burns, a mean of 6.3 man-days/ha of labor are spent on coivara (SD = 8.9, N = 200); for second growth burns the mean is 1.7 man-days/ha (SD = 2.3, N = 12) (Fearnside 1986a, 179). Coivara is most frequent when the burn is of intermediate quality: if the burn is good then coivara is unnecessary, while if it is very poor then it is impractical except perhaps for a very small patch to be used for subsistence production. The time between burning and planting (about 3 months) and the amount of labor available to the colonist usually limit the area to which coivara is done to a portion of the area felled.

Weed burns are done when preparing for a second year of planting on a slash-and-burn site. The residues from the previous crop plus any weeds that have grown up in the field are cut; often they are piled up in mounds prior to burning. Weed burn quality can be predicted from weather parameters, but the quality of weed burns is not significantly associated with the magnitude of soil fertility changes.

Pasture burns are done every 2–3 years in cattle pasture that is being maintained for grazing. All of the pasture in a given property cannot be burned in the same year, as the colonist or rancher must have somewhere to put the cattle when a part of the lot is burned. Usually the woody secondary forest species invading the pastures are cut prior to burning, although occasionally burns are attempted without this step. The logs left in the pasture from the original forest felling are often reduced to ashes when the pasture is burned; after about a decade the pastures are mostly free of the downed timber that is so evident in younger pastures.

7.4 Impacts of Burning on Amazonian Vegetation

Repeated burnings in deforested areas have a strong effect on the course of secondary succession. Burning eliminates most of the stock of seeds in the soil from the original forest (Brinkmann and Vieira 1971). Fields recently cleared from primary forest also have a strong contribution to regeneration from stump sprouts (Uhl 1987), which is also eliminated by repeated burnings. In Brazilian Amazonia most of the deforested areas are maintained in cattle pasture, beginning either directly after the clearing of primary forest (in the case of the large ranchers, who account for about 75% of the clearing) or after a year or two under annual crops (in the case of the small farmers, who account for about 25% of the clearing). In abandoned pastures the secondary vegetation grows much more slowly than it does in shifting cultivation fallows. Intensity of pasture use also has a strong effect. In Paragominas (Pará), Uhl et al. (1988) found that 8 years after abandonment lightly used pastures averaged twice the biomass accumulation of moderately used pastures and 17 times more than heavily used pastures. "Lightly used" pastures are defined as those never weeded, with no or little grazing, and "abandoned" (i.e., no longer weeded or burned) shortly after formation; "moderate" use refer to areas abandoned 6–12 years after formation, with grazing, and with weed cutting and burning every 2–3 years; "heavy use" refers to bulldozing and windrowing after several years under "moderate" use.

In certain parts of the Amazon region, repeated burning leads to dominance of the vegetation by fire-resistant palm species. This is true in Maranhão and southern Pará, where the babaçu palm (*Attalea speciosa* or *Orbignya phalerata*) forms solid stands completely eliminating both pasture grasses and other woody species. In areas cleared from dense forest in Roraima, the related inajá palm (*Attalea regia*) forms stands similar to those of babaçu.

Repeated burnings, combined with pasture degradation through soil compaction and nutrient depletion, can also contribute to deflecting ecological succession to a dysclimax of inedible grasses rather than the normal route to woody secondary vegetation. Tree dispersal and seedling establishment are impeded by the harsh site conditions and repeated burnings in the pasture (Nepstad et al. in press). In some highly degraded pastures in Acre the grass known as rabo de cavalo (*Andropogon* spp.) dominates, while in some other areas the sapé grass (*Imperata brasiliensis*) takes over; this also occurs in some shifting cultivation fallows (e.g., in the Gran Pajonal of Peru: Scott 1978). Amazonian successions are not presently subject to dominance by very aggressive grass species like the *Imperata cylindrica*, that outcompetes woody second growth in Southeast Asia (see UNESCO/UNEP/UN-FAO 1978: 224). It is not a God-given dictum that Amazonian successions will always tend to woody regrowth rather than grass, and the degradation of vast areas of cattle pastures in the region increases the danger that grass will become the predominant successional route in Amazonia as well.

The tendency to favor grasses over trees would be reinforced by changes in precipitation patterns that are expected to accompany large-scale deforestation in Amazonia. Approximately half of the rainfall in the region is the result of water that is recycled through the forest and returned to the air via evapotranspiration. Deforestation would reduce precipitation most during the dry season (Salati et al. 1979; Salati and Vose 1984). This is when rainfall is needed most to maintain the forest, which already suffers from drought stress approaching the limits of tolerance of the rain forest tree species during the occasional extended dry seasons that occur naturally even in the absence of massive deforestation. Grass is a particularly poor contributor to evapotranspiration during the dry season because the grass leaves are either eaten by the cattle or die, whereas the leaves of the forest remain green all year round and continue to transpire water. Deforestation could initiate a positive feedback loop leading to thinner, more xerophytic forests and more frequent and severe drought (Fearnside 1985a).

Fire is likely to be an increasing threat to the remaining forest as the climate of the region becomes dryer. Occasional droughts of unprecedented severity, rather than a gradual reduction of the mean precipitation, is the most likely and most dangerous result of reduced evapotranspiration. During these droughts, fires could escape into the standing forest without the trees having been felled and allowed to dry beforehand. This has already occurred in Borneo under similar circumstances during the 1982–1983 drought provoked by the Southern oscillation-El Niño phenomenon (Malingreau et al. 1985).

Logging disturbance greatly increased the danger of fires escaping into the standing forest burned in Borneo in 1982–1983 (although a large area of undisturbed forest was also burned). In Brazil, Uhl and Buschbacher (1985) have documented the increased probability of fire spreading from cattle pastures into surrounding forest where selective logging has occurred. Logging activity in Amazonia can be expected to increase in the future as the remaining stocks of South East Asian timber dwindle (Fearnside 1987a).

7.5 Indirect Effects of Burning

Burning of Amazonian forest has a variety of indirect effects, and these can be expected to increase as deforestation spreads. Burning has undoubtedly already altered the nutrient balance of the remaining forest through the deposition of nutrients in the form of aerosols. Smoke from Amazonian fires now blankets the entire Amazon Basin during the dry season, often resulting in closure of airports for extended periods. The smoke remains in the lower part of the troposphere where it can be removed by rainfall and deposited in the surrounding forest. In fact, burning often provokes localized rainfall events by supplying condensation nucleii and by the effect of rising air currents above the burn. Elements such as phosphorus, which is present in limited quantities in Amazonian soils and ecosystems, are likely to be scavenged from the solid and liquid deposition falling on the forest downwind of the fire. Amazonian forests are highly efficient in removing nutrients that are contributed through such deposition (Herrera et al. 1978).

One of the furthest-reaching effects of large-scale burning of Amazonian forest is the contribution to global warming through the greenhouse effect. Using an admittedly crude seven-category classification for Amazonian vegetation (based on Braga 1979), and existing information on the biomass and soil carbon between the natural vegetation and cattle pasture, conversion of the entire Legal Amazon to pasture would release approximately 50 billion metric tons (Gigatons = G tons) of carbon (Fearnside 1985b, 1986b, 1987b). Were this transformation to take place over a span of 50 years, then an average of one G ton of carbon would be released per year, or about 20% of the 5–6 G tons presently released from burning fossil fuels throughout the world. The present rate of deforestation in Brazilian Amazonia would represent a contribution of about 5–7% of the global fossil fuel total. The contribution to the greenhouse effect would be greater than the percentage of carbon release would indicate, however, because part of the carbon is released in the form of methane, which is far more potent in provoking the greenhouse effect per ton of carbon than is carbon dioxide. Fossil fuel combustion is more efficient and releases only an insignificant amount of methane. Reburning of cattle pastures produces a net release of methane because, unlike carbon dioxide which is absorbed by vegetation through photosynthesis, methane continues to build up in the atmosphere, without biological sinks. Although far from the entire problem, this represents a significant contribution to a major environmental concern. On a per-capita basis, a single deforester in Amazonia can make a contribution to the greenhouse effect equivalent to hundreds, and in the case of large ranchers to hundreds of thousands, of people living in cities burning fossil fuels. Because the cattle pastures in the region produce very little beef, this and other environmental costs are being borne by society in exchange for almost no benefit.

Acknowledgments. Studies of burning in Altamira were funded by National Science Foundation grants GS-422869 (1974–1976) and ATM-86-0921 (1986–1988), and in Manaus by World Wildlife Fund-US grant US-331 (1983–1985).

References

Braga PIS (1979) Subdivisão fitogeográfica, tipos de vegetação, conservação e inventário da floresta amazônica. Acta Amazonica (Suppl) 9(4):53–80

Brazil, Instituto de Pesquisas Espaciais (INPE) (1989) Avaliação da Cobertura Florestal na Amazônia Legal Utilizando Sensoriamento Remoto Orbital. INPE, São José dos Campos, São Paulo, 54 pp

Brinkmann WLF, Vieira AN (1971) The effect of burning on germination of seeds at different soil depths of various tropical tree species. Turrialba 21:77–82

Budowski G (1956) Tropical savannas, a sequence of forest felling and repeated burnings. Turrialba 6:23–33

Fearnside PM (1978) Estimation of carrying capacity for human populations in a part of the Transamazon Highway colonization area of Brazil. Ph.D. Diss. in biological sciences, Univ Michigan, Ann Arbor, Univ Microfilms Int, Ann Arbor, Michigan, 624 pp

Fearnside PM (1980) Land use allocation of the Transamazon Highway colonists of Brazil and its relation to human carrying capacity. In: Barbira-Scazzocchio F (ed) Land, people and planning in contemporary Amazonia. Univ Cambridge Cent Latin Am Stud Occasional Pap No 3, Cambridge, UK, pp 114–138

Fearnside PM (1984) Simulation of meteorological parameters for estimating human carrying capacity in Brazil's Transamazon Highway colonization area. Tropic Ecol 25(1):134–142

Fearnside PM (1985a) Environmental change and deforestation in the Brazilian Amazon, pp 70–89. In: Hemming J (ed) Change in the Amazon Basin: man's impact on forests and rivers. Manchester Univ Press, Manchester, UK, 222 pp

Fearnside PM (1985b) Brazil's Amazon forest and the global carbon problem. Interciencia 10(4):179–186

Fearnside PM (1986a) Human carrying capacity of the Brazilian rainforest. Columbia Univ Press, New York, 293 pp

Fearnside PM (1986b) Brazil's Amazon forest and the global carbon problem: Reply to Lugo and Brown. Interciencia 11(2):58–64

Fearnside PM (1987a) Causes of deforestation in the Brazilian Amazon. In: Dickinson RF (ed) The geophysiology of Amazonia: vegetation and climate interactions. Wiley, New York, pp 37–61

Fearnside PM (1987b) Summary of progress in quantifying the potential contribution of Amazonian deforestation to the global carbon problem. In: Athié D, Lovejoy TE, Oyens P de M (eds) Proc Workshop on Biogeochemistry of Tropical Rain Forests: Problems for Research. Univ São Paulo, Cent Energia Nucl Agric (CENA), Piracicaba, São Paulo, pp 75–82

Fearnside PM (1989) Burn quality prediction for simulation of the agricultural system of Brazil's Transamazon Highway colonists. Turrialba 39(2):229–235

Fearnside PM (in press) Deforestation in Brazilian Amazonia. In: Woodwell GM (ed) The earth in transition: patterns and processes of biotic impoverishment. Cambridge Univ Press, New York (in press)

Fearnside PM (unpublished a) Extent and causes of tropical forest destruction. Statement to the German Bundestag Study Commission on "Preventative Measures to Protect the Earth's Atmosphere." Bonn, 2–3 May 1989

Fearnside PM (unpublished b) Amazonian deforestation: a critical review of the "Our Nature" Program estimate. Manuscript, 11 pp

Fearnside PM, Keller MM, Leal Filho N, Fernandes PM (unpublished) Rainforest burning and the global carbon budget: biomass, combustion efficiency and charcoal formation in the Brazilian Amazon. Manuscript, 31 pp

Herrera R, Jordan CF, Klinge H, Medina E (1978) Amazon ecosystems: their structure and functioning with particular emphasis on nutrients. Interciencia 3:223–232

Malingreau JP, Stephens G, Fellows L (1985) Remote sensing of forest fires: Kalimantan and North Borneo in 1982–83. Ambio 17(1):314–321

Nepstad D, Uhl C, Serrão EA (1990) Surmounting barriers to forest regeneration in abandoned, highly degraded pastures (Paragominas, Pará, Brazil). In: Anderson AB (ed) Alternatives to deforestation: steps toward sustainable land use in Amazonia. Columbia Univ Press, New York (in press)

Prance GT (ed) (1982) Biological diversification in the tropics. Columbia Univ Press, New York

Robinson JM (in press) Fire from space: Global fire evaluation using IR remote sensing. Int J Rem Sens (in press)

Salati E, Vose PB (1984) Amazon Basin: a system in equilibrium. Science 225:129–138

Salati E, Dall'Olio A, Matusi E, Gat JR (1979) Recycling of water in the Brazilian Amazon Basin: an isotopic study. Water Resour Res 15:1250–1258

Sanford RL Jr, Saldarriaga J, Clark KE, Uhl C, Herrera M (1985) Amazon rain-forest fires. Science 227:53–55

Scott GAJ (1978) Grassland development in the Gran Pajonal of eastern Peru: a study of soil-vegetation nutrient systems. Hawaii Monogr Geogr No 1. Univ Hawaii at Manoa, Honolulu, 187 pp

Setzer AW, Pereira MC, Pereira Júnior AC, Almeida SAO (1988) Relatório de Atividades do Projeto IBDF-INPE "SEQE" – Ano 1987. Inst Pesquisas Espaciais (INPE), Pub No INPE-4534-RPE/565. INPE, São José dos Campos, São Paulo, 54 pp

Smith NJH (1980) Androsols and human carrying capacity in Amazonia. Ann Assoc Am Geogr 70(4):553–566

Sternberg H O'R (1968) Man and environmental change in South America. In: Fittkau EJ, Elias TS, Klinge H, Schwabe CH, Sioli H (eds) Biogeography and ecology in South America, Vol I. Junk, The Hague, The Netherlands, pp 413–445

Uhl C (1987) Factors controlling succession following slash-and-burn agriculture in Amazonia. J Ecol 75:377–407

Uhl C, Buschbacher R (1985) A disturbing synergism between cattle-ranch burning practices and selective tree harvesting in the eastern Amazon. Biotropica 17(4):265–268

Uhl C, Buschbacher R, Serrão EAS (1988) Abandoned pastures in eastern Amazonia. I. Patterns of plant succession. J Ecol 76:663–681

United Nations Educational Scientific and Cultural Organization (UNESCO)/United Nations Environmental Programme (UNEP)/United Nations Food and Agriculture Organization (UN-FAO) (1978) Tropical forest ecosystems: a state of knowledge report. UNESCO, Paris, 683 pp

8 Interactions of Anthropogenic Activities, Fire, and Rain Forests in the Amazon Basin

J.B. KAUFFMAN[1] and C. UHL[2]

8.1 Introduction

Evidence from soil charcoal studies indicates that fires have occurred since the Holocene in both Amazonian tropical rain forests (Saldarriaga and West 1986; Sanford et al. 1985) and Asian rain forests (Goldammer and Seibert 1989). However, it is likely that fire was an extremely rare event, because fire return intervals are many centuries in duration. The rarity of fire in this ecosystem is also reflected by the few apparent evolutionary relationships between rain forest vegetation and fire (i.e., the contribution of fire in shaping the structure and function of this ecosystem is not yet clear).

Prior to European settlement, it was only in those few areas modified by indigenous peoples through shifting cultivation that fire events were relatively common. However, in the last few decades of the 20th century, dramatic changes in the extent and intensity of anthropogenic activities have occurred. These activities have increased the incidences and extent of fire and, if left unchecked, could result in severe declines in the biological diversity and productivity of this ecosystem.

Estimates of the areal extent of Amazonia historically subjected to fire, and the concomitant increases in fire through the years are difficult, if not impossible, to ascertain accurately. Myers (1979) estimated that 36,000 km^2 of nonflooded forest was cleared by slash-and-burn agriculturalists during the period of 1966–1975. This was considered to be the most common form of land use in the region. By 1987, Setzer (in Booth 1989) reported that as much as 80,000 km^2 of forest was burned in the Brazilian Amazon in a single year (see also Kaufman et al. this Vol. and Fearnside this Vol.).

Forest harvesting and related land use activities have increased the likelihood and prevalance of fire in the Amazon Basin through three interrelated phenomena. First, deforestation increases fuel loads through the deposition of woody debris and the subsequent increase in fine fuels associated with increases in lianas and herbaceous species (Kauffman et al. 1988; Uhl and Kauffman 1990). In addition, deforestation alters microclimates (e.g., increased mean daily maximum temperatures, increased wind speeds, increased vapor pressure deficits, and decreased mean daily relative humidities), such that the rate of fuel dry down is more rapid and to a lower equilibrium moisture content. Finally, fire

[1] Department of Rangeland Resources, Oregon State University, Corvallis, Oregon 97331, USA
[2] Biology Department, Pennsylvania State University, University Park, Pennsylvania 16802, USA

is frequently used to clear slash for cropping, for conversion of forest to pasture, and for weed control in established pastures (Uhl and Buschbacher 1985). In a region characterized as a mosaic of pastures, regrowing forest, primary forest, and logged forest, this provides a widespread ignition source which results in frequent fires in those areas where logging or other disturbances have occurred.

Little is known on the science and ecology of fire in Amazon Basin rain forests. Given the potentially dire consequences associated with the widespread incidence of fire in this ecosystem, an increased understanding of the relationships between fire, humans, and the Amazonian biota is necessary. In this chapter we will review some recent advances in our knowledge of fire in the Amazon Basin with the understanding that tremendous gaps exist in our general understanding of fire in this ecosystem.

8.2 The Fire Environment

8.2.1 Fire History of Tropical Rain Forests

Charcoal originating from fires that occurred in the mid-late Holocene are commonly found in soils throughout the Amazon Basin. Based on radiocarbon dating of the charcoal, Sanford et al. (1985) and Saldarriaga and West (1986) reported that repeated fires have occurred in the north-central Amazon Basin during the past 6000 years. These data indicated that fire return intervals were 389–1540 years in Tierra Firme (upland) forests. Based on palynological evidence, these radiocarbon dates have been found to roughly correspond to what is believed to be drier periods of the Holocene (Absy 1982; Markgraf and Bradbury 1982). Although the widespread occurrence of charcoal indicates that fires played some role in the disturbance history of the Amazon Basin, little is known of the character of these fires, including the spatial extent of any given fire event, the fire behavior (low intensity surface fires or severe stand replacing fires), or ignition sources (humans or lightning). Recent findings of soil charcoal which date back to ca. 17,500 B.P. in the rain forests of eastern Borneo have raised similar questions about the evolutionary and ecological role of fire-induced disturbance in the rain forest biome (Goldammer and Seibert 1989; Goldammer and Seibert this Vol.).

Growth characteristics of tropical rain forest trees (i.e., a lack of annual rings) make traditional methods of fire history determinations impossible (e.g., pyrodendrochronology or stand-age analyses). However, knowledge of successional processes, vegetation adaptations and responses to fire, and current climatic patterns can provide a qualitative understanding of the contemporary fire regimes of any given ecosystem. In contrast to tropical savannas or sclerophyllous shrub communities of mediterranean-type climates, vegetation of tropical rain forests displays few evolutionary adaptations to fire survival. The extremely low capacity of primary forest flora to survive even a fire of low severity also suggests that fire occurrence was extremely rare in this ecosystem.

8.2.2 Fuel Biomass and Arrangement

8.2.2.1 Defining Tropical Rain Forest Fuels

Fuels in wildlands are most simply defined as all combustible materials. Theoretically, this includes all aboveground biomass in tropical rain forests (Table 1). However, severe crown fires similar to those that occur in many temperate forests have not been observed in mature Amazonian tropical rain forests. In those few documentations in which fires occurred in standing vegetation, these were largely restricted to surface fuels (Uhl et al. 1988b; Uhl and Kauffman 1990). Therefore, we have defined fuels in tropical rain forests to include all dead and downed woody debris and the organic horizons (i.e., litter layers and root mats).

8.2.2.2 Variability in Fuel Loads

If the areas are to be cut-and-burned, then "potential" fuels may equal total aboveground biomass (Table 1). Total aboveground biomass and, there-

Table 1. Biomass partitioning of living phytomass and necromass (fuels) in tropical evergreen rain forests. Data on the biomass of cutover sites before and after burning are conspicuously absent in the literature

Location	Above ground	Living phytomass ($Mg\ ha^{-1}$) Below ground	Total	Source
Panama	316	11	327	Golley et al. (1975)
Ghana	233	25	258	Greenland and Kowal (1960)
Brazil	264	35	299	Nepstad (1989)
Brazil	406	32	438	Klinge (1976)
Brazil	181	45	226	Fearnside (1985)
Venezuela	335	56	391	Sanford (in press)

Location	Fuel biomass ($Mg\ ha^{-1}$) Woody debris	Fine Fuels	Total	Source
Venezuela primary forest (Oxisols)	13	51*	64	Kauffman et al. (1988)
Venezuela primary forest (Ultisols)	27	81*	108	Kauffman et al. (1988)
Venezuela primary forest (Spodosols)	5.3	39	44	Kauffman et al. (1988)
Brazil primary forest	51	4	55	Uhl and Kauffman (1990)
Brazil selectively logged forest	173	6	179	Uhl and Kauffman (1990)
Brazil converted pasture	40	11	52	Uhl and Kauffman (1990)
Brazil second growth forest	23	4	28	Uhl and Kauffman (1990)

fore, potential fuels vary greatly among different rain forest communities in the Amazon Basin (Table 1). As a result, site-specific measurements are of particular importance in attempts to ascertain carbon storage by intact forests and the potential atmospheric inputs through biomass combustion or decomposition following fire.

In the Venezuelan Amazon, fuel loadings of forest communities ranged from 253 Mg ha^{-1} in floodplain forests to 13 Mg ha^{-1} in oligotrophic Low Caatinga (Bana) forests (Kauffman et al. 1988). When restricting comparisons to intact tall evergreen forests, fuel loads can vary by almost threefold (Table 1). Biomass partitioning of fuel loads will also vary among rain forest communities. In the eastern Amazon, the majority of the fuel load was woody debris (93% of the total fuel load) (Uhl and Kauffman 1990). A root mat was nonexistent in this tropical moist forest. In contrast, biomass of woody debris was much lower and comprised only 12–33% of the fuel loads in forest types of the Venezuelan Amazon (near San Carlos de Rio Negro) (Table 1). In the Venezuelan Amazon, a well-developed organic horizon (rootmat) primarily composed of live fine roots and decaying organic matter composed the majority of the fuel load.

With respect to the downed woody fuels and organic horizons (i.e., available fuels), we have found that biomass is seldom the limiting factor to sustained fire spread in tropical rain forests. In the Venezuelan Amazon, Kauffman et al. (1988) characterized all rain forest communities as having a continuous (> 98%) surface cover of organic materials (fuels) on the forest floor. Although fuel arrangement is continuous in primary rain forest, the majority of coarse woody debris is located in and around treefall gaps.

8.2.2.3 Effects of Disturbance on Fuel Biomass

Deforestation results in dramatic and long-term changes in the biomass, arrangement, and structure of fuels in tropical rain forest ecosystems. Uhl and Vieira (1989) reported the results of a typical mechanized logging operation in the eastern Amazon, where only 50 m^3 of bole wood per hectare were harvested. This amounted to 4.2 trees per hectare or 1.6% of all tree stems 10 cm or greater. In the process of extraction, 12% of the remaining trees lost their crowns, and 11% were uprooted by bulldozers. Logging slash and trees toppled by the logging operation were left on site. As a result, while only 50 m^3 ha^{-1} of wood was harvested, greater than 150 m^3 ha^{-1} of wood was deposited on the forest floor. Fuel loads increased from 55 Mg ha^{-1} in primary forest to 179 Mg ha^{-1} in these selectively logged forests (Table 1). This increase in woody debris on the forest floor represents approximately one-half of the total standing biomass of the primary forest.

Along a common disturbance/land use sequence of primary forest, to logged forest, to cattle pasture and finally to abandoned pasture, fire and decomposition will result in dramatic declines in aboveground biomass. Whereas biomass of woody debris in recently logged forests was 173 Mg ha^{-1}, biomass in an active cattle pasture and abandoned pasture (second growth forest) was only 40 and 23 Mg ha^{-1}, respectively (Table 1).

The quantity of woody debris consumed by wildfires may affect long-term site productivity. Residual coarse woody debris can be an important source of nutrients for regenerating forests following disturbance. Buschbacher et al. (1988) found that woody debris accounted for approximately one-half of the annual nutrient accumulation of regrowing forests following disturbance in the eastern Amazon. In addition to their ecological function as nutrient pools, residual coarse woody debris are also sites of important mycorrhizal activity and provide microsites for germination of woody species and will influence soil transport, erosion, and retention (Harmon et al. 1986).

Along with the decline of woody fuels in sites converted to cattle pasture are increases in highly combustible grass fuels. Total biomass of fine fuels (living and dead) in cattle pastures of the eastern Amazon was 8.9–11.3 Mg ha^{-1} (Nepstad 1989; Uhl and Kauffman 1990). These fine fuels are very flammable and increase the susceptibility of these communities to repeated fires.

Upon pasture abandonment and in the absence of additional fires, woody debris will eventually reaccumulate in second growth forest. In a 10-year-old second growth forest, Uhl and Kauffman (1990) reported that all of the forest litter and woody debris $<$ 7.6 cm in diameter originated from the regrowing stand (approximately 32% of the total fuel load). In addition, some of the coarse woody debris (i.e., that greater than 7.6 cm diam.) originated from the early successional-short-lived species of the second growth forest.

8.2.3 Microclimates and Fire in Tropical Rain Forests

In order for fire to occur in Amazonia, an ignition source must be present, fuels must be in a relatively continuous arrangement, and microclimate must be such that combustible materials can dry below a threshold of combustion. Ignition sources from either lightning or humans are abundant in the Amazon Basin. Because fuel loads and arrangement do not appear to limit surface fires from spreading in Amazonian rain forests, it is probable that it is climatic characteristics that limit the occurrence of large-scale fires in undisturbed forests of Amazonia. However, fires are very common in disturbed sites where microclimates are altered to drier, warmer, and windier conditions. Intact tropical rain forests create a microclimate that maintains fuel moisture contents of forest litter and woody debris above that point where sustained combustion is possible.

Relative humidity in intact tropical rain forests can remain at 100% for 20 h or more during the dry season (Uhl et al. 1988b; Uhl and Kauffman 1990). In tall closed canopy forest, it rarely falls below 70%. During the 6-month dry season of 1987, midday relative humidity in an eastern Amazonian primary forest fell below 65% on only seven occasions (Fig. 1). The mean midday relative humidity (i.e., the daily minimum) during the dry season in four different tall closed canopy forests of Venezuela and Brazil was found to remain above 80% (Table 2). This level of relative humidity minima is relevant in relation to fire

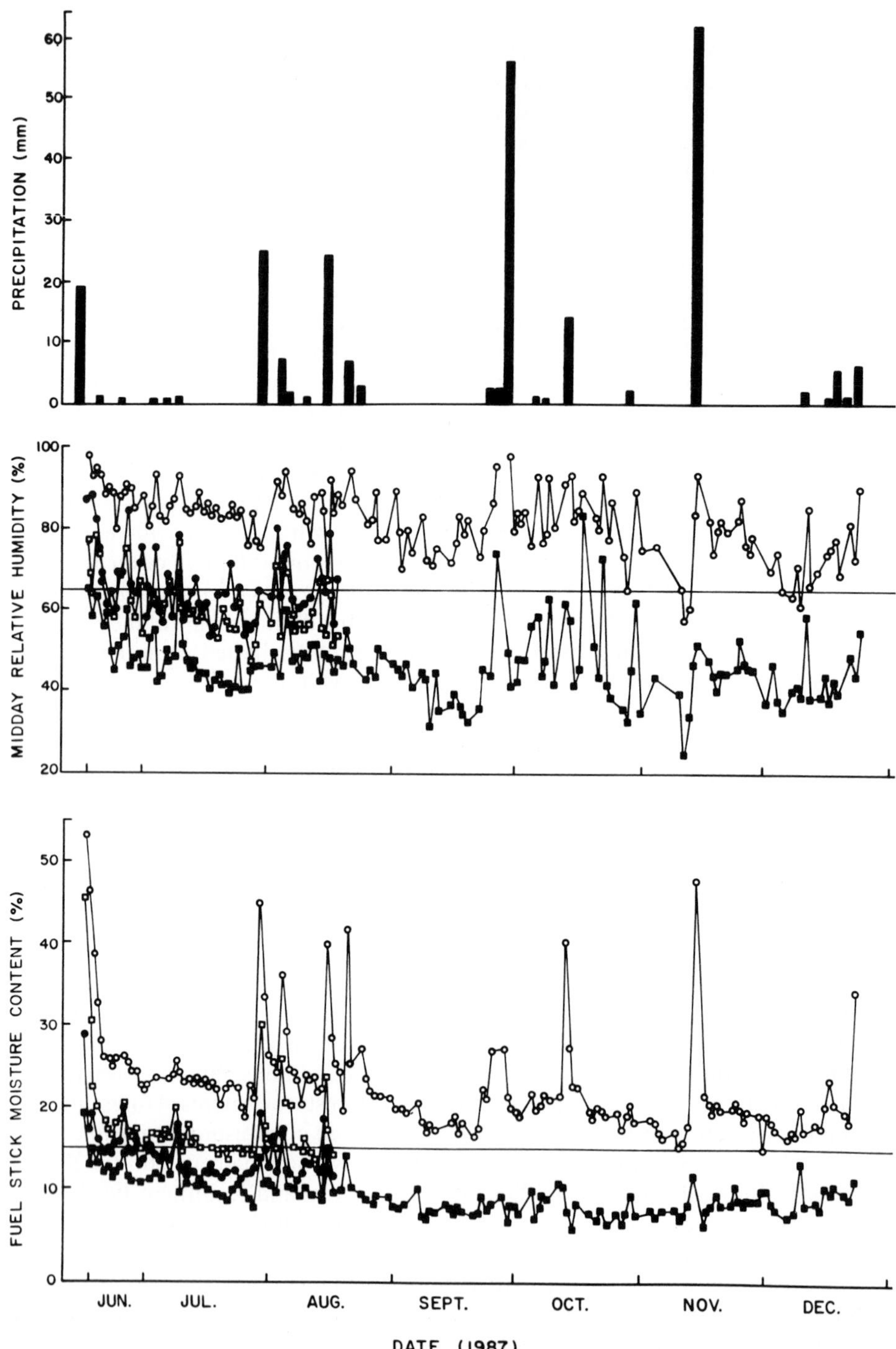
PRECIPITATION (mm)
0
10
20
30
40
50
60
MIDDAY RELATIVE HUMIDITY (%)
20
40
60
80
100
FUEL STICK MOISTURE CONTENT (%)
0
10
20
30
40
50
JUN.
JUL.
AUG.
SEPT.
OCT.
NOV.
DEC.
DATE (1987)

Table 2. Mean maximum temperatures, minimum temperatures, and midday relative humidity for intact primary rain forests and disturbed sites. Data are from the Venezuelan, Amazon near San Carlos de Rio Negro and the eastern Amazon, near Paragominas Pará, Brazil (Uhl et al. 1988b; Uhl and Kauffman 1990). Data are mean (± SE)

	Maximum temperature (°C)	Minimum temperature (°C)	Midday relative humidity (%)
Primary forest (Oxisols)-Venezuela	28 ± 0.2	23 ± 0.1	85 ± 2
Primary forest (Ultisols)-Venezuela	28 ± 0.2	23 ± 0.1	84 ± 2
Primary forest (Spodosols)-Venezuela	29 ± 0.3	23 ± 0.1	80 ± 2
Primary forest (Oxisols)-Brazil	28 ± 0.2	22 ± 0.1	86 ± 0.7
Logged forest-Brazil	38 ± 0.3	22 ± 0.1	65 ± 1
Degraded pasture-Brazil	38 ± 0.2	20 ± 0.2	51 ± 1
Second growth forest-Brazil	33 ± 0.5	21 ± 0.1	62 ± 1

susceptibility, as we have observed that sustained combustion rarely occurs unless relative humidity is below 65% in tropical rain forest communities.

Canopy removal either through natural treefalls or deforestation can create microclimates of warmer and drier conditions. In areas in which 50% or more of the canopy is removed, average daily temperature maxima are increased by as much as 10°C, while the mean midday relative humidity may decrease by as much as 35% (Table 2). Although midday relative humidity during the dry season in primary forest rarely falls below 65%, it is below this threshold for approximately one-half the time in selectively logged forest, and almost a daily occurrence in pastures and second growth (Fig. 1).

In uncut forests ecotonal to those that are disturbed, microclimates can be altered to warmer and drier conditions. This can create conditions that allow for sustained fires to occur in these fragmented standing forests. In the eastern Amazon, Kauffman (in press) reported an almost complete crown mortality from a fire that burned into a forest that was adjacent to an area converted to cropland. Canopy removal of the adjacent area created enough of a disruption in the microclimate of the standing forest to lower fuel moisture contents where fire spread was possible.

8.2.4 Susceptibility of Tropical Rain Forest Ecosystems to Fire

As a result of high levels of atmospheric moisture in primary forests, fuels are usually at or above the fiber saturation point [i.e., that point where cell walls are completely saturated through adsorption processes (Shroeder and Buck 1970)].

Fig. 1. Patterns of rainfall, midday relative humidity, and fuel stick moisture content during the 1987 dry season in a tropical evergreen rain forest ecosystem of the eastern Amazon Basin (near Paragominas, Pará, Brazil). Data are represented as -○- primary forest, -●- selectively logged forest, -□- second-growth forest, and -■- for degraded pasture

This level of fuel moisture content is well above that threshold where sustained combustion would be possible [i.e., the moisture of extinction (Albini 1976)]. The moisture of extinction for litter in temperate hardwood forests has been reported to be 25% (Albini 1976) and it is probably close to this value for litter in tropical hardwood forests. Uhl and Kauffman (1990) reported that leaf litter moisture content in a primary forest of the eastern Amazon never fell below 30% during the entire dry season of 1987; a moisture content well above the moisture of extinction.

The seeming immunity to fire in primary rain forest as a result of high moisture contents can be further evidenced by examination of woody fuel moisture contents (Fig. 1). Sustained combustion is rarely possible if moisture content of a 10-h fuel stick (i.e., a wood particle 1 cm in diameter) exceeds 15%. Reviewing the entire dry season of 1987, fuel stick moisture content in intact rain forests fell to 15% for brief midday periods on only two occasions (Fig. 1). Based on 9 years of weather records from the Venezuelan Amazon, Uhl et al. (1988b) determined that microclimatic conditions allowing for a sustained fire event never occurred in tall, closed-canopy Tierra Firme forests.

Although the moist microclimate in primary forest is such that fuel moisture contents are rarely low enough for sustained combustion to occur, small shifts in regional climates towards warmer or drier conditions could alter this condition. The present equilibrium of water and energy cycles in the Amazon Basin is related to the present vegetation cover. A decrease in forest area might lead to a decrease in the water vapor in the atmosphere and consequently also in the precipitation (Salati 1987). A decrease in atmospheric water vapor would also result in lower equilibrium moisture contents of fuels and hence greatly increase the probability of fire in primary forest. Increased incidences of fire under this scenario could result in dramatic and permanent changes in the composition and structure of Amazon Basin tropical forests.

Fire is now a common event in many anthropogenically altered ecosystems of the Amazon Basin (Uhl and Buschbacher 1985; Kauffman in press). In open cattle pastures and second growth forests, fuel moisture contents may dry to a combustible moisture content on a daily basis during much of the dry season (Fig. 1). Driving the process of fuel moisture loss is the differential gradient between the vapor pressure of the atmosphere and that of the fuel particle. In disturbed ecosystems, this vapor pressure gradient between fuels and the atmosphere is steeper because of the lower levels of atmospheric moisture and higher midday temperatures (i.e., lower relative humidities and higher vapor pressure deficits). Uhl and Kauffman (1990) reported that the vapor pressure deficit of the atmosphere surrounding degraded pastures was six times that in primary forest. Because of the accelerated rates of moisture loss and lower equilibrium moisture contents, fires are possible in second growth forests after only 8–9 rainless days during the dry season. In selectively logged forests (ca. 50% canopy removal), fires can occur after only 5–6 rainless days. Pastures are the most fire-prone ecosystems in the Amazon Basin; during the dry season, sustained combustion is possible within 24 h following a precipitation event.

8.2.5 Fire Behavior and Biomass Consumption in Tropical Rain Forests

In the Venezuelan Amazon, Uhl et al. (1988b) recorded fire behavior, biomass consumption, and bare ground exposure as a result of fire in two disturbed vegetation types and one open-canopied forest. Overall, the fires were characterized as being of very low intensity; flame lengths were less than 37 cm and fireline intensity was less than 34 kW m^{-1}. Flaming combustion was largely restricted to surface fuels. Slightly higher levels of fireline intensity were measured for experimental fires in second growth forests of the eastern Amazon (Kauffman and Uhl unpubl. data). These fires were also restricted to surface fuels; flame lengths ranged from 20–150 cm and fireline intensity was as high as 658 kW m^{-1}. Maximum bark surface temperatures that resulted from these fires (at 10 cm in height above the litter layer) ranged from 109 to 497°C on the windward side of the trees and from 140 to 560°C on the leeward side of the trees. These low-intensity fires often resulted in crown mortality of the second growth trees.

Although these test fires were of low fireline intensity and restricted to surface fuels, significant fractions of the fuel bed were consumed. In the Venezuelan Amazon, mean fuel biomass consumption from experimental fires ranged from 25% in treefall gaps to 47% in second growth forest (Uhl et al. 1988b). However, one fire in second growth forest consumed 89% of the fuels. In this plot, bare soil exposure increased from 0% prior to burning, to 62% after fire.

High levels of biomass consumption have been observed in fires that occurred in selectively logged forests and slashed areas that were burned for conversion to cropland or pasture. Fires in slashed Brazilian tropical dry forest commonly have flame lengths of 10–20 m and consumption may exceed 90% of all aboveground biomass (Kauffman et al. 1990). However, it is unfortunate that precise quantification of fire behavior or biomass consumption does not exist for fires in Amazon rain forest ecosystems. There is a tremendous gap in our knowledge concerning the quantification of fire behavior, levels of biomass consumption, and environmental conditions present at the time of burning. These data will be necessary in order to improve our understanding of fire effects on nutrient cycling, global carbon cycling, losses in biodiversity, and ultimately, the future composition and structure of forests in the Amazon Basin.

8.3 Vegetation Adaptations and Responses to Fire

When fires occur in disturbed tropical rain forest ecosystems, dramatic changes in the composition, vegetation structure, and successional pathways can result. The magnitude of the change is variable depending upon such factors as fire severity, size of the fire, disturbance history of the site, climate, the composition and location of propagules, and the composition of ecotonal vegetation. In contrast to the biota in ecosystems characterized by frequent fire return intervals, rain forest species have not developed specific evolutionary adaptations that ensure persistence following fire in their environment. Those species that will persist in anthropogenically modified ecosystems characterized by frequent fires possess what we define as "fortuitous adaptations": those adaptations that allow species to survive fires or quickly establish following fire regardless of evolutionary derivation. These adaptations include morphological, physiological, or reproductive characteristics of vegetation.

These "fortuitous adaptations" to fire can be broadly generalized to include those traits which facilitate survival of the individual, and those traits which facilitate persistence of the population through reproductive means (Kauffman 1990). Examples of those traits which enhance persistence of the individual include thick bark, anomalous arrangements of stem tissues, sprouting from subterranean meristematic tissues, and epicormic sprouting. Those traits which facilitate persistence of the species or population include long-term seed viability in soil seedbanks, seeds which possess efficient dispersal mechanisms to sites immediately following fire, and fire-enhanced flowering and seed production. In this section we will briefly review what little is known on vegetation adaptations to fire in the Amazonian rain forest.

8.3.1 Bark Properties

Bark tissues can protect vascular and meristematic tissues from fire-related injuries. The insulative properties of bark are often adequate to insure survival of trees following surface fires. Thermal properties of bark may be influenced by density, moisture content, chemical composition, surface characteristics, and thickness. However, only bark surface characteristics and thickness have been found to influence heat flux through tropical bark tissues studied by Uhl and Kauffman (1990).

Bark surface properties that will influence heat flux, and hence survival, include color (absorbance) and texture (as it influences heat dispersal). During simulated fires in which external bark temperatures on tropical rain forest trees were measured, the lowest temperatures occurred on those trees with highly fissured bark. In contrast, the highest surface temperatures occurred on those trees with bark composed of thin exfoliations that readily ignited when exposed to flames (Uhl and Kauffman 1990). Smooth-barked species were intermediate in surface temperature.

In both temperate and tropical forest studies, thickness has been reported to be the primary characteristic determining the capacity of bark to protect underlying cambial and other meristematic tissues (Martin 1963; Uhl and Kauffman 1990). Bark thickness varies greatly among tropical rain forest species. In a survey of all trees 20 cm or greater in a 5-ha stand of primary rain forest (n = 699), Uhl and Kauffman (1990) found that bark thickness at 1.5 m in height ranged from 1.5 to 28.9 mm. It would appear that those with thick bark tissues would survive low-intensity surface fires. However, the capacity for bark to protect living aboveground tissues is limited by high levels of fire severity.

Following a low-intensity surface fire in a selectively logged forest, Kauffman (in press) reported that 36% of the residual tree crowns survived. In contrast, crown survival was only 3% following a severe fire in another partially logged forest purposely burned for cattle pasture conversion. Bark tissues were adequate in insulating a significantly greater number of individuals in the low-intensity fire. However, they were ineffectual in protecting cambial tissues during fires of higher severity.

Uhl and Kauffman (1990) reported that a significant relationship existed between cambium temperatures and bark thickness during simulated fires in tropical rain forests. They predicted that during low intensity surface fires (mean flame length of 40 cm, residence time of 141 s) all trees with a bark thickness of less than 6.4 mm would be topkilled by fire. Crown mortality from fires of this severity would include 98% of all tree stems greater than 1 cm in diameter and approximately half of the total stand basal area.

In summary, thick bark is not a widespread characteristic among tropical rain forest trees. Presence of thin bark tissues for many tropical rain forest tree species results in high rates of mortality even following low-intensity surface fires. Moreover, it is doubtful that possession of thick bark tissues would necessarily ensure species persistence in areas of Amazonia where repeated fires of high intensity have become a common phenomenon.

8.3.2 Anomalous Arrangement of Stem Tissues

Remarkable rates of survival following fire have been observed for many of the Palmae (Kauffman in press). Following severe fires in areas purposely burned for pasture conversion, we have observed high rates of survival (> 85%) of mature palms where crown mortality was almost 100% for the associated dicotyledonous trees. In converted pastures subjected to numerous surface fires, palms are often the only residual surviving taxa from the original primary forest (see also Goldammer and Seibert this Vol.). In addition, lianas have been observed to increase in abundance following disturbance in the eastern Amazon. Explanations for the persistence of these taxa may include anomalous arrangements of tissues in their mainstems. Often these species will have protected segments of vascular cambium deeply embedded within stem tissues. In lianas, Dobbins and Fisher (1986) reported that the occurrence of living tissues within the xylem resulted in rapid and vigorous regeneration following wounding or even complete girdling. Hypothetically, this anatomical arrangement could also facilitate survival of a fire event. Whereas heat generated from a surface fire may kill cambial tissues close to the bark surface, those deeply embedded within xylem tissues may survive.

8.3.3 Vegetative Sprouting

Following fire, a proportion of the establishing vegetation will often arise as sprouts from surviving dormant meristematic tissues located beneath bark tissues or the soil surface. Plants with the capacity to sprout may have a competitive advantage on burned sites due to the presence of well developed root systems which facilitate rapid rates of regrowth. For example, sprouts 20 months old formed a dense canopy of 4 m or more in height (Kauffman in press).

Sprouting from subterranean organs following disturbance is an adaptation shared by many families of Amazonia's tropical rain forest trees. Uhl et al. (1988a) reported that 94 of 171 tree species identified in second growth forests in the eastern Amazon had a sprouting capacity. In a study in the same region, Kauffman (in press) reported that 80% of all plant families (n = 25) had individuals that sprouted following fire. Of the 123 species sampled, 65% possessed a sprouting capacity following fire.

Following disturbance, sprouts may arise via three pathways: (1) epicormic sprouting – sprouting which occurs from dormant meristematic tissues located beneath bark on mainstems and trunks; (2) subterranean sprouting – sprouting from dormant tissues located belowground (i.e., from roots, rhizomes, lignotubers, bulbs, etc.); and (3) stump sprouting (coppice) – sprouting that arises from buds on aboveground portions of stumps.

Epicormic sprouting occurs when dormant buds that are located on large branches and trunks are stimulated to grow as a result of fire-induced mortality of existing foliage. Sprouting from dormant tissues on aboveground plant parts is common following crown breakage in neotropical trees (e.g., Putz and Brokaw 1989). However, epicormic sprouting following fire has received little attention, largely because fires in standing rain forests are a relatively recent phenomenon. In a survey of sprouting trees following fires in the eastern Amazon, Kauffman (in press) reported that 52% of all plant families and 31% of all species sampled had epicormic sprouts.

Epicormic sprouting is an advantageous adaptation following fires of low severity where the level of heat release is great enough to kill foliage, yet vascular and cambial tissues as well as dormant buds survive. Following fire in Australia, Gill (1978) reported that a *Eucalyptus* sp. had reestablished prefire leaf areas within 1 year following a fire. Sprouting from epicormic tissues provides a competitive advantage with respect to structure and rapidity of regrowth compared to those species that reestablish from seed or basal sprouts.

When plants are topkilled, either through logging or fire, basal sprouts may arise via two pathways; from subterranean organs or from aerial portions of the stump (i.e., coppice). Soil is an extremely effective insulator from the high temperatures experienced during forest fires. As a result, the capacity to arise from subterranean sprouts is an efficient means of persistence following fire. In contrast, while sprouts arising from aerial portions of stumps are an efficient means of forest regrowth in the absence of fire, their vulnerable position greatly predisposes them to fire-induced mortality. Uhl et al. (1981) reported that 3 months following overstory removal, stump sprouts accounted for 87% of the stem density. However, nearly all of these sprouts were killed by a fire that occurred at this time.

Following fire in the eastern Amazon, Kauffman (in press) reported that 68% of all plant families and 56% of forest tree species had the capacity to sprout from subterranean tissues. However, among sprouting species there is a great variation in the percentage of individuals that survive and sprout following fire. For example, Kauffman (in press) reported that survival through sprouting of the most common species sampled ranged from 15% for *Cecropia* spp. to 67% for a Lecythidaceae.

The level of fire severity (e.g., levels of fireline intensity and biomass consumption) is a very important determinant influencing the number of individuals that sprout following fire. Following surface fires of low severity, Kauffman (in press) reported that 64% of the rain forest trees survived through sprouting. In a site purposely burned for pasture conversion with high levels of fuel consumption, survival of residual trees through sprouting was only 31%. If no other disturbances occur on these sites (i.e., herbicides, weeding, cropping, or use by livestock), sprouts may account for a significant proportion of the regrowing forest. For example, Uhl and Jordan (1984) and Uhl et al. (1988a) reported that following deforestation, fire, and minimal disturbances associated with postfire land use, between 12 and 30% of individuals in regrowing forests arose from sprouts. In contrast, nearly all sprouting individuals may be eliminated through subsequent intensive levels of disturbance on sites (e.g., repeated weeding, pasture fires, herbicides, and heavy equipment use).

8.3.4 Seed Banks

In addition to sprouts, many of the first woody plant colonizers following fire in Amazonia will originate from the soil seedbank. Because soils are an effective insulator to heat, numerous seeds present in surface horizons will survive and germinate.

In primary rain forests, seed bank populations have been reported to range from 180 to 860 viable seeds m^{-2} (Uhl and Clark 1983). In seed banks of primary forests, the majority of seeds are usually early seral woody species (e.g., *Cecropia* spp.). The density of viable propagules in the soil seed bank could be explained by a high seed rain in combination with maintenance of seed viability for many years. Uhl and Clark (1983) reported that the seed rain in primary forest of the Venezuelan Amazon was $\sim$50 seeds m^{-2} yr^{-1}. In experiments of seed longevity, several early seral woody species maintained viability when left on the forest floor under natural conditions for 1 year (Uhl and Clark 1983). Seed longivity for *Cecropia* spp. may exceed 15 years (Uhl and Clark 1983).

Fire may greatly reduce the abundance of viable propagules in the seedbank. Uhl et al. (1981) reported a mean temperature of 100° C at 1 cm below the soil surface during a slash fire. They reported that the seed bank in a primary forest consisted of 752 viable seeds m^{-2} and burning reduced this to 157 viable seeds m^{-2}.

Repeated fires, invasions by early seral plants or exotic species, and conversion to cropland or pasture will dramatically change the composition and size of soil seed banks. For example, in the Venezuelan Amazon, Uhl and Clark (1983) reported that a 6-year-old cattle pasture contained 1250 viable seeds m^{-2}. Almost all of these seeds were composed of grasses and forbs rather than successional woody species.

8.3.5 Dispersal Mechanisms as Adaptations for Fire Survival

The capacity for propagules to rapidly disperse onto recently burned sites could facilitate persistence under a scenario of increased fire frequency in Amazonia. This adaptation is advantageous in that the propagule is not subjected to the fire, but establishes in the nutrient-rich seedbed that characterizes recently burned forest sites. Seeds may disperse onto recently burned sites through four vectors: wind, water, animals, and humans.

Many graminoids and herbaceous dicots originate from offsite sources via wind dispersal. In contrast, many of the common successional woody species have berries or fleshy fruits and rely on animals for dispersal. Uhl et al. (1988a) reported that 88% of all tree species identified on second growth forests following pasture abandonment had animal-dispersed fruits.

There are a number of limitations to animal dispersal of propagules onto burned sites. If all aboveground vegetation has been cut or consumed by a severe fire (i.e., destruction of the animals' habitat), then dispersal potential can be eliminated. In contrast, when some residual vegetation is present, (i.e., habitat for potential seed-dispersing animals in the form of unburned patches of vegetation), an increased rate of seed dispersal will most likely accelerate both the rate of forest recovery as well as diversity of species present in the regrowing composition. For example, in an area cleared for agriculture, Uhl et al. (1982) found that soil under residual trees (cashews) had 932 viable seeds m^{-2} of woody vegetation, as compared to 126 m^{-2} under slash, and 74 m^{-2} in areas of bare soil.

The failure of many potential seed-dispersing animals to venture into burned sites is only one barrier to germination. In abandoned pastures or other disturbed areas, the prevalence of repeated fires will limit establishment. Even in the absence of fire, seed predation by rodents or insects can be high. In addition, numerous physiological limitations to survival in a hotter, drier environment exist (Nepstad 1989).

Seed survival and plant establishment are related to the microenvironment in which the seed is deposited. Uhl et al. (1981) reported that successional woody species established best under slash rather than exposed sites. Grasses tended to establish best in exposed microhabitats. Fire severity will influence the abundance of these microenvironments. In areas where almost all slash is consumed by one or several fires, establishment of woody vegetation may be retarded while graminoids will be favored. In contrast, woody species may be favored following low-severity fires where much woody debris remains.

8.3.6 Fire-Enhanced Flowering

Fire-enhanced flowering and increased seed production following fire has not been quantified in Amazonia, yet is a common phenomenon following fire in both North American and Australian ecosystems. However, enhanced flowering and fruiting is likely in burned environments of Amazonia as seed production has been reported to be rapid and very high following fire. Fire-enhanced flowering and fruit production may be related to nutrient availability and competitive interactions following fire.

8.4 The Winners and the Losers

There is little doubt that the frequent and widespread use of fire in Amazonia is dramatically changing successional patterns, forest structure, and composition. Fires will result in the elimination of some species even though forest regeneration will occur following all but the severest of disturbances (Uhl et al. 1988a). However, many of the complex interspecific interactions between coevolved species that make these tropical forests unique may never recur. Scientists do not yet have the ability to predict the future composition and structure of tropical rain forests under a scenario of frequent fires due to anthropogenic activities. Which species are likely to become extinct and which ones will persist or even flourish?

Those species with one or more of the adaptations mentioned in Table 3 are most likely to persist. This includes early seral tree species such as *Cecropia* spp. that presently establish in disturbed sites. These early successional species will most likely persist through a combination of rapid growth rates in a high-light environment, rapid rates of seed production, effective long-distance dispersal mechanisms, and long-lived seeds in the soil seed bank. Grasses and forbs with short life-spans and wind-dispersed seeds are likely to increase in abundance.

Few primary forest tree species are known to possess seed strategies that will insure persistence following one or more fires. However, those species with strong capacities to sprout from subterranean tissues may endure frequent fires. Epicormic sprouts and thick bark are adaptations suitable to facilitate survival following low-intensity surface fires. However, this adaptation would be of little value in an environment with severe and frequent surface fires. Many of the Palmae, with meristematic tissues protected by dense leaf bases, bracts, and anomalous arrangements of meristematic tissues, will most likely persist in a disturbed environment with frequent fires. However, will the residual components of tropical rain forests be destined to extinction as well? It is not known if those species that persist through sprouting will ultimately be capable of producing viable propagules. Further, will these propagules be capable of germinating and establishing in an anthropogenically altered environment?

It is of great importance to quantify who the losers will be (i.e., which species are likely to disappear under a scenario of increased forest fires). Those species

Table 3. Some vegetative adaptations that may influence species persistence following fire in tropical rain forests of the Amazon Basin

I. Adaptations that facilitate persistence of individuals		
Adaptation	Adaptative function	Species
Thick bark[a]	Insulation of cambium from excessive heat	*Manilkara huberi* (Duck) Standley (Sapotaceae) *Lecythis lurida* (Miars) Mors. (Lecythidaccae)
Anomalous secondary growth[b]	Protection of regenerative tissues from heat	Lianas, Palmae
Subterranean sprouting[c]	Sprouts from meristematic tissues located on subterranean organs (e.g., roots, rhizomes, burls, tubers, etc.)	Numerous neotropical forest trees
Coppice[d]	Sprouts originating from stumps (aboveground)	Numerous neotropical forest trees
Epicormic sprouting[e]	Regeneration from meristematic tissues beneath bark on main stems and trunks	Numerous neotropical forest trees
Dense leaf bases[f]	Protection of meristematic tissues by densely compacted leaf tissues and bracts	Palmae, Graminae
II. Adaptations that facilitate persistence of populations		
Viable seeds in soil seed banks[g]	Establishment from seeds capable of long-term viability in soils	*Cecropia* spp., *Solanum* spp. Many successional tree species
Wind-dispersed mechanisms[h]	Rapid invasion on site from winds following fire	Early successional grasses and dicots
Fire-enhanced flowering[i]	Enhanced flowering and seed production following fire	Terrestrial orchids, Graminae Many of the monocotyledons

[a] Uhl and Kauffman (in press).
[b] Dobbins and Fisher (1986), Kauffman (in press).
[c] Kauffman (in press).
[d] Uhl et al. (1981).
[e] Kauffman (in press), Putz and Brokaw (1989).
[f] Gill (1981), Kauffman (in press).
[g] Uhl and Clark (1983).
[h] Uhl and Clark (1983).
[i] Gill (1981). This trait has not been documented for tropical species in the Amazon Basin, but it is a widespread adaptation in many tropical and temperate ecosystems.

which lack any capacity to sprout following crown mortality by fire are likely to decline. Coupled with the loss of any forest tree species is the loss of those numerous organisms (e.g., insects, microorganisms, etc.) with an absolute fidelity to primary forest (Mueller-Dombois and Goldammer this Vol.).

In conclusion, deforestation and fire in the Amazon Basin is occurring at an unprecedented rate. Although these practices are recognized as among the most significant of global environmental problems, very little fundamental information exists on fuels, microclimate, and fire behavior in this biome. The variability in fuel biomass, as well as the variability in biomass consumption by fire, is largely unknown. Relationships between fuels, microclimate, and fire behavior need quantification in order to better ascertain the role of Amazon fires in global CO_2 inputs and as sinks for carbon storage. Data on fire in this environment are needed to quantify influences on vegetation diversity, nutrient losses, effects on the rhizosphere, and the overall long-term site productivity. The need for an improved understanding of the effects of fire on the Amazon environment is of importance if increased and intensified land use activities associated with increases in a desperately impoverished population occur (Fearnside this Vol.). Regional levels of deforestation may lead to altered hydrological patterns that result in a drier climate (e.g., Salati 1987); thereby creating conditions favorable for more fires of greater severity, and greatly exacerbating the effects mentioned above (Fosberg et al. this Vol.). If we are to change the current trend of degradation in the tropical rain forest biome, a fundamental understanding of fire effects and more importantly, potential alternatives to the widespread use of fire is of paramount importance.

References

Absy ML (1982) Quatenary palynological studies in the Amazon Basin. In: Prance GT (ed) Biological diversification in the tropics. Columbia Univ Press, New York, pp 67–73

Albini FA (1976) Estimating wildfire behavior and effects. USDA Forest Service, Gen Tech Rep INT-30. Intermountain Forest Research Station, Ogden, Utah, 97 pp

Booth WL (1989) Monitoring the fate of forests from space. Science 243:1428–1429

Buschbacher R, Uhl C, Serrao EAS (1988) Abandoned pastures in eastern Amazonia. II. Nutrient stocks in the soil and vegetation. J Ecol 76:682–699

Dobbins DR, Fisher JB (1986) Wound responses in girdled stems of lianas. Bot Gaz 147:278–289

Fearnside PM (1985) Summary of progress in quantifying the potential contribution of Amazonian deforestation to the global carbon problem. In: Workshop on biogeochemistry of tropical rain forest. Piracicaba, Sp Brazil, Sept 30-Oct 4, Proc CENA/USP, pp 75–82

Gill AM (1978) Crown recovery of *Eucalyptus dives* following wildfire. Aust For 41:207–214

Gill AM (1981) Fire adaptive traits of vascular plants. In: Mooney HA, Bonnicksen RM, Christensen NC, Lotan JE, Reiners WA (eds) Proc Conf Fire Regimes and Ecosystem Properties. USDA Forest Service, Gen Tech Rep WO-26, Washington DC, pp 208–230

Goldammer JG, Seibert S (1989) Natural rain forest fires in eastern Borneo during the Pleistocene and Holocene. Naturwissenschaften 76:518–520

Golley FB, McGinnis JT, Clements RG, Child GI, Duever MJ (1975) Mineral cycling in a tropical moist forest ecosystem. Univ Georgia Press, Athens, 248 pp

Greenland DJ, Kowal JML (1960) Nutrient content of the moist tropical forest of Ghana. Plant Soil 12:154–173

Harmon ME, Franklin JF, Swanson FJ, Sollins P, Gregory SV, Lattin GD, Anderson NH, Cline SP, Aumen NG, Sedell JR, Lienkaemper GW, Cromack K Jr, Cummins KW (1986) Ecology of coarse woody debris in temperate ecosystems. In: MacFadyen A, Ford ED (eds) Advances in Ecological Research. Academic Press, New York, 15:133–302

Kauffman JB (1990) Ecological relationships of vegetation and fire, chapter 4. In: Walstad JD, Radosevich SR, Sandberg DV (eds) Natural and prescribed fire in Pacific Northwest Forests. Oregon State Univ Press, Corvallis, Oregon

Kauffman JB (in press) Fire effects on standing tropical rain forest trees in the Brazilian Amazon. (Biotropica)

Kauffman JB, Uhl C, Cummings DL (1988) Fire in the Venezuelan Amazon: 1. Fuel biomass and fire chemistry in the evergreen rain forest of Venezuela. Oikos 53:167–175

Kauffman JB, Sanford RL, Cummings DL, Sampaio E (1990) Biomass burning in Brazilian tropical dry forest: carbon, nitrogen and phosphorous losses. In: Abstracts Chapman Conference on Global Biomass Burning. Williamsburg, Virginia, March 19–23, 1990. American Geophysical Union

Klinge H (1976) Root mass estimation in lowland tropical rain forests of Central Amazonia, Brazil. IV. Nutrients in roots from latosols. Tropic Ecol 17:79–88

Markgraf V, Bradbury JP (1982) Holocene climatic history of South America. Striae 16:40–45

Martin RE (1963) A basic approach to fire injury of tree stems. In: Proc 2nd Annual Tall Timbers Fire Ecology Conf. Tallahassee, Florida, pp 151–162

Myers N (1979) The sinking ark. Pergamon Press, Oxford. 308 pp

Nepstad DC (1989) Forest regrowth in abandoned pastures of eastern Amazonia: Limitations to tree seedling survival and growth. Ph.D. Diss., Yale Univ, New Haven, Connecticut, 234 pp

Putz FE, Brokaw NVL (1989) Sprouting of broken trees on Barro Colorado Island, Panama. Ecology 70:508–512

Salati E (1987) The forest and the hydrological cycle. In: Dickinson RE (ed) The geophysiology of Amazonia: Vegetation and climatic interactions. Wiley, New York, pp 273–296

Saldarriaga JG, West DC (1986) Holocene fires in the northern Amazon Basin. Quart Res 26:358–366

Sanford RL (1989) Root systems of three adjacent, old growth Amazon forests and associated transition zones. J Tropic For Sci 1 (in press)

Sanford RL, Saldarriaga J, Clark KE, Uhl C, Herrera R (1985) Amazon rain forest fires. Science 227:53–55

Schroeder MJ, Buck CC (1970) Fire weather. USDA For Serv Agric Handbook No 360

Uhl C, Buschbacher R (1985) A disturbing synergism between cattle ranch burning practices and selective tree harvesting in the eastern Amazon. Biotropica 17:265–268

Uhl C, Clark K (1983) Seed ecology of selected Amazon Basin successional species. Bot Gaz 144:419–425

Uhl C, Jordan CF (1984) Succession and nutrient dynamics following forest cutting and burning in Amazonia. Ecology 65:1476–1490

Uhl C, Kauffman JB (1990) Deforestation effects on fire susceptibility and the potential response of tree species to fire in the rain forest of the eastern Amazon. Ecology 71:437–449

Uhl C, Vieira ICG (1989) Ecological impacts of selective logging in the Brazilian Amazon: a case study from the Paragominas region in the State of Para. Biotropica 21:98–106

Uhl C, Clark KE, Clark H, Murphy P (1981) Early plant succession after cutting and burning in the upper Rio Negro Region of the Amazon Basin. J Ecol 69:631–649

Uhl C, Clark H, Clark K, Maquirino P (1982) Successional patterns associated with slash-and-burn agriculture in the Upper Rio Negro region of the Amazon Basin. Biotropica 14:219–254

Uhl C, Buschbacher R, Serrao EAS (1988a) Abandoned pastures in eastern Amazonia. I. Patterns of plant succession. J Ecol 76:663–681

Uhl C, Kauffman JB, Cummings DL (1988b) Fire in the Venezuelan Amazon. 2: Environmental conditions necessary for forest fires in the evergreen rain forest of Venezuela. Oikos 53:176–184

9 Social and Ecological Aspects of Fire in Central America

A.L. Koonce and A. González-Cabán[1]

9.1 Introduction

The seven countries included in Central America – Belize, Honduras, El Salvador, Nicaragua, Costa Rica, and Panama – cover a small area, 537,840 km^2. The region was heavily forested in the past with tropical wet and dry forests. These forests, however, have been extensively exploited for their timber and for a spectrum of agricultural purposes, resulting in a drastically reduced forest cover throughout the region. The amount of tropical dry forest that remains is less than 2% of the original cover (Janzen 1986b), and the moist forest has probably been reduced by two thirds. The area of pasturelands and the number of beef cattle in Central America have more than doubled since 1950, principally at the expense of primary forest (Myers 1980).

Central America contains approximately 300,000 km^2 of the approximately 5 million km^2 of tropical moist forests remaining in Latin America (Myers 1980). This is a fraction of the total original forest cover of the region at the time European settlers arrived in the early 1500's. An example of the rapid conversion of the tropical forests in Central America is Costa Rica (Fig. 1) where from 1940 to 1983 more than 50% of the land was deforested (Caulfield Vasconez 1987; Janzen 1986b).

Increasing population, forest farming, cattle raising, and timber trading, all of which involve the use of fire, are the factors contributing most to the problem of tropical forest conversion and environmental quality degradation in Central America (Myers 1980; Batchelder 1967).

Fire is a human artifact in this region. It is used to create and maintain farm, pasture, and savanna vegetation. As reported by Janzen (1986b), Budowski (1966), and Schwab (1988), plant adaptations to xeric conditions confer fire resistance as well; but the native vegetation has not evolved with fire, and the habitats created by fire are unnatural. Most of the living primary forest is unburnable. The high moisture content of the vegetation and the absence of understory vegetation do not allow the forest to carry fire. There is no evidence of fire adaptation in plants or animals, nor is there evidence of natural ignitions. All lightning occurs during the rainy season and no evidence of holdover fires can be found in old trees, snags, or surface debris. This fact is central to understanding the fire ecology of Central America.

[1]USDA Forest Service, Pacific Southwest Forest and Range Experiment Station, Riverside, California 92507, USA

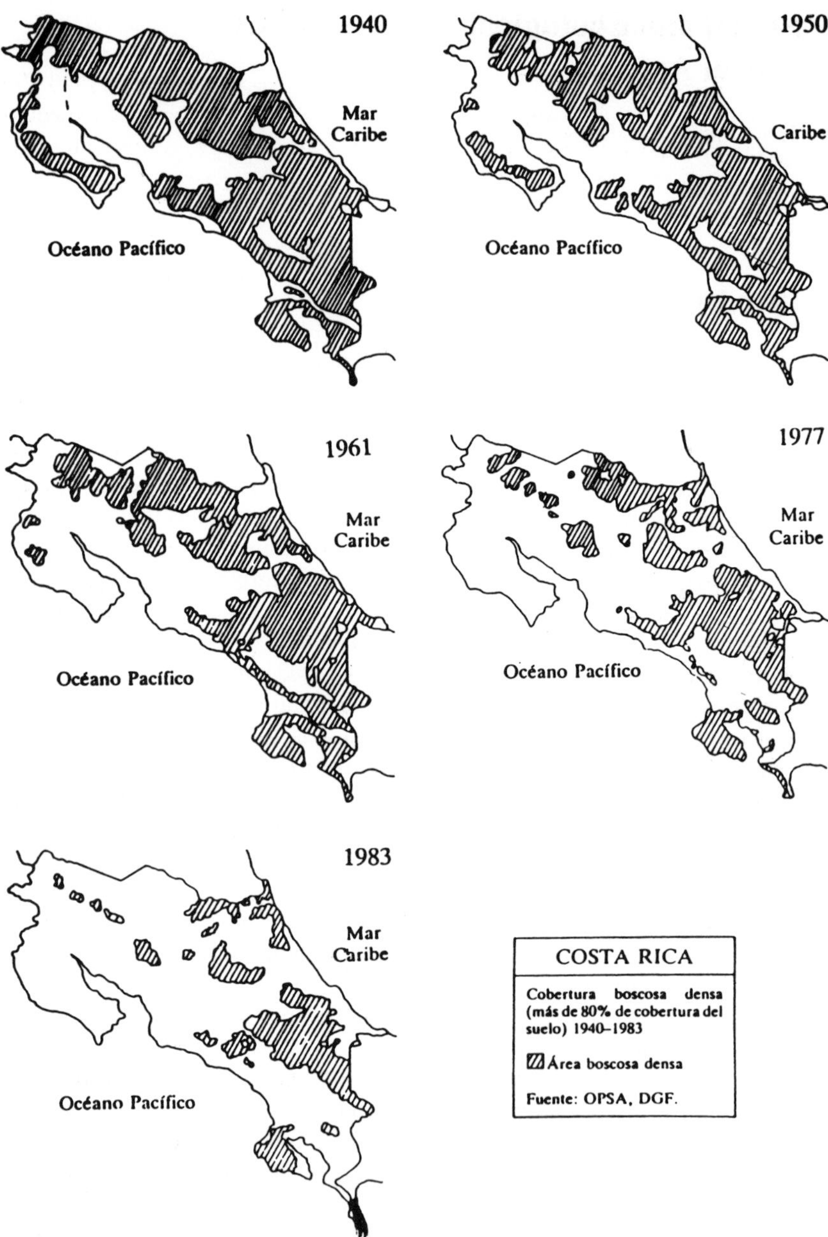

Fig. 1. The closed canopy forest of Costa Rica has shrunk over 75% from 1940 to present. (Janzen 1986b)

9.2 Socio-Economic Factors

A critical factor in Central America is human population pressure. In 1920, the region's population was 4.9 million. Today it is almost 28 million and is expected to reach 40 million by the end of the century. The percentage of the population located in urban centers varies from about 30% in Guatemala, Belize, El Salvador, and Honduras to 50% in Costa Rica, Nicaragua, and Panama. This is larger than in most African and South East Asian countries, but lower than in most Latin American countries. An increase in urbanization is usually associated with a reduction in agriculture as the main source of income, employment, and hard currency for foreign exchange, but this is not the case in Central America. Although fewer people are actually living in rural communities, agriculture and mineral primary products account for over 50% of all exports of six of the seven countries of the region – Panama being the exception (International Monetary Fund 1986). A higher urbanization rate has not resulted in a reduction in the rate of forest land converted for subsistence agriculture or commercial exploitation.

The pressure of population growth directly affects use of forest resources by increasing needs for timber for housing, wood and charcoal for heating and cooking, and agricultural and pasture land for growing food. Thus, the forest farmer becomes the most important factor in conversion of tropical forests.

Of the various forms of forest land agriculture, the main ones are traditional shifting cultivation, small holder agriculture, and sundry types of squatter colonization (Myers 1980). All through the region farmers often find themselves in a vicious cycle once the land is cleared. The wood is burned, except some larger logs, which may be sold. Next, they plant annual crops such as corn or beans. Some go on to plant coffee or cacao for export. Soon the topsoil washes away or the area is overgrown with unmanageable weeds, and the land will no longer profitably support crops. Farmers then turn the plot to a pasture for livestock production; but unless grazing is carefully managed, weeds that the animals will not eat proliferate and land quality is degraded further. Eventually the plot cannot even be used for livestock production and more land needs to be cleared (Caulfield Vasconez 1987).

The second most important factor in the conversion of tropical forests in Central America is cattle raising (Fig. 2). In the 25 years between 1950 and 1975, the amount of forest land cleared for pasture more than doubled. During the same period the total number of beef cattle more than doubled, while the per capita consumption of beef by Central America citizens actually declined. The surplus meat was exported to developed countries.

Acquisition of beautiful and exotic tropical hardwoods by developed countries is another significant factor in the conversion of the tropical forests. This factor will become more relevant in Central America as the hardwoods of the tropical forests of South East Asia are depleted. By 1976, the export value of tropical wood exceeded 4.2 billion U.S. dollars, making it one of the five most important export earners among the major commodities produced by the developing world.

Fig. 2. Cattle raising in converted forest land in Copan, Honduras. (Photo by A. Koonce)

9.3 Tropical Forest Resources

The tropical forest resources of the Central America region vary significantly among the countries of the region. The percent of forest cover varies from a low of 10% in El Salvador to almost 63% in Honduras. Likewise, the information available about their forest resources and the effect of fires on those resources is very uneven. Most of the information available is for Belize, Honduras, Costa Rica, and Panama.

Guatemala

Almost 49% of Guatemala's 108,889 km^2 is reported to be forested (Myers 1980). Approximately 10,000 km^2 of the forested area consists of temperate coniferous montane forests. The largest area of tropical moist forest in Guatemala is found in the Department of Peten in the northeast, occupying one third of the national territory (Myers 1980).

Belize

Out of Belize's 22,965 km^2, close to 91% is officially listed as forest lands, though a large portion consists of open woodland and various forms of savannas. Moist forests account for no more than 16,300 km^2, of which only 75% supports vigorous communities (Myers 1980).

Most of Belize is still covered by pine forests. Around 4950 km^2 has been declared forest reserve. About 25% of this reserve is in pine savannas (Hudson 1976). These pine savannas are part of the native range of pine savannas in Central America that cover an estimated 45,000–50,000 km^2. The number of pine species in Central America from Guatemala to Nicaragua is only three; of these, Belize only has two, *Pinus caribaea* Morelet and *Pinus oocarpa* var. *ochoteranai* Mart.

The factors most seriously affecting the ecology of these pines are the nature of the bedrock, rainfall distribution, and fire frequency (Hudson 1976). Human ignitions are the principal cause of fire in Belize, except in the highlands, where population density is low and lightning strikes are frequent.

Honduras

Honduras has the largest forest reserves in Central America (Myers 1980). Of the country's 112,044 km^2 around 63% are forested. More than 40,000 km^2 of the total forested area is classified as tropical moist forest. The eastern section of the country contains the relict Mesquitia Forest, one of the three major expanses of rain forest remaining in Central America (Myers 1980). The rest of the forested area is composed of 27,000 km^2 of conifers and savannas, and 3500 km^2 of mangrove and swamplands.

With the largest forest reserves in Honduras, forestry products are potentially the greatest exportable natural resource. Most of the lumber produced is exported, despite domestic scarcity and high prices for local wood. Timber production has shown an upward trend since 1975, from 203,000 board feet in 1975 to 260,000 board feet in 1979. Because of increased domestic demand, exports have decreased from 95% in 1975 to only 54% by 1979 (Evaluation Technologies 1981). Nearly all commercial wood comes from coniferous forests.

The practice of shifting agriculture, widespread burning of forests, and fuelwood cutting have caused rapid depletion of forest resources and increasingly serious soil erosion problems. These problems plus wasteful timber exploitation practices will eventually threaten the timber industry. The present timber exploitation practices are also wasteful. Only about 70% of the usable wood reaches the mills, and only 33% of that ends up as lumber (Evaluation Technologies 1981). Little additional processing of lumber occurs.

In 1974, the Honduran Forest Development Corporation (COHDEFOR) was formed to oversee and control the future development of the forestry industry in the country. The National School of Forest Sciences (ESNACIFOR) was also created to educate Central American students in forestry and natural resources. The school is developing greater technical expertise to manage the declining forest resources rationally.

El Salvador

El Salvador is the smallest country in Central America but it is the most densely populated in all of Latin America. The estimated population, as of 1984, was 5.1 million with a population density of over 238 inhabitants per km^2. Only 2600 km^2 of El Salvador is forested and that is considered degraded forest. Eighty percent of the population depends upon agriculture for their livelihood, and up to 75% requires charcoal for fuel. The forests in El Salvador are not likely to last until end of the century (Myers 1980).

Nicaragua

Nicaragua is the largest of the Central American countries, with a land base of 147,943 km^2. In early 1970, it was estimated that 43% of the total territory was forested. Tropical forest covered 50,000 km^2 and the rest was mainly montane coniferous forest. By the early 1980's, the tropical forest in the country had been reduced to about 35,000 km^2 (Myers 1980). Even this large forest cover is only a fraction of what was supposed to have existed only 50 years ago.

Costa Rica

Of Costa Rica's 49,132 km^2, a 1967 inventory estimated the forest area to be only 45%, most of it tropical forest together with some deciduous and montane cloud forest (Myers 1980). More recent surveys suggest that the forested area amounts to no more than 16,000 km^2. The remaining forest cover represents less than 35% of the original moist forest cover, and this is being converted at a rate of 550 km^2 every year (Caulfield Vasconez 1987). Figure 1 shows the dramatic reduction in tropical dry forest cover of Costa Rica from 1940 to 1983 (Janzen 1986b).

In Costa Rica, as a whole, the remaining forested area comprises 33.5% of the land area. Of this, 4.2% is "moderately intervened" forest, meaning that between 60 and 90% of the original forest vegetation remains, 9.8% is strongly intervened, 30–60% original vegetation remains, and only 17% of the country remains in primary forest (as of 1983). Few forest tracts remain in proximity to road or railroad rights-of-way, which provide the principal access for agricultural and forest clearing (Sader and Joyce 1988). By 1961, essentially all the dry forest was gone, and the low elevation moist forest became the next target for exploitation. Forest clearing accelerated between 1977 and 1983 from less than 2% per year to more than 7% per year. Of the remaining primary forest, only 8.5%, or roughly half, is located in parks or reserves. The remaining forest habitat is located on the steepest slopes, and is the most critical to manage from the standpoint of potential soil erosion and degradation.

Panama

Of Panama's 75,474 km^2, around 54% is officially classified as forests. At least over 10,000 km^2 of these forests have been seriously disrupted through slash-and-burn agriculture (Myers 1980). This disruption is evident in the tributary watersheds surrounding the Panama Canal, which were covered with dense rain forests. As recently as 1952, the area was still over 85% forested, but in the 20 years from 1960 through 1980, most of the forest was converted to agricultural uses.

The largest remaining tract of tropical forest in Panama is found in the Darien Province in the southeastern part of the country along the Caribbean Sea. Now this area is experiencing the pressure of a population explosion. Landless peasants are searching for a place to grow crops, and cattle ranchers are seeking to expand production of beef destined for international markets.

Expansion into the forest land is cyclic. Peasants from the increasingly arid interior of the country burn the forest for subsistence agriculture. They are driven away from their homes by cattle ranchers, who continually move into the recently cleared areas. In 3 to 4 years, when the newly cleared and degraded land no longer supports crops like bananas, rice, or corn, farmers plant pasture and try to sell their holdings to ranchers. In a few more years the ecologically fragile soils are so overused that cattle-ranching will also fail (Breslin and Chapin 1984). Opening of the inter-American highway through the Darien Province is facilitating the acceleration of the tropical forest conversion by providing access where none existed before.

9.4 Fire in the Tropical Forests of Central America

Fires in the tropical lowlands of Central America are greatly affected by climatic conditions, particularly temperature and rain (Budowsky 1966; Munro 1966). The average temperature in the tropical lowlands is generally above 24°C, but rainfall varies tremendously. By and large, the most intense and frequent wildland fires occur in regions where the seasons alternate between a dry and a wet season. This condition is typical of the Central American region.

All areas experience a pronounced and variable dry season early in the year, normally from February to March (Hudson and Salazar 1980). The wet season starts in May and June, preceded by electrical storms, and continues until the end of October with a 2–4-week dry spell in July through August. The months of greatest rainfall are normally September and October.

In areas of heavy rainfall, the montane forests will not carry fire under normal conditions. By contrast, secondary woody vegetation and grass often burn after short dry spells make these fuels more flammable. The adaptations of the secondary plants and the low water-holding capacity of the soils are more conducive to fire ignition (Budowsky 1966). Pioneer species, those that colonize areas devoid of vegetation, adapt to drought by shedding their leaves, which provide enough surface fuel to carry fire (Budowski 1966), and, although there is no marked dormant period of vegetative growth in Central America, the grasses cure in the dry season (Hudson and Salazar 1980). Undisturbed forests do not burn because the environment at the soil surface remains sufficiently moist.

Successional species may become established with the aid of light annual fires. These reduce the fuel that could support more intense fires. Grazing also results in fuel reduction and may promote invasion by trees not browsed by cattle. According to Budowski (1966, 17): "The general rule is that these species originated from much drier areas and become invaders. This has often led to the erroneous belief that climate has changed. Actually, a vegetation adapted to drought has invaded the area because of the combination of fire and grazing".

However, it is now known that the land clearing and burning can affect both micro- and meso-scale climatic change (Fig. 3), and the effects of local warming and reduction of rainfall will have much more profound effects on long lived forest vegetation than on short-rotation agricultural crops (Janzen 1986a). The forests that are broken up into habitat fragments are more vulnerable to the drying effects at the forest edge. If long-term changes in climate result in species mortality in these fragments, there will be no large forested reserve of similar species to reinvade the habitat.

This phenomenon can already be seen in the decline of *Quercus oleoides*, lowland tropical oak, in Santa Rosa, Costa Rica. The acorn requires moist soil to germinate and grow. If the soil or litter layer is dry, the seed remains dormant and eventually dies. Forest clearing, pasture development, and frequent fires have created a habitat vulnerable to the drying influence of winds, which prevent acorn germination. If the acorns do manage to produce seedlings, they are generally killed by dry season fires. Even the mature oaks are gradually killed from basal girdling caused by repeated surface fires. As a result of this phenomenon the entire oak forest is receding (Janzen 1986a).

Aerial photography and satellite imagery are increasing the accuracy of predicting gross rates and extent of tropical forest clearing (Sader and Joyce 1988). These methods are unable to suggest specific information about species removal or recovery. Some partial habitats and habitat fragments can con-

Fig. 3. The impact of large fires like this one in the outskirts of Tegucigalpa, Honduras, can contribute to micro- and meso-scale climatic changes. (Photo by A.L. Koonce)

tribute to wildlife, tree and crop production, and to the desirable gas-exchange which offsets the trends of the greenhouse effect and global warming.

The contribution of CO_2 and particulate emissions by agricultural burning in developing countries is critical (Pasca 1988). Over 80% of the phytomass burned annually is in these countries (Seiler and Crutzen 1980). Little scientific information is available on the amount of carbon that tropical fires contribute to aerosol emissions. An estimated 0.9 to 2.5×10^{15} g CO_2-carbon are released into the atmosphere from forest clearing each year (Hao et al. this Vol.).

Suman (1984, 1986, 1988) monitored fires in Panama during the dry season of 1981, during which over 10% of the land area was burned. During the peak, charcoal concentrations in the aerosol reached 3.1 μgC m^{-3} and one third of that amount during off-peak times. Most of the charcoal mass was in the fine-sized fraction ($< 2\ \mu$m), suggesting that these particles can be advected great distances over the Pacific Ocean by the northeasterly Trade Winds. During burning this advection results in hazy conditions and poor visibility.

9.4.1 Fire Effects on Soils

Nutrients

While forest farmers are blamed for increasing loss of forest habitat, intensification of land use also has a negative impact on soil fertility. In dry forest habitats in Belize, soil fertility declined in milpa plots (Arnason et al. 1982).

Macronutrients, particularly P and Ca, decreased after burning for site preparation and during the first 3 years of cropping. The first year after burning K increased and the soil pH decreased. During the 3-year study period P became a limiting factor to plant growth. Ten years were required for levels of P to recover after only 2 years of cropping. *Gynerium*, *Heliconia*, and *Piper amalago* are efficient at concentrating P and may be planted during the fallow periods to enhance recovery.

Savanna soils are notably low in nutrients, so nutrient losses could have a critical impact on site productivity. Kellman and Sanmugadas (1985) examined the nutrient status of soil solutions in Belize savannas between dry and wet seasons. In the absence of fire, the soil was able to rapidly immobilize nutrients in solution, indicating resistance to leaching. Since root uptake was not an important mechanism in short-term nutrient retention, immobilization of nutrients in large fluctuations associated with fire occurrence should occur in the soil itself. To test this hypothesis, Kellman et al. (1985) also studied nutrient retention after fire. As expected, after fire they found large increases of nutrients in the rooting zone within 1 week. Although percolation was considerable, these nutrients were not recorded in deeper soil layers. After 1 year Ca and P were effectively retained in the topsoil, Mg and K retention was less, and Na retention was lowest. Nutrients are believed to be lost principally in surface runoff. However, burning does not significantly reduce the fertility of the top soil, and nutrient volatilization is not the critical factor in soil deterioration associated with intensive land use.

Since runoff is most critical for nutrient loss, soils that have considerable vertical or lateral drainage due to climate or slope are most vulnerable to nutrient depletion (Fig. 4) (Furley 1987). This was corroborated by Furley (1987) in a lowland evergreen forest in Belize, where forest clearing, burning, and cultivation of 2–3 years were most influential on the summit and upper slope regions. At the summit and upper slopes Mg and K were reduced. Soil depth indicated increased erosion from summit areas and deposition at the foot of the slope. Total carbonates and pH were also affected.

In *Pinus oocarpa* stands in Honduras, Hudson et al. (1983a,b) reported the effects of prescribed burning on surface runoff, sediment loss, and nutrient cycling. Over a period of 2 years runoff was 1.73% on the control plots and 5.03 on the burned plots. Mean sediment loss was 80 kg ha^{-1} and 1732 kg ha^{-1} respectively. Sediment loss was responsible for the majority of nutrient depletion on the site. Net losses of Ca, P, and Mg were recorded for the first 16½ months after burning, and levels returned to normal within 2 years. Recovery was due to rapid revegetation in the plots. Burning should be conducted early in the dry season (November-December), or just after the onset of the wet season, to allow for the vegetation to recover as much as possible before the rainy season begins in late May or early June. Following these guidelines, burning cycles of 3–7 years will not deplete the nutrient capital of the site.

In *Pinus caribaea* stands growing on savanna soils in Belize both shrub and graminoid stata resprouted vigorously after prescribed fires (Kellman et al. 1987). Nutrient concentrations were significant in shrub, grass, and sedge

Fig. 4. During the rainy season this steep burned area in northern Honduras most likely will suffer severe nutrient losses. (Photo by A.L. Koonce)

tissues, suggesting a storage function for these plants. Since nutrients were higher in the shrub tissues, and since they are not completely consumed in cool to moderate fires, shrubs can be managed as part of a nutrient conservation strategy for the site.

More fertile soils on tropical forest sites followed a similar pattern (Matson et al. 1987). Ammonium and nitrate N increased dramatically in surface soils after clearing and burning, but returned to levels similar to secondary forest soils within 6 months. Plant N uptake, immobilization, and nitrate adsorption are all important avenues for retaining on-site nitrogen.

Stored Seeds

Fire has more impact on viability of seed stored in the soil, affecting species composition and population density in post-burn succession. Fire effects on a wet forest site in Costa Rica were compared to forest harvesting and dry mulching (Ewel et al. 1981). From forest soils 8000 seeds m^{-2} and 67 species germinated. From soils collected below slash 6000 seeds m^{-2} and 51 species germinated, and from burned soils less than 3000 seeds m^{-2} and 37 species germinated. Post-burn soils had increased pH, P, and cation availability.Ca and P were readily available for plant uptake but were susceptible to leaching, runoff, and removal by blowing ash. Forty to fifty percent of these elements were lost after burning compared with 11–16% after harvest. K was readily leached and 31% of the C, 22 percent of the N, and 49% of the S were volatilized by burning. The slash-and-burn combination decreased the seed storage capacity of the soil by 63%, reducing the number of individuals more than the number of species, but it did not retard succession. Within 3 months, the site was revegetated by more than 100 species. Most significant was the loss of S by the removal of the overstory vegetation. This nutrient is not easily restored and may have long-term implications for site deterioration with increased tree cropping.

9.4.2 Fire Effects on Pine Forests

There are 45,000 to 50,000 km^2 of naturally occurring pine forests in the Central American Isthmus, from Guatemala and Belize in the north to Nicaragua in the south (Hudson and Salazar 1980). The number of species declines with latitude but *P. caribaea* and *P. oocarpa*, the species of greater commercial importance, are found in all countries. The natural distribution of *P. caribaea* is confined to the Atlantic drainages of Central America and to some islands of the Caribbean. The inland form occurs in Nicaragua, Honduras, Belize, and possibly Guatemala up to an elevation of 1000 m. The lowland form occurs in Nicaragua, Honduras, Belize, Guatemala, Cuba, Isle of Pines, Bay Islands, and the Bahamas (Munro 1966).

Establishment of the Pine Forest

The role of fire in the perpetuation and distribution of these pine species relative to that of surrounding broadleaved forests is widely accepted (Denevan 1961; Munro 1966; Budowsky 1966; Guevara and Hudson 1979; Hudson and Salazar 1980). Munro (1966, 72) states that "The effect of fire on the distribution, establishment and growth patterns of Caribbean pine is profound and it can be said there are no natural stands of this species in Central America which, even though there is no apparent relationship, have not been affected by fire at least once, if not numerous times, in their life times. . . . Caribbean pine savannas are not permanently fixed in one place but are expanding or contracting their boundaries according to fire frequency". He identified seven phases, or seres, which follow the evolution of an area from hardwood, to pine, and back to hardwood (Fig. 5). The following description is summarized from Munro's work.

Phase 1 is a climatical climax tropical hardwood stand which has probably not burned in the last 500 years or more. This stand is over 40 m high and could contain 100 or more species which reach 50 cm DBH. Phase 2 is the most complex of the seven and can be considered the buffer or transitional stage between the hardwood stand and the true savanna. This zone is usually narrow and is composed of an overstory of mature pine and an understory, often present in several layers, of pine and hardwood. The composition of this phase is often the

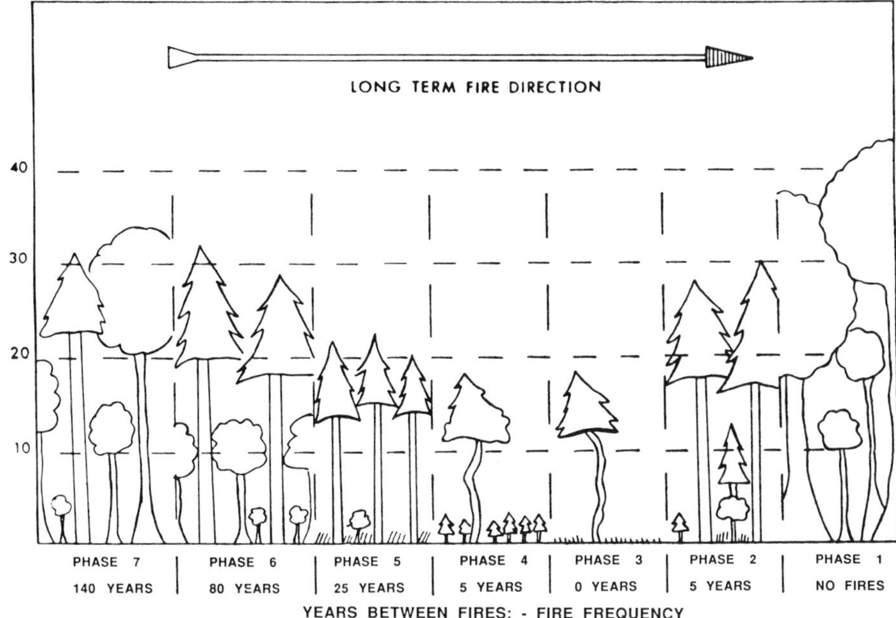

Fig. 5. Diagram of fire frequency and Caribbean pine (*Pinus caribaea*) succession relationships. The *long arrow* above the diagram indicates the direction of savanna movement, or savanna development, over centuries. (Munro 1966)

only way it can be separated from phases 5 and 6 in the field. This area is not burned annually and in many cases the fires are only light grass fires sufficient to kill fire-susceptible species. It is through this area that fires penetrate in exceptional dry years to the tropical hardwood and old garden areas to expand the pine area.

Phase 3 is the most common phase presently existing in Nicaragua and is characterized by crooked, fire-scarred, mature pine trees. This phase is burned annually, often twice a year, and generally has no understory or regeneration. The herbaceous layer is grass and sedges. Phase 4 represents an area which was phase 3 but has not been burned for the last 5 years.

Phase 5 is an old pine savanna area which has not been burned for 25–30 years. The regeneration from phase 4 is now 15–20 m tall and forms a complete cover which prevents the development of more pine regeneration because Caribbean pine is not tolerant of heavy shade. Small hardwoods are now coming up through the heavy herbaceous layer and, unless halted, will eventually take over the site. The herbaceous layer usually consists of tall grasses and bracken fern which present a high fire hazard in the dry seasons.

Phase 6 is approximately 80 years after the last fire and the pine have now reached a height of 30 m or more and the invading hardwoods have formed a well-established second story up to 12 m high. There is no pine regeneration and the life of the area as a pine site is limited.

The final phase, phase 7, contains only the remnants of the pines and is almost completely taken over by the hardwoods. The pines are decadent and it is not unusual to find they have little commercial value because of decay and termite attack. It is not uncommon to find pine windfalls and old stumps in this type, which is only one step away from phase 1.

Lightning-caused fires are common in some parts of the region (an average of four per year in the 650 km^2 of the Mountain Pine Ridge Reserve in Belize), but human-caused fires are of much greater significance. Fire occurrence has greatly increased in recent years, and fire and uncontrolled logging in accessible areas have degraded vast areas of the Central American forests (Hudson and Salazar 1980).

Pine Fuels

The principal fuels in the pine forests of Central America are grasses and sedges. Wire grasses (similar to those in the southern part of the U.S.A.) are dominant in the coastal and lower montane areas, and broadleaved grasses are dominant at higher elevations. These grasses are highly adapted to fire, and many flower only after burning. The wire grasses are particularly responsive to weather changes, ignite more easily than broadleaved grasses, and support a faster rate of spread (Hudson and Salazar 1980). The broadleaved grasses attain higher fuel loadings and present greater resistance to control. Burning in the period from November to February, while a large proportion of the grass fuels are still actually growing, is not as favorable for forage improvement as burning in the late dry season, when grasses would naturally burn. Early seasonal burns, however, reduce damage to the pine overstory.

At higher elevations, ferns and forbs constitute a significant portion of surface fuels. For example, *Dicranopteris pectinata*, a fern common in Belize, attains high fuel loadings (up to 25 ton ha^{-1}) as a continuous groundcover in the open and under the pine canopy (Hudson and Salazar 1980). It is extremely flammable and is considered the most problematic fuel in the area. At higher elevations in Honduras, *Pteridium aquilinum* is often found in association with brambles (*Rubus* spp.). In Costa Rica, Gliessman (1978) found that bracken (*Pteridium aquilinum* L. Kuhn) is encountered anywhere from sea level on well-drained soils up to 3000 m elevation. He claims that the extensive presence of bracken is mainly due to the use of fire in the clearing of dense tropical forests. Burning of the down material right before the rainy season (as commonly done under the slash-and-burn agricultural system) increases soil pH level, creating the most favorable conditions for bracken establishment. In cutover areas that have escaped burning, he found no bracken establishment. Gliessman (1978, 44) concluded that "... fire should not be used after clearing the forest ... [because] bracken rapidly takes advantage of conditions created after fire in the tropics". The latter is important because, once having established, the fern is difficult to eradicate manually or mechanically, and only the widespread application of new chemical fernicides offers control (Martin 1976, as cited by Gliessman).

Throughout the pine forests of Central America, fire-tolerant broadleaved trees and shrubs occur as scattered populations. In frequently burned areas their contribution to available fuel is negligible. Fire exclusion results in the increase in density of fire-tolerant species from existing root stocks, and the establishment of seedlings, and in the invasion of fire-susceptible species (Hudson and Salazar 1980). Some of the live oaks spread by root suckering to form dense, but discontinuous patches of undergrowth in the absence of fire. The live oaks are highly flammable and can result in considerable damage to a pine overstory. For example, many years of fire protection in the Mountain Pine Ridge Forest Reserve in Belize has resulted in the formation of a dense understory of melastomes which, when ignited, can cause considerable damage to the associated pines, if they are small (Hudson and Salazar 1980).

The pines themselves can rarely be considered as available fuel except as young regeneration within a well-developed understory or in dense young plantations which, as yet, are negligible in Central America. Both species, *P. caribaea* and *P. oocarpa*, cast dead branches rapidly and, in a natural condition, rarely achieve high stockings.

Adaptation to Fire

Tropical pines are well adapted to forest fires and some, like *P. oocarpa* Schiede var. *ochotenarai* Mart. from Honduras, when young, can resprout when killed by fire (Hudson and Salazar 1981). *P. caribaea* is intolerant (Munro 1966), and its seedlings will die in the understory of broadleaved trees. If fire is totally excluded from the pine forest, pine regeneration will not occur, and eventually the forest will become a broadleaved forest (Wolffsohn 1988). On the other hand, frequent forest fires over a long period of time can change the forest into a tropical savanna without trees (Hudson and Salazar 1981; Wolffsohn 1988). A high fire frequency may also lead to the impermeability of the soil (Wolffsohn 1988).

Fire may be the dominant factor in the established savanna environment in Latin America (Alexander 1973; Blydenstein 1968; Munro 1966). For example, in northeastern Nicaragua the savanna is perpetuated by repeated burning of the cover on infertile soils (Alexander 1973). Herbaceous savanna vegetation is excellently adapted to repeated burning and has several advantages over forest vegetation under such conditions. The accumulated vegetable material that is burned off is dead already and expendable. New growth may actually be stimulated by letting more light enter at the level of the basal growing points of bunchgrasses. Although yearly fires do not permit pine forest regeneration, a 3- to 6-year period without fires is enough to permit the forest to regenerate.

The increase in density of broadleaved trees and shrubs in the Mountain Pine Ridge Forest in Belize, and to a lesser degree in northeastern Nicaragua, will make regeneration of pine difficult and more expensive if allowed to continue (Hudson and Salazar 1980). Elimination of these broadleaved species by the use of fire is not practical, because of their persistent sprouting behavior, nor is it necessarily desirable. For example, Kellman (1976, 1979) found no evidence of competition between pine and broadleaved species in the Mountain Pine Ridge Forest. These broadleaved species act as nutrient sinks, enriching the surface soil beneath them by capture of precipitation inputs. Blydenstein (1968) also found that the only discernible effect of fires burning on the savanna soils, with respect to forest soils, is in its pH. The lower acidity of the savanna soils can be attributed to the accumulation of ashes, high in K content, after grassland fires. The periodic application of fire could be used to maintain these broad-leaved species in a low coppicing condition, thereby retaining their utility as soil-enrichers, while improving conditions for the regeneration of pine at the end of the rotation (Hudson and Salazar 1980).

The time of year in which a fire occurs can be crucial to seed germination and pine regeneration. For example, *P. oocarpa* drops its seeds during the dry season. A dry season fire will kill most of the seeds that fell before the fire. The best chances for seed germination and survival is when the seeds fall in areas in which the vegetation has recuperated from the fires enough to hide the seeds but not long enough to permit insect recolonization (Hudson and Salazar 1981; Lara Marquez 1977). This situation usually happens in *P. caribaea* Morelet var. *hondurensis* Ban and Golf, which releases its seeds during the rainy season, but rarely in *P. oocarpa*, which releases the majority of its seeds during the dry season (Hudson and Salazar 1981).

P. pseudostrobus Lindl. is not a tropical pine, although present in the tropical pine forest, and it is not well adapted to fires. Its presence in the tropical forest is due in part to fire (Hudson and Salazar 1981) because it is intolerant and needs very fertile soils to grow (Lara Marquez 1977). *P. oocarpa* can successfully compete with *P. pseudostrobus* below 1200 m, especially because of the higher incidence of wildfires below the 1200 m mark. Below this altitude most fires are human-caused and population concentrations are large. Frequent fires may reduce soil fertility in this range, further benefiting *P. oocarpa*.

9.4.3 Fire Effects in Dry Forest

General Effects

Separating the effects of fire from those of other land-clearing practices, such as farming, grazing, and logging, is difficult. These activities are always practiced in conjunction with each other. Slash-and-burn agriculture is characterized by clearing land, either land that has previously been cleared by farmers or loggers or by moving into primary forest. The resulting debris is burned prior to the wet season to open the sites for planting and to provide fertilizer for the crops which will be planted. Burning temporarily suppresses competition from native species, but after several years of cultivation the plot is invaded by unmanageable weeds and pests, which cause it to be abandoned (Lambert and Arnason 1986). Under the population pressures of immigration and translocation, the land is not allowed to fallow for a sufficient length of time. With continued burning, the forest borders become more distant and cannot provide the seed or the microclimate which will aid in site recovery, and the land becomes increasingly impoverished.

Janzen (1973a,b, 1986a,b, 1988a,b,c) has described the effects of fire in dry forest habitats. Fires burn continually throughout the dry season (Edgar 1989; Janzen 1989 personal commun.). They are started deliberately for reasons related to land clearing for agriculture, pasture, hazard reduction, clearing of debris and unwanted vegetation, reduction of insect pests, or other management reasons. Some are escapes from camping and cooking fires, and occasionally they are set intentionally for pleasure, mischief, or revenge (Janzen 1988b).

Twenty to 60% of the Pacific coastal area is burned annually. Little thought is given to controlling these fires unless they are posing some immediate threat to valuable resources such as buildings or land improvements (Janzen 1986a). Due to the increased population presence in grass and shrubland habitats, the wildland-urban fire threat is becoming a more recognized problem.

Fires are started in flammable vegetation outside forest boundaries. The external vegetation is characterized by agricultural residue, African "jaragua" grass (*Hyparrhenia rufa*) (Pohl 1983), annual grasses, savanna vegetation, or shrubs and woodlands of native and exotic species. Jaragua grass is the most commonly cultivated pasture grass as it is resistant to grazing pressure. It is abundant along roadsides and grows from 1 to 2 m tall in the absence of grazing (Pohl 1983).

Fire behavior varies depending on the fuel conditions that exist outside the forest (Janzen 1986b). If the vegetation is composed of annual grasses, if the moisture content of the vegetation is high, and if the wind is not too strong, the fire will burn up to the edge of the forest and go out. A "cool" fire of this type will damage only the trees at the forest boundary. However, the increased light at the forest edge will support new herbaceous vegetation, increasing the ability of fire to spread into the forest, and the exposed wood in the fire scars of the bordering trees will be increasingly vulnerable to ignition and consumption by future fires. The margin of the forest thus becomes more flammable and will support an even more damaging fire spread at its edge the following year.

If a fire is spreading in tall jaragua grasses, if it is pushed by a strong wind, or if the year has been unusually dry, the heat and flames may penetrate 10 to 15 m into the forest edge. It will consume the forest floor litter, causing damage from the heat generated to the soil and to the cambium of the trees in its path. If a firebrand lands in the forest interior and manages to ignite dead vegetation (generally dead branchwood or snags in the upper canopy) it will open a small hole, allowing the entry of light, again promoting the growth of surface fuels. These fuels ignite easily, will carry future spot fires, and will provide kindling for ignition in the dead or injured woody vegetation resulting from the first spot fire (Janzen 1986a).

The first fire in forest vegetation is a surface fire burning in deciduous leaf litter and in grasses that may penetrate the borders of the forest where light can penetrate past the forest edge (Janzen 1986b). Fires die down at night because of the increasing humidity and are fanned again the following day with the return of warmer, drier conditions, and wind. A fire burning in a deciduous forest that borders the moist evergreen forest will generally stop burning at the intersection of the two, creating a sharp boundary between them. A fire burning in dead trees or downed logs will smolder for several days. The heat delivered to the soil surface results in severe soil sterilization, baking it into a hard red clay, which remains visible for 3 to 5 years in the understory. Burned-out roots serve as tunnels or burrows for small mammals and other forest life. Trees that survive or that replace fire-killed vegetation are generally those that can pioneer in dry habitats or that can seed in and grow quickly.

The earlier in the dry season that a fire burns, the lower its intensity. Early fires will tend to leave patches or individual grass culms unburned in their wake. The grass and herbaceous vegetation at the boundaries of forest and woody vegetation invading the grasslands will stay green and moist longer into the season. A field of jaragua grass left ungrazed for one season will result in headfire flame heights of 1 to 4 m. If ungrazed for more than one season, the fire will burn with greater intensity, and will kill all woody vegetation in the field and into the forest boundaries which it comes up against. Ungrazed native grass stands will also carry fire but will not kill resistant trees such as *Curatella americana*, and *Byrsonima crassifolia* (Janzen 1988b).

Grassland Succession

Repeated burning will convert the native forested habitat to grassland and that grassland will be maintained, promoted, and increased in its extent by annual or repeated burning. The succession of grassland back to forest will also depend on the nature of its disruption and its location with respect to seed sources and to microclimatic influences of forest edges or remnants. The rate of woody succession in pastures will depend on the following: (1) whether the pasture was jaragua or native grasses, (2) the proximity and configuration of the forest that can supply seed and microclimatic buffers, (3) the availability and access to the site for dispersal agents, (4) the environmental conditions on the site, (5) the length of time in grassland vegetation, and (6) whether or not grazing continues (Janzen 1988b,c).

Jaragua grass provides a dense and competitive block of vegetation that restricts woody encroachment, but jaragua grasslands are invaded more quickly than their native grass counterparts, which are more open and sparse. This difference is attributed to drier or harsher sites than the native grasses are able to occupy, which are less conducive to the growth of forest vegetation. The seed source of the forest vegetation may be more distant from native grassland sites, and the woody pioneers are slower to achieve habitat dominance. It may take 20 to 50 years for a pasture to revert to woody vegetation, 100 to 200 years to resemble the primary forest, and 1000 years to fully recover to a balanced and dynamic natural system.

Woody invasion progresses most rapidly at the forest pasture-boundary. There is a strong seed rain 20 to 100 m downwind from a forest margin and a major part of the initial forest invasion may be by wind-dispersed seeds. Unfortunately, the number of trees with wind-dispersed seeds represent only a quarter of the total species in the primary forest. They are usually tall, fast-growing, and adapted to fill quickly the forest openings as they occur. Representative species include *Enterolbium cyclocarpum*, *Pithecellobium saman*, *Guazuma ulmifolia*, *Spondia mombin*, and *Ficus* spp. A forest comprised of these species alone is an artifact of a natural association. In the best cases, however, they can serve to shade out the fire-prone grass. They provide roosting habitats, shade, and protection for birds and animals which disperse other types of tree seeds through regurgitation, defecation, or bodily transport in their fur or feathers (Janzen 1988b,c).

Isolated trees in a pasture can serve as loci for the growth of animal-dispersed plants which, in turn, can spread and fill in from these margins. Some "living dead" trees also may exist in the pasture. These trees are "dead" because they are not reproducing, their seed is not being effectively dispersed, or they are unable to germinate or grow. The ultimate fate of these trees depends on site development. Their primary function may be to serve as nurse trees for the growth of other plants.

The probability that wildlife will sufficiently disperse seeds throughout the pasture area is a function of the size of forested block adjacent to the grassland. Patches of forest may not be large enough to support the necessary birds and animals that disperse seeds, or the wildlife species may not find the seeds. If the pasture is large, wildlife may not go into or cross the pastures, and would not deposit seeds in the open.

Woody invasion is fastest on moist to mesic sites. In swampy areas the woody reinvasion may be slowed by the formation of dense stands of *Mimosa pigra* and *Sesbiana emerus*, both of which are flammable in the dry season. In the absence of fire, however, succession to forest vegetation continues. The dry forest pastures are quicker to return to forest vegetation than the rain forest pastures. This may be due to the fact that dry forest pioneer species are more drought- and heat-tolerant, or that the dry forest pastures have richer soils with mycorrhizal fungi that can facilitate the growth of trees. The slowest to revegetate are the rocky and serpentine soils. The habitats are located upwind from seed sources and offer little forage or habitat to animals that would drop seeds in passage.

In the first few years after a site is cleared the soils are rich, some seed bank remains in the soil, there are roots and stumps which can sucker and sprout, and grass growth is relatively poor. These newly cleared sites can return to forest vegetation very quickly. Old pastures may recover quickly as well, but their recovery is a function of a number of factors that have not been studied.

Moderate grazing can also facilitate succession by keeping the fuel loading and flammability of grasses to a minimum and by dispersing some seeds. The extent to which grazing animals browse woody vegetation and trample and compact delicate sites is not known.

9.4.4 Fire Effects on the Aripo Savannas

Fire is considered a key factor in the maintenance of most neotropical savannas. The Aripo Savannas complex located in the proposed scientific reserve, in north-central Trinidad however, is one notable exception (Schwab 1988). These savannas are maintained by edaphic rather than pyric conditions. Infertile soils accentuated by extreme seasonal conditions support sparsely vegetated primary savannas, regardless of the presence of fire. Fires that do occur in the area are set illegally by local residents to create fresh forage for cows and goats, or to smoke out agouti, a favored game animal.

Some common adaptations to drought also confer fire resistance (Schwab 1988). *Brysonima crassifolia*, an infrequent inhabitant of the Aripo Savannas, has thick, corky bark and roots that penetrate deep into the soil. They do not die unless severely burned because a large portion of their reserves are found below ground level. Repeated burning, however, can affect even the most fire-resistant species, so that gradual elimination of some species is expected. For example, the orchid *Cyrtopodium parvifolium* is now considered rare in areas subject to recurrent fires. Elimination of fire-sensitive species would then favor the development of a community of species coincidentally adapted to fires as well as to the environmentally drier conditions. Certain grasses – *Andropogon leucostachyus*, *Paspalum pulchellum*, and *Leptocorphium lanatum* – are common residents in the regularly burned savannas of Venezuela and Colombia. Studies of postfire succession in the Aripo Savannas reported these species as being common, whereas earlier studies reported them either infrequently or confined to the edge between open savanna and palm marsh communities. Plants from disturbed areas such as nearby access roads are encroaching on the Aripo Savannas, supporting the theory that recurrent fires are affecting the floristic composition.

Fortunately, the Aripo Savannas appear to be resilient, but no ecosystem can be adapted to fires that occur more frequently than their ability to replace the majority of their original species to their former proportions. Recurrent fires in the area may change the balance of this ecosystem in favor of a fire-conditioned community.

9.4.5 Fire Effects on Montane Forests

The rate of regeneration after a tropical high elevation fire has also been investigated on the south side of the Cerro Asuncion, Costa Rica (3300 m elevation) by Janzen (1973b). The fire occurred during the dry season in "paramo" shrubby, chaparral-like vegetation. Three years after the fire the plants had recovered only 24 to 53% of their aboveground biomass. There were numerous patches of bare ground where cinders and burned wood could be found. The soil patches were being colonized by liverworts, mosses, and roseaceous herbs. Succession to the original vegetation was very slow owing to lack of seed source or environmental stress. The plant species which are the best competitors in this habitat form nearly pure stands on a given exposure and soil type. Litter breakdown is slow, and typical lowland decomposers are absent at these elevations. The low rates of plant tissue replacement are believed to be responsible for the presence of only a small community of herbivores.

9.5 Closing Remarks

The fire ecology of Central America is evolving in response to human, social, and economic pressure. Population increase results in greater needs for subsistence agriculture in a region poorly suited for farming (Janzen 1988a). Nutrient runoff, erosion, and rapid weed proliferation are detriments to both intensive and rotational cultivation patterns. At the same time, political expediency, or shortsightedness, tend to support development of export commercial commodities over development of workable agricultural reforms and their implementation and enforcement.

A complex web of theories, policies, and beliefs restricts progress towards halting deforestation and reversing the "the ecological fitting of fire tolerant parts" (Janzen 1988a). Fire is viewed as a natural and beneficial force in the ecosystem. Temperate ecosystem bias has been installed with its European and American immigrants and technical advisors. The human need to win in the fight against nature has justified an ever-increasing persistence and sophistication in modifying the landscape until little that is natural is left.

The solution to slash-and-burn practices is not clear. The strategy of Janzen is to "save it; figure out what you have saved; and then put . . . [it] to work for society" (Lewin 1988). Nongovernmental organizations have been successful in financing and facilitating reforestation programs and appropriate agricultural reforms (Talbot 1987)."Sustainability" has become the goal of current natural resource experts (Gradwohl and Greenberg 1988; Budowski 1988; Wadsworth 1988; Gyllenhaal 1984). Modern forestry practices abandon clearcutting in favor of selective logging and intensive culture of mixed species, age, and functional individuals.

Sustainability is appropriate for productive habitats. The rest must be "restored," through the practice of rehabilitating degraded, deforested, and depleted lands so that productivity and usefulness can be enhanced. Most programs depend on fast-growing, nitrogen-fixing trees because of their proven effectiveness in some habitats. While these may be useful in agro-forestry or sustained yield agricultural system, their proliferation is suspect. These introduced trees may take over habitats from better-suited native species and are often susceptible to pests and biotic and abiotic diseases. Gradual diversification and reestablishment of natural vegetation is recommended in both sustained yield and restoration programs (Grawohl and Greenberg 1988).

To be successful, programs must also support local ownership and be compatible with local customs and habitat. In Panama, reestablishment of the green iguana (*Iguana iguana*) as a protein source has been successful in reducing deforestation and increasing protection of wildlife. The iguanas are naturally adapted to forest life, and campesinos are enthusiastic about iguana ranching (Gradwohl and Greenberg 1988).

Other reforms are needed outside the Central American countryside. International organizations such as the World Bank, the Food and Agricultural Organization of the United Nations, and the Agency for International Development of the United States of America need to assess their programs on the basis of environmental preservation and restoration. In the past, big visible projects have received the most funding support. Now community focus and involvement are considered to be the most manageable and productive in terms of environmental fitness (Pasca 1988).

References

Alexander EB (1973) A comparison of forest and savanna soils in northeastern Nicaragua. Turrialba 23:181–191

Arnason T, Lambert JDH, Gale J, Cal J, Vernon H (1982) Decline of soil fertility due to intensification of land use by shifting agriculturists in Belize, Central America. Agro-Ecosystems 8:27–37

Batchelder RB (1967) Spatial and temporal patterns of fire in the tropical world. In: Proc Ann Tall Timbers Fire Ecol Conf. Tall Timbers Res Stat, Tallahassee, Florida 6:171–208

Blydenstein J (1968) Burning and tropical American savannas. In: Proc Ann Tall Timbers Fire Ecol Conf. Tall Timbers Res Stat, Tallahassee, Florida 8:1–14

Breslin P, Chapin M (1984) Conservation Kuna Style. Grassroots Develop 8:26–35

Budowski G (1966) Fire in America tropical lowland areas. In: Proc Ann Tall Timbers Fire Ecol Conf. Tall Timbers Res Stat, Tallahassee, Florida 5:5–22

Budowski G (1988) Is sustainable harvest possible in the tropics? Am For 94(11 and 12):34–37, 79–81

Caulfield Vasconez K (1987) Costa Rica, a country profile. Arlington, Virginia: Evaluation Technologies, under contract AID/SOD/PDC-C-3345, Office of Foreign Disaster Assistance, Agency for International Development, Dept State, Washington DC, 128 pp

Denevan WM (1961) The upland pine forests of Nicaragua, a study in cultural plant geography. Univ Cal Publ in Geography 12:251–320

Edgar B (1989) No title. Re: Costa Rica's dry forests. Pac Discovery Mag, 6 pp

Evaluation Technologies (1981) Honduras a country profile. Arlington, Virginia: Evaluation Technologies, under contract AID/SOD/PDC-C-0283, Office of Foreign Disaster Assistance, Agency for International Development, Dept State, Washington DC, 86 pp

Ewel J, Berish C, Brown B, Price N, Raieh J (1981) Slash and burn impacts on a Costa Rican wet forest site. Ecology 62:816–829

Furley PA (1987) Impact of forest clearance on the soils of tropical cone karst. Earth Surface Processes Landforms 12:523–529

Gliessman SR (1978) Establishment of bracken following fire in tropical habitats. Am Fern J 68:41–44

Gradwohl J, Greenberg R (1988) Sustainable agriculture. In: S ving the tropical forests. Earthscan, London. Chapter 2, pp 102–137 and Chapter 4, pp 163–189

Guevara J, Hudson J (1979) Los Incendios Forestales en la Zona de Estudio de la ESNACIFOR durante 1979. (Forest Fires in the Study Zone of the ESNACIFOR during 1979). COHDEFOR, Honduras (Mimeograph) 29 pp

Gyllenhaal C (1984) Cultural and ecological factors affecting nutrient dynamics and crop yields in shifting cultivation. Diss, Univ Alabama. Ecology 45:2026-B

Hudson J (1976) Proteccion Contra Incendios de Pino en Belice. (Pine Forest Fires Protection in Belize). Curso Intensivo Sobre Manejo y Aprovechamiento de Bosques Tropicales. CATIE (Mimeograph) 13 pp

Hudson J, Salazar M (1980) Prescribed fire study tour of the United States of America. Jan 16–Feb 8, 1980 (Mimeograph) 44 pp

Hudson J, Salazar M (eds) (1981) Las Quemas Prescritas en los Pinares de Honduras. (Prescribed burning in Honduras pine forests). Serie Miscelanea No 1, Escuela Nacional de Ciencias Forestales, Siguatepeque, Honduras, 58 pp

Hudson J, Kellman M, Sanmugadas K, Alvarado C (1983a) Prescribed burning of *Pinus oocarpa* in Honduras. I. Effects on surface runoff and sediment loss. For Ecol Manage 5:269–281

Hudson J et al. (1983b) Prescribed burning of *Pinus oocarpa* in Honduras. II. Effects on nutrient cycling. For Ecol Manage 5:283–300

International Monetary Fund (1986) World economic outlook, a survey by the staff of the International Monetary Fund. Washington DC, Int Monetary Fund, 268 pp

Janzen DH (1973a) Tropical agroecosystems. Science 182:1212–1219

Janzen DH (1973b) Rate of regeneration after a tropical high elevation fire. (Abstr) Biotropica 5:117–122

Janzen DH (1986a) The eternal external threat. In: Soule ME (ed) Conservation biology: The science of scarcity and diversity. Sinauer Associates, Sunderland, MA, pp 286–303

Janzen DH (1986b) Guanacaste National Park: Tropical ecological and cultural restoration. 1st edn San Jose, CR: EUNED-FPN-PEA, 104 pp

Janzen DH (1988a) Complexity is in the eye of the beholder. In: Almeda F, Pringle CM (eds) Diversity and conservation. Calif Acad Sci AAAS, San Francisco, CA, pp 29–51

Janzen DH (1988b) Guanacaste National Park: Tropical ecological and biocultural restoration. In: Cairns J Jr (ed) Rehabilitating damaged ecosystems, Vol II. CRC, Boca Raton, FL, pp 143–192

Janzen DH (1988c) Management of habitat fragments in a tropical dry forest: growth. Ann Missouri Bot Gardens 75:105–116

Kellman M (1976) Broad-leaved species interference with *Pinus caribaea* in a managed pine savanna. Commonw For Rev 55:229–245

Kellman M (1979) Soil enrichment by neotropical savanna trees. J Ecol 67:565–577

Kellman M, Sanmugadas K (1985) Nutrient retention by savanna ecosystems. I. Retention in the absence of fire. J Ecol 73:935–951

Kellman M, Miyanishi K, Hiebert P (1985) Nutrient retention by savanna ecosystems. II. Retention after fire. J Ecol 73:953–962

Kellman M et al. (1987) Nutrient sequestering by the understory strata of natural *Pinus caribaea* stands subject to prescription burning. For Ecol Manage 21:57–73

Lambert JDH, Arnason JT (1986) Nutrient dynamics in milpa agriculture and the role of weeds in initial stages of secondary succession in Belize, Central America. Plant Soil 93:303–322

Lara Marquez H (1977) Estudio del Efecto del Fuego Sobre la Germinacion de Semillas de *Pinus caribaea*, *Pinus oocarpa* y *Pinus pseudostrobus* Ubicados en Diferentes Partes del Suelo Forestal y su Supervivencia Subsiguiente. (A Study of the Effect of Fire on *Pinus caribaea*, *Pinus oocarpa* and *Pinus pseudostrobus* Seeds Germination and Survivability Depending on Their Location on The Forest Soil). Bachelor's thesis, Escuela Nacional de Ciencias Forestales, Siguatepeque, Comayagua, Honduras, 41 pp

Lewin R (1988) Costa Rican biodiversity. Science 242:1637

Matson P, Vitousek PM, Ewel JT, Mazzarino MJ, Robertson GP (1987) Nitrogen transformations following tropical forest felling and burning on a volcanic soil. Ecology 68:491–502

Munro N (1966) The fire ecology of Caribbean pine in Nicaragua. In: Proc Ann Tall Timbers Fire Ecol Conf. Tall Timbers Res Stat, Tallahassee, Florida 5:67–83

Myers N (1980) Conversion of tropical moist forests. Nat Acad Sci, Washington DC, 205 pp

Pasca TM (1988) The politics of tropical deforestation. Am For 94 (11 and 12):21–24

Pohl RW (1983) *Hyparrhenia rufa* (Jaragua). Description. Univ Chicago, IL. CR Natural History, pp 256–257

Postel S (1988) Global view of a tropical disaster. Am For 94(11 and 12):25–29, 69–71

Sader SA, Joyce AT (1988) Deforestation rates and trends in Costa Rica, 1940 to 1983. Biotropica 20:11–19

Schwab S (1988) Vegetation response to arson fires in the Aripo Savannas. Scientific Reserve, Trinidad, WI, Master's thesis, Univ Wisconsin, Stevens Point, Chapter 2, pp 68–132

Seiler W, Crutzen PJ (1980) Estimates of gross and net fluxes between the biosphere and the atmosphere from biomass burning. Climatic Change 2:207–247

Suman DO (1984) The production and transport of charcoal formed during agricultural burning in Central Panama. Interciencia 9:311–313

Suman DO (1986) Charcoal production from agricultural burning in Central Panama and its deposition in the sediment of the Gulf of Panama. Environ Conserv 13:51–60

Suman DO (1988) The flux of charcoal to the troposphere during the period of agricultural burning in Panama. J Atmos Chem 6:21–34

Talbot J (1987) Tropical forests, biological diversity and preservation: Who's on call for action? Ecol Soc Am 68:497–500

Wadsworth FH (1988) Finding forestry alternatives. Am For 94(11 and 12):36–77

Wolffsohn A (1988) El Papel del Fuego en el Bosque Humedo Neotropical. (The Role of Fire on Neotropical Forests). Nota Tech No 10, Escuela Nacional de Ciencias Forestales, Siguatepeque, Comayagua, Honduras, 3 pp

10 Fires and Their Effects in the Wet-Dry Tropics of Australia

A.M. Gill[1], J.R.L. Hoare[2], and N.P. Cheney[2]

10.1 Introduction

In the Northern Territory, eucalypt forests and woodlands tend to have sharply separated components of tree canopies and a grass-shrub understorey. Where the grassy understorey species are predominantly annual *Sorghum* spp., fires early in the dry season are patchy, of low intensity and tend to go out at night; they disturb large numbers of grasshoppers and other invertebrates which become the object of predation by birds; some *Sorghum* seed may be killed, then consumed by cockatoos on the ground. Fires later in the dry season are of higher intensity, remove grassy fuels over extensive areas and persist overnight; if fuels are substantial, e.g. 20 t ha^{-1} of litter, cambial damage to overstorey trees is likely; any eucalypt seedling present is likely to be killed thereby preventing recruitment. Fire behaviour in the wet season has not been reported, but fires can occur given an accumulation of dead material from the previous year or years. Because all *Sorghum* regenerative potential is found in seedlings at that time, fire will effectively eliminate *Sorghum* from the stand, leaving the shrubs, perennial grasses and herbs to predominate. In the absence of fire, the understorey may grow up such that the shrub and tree canopies become vertically continuous.

Fires may have deleterious effects on the extent of *Callitris* forests and mesic and xeric rain forests. Severe fires in adjacent grassy communities may kill trees on the edges of these fire-sensitive communities; during severe droughts, in fuels augmented by drought-induced leaf fall, fires may burn right through rain forest patches; after cyclones have felled forests, desiccation of debris and invasion of grasses may predispose the area to severe fires. After the community has been broached by fire, it is more vulnerable to further fires because of the general invasion of flammable grasses.

Fire management of eucalypt forests and woodlands in the wet-dry tropics of Australia has: historical importance in relation to Aboriginal burning practices; ecological importance in relation to tourism and the conservation of flora and fauna; contemporary economic importance in relation to the production of cattle and timber (*Callitris*); and potential global importance in relation to the production of emissions which affect atmospheric composition and climate.

[1]CSIRO Division of Plant Industry, Canberra, A.C.T. 2601, Australia

[2]CSIRO Division of Forestry and Forest Products, Queen Victoria Terrace, A.C.T. 2600, Australia

10.2 Location and Landscape

The wet-dry tropics of Australia form an arc from the Kimberley Region of northwestern Western Australia, east to Cape York Peninsula in Queensland, then south parallel to the eastern coast to the vicinity of Brisbane in southeastern Queensland (Fig. 1). The region has not been tightly defined but represents a convergence of regional distributions defined for various purposes. It is largely represented by the megatherm seasonal region of Nix (1983), where it occurs north of Brisbane, Queensland. The region has a seasonally wide variation in rainfall and "is dominant where mean annual temperature exceeds 24°C". The average annual rainfall of the region is generally greater than 600 mm (Ridpath 1985); its seasonality is at a maximum in the northwest and a minimum in the southeast. The soils are generally infertile (Ridpath 1985) having arisen in situ over long periods or from the dismantled and transported products of such soils; ferricretes (e.g. "laterites") are widespread (Story et al. 1969). Topographic relief is modest, the most noticeable relief being found in the Kimberley region, Arnhem Land and the Eastern Highlands (Fig. 1).

Biologically, the region has a predominance of forest and woodland plant communities but other communities such as hummock grasslands and even heaths occur. Because of the low-lying topography in many areas and the strongly seasonal rainfall, seasonal inundation of portions of the landscape occurs during the wet season. Tidal swamps and marshes fringe the coast but where fresh water predominates adjacent to the rivers and streams, open graminoid marshes occur. Where inundation is less extensive, swamps dominated by *Melaleuca* spp. are typical; the distribution of *M. leucadendra* encompasses the tropical (cf. subtropical) portion of the region (Barlow 1988). On higher ground – where inundation is rare, limited or absent – forests and woodlands of *Eucalyptus* are typical. (Taylor and Dunlop 1985 and Bowman

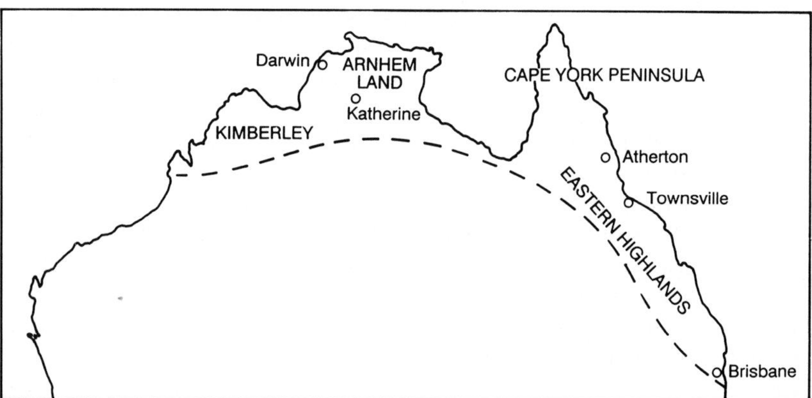

Fig. 1. The region of interest is on the wetter, coastal side of the *dashed line* – where the annual rainfall is mostly greater than 600 mm. The distance from Darwin to Townsville is approximately 1900 km

and Minchin 1987 provide much greater detail on aspects of vegetation catenas).

The region of interest has considerable biogeographic significance in being similar to the "tropical zone" of Burbidge (1960), the sum of two major regions with internally similar eucalypt floristics (Gill et al. 1985), and the region in which xeric or "dry seasonal rain forests" are found (Gillison 1987). Along the wetter more coastal margins of the region are the mesic tropical rain forests (Winter et al. 1987).

All the plant communities mentioned above – with the likely exception of those of tidal swamps and marshes – are subject to fires. Below, our emphasis is on the extensive eucalypt forests and woodlands of the region, especially those of the World Heritage Area known as Kakadu National Park about 200 km east of Darwin (Fig. 1). We focus on the plants and plant communities of the region along with their responses to fires, and largely put aside the increasing interest in, and knowledge of, the fauna of the region and the affects of fire regime upon it.

10.3 Proneness to Fires and the Fire's Characteristics

10.3.1 Fuels

Bowman and Wilson (1988) provide a "snapshot" of fuel types and their characteristics in an area near Darwin. Eucalypt communities had grassy fuels over litter (averaging 2.2 t ha^{-1} and 2.7 t ha^{-1} respectively of the total 6.3 t ha^{-1}) while *Callitris* forest (4.8 t ha^{-1} total), mixed thicket – a mixture of eucalypts and monsoon rain forest (8.1 t ha^{-1} total), and pure thicket, i.e. without eucalypts (7.5 t ha^{-1}) – had little grass and mostly litter as the major fuel component. Litter moisture contents decreased from pure thicket to mixed thicket to *Callitris* forest to eucalypt forest in proportion to decreases in canopy cover both in the early and late dry season.

Grass (or graminoid material) is particularly important as a fuel in the region because of its flammability, especially in the dry season. As the dry season progresses, the soils desiccate and the grasses either die completely (the annuals) or die back to the root stock (the perennials). Soil desiccation depends on topographic position which affects the extent of wet season flooding, the timing of the floods ebbing away and exposure to the sun: thus the nonflooded ridges dry first, then various locations in lower topographic locations. Just how much grass is present in areas not subject to inundation may be influenced by the dramatic variations in wet-season rainfall that the area experiences, e.g. near Darwin (Taylor and Tulloch 1985).

Grassy fuels in the region typically reach 5–10 t ha^{-1} 4 to 7 years after fire (Walker 1981; Mott and Andrew 1985; Sandercoe 1986). First-year production may be as high as 6–8 t ha^{-1} (Walker 1981). In studies where the time since fire was unknown, Tunstall et al. (1976), Gill et al. (1987) and N.P. Cheney and J.S.

Gould (pers. commun.) quote figures for grassy fuels between 1.1 and 9.0 t ha^{-1}. Annual *Sorghum* growing in relatively moist, open areas can create fuel loads up to 14 t ha^{-1} (Rowell and Cheney 1979).

Annual *Sorghum* fuels in eucalypt communities, common in the Northern Territory (Mott and Andrew 1985), increase as expected from the general rules above up to 5 or 6 years after fire, then begin to decrease until they became quite sparse. For example, J.R.L. Hoare (unpubl.) found *Sorghum* cover to decline by 24% with 5 years' protection from fire. Perennial grasses may decline too, but at a slower rate. During the whole period until the next fire, litter fuels (leaves and twigs of woody plants) may keep increasing, perhaps up to the 23.8 t ha^{-1} level reported by Gill et al. (1987) in Kakadu National Park. J.R.L. Hoare (unpubl.) found that, after 13 years of fire protection in a tall eucalypt forest, litter fuels reached an average of 12.1 t ha^{-1} with a maximum value of 21.5 t ha^{-1}.

10.3.2 Fire Climate

A quantitative measure of fire climate is the fire danger index (McArthur 1966, 1967). The index is readily linked to fire behaviour variables and is used as an expression of fire danger by the Australian Bureau of Meteorology. Two indexes are used – one for grassland (McArthur 1966) and one for eucalypt litter fuels (McArthur 1967). Though designed for fuel types in a quite different area (Canberra, Australian Capital Territory), they can be used as an *index* of seasonal and yearly variation of fire weather. Both indexes have a set maximum of 100 which was to represent the near worst-possible fire weather conditions to be experienced in Australia (Luke and McArthur 1978). The indexes are calculated from equations combining the effect of air temperature, relative humidity and windspeed together with a drought factor for litter fuels and a curing factor (or percentage dead grass) for grassy fuels (Noble et al. 1980). Figure 2a shows the average monthly "forest fire danger index" for Darwin, while Fig. 2b shows the calculated grassland fire danger index for the same locality, assuming different degrees of grass curing. The dotted lines show the trajectories of grassland fire danger for grassy fuels in different topographic positions viz. (1) highest topographic position and (2) and (3) for progressively moister sites. Coastal Darwin's forest fire danger index reached a peak in July while grassland fire danger index peaked in September; further east at a more inland site, Jabiru in Kakadu National Park, both fire danger indexes peaked in September (Gill et al. 1987), the height of the dry season.

10.3.3 Ignition Sources and Fire Frequencies

At the end of the dry season, lightning is abundant and fires may be ignited. Soon, however, relative humidities increase and rainfall begins so the chances of ignition decrease. In the pre-Aboriginal period, the late dry season may be

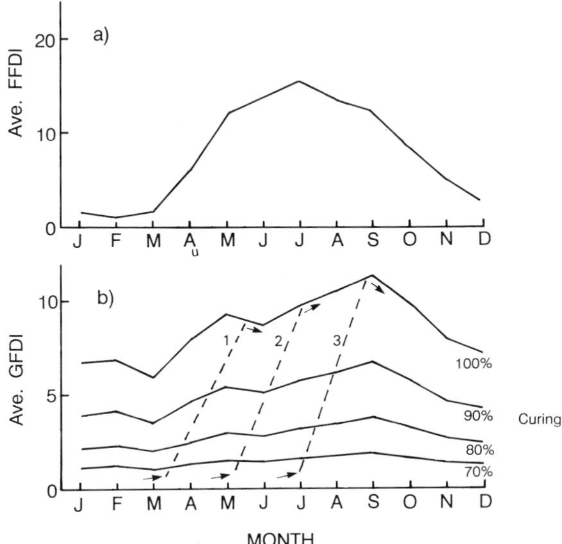

Fig. 2. a Average monthly 3 pm forest fire danger indexes (FFDI) for Darwin for the period 1963–1986. **b** Average monthly 3 pm grassland fire danger indexes (GFDI) for Darwin for the period 1963–1986 using various grass curing percentages. The *dashed lines* (1) to (3) indicate hypothetical trajectories for different grass species and locations as grasses cure during the dry season

regarded as the major time for fires. Braithwaite and Estbergs (1985) suggest that this time is from September to December in the Kakadu National Park.

Aborigines were believed to have arrived in Australia 32,000 years ago (Barbetti and Allen 1972) although some authorities now suggest the possibility of up to 130,000 years (Singh and Geissler 1985). In the wet-dry tropical area of the Northern Territory evidence of Aboriginal antiquity dates to 20,000 years ago (Russell-Smith and Dunlop 1987).

Aborigines in northern Australia defined a number of seasons and sub-seasons which regulated their activities (e.g. Haynes 1985; Stephenson 1985). Haynes (1985) studied the pattern of Aboriginal burning in a 9000-ha area of Arnhem Land. He found that deliberate Aboriginal burning was controlled almost entirely by the timing and placement of fires. Thus, strips were burnt around resource-rich monsoon forests (rain forest) and around hunting grounds where intense fires would be used to drive game later in the dry season. Fires were very small in the early dry season and were in areas near camps and walking tracks; fires up to hundreds of hectares may burn later in the dry season only to go out at night; in the late dry season effective control is lost and extensive areas may burn (Haynes 1985). The details of Haynes' (1985) study area are not necessarily the same for others. Stephenson (1985) and Russell-Smith and Dunlop (1987) have noted that variations occur due to different tribal customs. It is likely that Aboriginal fire regimes varied spatially and temporally

(Russell-Smith and Dunlop 1987). Braithwaite and Estbergs (1985) suggested that Aboriginal burning in eucalypt forests was mainly in the period May to September with some further burning post-October, while in eucalypt woodland, burning began in April, peaked in July, was equally prevalent from July to November and then declined to December.

Press (1988) examined LANDSAT imagery for a large area around, and including, Kakadu National Park. He found that there was a correlation between areas burned in the early dry season with those burned in the late dry season, implying that if the area did not burn early it was burned late. His observations of burning in the Arnhem Land Aboriginal Reserve supported Hayne's (1985) observations. In leaseholds for cattle raising, early dry-season burning was the norm. However, this pattern of early dry-season burning by Europeans may not have always been the case. Between 50 and 5 years ago, as Europeans displaced Aborigines, the season of burning in forest was shifted to reach a peak in September while that in woodlands may have been spread largely across the June-September period (Braithwaite and Estbergs 1985). A large proportion of fires now are caused by escapes from burning-off operations and the careless use of fire by the public.

Throughout the region some burning is annual (Pedley and Isbell 1971; Braithwaite and Estbergs 1985) but the return time varies from place to place. Haynes (1985) recorded that some floodplains were burned several times per year by Aborigines. Eucalypt forest had a fire in 2 out of 3 years in the Kakadu National Park area. Walker (1981) in a review suggested return times of 1–5 years for the region.

10.3.4 Fire Characteristics

The most important fire characteristics in relation to biological effects are flame dimensions, temperature profiles and intensity. A case could be made for smoke as an important characteristic because of its apparent signalling affect on birds of prey (such as the fork-tailed kite, *Milvus migrans*), which then feed on invertebrates and small vertebrates disturbed by the passage of fire. Further, the fine-grained patchiness left by the fire is important to niche differentiation in lizards (Braithwaite 1985).

Fine-grained patchiness is marked in early dry-season fires. At that time, there is a variety of moisture conditions with some fully cured grasses (especially annuals), some greener grasses (e.g. perennial grasses like *Heteropogon triticeus*) and low-lying shrubs (e.g. *Xanthostemon* sp.). Coarse dry annuals like *Sorghum intrans* may have bare ground between plants and very little leaf material at ground level: even the few centimetres between stalks of grass may be sufficient to cause the fire to go out when winds are calm and evening approaches. Early dry season fires burn in a mosaic pattern and go out at night. As the dry season progresses, however, the grasses desiccate further and semi-deciduous or deciduous shrubs and trees like species of *Terminalia*, *Eucalyptus*, *Xanthostemon*, *Brachychiton* and *Cocholspermum* shed leaves which not only add to the litter fuel weight but also promote a continuous layer of fuel. Fires become more intense and more continuous and burn throughout the night at the height of the dry season. Towards the end of the dry season, a single fire may have burned hundreds of square kilometres.

Flame dimensions have been linked with leaf damage on trees (McArthur 1962) and the relationship has been depicted for forests and woodlands in Kakadu by Gill et al. (1987). They showed that peaks in flame heights were an appropriate measure to correlate with the response (especially in conjunction with fuel weight).

Although temperature measurement in fires is fraught with difficulty (Martin et al. 1969), thermocouple values do reveal interesting phenomena. Tunstall et al. (1976) were able to explain how fire scars could form on the lee side of trees during fires using thermocouple-lined cylinders set in burning tropical grasslands. They found that maximum temperatures were reached near canopy level in their grassy fuels, a result recorded also by Gillon (1983) in fires burning in a similar fuel in Africa.

Intensity is the product of: the heat value of fuels (say 16 to 22,000 kJ/kg^{-1} Bowman and Wilson 1988 for dead materials); fuel quantity (say 1 to 24 t ha^{-1} or 0.1 to 2.4 kg m^{-1}); and rate-of-spread of the fire perimeter (Byram 1959). Fire intensities in tropical forests have been reported up to 7000 kW m^{-1} (Gill et al. 1987) although much higher values may be expected, especially where heavy litter fuels occur. In open grasslands, Cheney and Gould (pers. commun.) found rates of spread largely independent of fuel weight, up to 3.2 m s^{-1}: they recorded intensities of 17,000 kW m^{-1}.

10.4 Fire's Impact on Plants: Demographic Aspects

Grass fires seem to cause surprisingly little direct damage to trees perhaps because of short residence times and relatively low fuel quantities. Bark losses from gum-barked trees (Gill et al. 1986) seem minor, although scorch of canopy leaves can be complete. Leaf loss by fire in the dry season may be functionally similar to dry-season deciduousness and have no apparent influence on the restoration of canopies in the wet season. Yearly death of canopy leaves and buds could prevent flowering and seed set in some species. Such scenarios of relatively minor damage to canopy trees are possible although, in the absence of careful study, they should be regarded as tentative. The presence of fire scars on trees (Tunstall et al. 1976) and the record of population decline of eucalypts in Queensland (Sandercoe 1988) suggest caution. If long periods without fire occur and fuels become litter-based and heavy, late dry-season fires could be very damaging.

The absence of recruitment among eucalypts in many stands in the Northern Territory could be due to fires killing young cohorts of seedlings, long-term episodic establishment events (like a combination of particularly abundant seed set and an extended wet season) or competitive exclusion by other plants. Hoare et al. (1980) found an absence of seedling recruitment in all stands, even those unburned. However, in the unburned stand, there was a release from fire-caused stunting of the common small eucalypt plants vegetatively propagated from underground tissues of larger plants (Lacey et al. 1982).

The situation with trees flagged as relatively fire-sensitive earlier in this chapter represent a separate category. *Callitris* is likely to be killed if all of its foliage is scorched, while many rain forest species may die or resprout from the base. Rain forest trees appear to be relatively fire-sensitive (Gillison 1987; Kahn and Lawrie 1987) cf. eucalypt trees, although survival of some does occur; given the diversity of rain forest trees in both mesic and relatively xeric situations, generalizations would be better served by an expanded data set. In Queensland, there is evidence that a wide range of rain forest species can recover vegetatively after fire (Stocker 1981; Unwin et al. 1985).

All trees and shrubs of eucalypt forests and woodlands recorded by Hoare et al. (1980) have vegetative means of survival of fire. Large plate-like structures, lignotubers, of some eucalypts provide an extreme case of at least partly buried and therefore protected regenerative tissue (Lacey et al. 1982). Table 1 presents the results of Hoare et al. (1980) according to life-form and leaf persistence. Apart from the annual grasses and herbs, all species are able to regenerate by vegetative means after fire. The general absence of woody species depending on seed for recovery after fire contrasts with the situation in forests in southern Australian eucalypt forests and sclerophyll shrub woodlands where such species are often understorey dominants (e.g. Gill 1981). The main difference between mechanisms of plant persistence in temperate and tropical eucalypt forests may be explained by the very high frequency of fires in the tropical eucalypt forests and woodlands.

Many woody plants of the "monsoon forests" may recovery vegetatively after fire (Russell-Smith and Dunlop 1987), but little systematic survey has been carried out. Stocker (1966) points out that many woody species of these forests are

Table 1. Classification of species regenerative strategies by life form for forest and woodland plots in Kakadu National Park. (Hoare et al. 1980)

Category	Species numbers Forest	Woodland
Annuals	28	82
Perennial grasses	4	19
Perennial herbs	31	54
Evergreen shrubs	6	6
Evergreen trees	14	14
Deciduous trees	5	9
Deciduous vines	1	1
Evergreen vines	0	2
Deciduous shrubs	0	2
Total	89	189

"fire-sensitive" in the sense that they can be entirely eliminated by *repeated* burning (our emphasis).

The classification of species according to mechanisms of persistence during fires does not underscore the interesting and varied biology of the species concerned. While the biology is as yet mostly unstudied, some anecdotal information from Kakadu National Park is presented to promote the subject. Two examples are mentioned. The first is *Planchonia careya* (family Lecythidaceae), which often occurs as a short shrubs in eucalypt woodlands but which can grow into a tree on the edges of flood plains and coastal forests (Brennan 1986). In the shrubby stage, at least, this species resprouts following early dry-season burning and may flower on the short stems produced by September – even in the absence of rain. The second example is *Cochlospermum fraseri* (family Bixaceae), a medium to tall shrub found commonly in eucalypt woodlands. This species appears quite resistant to early dry-season fires (unlike short *Planchonia* shrubs) and will flower and fruit before the wet season begins.

Grasses are the group of plants best studied in the wet-dry tropics in relation to their reactions to various fire regimes. Mott and Andrew (1985) have reviewed the subject recently. If annual *Sorghum* (e.g. *S. intrans*) is compared with perennial *Themeda australis*, the similarities are more surprising than the contrasts. Thus, it was found that both species have similar transient seed dormancies which led to all soil seed germinating in the wet season following dry season dispersal. Contrasting behaviour of seedlings was evident, however: *Sorghum* seedlings established, grew up to 2 m tall and produced seed; *Themeda* seedlings failed to establish in unburnt communities. In the absence of fire, *Themeda* basal area reached a peak in year 2 then declined to year 6, the longest time studied; biomass was maintained during years 3 to 6, however. With biennial burning in the dry season, *Themeda* tussocks were stable.

Andrew and Mott (1983) found nearly 400 seeds m^{-2} germinating in stands of annual *Sorghum*, of which 40% survived as seedlings. A similar percentage of seedlings arose from soil seed in burnt plots, but only 250 seeds m^{-2} were present. Details of seed survival during fires are lacking but it has been noticed that direct seed death and bird (red-tailed black cockatoos, *Calyptorhynchus magnificus*) predation of seed does occur in early dry season fires at least (Gill et al. 1987). Presumably, enough seed is buried through use of hygroscopic awns, then protected from the heat of the fire by the soil covering, to provide a crop the following year. In areas protected several years from fire, production and transient storage of seed declines presumably due to a reduction in plant density following canopy closure and build up of litter (Hoare et al. 1980; Mott and Andrew 1985).

10.5 Fire and Communities

10.5.1 Eucalypt Forests and Woodlands

What happens to whole communities of plants depends on the species present, the stage in their life cycle when they are exposed to fire, the time of year, the intensity of fire and any post-fire grazing effects. In this section, the results of a long-term burning experiment (set up in 1972, Hoare et al. 1980) are described. The location of the experiment was Munmarlary, now in Stage 2 of Kakadu National Park. Two sites a short distance apart were chosen, one in eucalypt forest, the other in eucalypt woodland: both sites had understories dominated by *Sorghum intrans*.

Each experiment had four treatments: (1) no burning; (2) early dry-season annual burning; (3) early dry-season biennial burning; and (4) late dry-season annual burning. The treatments were replicated three times.

After 5 years of experimentation, species composition had changed very little. The few changes that had been detected could be ascribed to small cryptic populations or to a flux of annuals perhaps associated with seed redistribution during overland flow of water during the wet season. The changes could not be ascribed to fire regime differences. However, population decline of the annual grass *Sorghum intrans* in the unburnt plots was consistent with the absence of fire.

The major change due to fire regimes after 5 years was the release of fire-stunted shrub and tree species in the no-fire treatment thereby causing a dramatic structural change particularly in the 2–8 m tall mid storey (Figs. 3 and 4). Species with such a growth response included: *Eucalyptus tetrodonta* (family Myrtaceae), a canopy tree; two *Acacia* spp. (family Mimosaceae); *Grevillea heliosperma* and *Stenocarpus cunninghamii* (family Proteaceae); *Erythrophleum chlorostachys* (family Caesalpinaceae); *Alphitonia excelsa* (family Rhamnaceae); and, *Pandanus spiralis* (family Pandanaceae). This structural change seemed to be reflected in an increased abundance and in situ breeding of birds.

There was little difference between height growths of woody plants across all *burning* treatments, thereby emphasizing the suppressant effect of fires irrespective of their intensity. Before experimental fire regimes were applied, no plants were recorded in the 2–8-m height category in any treatment, yet after 5 years, all treatments registered plants in this stratum. This result suggests that previous fire regimes were somehow more severe or that the 5-year period of review was more favourable to growth.

This experiment did not include the effects of wet-season burning – which presumes some dry weather and a carry-over of dead grass from the previous season – but various investigators near Katherine (Fig. 1), which has an annual rainfall of about 900 mm, have examined the effects on pastures. Smith (1960) found that the percentage of *Sorghum plumosum*, and another perennial grass *Themeda australis*, fell by 30 and 50% respectively due to wet-season burning, while the percentage of a further perennial grass, *Chrysopogon fallax*, increased. Norman's (1963) results elicit a note of caution in interpretation because he

Fig. 3. Annually burnt forest of *Eucalyptus tetradonta-E. miniata* in Kakadu National Park, 200 km east of Darwin. The grassy understorey is largely composed of *Sorghum intrans.* (Experimental plot of Hoare et al. 1980)

suggested that a shift he observed from a dominance of *S. plumosum* to one of *T. australis* in unburned plots was due to a short-term cycle of above-average rainfall. Despite this, wet-season burning decreased the proportion of *Themeda* and increased the proportion of *Chrysopogon*, as found by Smith (1960).

Norman regarded the most significant shift in composition in his experiments to be due to late dry season fires which caused a proportional increase in annual grasses and forbs. Further north, where the annual grass *S. intrans* is so prominent, Stocker and Sturtz (1966) made the discovery that wet-season burning virtually eliminates that species, a result which can be explained now in relation to its having no soil seed store in the wet season and no viable seed on the standing crop (Mott and Andrew 1985).

Fig. 4. Similar forest to Fig. 3, but unburnt for 5 years. Notice the growth of the previously fire-stunted understorey. (Experimental plot of Hoare et al. 1980)

10.5.2 "Fire-Sensitive" Communities

Earlier in this review, we grouped a number of rain forest types together with *Callitris* forests and declared them a "fire-sensitive" group. While such grouping is a convenience to provide the contrast they present with eucalypt communities, the group is a diverse assemblage of non-eucalypt communities which are rarely subject to fire. Apart from being relatively fire-sensitive, perhaps the only common feature of this non-eucalypt collation is their patchy distribution.

While the more mesic rain forests are found in the coastal areas of Queensland, the relatively xeric "vine thickets" (Kahn and Lawrie 1987), "monsoon vine forests" (Russell-Smith and Dunlop 1987), "dry rain forests" (Gillison 1987) and "coastal monsoon forests" (Bowman and Wilson 1988) are found scattered throughout the drier areas; *Callitris* forests are found in patches throughout northern Australia (Stocker and Unwin 1986). Interestingly, without fire, a successional sequence can be from eucalypt community to *Callitris* to "monsoon forest" (Stocker and Unwin 1986).

The patchy distribution of the non-eucalypt forests – set in a matrix of eucalypt forests and woodlands – elicits a number of theories which can be complementary: (1) that patches occur where they do because physical site characteristics are favourable to these patches only (Beard 1976); (2) that patches are the remnants of previously continuous vegetation that has been fragmented by adverse climatic change (Specht 1958); and (3) that patches represent relics created by the spread of more flammable vegetation types (e.g. Clayton-Greene and Beard 1985; Stocker and Unwin 1986).

The interactions of the non-eucalypt communities with fires may be through the fire's effects on the edges of the communities, through burning generally within the community, and through the compounding effects of cyclones and fires. These three interactions will be discussed in turn.

That edges of "rain forest" have retreated with consequent expansion of grasslands or eucalypt forests has been shown by occurrences of abandoned mounds of exclusively "rain forest"-inhabiting scrub fowls (*Megapodes reinwardt*) on Melville Island near Darwin (Stocker 1971), on Cape York (Lavarack and Godwin 1987) and in Kakadu National Park (Russell-Smith 1985). Fire may have been the principal cause of these changes, but feral water buffalo could also have been involved in Russell-Smith's area (through direct structural damage) and climate change could have been involved in the Melville Island area where the mounds are older (Stocker 1971). However, climatic change and buffalo damage cannot explain the direct evidence of fire-killed "rain forest" seedlings in grassland areas adjacent to "rain forest" patches (Clayton-Greene and Beard 1985) and are less attractive as hypotheses in the explanation of sharp vegetation boundaries (between eucalypt and non-eucalypt communities) where no discernible physical site variation occurs, where burnt-out shells of "rain forest" species are found marginal to patches, and where direct observation of death of trees at margins have been observed (Russell-Smith and Dunlop 1987). Further support for the fire theory comes from the observation that rain forest margins expand when fire is excluded (e.g. Clayton-Greene and Beard 1985; Stocker and Unwin 1986; Russell-Smith and Dunlop 1987). There are a variety of reasons for "rain forest" distribution, one of which is fire occurrence.

Direct observation of "rain forest" burning have been reported for "monsoon vine thickets" of the Northern Territory (Russell-Smith and Dunlop 1987) and of Queensland (Ridley and Gardner 1961; Stocker and Unwin 1986). In severe droughts, partial deciduousness may occur so that fuels on the forest floor are augmented (Ridley and Gardner 1961), and because of the more open canopy, desiccation of litter proceeds to more extreme levels. Detailed studies of the effects of such fires are lacking but it has been observed that many species may resprout (Russell-Smith and Dunlop 1987). With the entry of fire into the community and further defoliation, grass invasion may occur thereby predisposing the area to further fire (Ridley and Gardner 1961). Repeated fires at short intervals are likely to have a devastating effect on many "rain forest" species.

A similar story, though more extreme, is that of the compounding effects of cyclones and fires (Webb 1958; Stocker and Unwin 1986). The cyclone (in its extreme form) smashes down the forest, exposes the community to desiccation, creates an unusually heavy fuel load and allows grasses to establish. Thus the forest is more predisposed to fire. In less extreme forms, cyclonic winds may increase the fuel load on the floor of eucalypt forests and woodlands as well as "rain forest" by exacerbating leaf and twig fall (Stocker 1976).

10.6 Management

10.6.1 National Parks

Kakadu National Park, already referred to above, is the most significant terrestrial national park in the region in terms of visitor use, size, controversy and status. The park, situated about 200 km east of Darwin, has an area of nearly 20,000 km^2. It attracts large numbers of visitors who are interested in Aboriginal rock-art and wildlife. The area is controversial because of the mining for uranium that occurs in a designated zone within the park. The park enjoys special status as a World Heritage Area. It is leased by the Australian government on a long-term basis from the traditional Aboriginal land owners.

Fire management of any national park must be considered in relation to the general aims of management. In part, only, the general aims of management are: "to conserve the natural, cultural, scientific and scenic resources of the Park whilst developing an Aboriginal perspective for management"; "to protect Park resources from the adverse consequences of fire, erosion, environmental change, pollution and misuse by people" (Anonymous 1986). Twelve other "key" management aims are listed.

Fire management policy for the Park is based on the assumption that the biota is adapted to the fire regimes of the Aboriginal people. Aboriginal fire regimes have been disrupted to various degrees since European settlement, particularly over the last 100 years (Anonymous 1986), but especially between 50 and 5 years ago (Braithwaite and Estbergs 1985). Today, fire management by Park staff is aimed at early dry-season burning in order to prevent extensive late dry-season fires; ignition may be carried out on the ground or from the air.

Around Aboriginal settlements and in areas used extensively for hunting, burning is carried out by Aborigines. Particular sites occupied by "monsoon forests" or rare palms (*Gronophyllum ramsayii*) are given special protection by Park staff (Anonymous 1986). Overall, the aim of fire management has been stated in the Plan of Management as "to re-establish so far as possible the traditional Aboriginal patterns of burning . . . to reduce the frequency, extent and intensity of wild fires . . . to prevent late dry season fires entering or leaving the Park . . . and protect Park assets and ensure visitor safety".

Prior to being national park, much of the area was used for raising cattle, the subject of the next section. While cattle raising has been lost from the park, feral water buffalo have been a particular problem. Buffalo are now being eradicated from the Park as part of a tuberculosis and brucellosis eradication campaign. Other changes from Aboriginal times are the shifts in Aboriginal culture, the presence of roads and built assets, large numbers of visitors, and the introduction and expansion of exotic plant distributions.

Of particular concern to managers is the unauthorized ignition of fires in the Park in the late dry season, a time of year when resources are inadequate for fire control (J Day pers. commun.). This situation makes difficult the implementation of policies other than frequent early dry-season burning. As an example of alternative policies, Hoare et al. (1980) have suggested, on plant ecological grounds, that fires every 4 years or so early in the dry season would provide ideal circumstances for the optimal development of eucalypt forest and woodland.

A recent innovation in Kakadu National Park has been the installation of an "expert system" with an appropriate geographical information system (Davis et al. 1986). "Expert systems" are rule-based systems – if X, then Y – which can assist decision-making by managers on the basis of the opinions of experts. Because they run on computers which sort through the appropriate rules and information required, quite complex questions can be answered on the basis of many rules. Already, the "system" is being used to explore what consequences would follow certain potential management options.

Fire management is complex and involves many issues and the consideration of many operational and administrative techniques. Here we have not mentioned the important topics of survey, monitoring and research. Much is new in management at Kakadu: the Park was established in 1979 only, doubled in size in 1984 and expanded again in 1987; visitor numbers are growing by thousands each year; and new information from survey and research is becoming available annually. Developments in fire management in this unique situation will be interesting to follow.

10.6.2 Cattle Raising

The wet-dry tropics of Australia are largely used for cattle raising. Stocking rates are low on large properties utilizing native pastures. British breeds of cattle have been traditional, but Asian and Asian-cross cattle are now widely found.

A simple view of cattle raising in the area is of vast paddocks, little stock handling, and widespread dry season burning. Cattle have been found to graze patches of vegetation in the wet season and to continue grazing there presumably because of more ready access to short palatable pasture (Andrew 1986). In the Northern Territory, persistent grazing of patches for several years has resulted in local death of pastures (Mott 1986). In northeast Queensland, *Heteropogon contortus* may be replaced by the introduced naturalized grass *Bothriochloa pertusa* (Tothill and Mott 1985). In southeast Queensland, burning and grazing seem responsible for the shift in pasture dominance from *Themeda* to *H. contortus* (Shaw 1957).

The apparently universal observation that herbivores prefer to feed on the "green pick" after burning of pastures has been put to good use in experiments designed to achieve a more uniform and less damaging system of pasture management (Andrew 1986). What is done is to burn half of the set stocked paddock in the first year in order to attract animals to it. Then, in the following year the animals move to the other half after it has been burnt. In this way half the pasture is burnt and half spelled each year and pasture degradation is arrested (Andrew 1986).

With gradual intensification of pasture use, the fire-grazing manipulation may have to be modified from the above because the grazing animals reduce the quantity of fuel (Andrew 1986). Intensification of grazing may occur with the use of mineral supplements and the introduction of exotic plant species (Tothill 1971).

10.6.3 Invasive Plants

Invasive plants have particular significance to conservation management. *Lantana* spp. have come from South America to become species widely spread in disturbed areas. Their spread has been exacerbated by fires in rain forests in Queensland (Ridley and Gardner 1961). Another species in tropical Queensland appears to have the potential to significantly alter fire regimes: *Melinus minutiflora*, or molasses grass, is a productive naturalized grass which is highly flammable (Tothill and Hacker 1973).

In the Northern Territory, another grass is threatening a change in fire regimes. *Pennisetum polystachyon*, mission grass, is a productive grass (Skeat 1986) which cures late and therefore has the potential to support severe late dry-season fires. This species could come to dominate the ground flora of forests and woodlands (R.W. Braithwaite et al. quoted by Anonymous 1986), presumably at the expense of many other species.

Perhaps the most important invasive species in Kakadu National Park is *Mimosa pigra* (family Leguminosae). This species spreads rapidly, stores large quantities of seed in soil, and can form dense thickets which shade out the understory (Lonsdale and Braithwaite 1988), thereby implying that fire frequency would be reduced by the presence of the species. Of course, the conservation consequences of the presence of thickets of this species spread far wider than any effect on fire regimes.

Detailed studies of the implications of species like those mentioned above have not been made in relation to fires. Such studies seem particularly germane to conservation management.

10.6.4 Emissions

Large areas of the wet-dry tropics are burnt every year, releasing vast quantities of smoke which have the potential to contribute to changes in global atmospheric composition. Cheney et al. (1980) estimate 25×10^6 ha of tropical and subtropical grasslands are burned each year, contributing 10^8 tonnes of fuel and about 45×10^6 tonnes of carbon.

What the contribution of fire emissions to the global atmospheric composition is, is a difficult problem to solve because of scale and the various feedback and buffering systems involved. In a management sense, the effects of decay, grazing and burning could be compared. In terms of the thermal infrared absorption of various gases, pertinent to the greenhouse effect, perhaps carbon dioxide and methane are of primary interest. Methane occurs in relatively small quantities in the atmosphere but has an important effect on warming (Crutzen et al. 1986).

Aerobic decay of organic matter might be expected to release carbon dioxide and no methane while anaerobic digestion by cattle releases considerable quantities of methane (Crutzen et al. 1986). Combustion in well-aerated fuels would be expected to produce carbon dioxide and little methane: Crutzen et al. (1985) suggest that the ratio of CH_4 to CO_2 in biomass burning is about 1%.

Apart from composition of emissions, the amount of material involved is most important. With time since fire, annual increments of growth usually decline each year such that the lower annual productivity leads to lower annual emission of gases. If fires are annual productivity is relatively high and emissions from fires are maximized. However, grazing in this region consumes relatively little pasture at present (Andrew 1986). Burning at 4- to 5-year intervals early in the dry season (relatively mild patchy fires) – as suggested for National Park management (Hoare et al. 1980) – is a case in point: this regime would produce less carbon dioxide and methane than one of annual late dry season fires (often intense and consuming all surface fuels).

10.7 Conclusions

The vast wet-dry tropics of Australia are clothed mostly by native plants. Fire is almost universally part of the environment of this large region but different fire regimes may occur in different areas, different vegetation types and in different management systems. Historical shifts in fire regimes appear to be altering the balance between eucalypt forests and woodlands, on the one hand, and relatively fire-sensitive forests ("rain forests' and *Callitris* forests) on the other. Most fire studies in the wet-dry tropics have been either in the Darwin-Katherine-Kakadu region, in southeastern Queensland or in the Atherton region of Queensland. Very little is known of the effects of fire regimes on tropical flood plains, *Melaleuca* swamps, tropical shrublands, various "rain forests", hummock grasslands and other non-eucalypt communities.

Acknowledgements. The senior author acknowledges the able technical assistance of Mr P.H.R. Moore. All the authors have received welcome financial assistance from the Australian National Parks and Wildlife Service while the junior authors have been supported also by the Northern Territory Conservation Commission.

References

Andrew MH (1986) Use of fire for spelling monsoon tallgrass pasture grazed by cattle. Tropic Grasslands 20:69–78

Andrew MH, Mott JJ (1983) Annuals with transient seed banks: the population biology of indigenous *Sorghum* species of tropical north-west Australia. Aust J Ecol 8:265–276

Anonymous (1986) Kakadu National Park Plan of Management. Aust Nat Park Wildlife Serv, Canberra

Barbetti M, Allen H (1972) Prehistoric man at Lake Mungo, Australia, by 32,000 years B.P. Nature (Lond) 240:46–48

Barlow BA (1988) Patterns of differentiation in tropical species of *Melaleuca* L. (Myrtaceae). Proc Ecol Soc Australia 15:239–247

Beard JS (1976) The monsoon forests of the Admiralty Gulf, Western Australia. Vegetatio 31:177–192

Bowman DMJS, Minchin PR (1987) Environmental relationships of woody vegetation patterns in the Australian monsoon tropics. Aust J Bot 35:151–169

Bowman DMJS, Wilson BA (1988) Fuel characteristics of coastal monsoon forests, Northern Territory, Australia. J Biogeogr 15:807–817

Braithwaite RW (1985) Fire and fauna. In: Braithwaite RW (ed) Kakadu fauna survey: final report. CSIRO Australia, pp 634–650

Braithwaite RW, Estbergs JA (1985) Fire patterns and woody vegetation trends in the Alligator Rivers region of northern Australia. In: Tothill JC, Mott JJ (eds) Ecology and management of the world's savannas. Aust Acad Sci, Canberra, pp 359–364

Brennan K (1986) Wildflowers of Kakadu. Brennan, Jabiru

Burbidge NT (1960) The phytogeography of the Australian region. Aust J Bot 8:75–212

Byram GM (1959) Combustion of forest fuels. In: Davis KP (ed) Forest fire: control and use. McGraw-Hill, New York, pp 61–89

Cheney NP, Raison RJ, Khanna PK (1980) Release of carbon to the atmosphere in Australian vegetation fires. In: Pearman GI (ed) Carbon dioxide and climate. Aust Acad Sci, Canberra, pp 153–158

Clayton-Greene KA, Beard JS (1985) The fire factor in vine thicket and woodland vegetation of the Admiralty Gulf region, north-west Kimberley, Western Australia. Proc Ecol Soc Aust 13:225–230

Crutzen PJ, Delany AC, Greenberg J, Haagensen P, Heidt L, Lueb R, Pollock W, Seiler W, Wartburg A, Zimmerman P (1985) Tropospheric chemical composition measurements in Brazil during the dry season. J Atmos Chem 2:233–256

Crutzen PJ, Aselmann I, Seiler W (1986) Methane production by domestic animals, wild ruminants, other herbivorous fauna and humans. Tellus 38B:271–284

Davis JR, Hoare JRL, Nanninga PM (1986) Developing a fire management expert system for Kakadu National Park, Australia. J Environ Manage 22:215–227

Gill AM (1981) Coping with fire. In: Pate JS, McComb AJ (eds) The biology of Australian plants. Univ Western Australia Press, Nedlands, pp 65–87

Gill AM, Belbin L, Chippendale GM (1985) Phytogeography of *Eucalyptus* in Australia. Aust Flora Fauna Ser, Bureau Flora Fauna 3:53 pp

Gill AM, Cheney NP, Walker J, Tunstall BR (1986) Bark losses from two eucalypt species following fires of different intensities. Aust For Res 16:1–7

Gill AM, Pook EW, Moore PHR (1987) Measuring fire properties for ecological effects at Kakadu National Park. Unpublished Report to Australian National Parks and Wildlife Service, Canberra

Gillison AN (1987) The 'dry' rain forests of Terra Australis. In: The rain forest legacy, Vol 1. Aust Gov Publ Serv, Canberra, pp 305–321

Gillon D (1983) The fire problem in tropical savannas. In: Bourlière F (ed) Tropical Savannas. Elsevier, Amsterdam, pp 617–641

Haynes CD (1985) The pattern and ecology of munwag: traditional Aboriginal fire regimes in north-central Arnhem Land. Proc Ecol Soc Australia 13:203–214

Hoare JRL, Hooper RJ, Cheney NP, Jacobsen KLS (1980) A report on the effects of fire in tall open forest and woodland with particular reference to fire management in Kakadu National Park in the Northern Territory. Unpublished report to Australian National Parks and Wildlife Service, Canberra

Kahn TP, Lawrie BC (1987) Vine thickets of the inland Townsville region. In: The rain forest legacy, Vol 1. Aust Gov Publ Ser, Canberra, pp 159–199

Lacey CJ, Walker J, Noble IR (1982) Fire in Australian tropical savannas. In: Huntley BJ, Walker BH (eds) Ecological Studies, Vol 42: Ecology of tropical savannas. Springer, Berlin Heidelberg New York Tokyo, pp 246–272

Lavarack PS, Godwin M (1987) Rain forests of northern Cape York Peninsula. In: The rain forest legacy, Vol 1. Aust Gov Publ Serv, Canberra, pp 201–226

Lonsdale M, Braithwaite R (1988) The shrub that conquered the bush. New Sci 120(1634):52–55

Luke RH, McArthur AG (1978) Bushfires in Australia. Aust Gov Publ Serv, Canberra

Martin RE, Cushwa CT, Miller RL (1969) Fire as a physical factor in wildland management. Proc Ann Tall Timbers Fire Ecol Conf 9:271–288

McArthur AG (1962) Control burning in eucalypt forests. Aust For Timber Bureau Leaflet 80

McArthur AG (1966) Weather and grassland fire behaviour. Aust For Timber Bureau Leaflet 100

McArthur AG (1967) Fire behaviour in eucalypt forests. Aust For Timber Bureau Leaflet 107

Mott JJ (1986) Patch grazing and degradation in native pastures of the tropical savannas of northern Australia. In: Horne PF (ed) The plant-animal interface. Winrock Int, Arkansas (from Andrew 1986)

Mott JJ, Andrew MH (1985) The effect of fire on the population dynamics of native grasses in tropical savannas of north-west Australia. Proc Ecol Soc Australia 13:231–239

Nix H (1983) Environmental determinants of biogeography and evolution in Terra Australis. In: Barker WR, Greenslade PJM (eds) Evolution of the flora and fauna of arid Australia. Peacock, Adelaide, pp 47–66

Noble IR, Bary GAV, Gill AM (1980) McArthur's fire danger meters expressed as equations. Aust J Ecol 5:201–203

Norman MJT (1963) The short term effects of time and frequency of burning on native pastures at Katherine, NT Aust J Agric Anim Husbandry 3:26–29

Pedley L, Isbell RF (1971) Plant communities of Cape York Peninsula. Proc Roy Soc Queensland 82:51–74

Press AJ (1988) Comparisons of the extent of fire in different land management systems in the Top End of the Northern Territory. Proc Ecol Soc Australia 15:167–175

Ridley WF, Gardner A (1961) Fires and rain forest. Aust J Sci 23:227–228

Ridpath MG (1985) Ecology in the wet-dry tropics: how different? Proc Ecol Soc Australia 13:3–20

Rowell MN, Cheney NP (1979) Firebreak preparation in tropical areas by rolling and burning. Aust For 42:8–12

Russell-Smith J (1985) A record of change: studies of Holocene vegetation history in the South Alligator River region, Northern Territory. Proc Ecol Soc Australia 13:191–202

Russell-Smith J, Dunlop C (1987) The status of monsoon vine forests in the Northern Territory: a perspective. In: The rain forest legacy, Vol 1. Aust Gov Publ Ser, Canberra, pp 227–288

Sandercoe CS (1986) Fire management of Cooloola National Park – fuel dynamics of the western catchment. In: Roberts BR (ed) Third Queensland Fire Research Workshop. Darling Downs Inst Adv Education, Toowoomba, pp 139–158

Sandercoe CS (1988) An aerial photographic study of the long-term effects of wild fires on Magnetic Island. Proc Ecol Soc Australia 15:161–165

Shaw NH (1957) Bunch spear grass dominance in burnt pastures in south-eastern Queensland. Aust J Agric Res 8:325–334

Singh G, Geissler EA (1985) Late Cenozoic history of vegetation, fire, lake levels and climate at Lake George, New South Wales, Australia. Phil Trans Roy Soc Lond Ser B 311:379–447

Skeat A (1986) Kakadu National park weed control. Aust Ranger Bull 3(4):16

Smith EL (1960) Effects of burning and clipping at various times during the wet season on tropical tall grass range in northern Australia. J Range Manage 13:197–203

Specht RL (1958) The geographical relationships of the flora of Arnhem Land. In: Specht RL, Mountford CP (eds) Records of the American-Australian Scientific Expedition to Arnhem Land. Melbourne Univ Press, pp 415–478

Stephenson PM (1985) Traditional Aboriginal resource management in the wet-dry tropics: Tiwi case study. Proc Ecol Soc Australia 13:309–315

Stocker GC (1966) Effects of fires on vegetation in the Northern Territory. Aust For 30:223–230

Stocker GC (1971) The age of charcoal from old jungle fowl nests and vegetation change on Melville Island. Search 2:28–30

Stocker GC (1976) Report on cyclone damage to natural vegetation in the Darwin area after Cyclone Tracey, 25 December 1974. Aust For Timber Bureau Leaflet 127

Stocker GC (1981) Regeneration of a north Queensland rain forest following felling and burning. Biotropica 13:86–92

Stocker GC, Sturtz JJ (1966) Use of fire to establish Townsville lucerne in the Northern Territory. Aust J Exp Agric Anim Husbandry 6:277–279

Stocker GC, Unwin GL (1986) Functioning of tropical plant communities. 3. Fire. In: Clifford HT, Specht RL (eds) Tropical plant communities. Univ Queensland, Brisbane, pp 91–103

Story R, Williams MAJ, Hooper ADL, O'Ferrall RE, McAlpine JR (1969) Lands of the Adelaide-Alligator area, Northern Territory. CSIRO Land Res Ser No 25

Taylor JA, Dunlop CR (1985) Plant communities of the wet-dry tropics of Australia: the Alligator Rivers region, Northern Territory. Proc Ecol Soc Australia 13:83–127

Taylor JA, Tulloch D (1985) Rainfall in the wet-dry tropics: extreme events at Darwin and similarities between years during the period 1870–1983 inclusive. Aust J Ecol 10:281–295

Tothill JC (1971) A review of fire in the management of native pasture with particular reference to north-eastern Australia. Tropic Grasslands 5:1–10

Tothill JC, Hacker JB (1973) The grasses of Southeast Queensland. Univ Queensland Press, Brisbane

Tothill JC, Mott JJ (1985) Australian savannas and their stability under grazing. Proc Ecol Soc Australia 13:317–322

Tunstall BR, Walker J, Gill AM (1976) Temperature distribution around synthetic trees during grassfires. For Sci 22:269–276

Unwin GL, Stocker GC, Sanderson KD (1985) Fire and the forest ecotone in the Herberton highland, north Queensland. Proc Ecol Soc Australia 13:215–224

Walker J (1981) Fuel dynamics in Australian vegetation. In: Gill AM, Groves RH, Noble IR (eds) Fire and the Australian biota. Aust Acad Sci, Canberra, pp 101–127

Webb LJ (1958) Cyclones as an ecological factor in tropical lowland rain forest, north Queensland. Aust J Bot 6:220–228

Winter JW, Atherton RG, Bell FC, Pahl LI (1987) An introduction to Australian rain forests. In: The rain forest legacy, Vol 1. Aust Gov Publ Serv, Canberra, pp 1–7

11 Fire Management in Southern Africa: Some Examples of Current Objectives, Practices, and Problems

B.W. VAN WILGEN[1], C.S. EVERSON[2], and W.S.W. TROLLOPE[3]

11.1 Introduction

This review covers the use of fire in the management of vegetation in selected areas of southern Africa. Major existing reviews of the ecological effects of fire in Africa have appeared in the recent past (Booysen and Tainton 1984; Komarek 1972; Phillips 1974; Trollope 1980b). It is not our intention to repeat what has been covered by earlier reviews, but rather to concentrate on the use of fire in the management of southern African ecosystems. Information on the objectives of management in areas of natural vegetation and the means by which they are achieved through the use of fire is widely scattered. The research results on which such policies are based are also scattered, and a synthesis would seem timely. The areas we have selected as examples cover the major vegetation types as well as the major forms of land use (agricultural, conservation, and catchment areas).

11.2 Major Vegetation Types of Southern Africa

White's (1983) description of the vegetation of Africa provides the broadest basis for discussion of the vegetation of the subcontinent. For the purposes of this review, we recognize five major vegetation types in southern Africa: fynbos (White's type 50, Cape shrubland), forest (White's type 19, Afromontane forest), grasslands (White's types 20, Highveld grassland-Afromontane forest transition, 57, Karoo grassy shrubland, and 58, Highveld grassland), moist savannas (White's types 16, East African coastal mosaic, and 29, undifferentiated woodland), and arid savannas (White's types 28 Mopane woodland and scrub woodland, and 44, Kalahari *Acacia* wooded grassland). Although desert and semi-desert vegetation covers a great deal of the subcontinent, these seldom support fire, and are therefore not considered here.

Fynbos comprises evergreen sclerophyllous heathlands and shrublands in which fine-leaved low shrubs and leafless tufted reed-like plants are typical.

[1] Jonkershoek Forestry Research Centre, Stellenbosch, 7600, South Africa
[2] Cathedral Peak Forestry Research Station, Winterton, 3340, South Africa
[3] Department of Agronomy, Faculty of Agriculture, University of Fort Hare, Alice, 5700, Ciskei, South Africa

Broadleaved sclerophyllous shrubs, mainly of the family Proteaceae, dominate in many places, but trees and evergreen succulent shrubs are rare, and grasses form an insignificant part of the biomass.

Grasslands occur where the vegetation is dominated by grasses and occasionally by other plants of grassy appearance (e.g., sedges) and in which woody plants are absent or rare (Huntley 1984). Grasslands occupy most of the higher eastern regions of the southern African continent, particularly those areas which are relatively cool, and with annual rainfall between 600 and 1200 mm (Tainton 1981).

Two main grassland types ("true" and "false") are recognized. True grasslands are those where succession does not proceed beyond the grassland climax. These occur as two ecologically distinct types, distinguished by the elevations at which they are found. At low elevations (extensive areas of the Orange Free State and southern Transvaal) the vegetation comprises tufted grass species dominated by the tribe Andropogoneae. This area is inherently stable but has been degraded considerably during the last century. It is, relative to natural grasslands elsewhere in the world, extremely productive (Tainton 1981). At higher elevations (the Lesotho highlands) the second type, comprising temperate C_3 grasses, occurs. The vegetation is characteristically short and seasonally unpalatable ("sour") and is dominated by the tribes Andropogoneae, Chlorideae, Eragosteae, Arundinelleae, and Paniceae.

False or fire climax grasslands occupy areas climatically suited to the advance of plant succession beyond the grassland stage into forest or savanna, but where grassland has persisted through the restraining effects of fire and grazing. These grasslands are often interspersed with patches of forest or bush clumps. They are of two basic types: (1) those which occur in areas of potential forest climax, where conditions for growth (particularly rainfall) are very good in summer, and (2) those occupying areas of potential savanna climax, where conditions are drier and less suitable for growth.

Savannas can be broadly regarded as including all ecosystems in which C_4 grasses potentially dominate the herbaceous stratum and where woody plants, usually fire-tolerant, vary in density from widely scattered individuals to a closed woodland, broken now and again by drainage-line grasslands. Rainfall occurs in the warmer, summer months with a dry period of between 2 and 8 months duration during which fire is a typical phenomenon at intervals varying from 1 to 50 years (Huntley 1982). In this chapter we will differentiate between arid and moist savannas.

The arid savannas are physiognomically diverse and include open sparse grassland with scattered shrubs and short trees to dense thorn thickets in which the herbaceous layer may be insignificant. The woody component is dominated in particular by the genera *Acacia* and *Commiphora*. Common grasses include *Eragrostis* and *Pancium* species with a large percentage of annuals in the driest areas. Rainfall is usually restricted to 5 or 6 months of the year and ranges typically from 250–650 mm per annum (Huntley 1984).

The moist savannas comprise a catena of tall closed deciduous woodland and open drainage-line grasslands. Common trees and shrubs include *Burkea*,

Terminalia, and *Ochna*, while the grass layer is dominated by tall perennial mesophytic species mainly of the tribe Andropogoneae. Rainfall occurs over 6 to 9 months and varies from 500–1100 mm per annum (Huntley 1984).

11.3 Management of Southern African Areas Using Fire

We have selected seven areas (one in fynbos, one in grasslands, and the rest in savannas) to serve as examples of how fire is used in natural vegetation to achieve stated objectives. For each area, we list the management objectives, and review the scientific basis for the current management regimes. We then discuss the current fire management, and highlight some problems. The distribution of the areas is shown in Fig. 1, and their salient features are listed in Table 1. Climatic features of the areas are presented in Fig. 2.

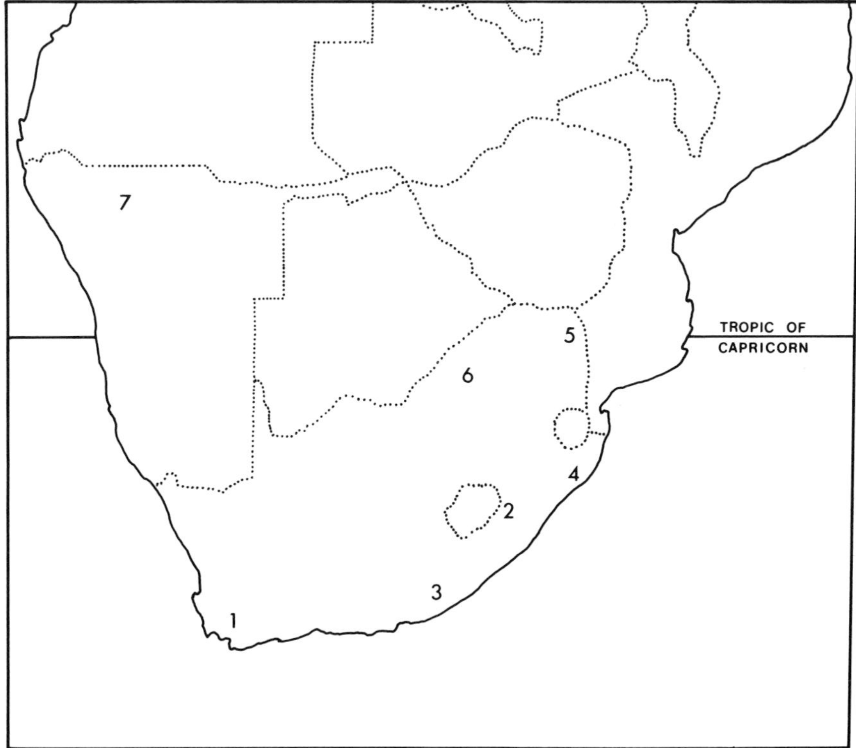

Fig. 1. Map of southern Africa showing the position of seven areas selected to illustrate contemporary fire management. These areas are: *1* Fynbos catchment areas; *2* Drakensberg catchments; *3* Eastern Cape; *4* Hluhluwe Game Reserve; *5* Kruger National Park; *6* Pilanesberg National Park; *7* Etosha National Park

Table 1. Salient features of seven areas in southern Africa, selected to illustrate the range of fire regimes and management objectives in the subcontinent

Area	Vegetation	Size (ha)
Mountain catchment areas in the western Cape Province	Fynbos	813 780
The Natal Drakensberg catchment areas	Montane grassland	243 000
The Hluhluwe-Umfolozi Game Reserves Complex	Moist savanna	96 453
The eastern Cape agricultural areas	Moist and arid savanna	*c* 3 000 000?
The Kruger National Park	Moist and arid savanna	1 948 528
The Pilanesberg National Park.	Arid savanna	55 000
The Etosha National Park	Arid savanna	2 200 000

11.4 Fynbos Catchments in the Western Cape Province

11.4.1 Aims of Management

Fynbos shrublands cover the mountain catchment areas in the western Cape Province of South Africa. They are managed for a variety of goals, the most important of which include maintaining sustained yields of high quality streamflow, nature conservation, fire hazard reduction, afforestation, grazing, and the provision of tourist and recreational opportunity. Many areas may be managed for more than one of these goals simultaneously. Management for these goals centers largely on the application of fire. Each of the major aims of management is discussed briefly below.

Sustained Streamflow and Water Quality

Streamflow can be increased by frequent burning in some cases. The actual magnitude of such increases depends largely on structural and other features of the vegetation. Fire causes an initial increase in streamflow by reducing evapotranspiration and eliminating water loss due to interception of rainfall by the vegetation canopy. The magnitude of this increase is dependent on pre-fire vegetation structure, or biomass. Such increases vary from 180 mm rainfall equivalent in the first year after fire in tall closed shrublands to almost nothing in very sparse vegetation (Bosch et al. 1986). Secondly, the rate of recovery of the vegetation to its pre-fire condition will determine the rate at which streamflow returns to pre-fire levels. The streamflow recovery rate can vary from 4 to 60 mm per year depending on vegetation recovery rate (Bosch et al. 1986). Under certain conditions, such as where the mature vegetation has a high biomass and

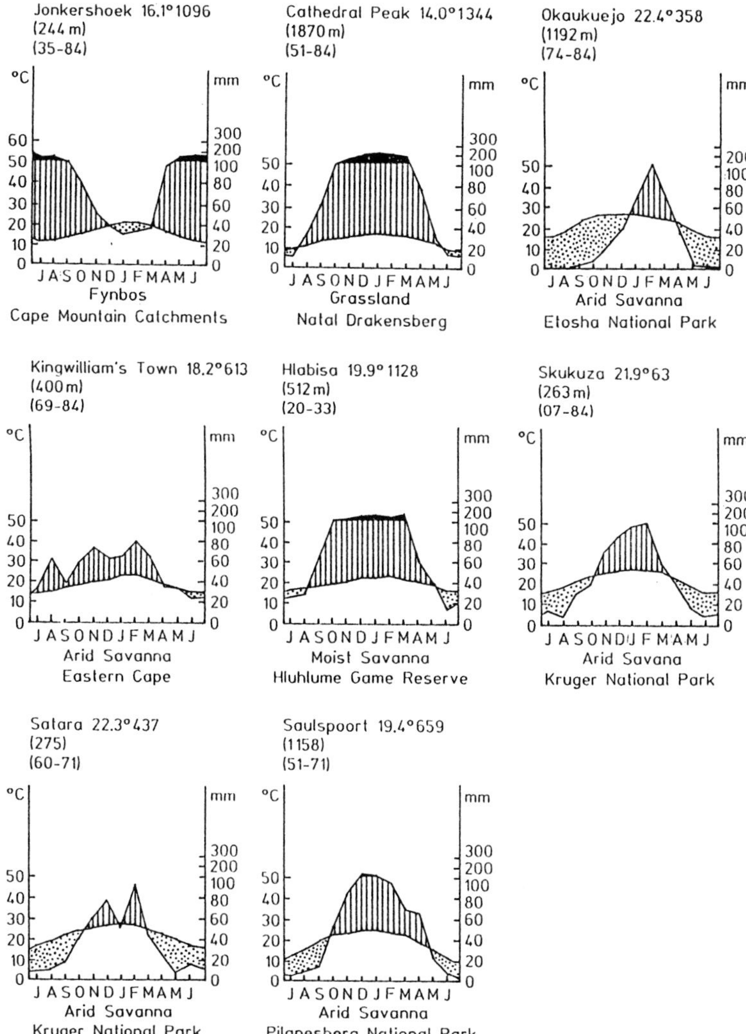

Fig. 2. Representative climate diagrams for seven areas in southern Africa, selected to illustrate contemporary fire management

the post-fire recovery is relatively slow, streamflow from fynbos catchments could be augmented by more frequent burning.

Nature Conservation

The fynbos mountain catchments form a major part of the Cape flora, one of six plant kingdoms of the world. The Cape flora consists of 8574 species of flowering plants, gymnosperms, and ferns. There are 989 genera of plants, of which 19.5% are endemic; 5847 (68.2% of the total) species are endemic. A detailed analysis of the flora is given by Bond and Goldblatt (1984). The degree of endemism in faunal components of the Cape Floral Kingdom is far less marked. The list of endemic fauna in the region nonetheless includes 20 reptiles, nine amphibians, eight mammals, and six birds (Bigalke 1979). The contribution of fynbos catchment areas to the conservation of this biotic diversity, in terms of the total number of species and endemism, is not known, but the catchment areas are undoubtedly important as conservation areas, as the vast majority of extant natural areas are in the mountains.

The aims of nature conservation in catchments are, among others, to maintain species diversity, and should be achieved by ensuring that the natural processes necessary for the maintenance of the full complement of species are allowed to operate. Fire is the major natural process which affects species in catchments. Secondly, unnatural or disruptive processes such as invasion by alien biota need to be combated. Management for nature conservation therefore consists largely of prescribed burning and various weed control programs.

Prescribed Burning. Prescribed burning is defined as the deliberate application of fire under specified (prescribed) conditions to the vegetation so as to achieve specific objectives safely. The frequency, season, and intensity at which fires occur in a given region is termed the fire regime. Burning operations are prescribed in terms of the fire regime, and are based on a knowledge of the effects of fire at different frequencies, in different seasons, and at different intensities on the vegetation. Fires in fynbos are prescribed to occur at 12–15-year intervals, preferably between November and April. They are usually intended to be of moderate intensity.

Clearing of Invasive Alien Plants. Invasion of catchments by exotic woody plants results in the displacement of indigenous biota, a reduction in species diversity, and an impact on the biological and physical processes necessary for the maintenance of the indigenous biota (for reviews, see Macdonald et al. 1986). Alien invasives therefore need to be cleared to achieve the aims of nature conservation.

In fynbos there are two broad groups of woody alien weeds. The most important group includes plants such as *Hakea sericea* and *Pinus pinaster*, that are killed by fire and release their seed stores on the death of the parent plant. These are felled and left for a period before burning. The fire kills seedlings, but regular follow-up is necessary to ensure that survivors are eliminated. In fynbos, follow-up weeding operations are generally carried out 2½ and 10 years after fire (Macdonald et al. 1985). The second group includes those species with continual seed production and release (for example the Australian *Acacia* species). Large quantities of hard-coated and persistent seeds accumulate in the soil. Here, control is more problematic. Felling and burning results in abundant seedlings which need to be cleared by hand-pulling, a time-consuming and labor-intensive operation.

Reduction of Fire Hazard

Increases in fuel loads in natural vegetation are responsible for increases in fire hazard, which in turn pose a threat to neighboring crops and property. Reduction of such hazard is usually achieved by regular burning. The effects of increasing post-fire age on biomass, fuel bed depth, and simulated fire behavior are shown in Fig. 3, using biomass data from Kruger (1977), van Wilgen (1982), van Wilgen et al. (1985) and from interpolation. Simulations of fire behavior were done using Rothermel's (1972) fire model, assuming fuel moistures of 10 and 100% for dead and live fuel respectively, and a 10 km h^{-1} wind. Van Wilgen's (1984) fuel model, and modifications of this based on Fig. 3, were used as inputs. Increasing post-fire age clearly increases fire behavior parameters. Reducing fire frequency from longer periods to once every 12 to 15 years will reduce maximum fuel loads to around 1200 g m^{-2}, thus considerably reducing the severity and increasing the controlability of fires. In addition, a mosaic of different post-fire ages of vegetation created by regular burning would break up continuous fuel beds.

Invasion can also increase fire hazard, as biomass and fuel loads are increased when alien trees or shrubs replace indigenous vegetation. For example, invasion of fynbos by the alien shrubs *Acacia saligna* and *Hakea sericea* increased fuel loads (all particles > 6 mm diameter) from 1200 g m^{-2} to 1800 and 2000 g m^{-2} respectively (Van Wilgen and Richardson 1985a). Alien invasions need to be controlled for this reason as well.

11.4.2 Fire Frequency

Fynbos communities accumulate enough fuel to readily sustain a running fire under average summer conditions once they have reached a post-fire age of 4 years, and may burn at 3 years under severe fire danger conditions (Kruger 1977). Thus fire cycles of less than 4 years are not possible (see Fig. 3). On the other hand, most attempts to exclude fire from fynbos catchments for longer than 45 years have met with failure, indicating that exclusion of fire for longer than this should, for practical purposes, not be considered feasible. For example, 50% of the Cederberg catchment area in the western Cape was burnt within 12 years after fire, and 95% within 30 years (Fig. 4).

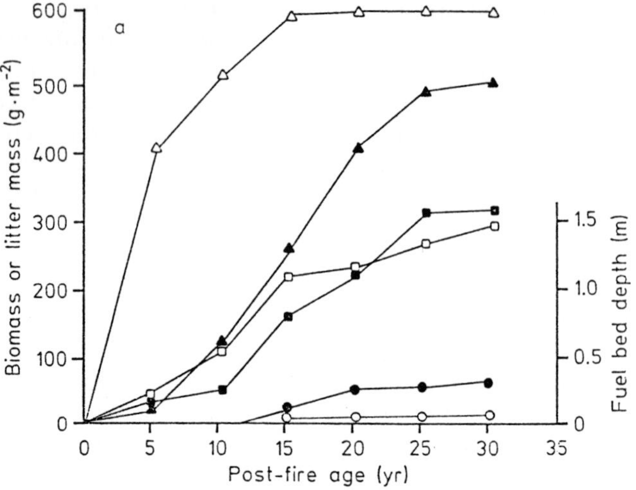

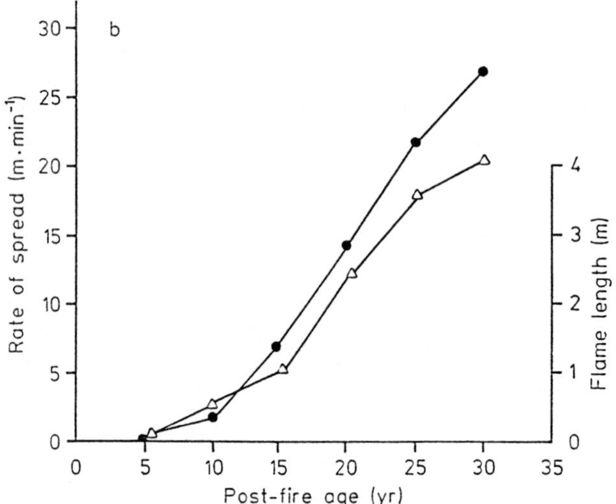

Fig. 3. a Post-fire development of biomass, litter mass, and mean vegetation height (fuel bed depth) in fynbos, used as inputs to simulate fire behavior. △ = live herbaceous fuel; ▲ = live woody fuel; ■ = litter fuels < 6 mm diameter; ● = litter fuels 6–25 mm diameter; ○ = litter fuels > 25 mm diameter; and □ = fuel bed depth. **b** Predictions of fire behavior under similar conditions (see text) at different post-fire stages in fynbos vegetation. ● = rate of fire spread; and △ = flame length

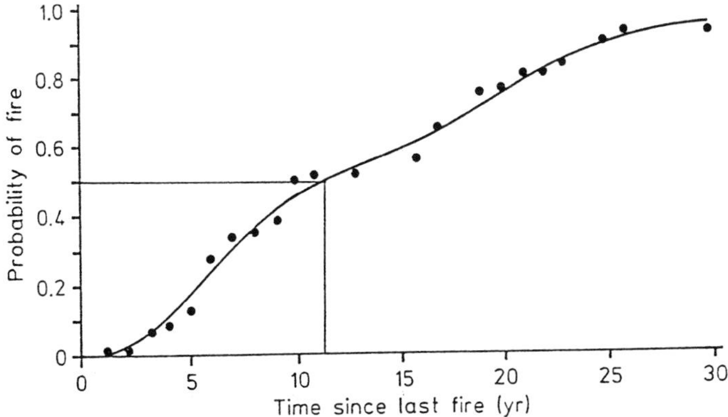

Fig. 4. The probability of a fire occurring in fynbos, expressed as a function of time since the previous fire. Data are from the Cederberg catchment area in the western Cape

Ecological considerations require that fynbos be burnt within narrower limits of frequency than the limits set by fuel dynamics alone, if the aim of maintaining species diversity is to be met. A useful way to determine the response of plant communities to different fire frequencies could be to use the vital attributes scheme proposed by Noble and Slatyer (1980). Although data for most fynbos species are lacking, several indicator species can be examined in this context.

Serotinous Proteaceae, dominant shrubs in many fynbos areas, fall into Noble and Slatyer's CI class of species. These are killed by fire and regenerate from seed stored in the canopy. CI species are "intolerant" and recruitment is confined to the period immediately after fire and is negligible at other times. The seed is short-lived after release and no effective seed reserves persist after the death of the parent and cone opening. Populations of CI species may become locally extinct either from disturbance before they are reproductively mature (no seed reserves) or as a result of intervals between fires exceeding the life span of the individuals in the even-aged population. High frequency fires are a well-known cause of local extinction of slow-maturing Proteaceae populations (Kruger 1977; van Wilgen and Kruger 1981). Old stands of fynbos are rare, but Bond (1980) has reported poor regeneration of CI Proteaceae after long (40–50-year) intervals between fires, interpreted as the result of reduction in seed reserves with senescence of the parents. An analysis of the flora would thus show, for species of *Protea* at least, that fire intervals of between 8 and 30 years are needed to ensure survival.

Populations of shrubs of the genus *Mimetes* (Proteaceae) are found in many mountain fynbos areas. Some data for *Mimetes splendidus* (a GI species) are available (A.J. Lamb pers. commun.). GI species build up considerable seed stores, with longevities exceeding the life span of adult plants (> 12 years), in the soil. The species are intolerant and establish only after fire, at which time the seed pool is exhausted. Data on the longevity of the soil-stored seed pool would be necessary to determine maximum periods between fires. Minimum periods are set by the age at which these plants attain maturity (about 5–6 years).

Most species in fynbos are able to sprout (Noble and Slatyer's groups, V, U, and W). Such species are far less affected by fire frequency, as they do not have to reach maturity to survive fires; they are also generally longer-lived than re-seeding plants. Thus re-seeding plants are far more sensitive to changes in fire frequency and such plants therefore dictate the fire frequencies required to maintain species diversity.

The only data set available to test the effects of different fire frequencies on fynbos is a list of 174 species from the Jakkalsrivier catchment in the western Cape (Kruger 1987). Fire frequencies of 5 years were found to eliminate only four species from this list. Fire frequencies of 10 and 15 years maintained all of the species. Progressively more species are eliminated as the mean fire frequency increases beyond 20 years (Fig. 5). Nonetheless, fire frequencies of 25 years eliminate only three species, but frequencies of 50 years eliminate 48 species. This analysis indicates that fire frequencies of between 10 and 15 years will be necessary to maintain as many species as possible in the fynbos. This result must be considered with the drawbacks of Noble and Slatyer's scheme in mind. These authors emphasize the recurrence time of disturbance to predict the composition of the vegetation. However, fire frequency is only one element of the fire regime. Fire season, for example, can radically alter the response of CI Proteaceae (Bond et al. 1984). For the GI species *Mimetes splendidus*, re-establishment after fire is very good following high-intensity fires, and almost nonexistent following low-intensity fires (A.J. Lamb personal commun.). In order to conserve elements of the biota, managers are therefore constrained by all three elements of the fire regime: frequency, season, and intensity.

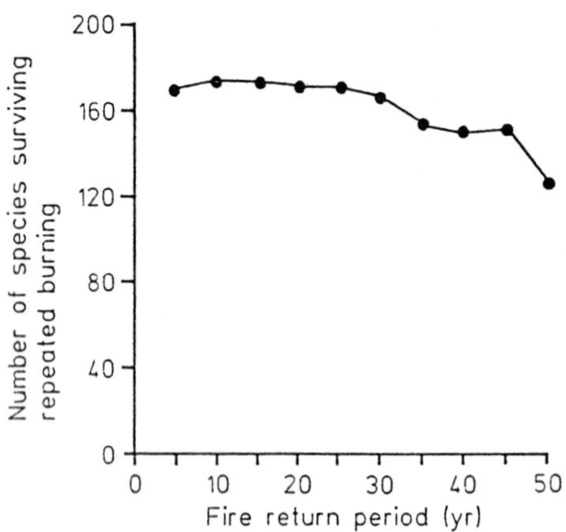

Fig. 5. The number of species, from a list of 174 species (Kruger 1987), surviving repeated burning expressed as a function of the fire return period. The curve was obtained using Noble and Slatyer's vital attributes scheme. (Noble and Slatyer 1980)

11.4.3 Fire Season

As far as the biological effects of fire season is concerned, fires in winter and spring are undesirable in fynbos. The maximum flowering activity occurs in late winter and spring (Kruger 1981), which implies that the maximum seed loads will be available in late summer or early autumn. Shrubs and trees that are killed by fire show maximum seedling recruitment after late summer and early autumn fires (Bond et al. 1984; Manders 1985; van Wilgen and Viviers 1985). Regular prescribed burning outside the late summer-early autumn period could result in the local extinction of species (Bond et al. 1984), and is therefore not usually applied.

The time of year at which fires are likely or able to occur is determined by climatic factors and by seasonal variations in fuel properties. Seasonal curing is not a feature of fynbos vegetation (van Wilgen 1984) and fire season therefore depends largely on climatic factors. Due to the mediterranean climate over much of the fynbos biome, fires occur mainly in the summer, but can occur in all months. For optimum ecological effects, fires should occur between late November and mid-April. Burning in the summer months (November to February) is effectively prevented due to hot and dry weather conditions at these times (Fig. 6). The risk of prescribed burns becoming uncontrollable is simply too great to allow such burns to take place. Prescribed burning is therefore only feasible in March and April. However, only about 12 days (on average) in March and April have weather suitable for burning in the western Cape Province (van Wilgen and Richardson 1985b). Given the large areas to be managed, prescribed burning programs need to be carefully planned if they are to be completed. Table 2 gives data on the areas that are managed by prescribed burning. In order to complete the 52 burning operations required each year in the estimated 12 available days, 4.3 burning operations must be carried out per suitable day in the western Cape alone. Managers will have to ensure that every

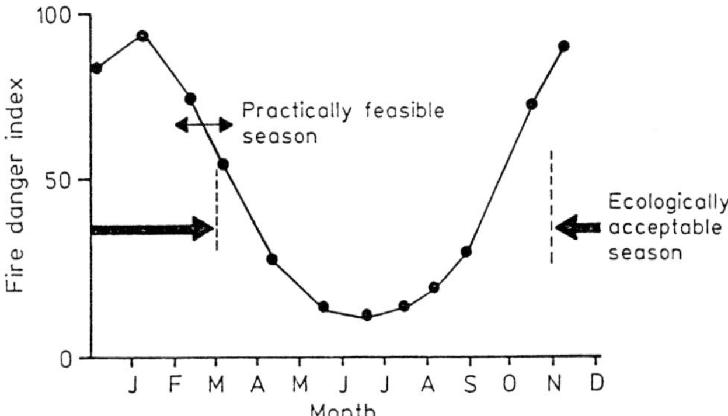

Fig. 6. The annual cycle of fire danger at a typical inland mountain fynbos site in the Western Cape. The *large arrows* show the ecologically acceptable burning season. Due to dangerous conditions for most of this time, the practically feasible burning season (*small arrows*) is much shorter

Table 2. Areas of mountain catchments managed by prescribed burning in two forestry regions in South Africa

Region	Total area (ha)	Mean fire cycle (yr)	Mean area burnt per fire (ha)	Number of fires per year
Western Cape	813 780	15	1031	52
Drakensberg	243 300	3[a]	500	162

[a] A mean based on a 2-yr cycle in seasonal grasslands, 3-yr cycle in *Protea* woodlands and a 4-yr cycle in subalpine and alpine heath communities. (Bainbridge 1987).

Table 3. The actual number of blocks burnt per year in the nine catchment areas in the Western Cape forestry region from 1981 to 1988. Departmental records

Catchment area	Number of blocks burnt							
	1981	1982	1983	1984	1985	1986	1987	1988
Cederberg	2	9	0	0	0	0	1	0
Groot Winterhoek	0	2	4	0	1	0	2	0
Kouebokkeveld	6	5	0	0	1	3	0	0
Matroosberg	1	2	1	0	1	0	0	0
Hawequas	10	9	5	4	3	1	3	3
Riviersonderend	6	3	3	2	1	2	2	1
Hottentots-Holland	5	2	6	4	1	0	1	5
Langeberg-West	6	5	5	2	4	2	1	0
Langeberg-East	3	4	1	1	1	1	0	2
Total	39	41	25	13	13	9	10	11

suitable day is utilized if the planned burns are to be completed. Data on the actual number of burns completed in the Western Cape over the past 6 years are given in Table 3. The target of about 50 burning operations per year has never been realized.

11.4.4 The Control of Alien Woody Weeds in Fynbos Catchments

Areas that have not been cleared of woody weeds can hold up the burning schedules. Accidental fires in untreated areas may also necessitate additional follow-up operations. Taking *Hakea sericea* as an example, almost 20% of the area of catchments in the western Cape (ca. 156,000 ha) is infested. Assuming that an average area of 1000 ha is cleared as a unit, and that a fire frequency of 15 years is applied, ten blocks of 1000 ha have to be cleared annually. In any given year, another ten more will come up for follow-up clearing of seedlings at 2 years and another ten for follow-up at 10 years. Thus almost 30,000 ha must be worked over annually, for this species alone.

Clearing operations cause changes to the fuel properties of the vegetation. Dead plant material adds to the fuel load and such loads are concentrated close to the ground. This poses new constraints on managers. Unnaturally high fuel loads lead to detrimental effects on the vegetation following fire, and these have to be prevented. For example, where fuel loads are high following clearing operations, it may be necessary to burn under moderate conditions to reduce impact of higher-intensity fires (Richardson and Van Wilgen 1986). This further restricts the number of days available for burning, or forces managers to burn in spring to complete programs.

In order to prevent wasted effort, control operations, once started, must be regularly followed up. Experience has shown that single, isolated clearing operations are simply a waste of time (Macdonald 1989). Secondly, once areas have been cleared, re-infestation from nearby, uncleared areas poses a problem. This is particularly the case for species which have good dispersal properties, such as the wind-dispersed seeds of *Hakea* and *Pinus* species. The effective control of a species on a landscape basis is therefore necessary, and can probably only be achieved through co-operation by neighboring landowners, in combination with biological and mechanical control.

11.4.5 Wildfires as a Complicating Factor in Prescribed Burning

Wildfires (unplanned fires) are frequent in the Cape mountains (Table 4). About 74% of the area burnt in wildfires in catchments over the past 8 years, which upset buring programs intended to produce mosaics of vegetation of different ages. Wildfires also result in many areas being burnt too frequently, and they burn areas where the necessary pre-burn weed control has not been carried out. An improved ability to predict the likelihood of wildfire occurrence and thus to be in a position to prevent them, would possibly help managers. Fire danger rating

Table 4. Areas burnt in planned and unplanned fires in the Western Cape (1.4.1981–31.3.1989) and Natal Drakensberg (1.4.1980–31.3.1984). (Departmental records)

Catchment area	Total area burnt (ha.) Unplanned fires	Planned fires
Cederberg	74 655	103 22
Groot Winterhoek	28 085	3 110
Kouebokkeveld	17 023	13 160
Matroosberg	71 075	2 493
Hawequas	75 642	37 329
Riviersonderend	18 970	12 827
Hottentots-Holland	35 670	20 747
Langeberg-West	38 623	19 090
Langeberg-East	22 158	14 799
Total (Western Cape)	381 901	133 877
Cathedral Peak (Natal)	4 152	23 925

offers an opportunity to achieve this (Van Wilgen and Burgan 1984; Everson T.M. et al. 1988b).

11.5 Grassland Catchments in the Natal Drakensberg

11.5.1 Aims of Management

Grassland catchment areas are managed primarily to conserve water resources, for nature conservation, and to reduce fire hazard. Grazing and afforestation have been specifically excluded as objectives. Fire is the only practical means manipulating large areas of grassland, and is consequently widely used to achieve the major aims.

The Natal Drakensberg is an important conservation area. It constitutes the largest single natural area in Natal. There are about 1390 plant species in the southern Natal Drakensberg (Hillard and Burtt 1987), of which 394 (29.5%) are endemic. A further 317 species are restricted to montane and submontane areas in southern Africa, indicating that 53% of the southern Drakensberg flora is endemic to montane and submontane areas south of the Limpopo River. Hillard and Burtt (1987) give a detailed analysis of the flora of the region. The vertebrate fauna of the Drakensberg catchments includes 232 species of birds, 49 mammals, 44 reptiles, and 24 amphibia (Little and Bainbridge 1985; Bainbridge et al. 1986). Although none of these is endemic to the reserves (except possibly the Drakensberg dwarf chameleon *Bradypodion dracomontanum*), many are rare or have restricted distributions, and the Drakensberg catchments are therefore important to their conservation.

Approximately 3520 km^2 of grassland are conserved in southern Africa, or about 1% of the biome (Rutherford and Westfall 1986). This is an alarmingly small area considering the many grassland communities recognized in this area. In the Transvaal, for example, only two of the seven pure grassland and five false grassland types are regarded as adequately conserved (Clinning 1986). The largest conserved areas of grassland fall within mountain catchment areas which are considered marginal or unsuitable for most forms of agriculture. Many of the large rivers, on whose water both industry and agriculture depend, arise in these catchments. Their stability is therefore a primary objective.

11.5.2 Fire Regime

Grassland areas are prone to fire and have many plant and animal species which have obviously adapted to fire (Bews 1925; Bayer 1955; Levyns 1966; Gordon-Gray and Wright 1969; Mentis and Bigalke 1979). Natural fires are caused by lightning, and grassland areas coincide with regions where the ground lightning flash density is greater than 4 km^{-2} yr^{-1} (Edwards 1984). In addition to natural fires, man has increased the frequency of fires. Man-induced fires are of considerable antiquity and there is general agreement that primitive man was a significant fire agent throughout the Holocene, when the plants and animals in these systems were adapting to contemporary climatic conditions (Hall 1984). Fire is therefore regarded as a central component in the grasslands, and not as an extraneous factor.

Although it is uncertain to what extent early humans modified the natural fire regime, there is evidence that by the early 18th century regular burning was practiced by the Black peoples to stimulate a green flush of new growth to improve grazing and to aid hunting. More recently, European settlers of the early 19th century burned extensively from mid-August to early spring to promote early growth for grazing by sheep and cattle.

The fire regimes for the grassland areas are determined principally by climate and available fuel. In general, the grassland fire regime is one of regular fires occurring chiefly during late autumn, winter, and spring (Edwards 1984).

11.5.3 Fire Frequency

Regular burning is necessary to reduce fire hazard by keeping fuel loads low in grassland catchments. Figure 7 depicts a post-fire fuel dynamics model for grasslands, similar to that presented for fynbos in Fig. 3, using data from Everson (1985). Conditions used in the simulations are the same as for Fig. 3. The fuel model, of Everson T.M. et al. (1988b), and modifications of this based on Fig. 7, were used as inputs. Increases in post-fire age cause increases in simulated fire behavior, particularly between the first and second years post-fire. Thereafter, increases are small.

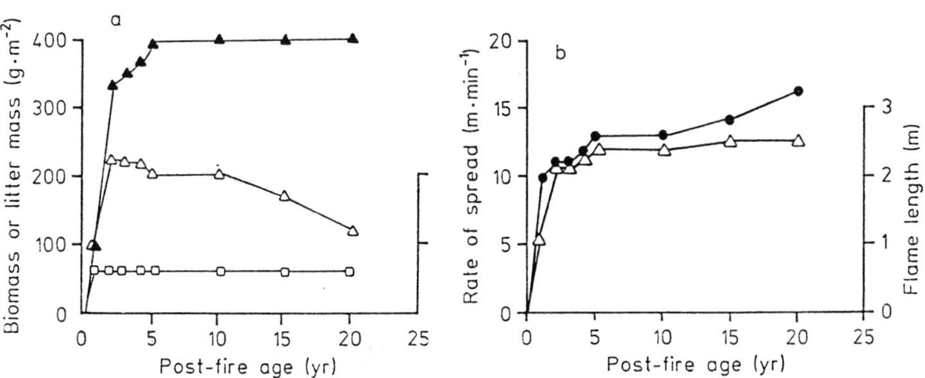

Fig. 7. a Post-fire development of biomass, litter mass, and mean vegetation height (fuel bed depth) in montane grassland, used as inputs to simulate fire behaviour. △ = live herbaceous fuel; ▲ = litter fuels $<$ 6 mm in diameter; and □ = fuel bed depth. **b** Predictions of fire behavior under similar conditions (see text) at different post-fire stages in fynbos vegetation. ● = rate of fire spread; and △ = flame length

Rapid fuel accumulation in montane grasslands makes annual burning possible (Fig. 7). On the other hand, it is unlikely that fire frequencies in excess of 15–20 years would be practically feasible. The absence of fire results in a decrease in the vigor of plants (Everson 1985). This is due to a decrease in light intensity at the base of the plant through the continual accumulation of dead material, which in turn results in a reduced rate of tillering (Everson C.S. et al. 1988). Regular burning is therefore considered essential for stimulating vegetative reproduction and maintaining grassland condition.

The condition of grasslands is often expressed (as a percentage) in terms of species composition by comparing species composition on a site to that on a benchmark site considered to have a desirable composition (Foran et al. 1978). Vegetation condition in grassland protected from fire for 18 years was significantly lower (35%) when compared with a biennial spring burn treatment (57%) (Everson and Tainton 1984). In this study the benchmark site had received 30 years of biennial spring burning. Moribund swards also have lower levels of alpha diversity than grass swards that are regularly burnt. For example, at Cathedral Peak a total of 58 species were recorded on 25 × 25 m plots in regularly burnt grassland, compared to only 48 species in an adjacent area protected from fire for 20 years. The maintenance of woody species in the ecosystem may well be adversely affected by frequent burning, although quantitative data to test this, for example in terms of Noble and Slatyer's (1980) vital attributes, are lacking. Granger (1977) noted major seral changes in an area protected from fire for 20 years. In areas previously dominated by *Themeda triandra* there was invasion by woody species such as *Philippia evansii*, *Leucosidea serecia*, *Widdringtonia nodiflora*, and *Buddleja salviifolia*. Following protection from fire for 40 years, grass species were replaced by shrub and forest species (Westfall et al. 1983), and higher levels of alpha diversity were recorded here than in grassland. Long inter-fire intervals will therefore increase species diversity (a major aim of nature conservation), but under such regimes grasses become moribund and lose reproductive vigor. In addition, the open nature of the vegetation would change. In general, there is little knowledge of the effects of fire on woody species in grassland, and consequently little on which to base management prescriptions. Woody species in montane grassland are typically confined to refuge sites where fires are infrequent, absent, or burn at low intensity (Tainton and Mentis 1984). Forest patches in the Drakensberg are sensitive to fire and concern has been expressed that they may be progressively eliminated by frequent fires (Everard 1986). Nevertheless, species diversity at a landscape level (gamma diversity) will not be reduced, as fire-sensitive species survive in forest patches and refuge sites. Protection from fire also has implications for the control of alien invasives, since it enables woody alien weeds to establish. For example, a site at the Cathedral Peak State Forest was protected from fire to encourage the spread of indigenous forest. Here *Pinus patula* invaded at densities of up to 160 stems ha^{-1} after 6 years. Adjacent, regularly burnt areas were free of pines, despite a similar proximity to seed sources.

The proportional species composition of grasses varies according to frequency of burning. *Themeda triandra*, *Heteropogon contortus*, and *Tra-*

chypogon spicatus increased in the sward with regular burning, while *Tristachya leucothrix*, *Alloteropsis semialata*, and *Harpechloa falx* decreased (Everson 1985). The reasons for such differential responses to fire were elucidated in demographic studies (Everson 1985). The results revealed some interesting generalities in the population dynamics of the perennial grasses studied (*T. triandra*, *H. contortus*, *T. spicatus*, *T. leucthrix*, and *H. falx*). These include:

1. In all species studied the majority of tillers remained vegetative until death, suggesting that flowering is of minimal importance for their propagation.
2. Shoot apices remained close to the soil surface, enabling all species to survive frequent defoliation.
3. In all examples described, recruitment of secondary tillers was stimulated by regular burning. Annual burning produced on average more secondary tillers per primary tiller than biennial spring burning.
4. All species exhibited smooth survivorship curves, suggesting that dramatic fluctuations in climate and severe defoliation, as by fire, have little impact on mortality.
5. The differential responses of these five grass species to burning was best explained by the combined effects of their different reproductive capacities and mortality rates.
6. A biennial burning regime would maintain the most important grass species at present levels of abundance in the Natal Drakensberg.

Since biennial burning maintains the dominant grass species in high abundance, increases vegetation vigor and reduces fuel loads and concomitant fire hazard, it is viewed as the most suitable burning regime for montane grasslands. Annual burning would also achieve this, but would not be cost-effective.

11.5.4 Fire Season

The climate of the Natal Drakensberg is characterized by dry winters and wet summers. Grasses grow actively in summer, resulting in high moisture contents in the fuel bed. Once the aboveground parts of the grasses are killed by the first winter frosts, the herbage gradually dies. This process, termed curing, is of prime importance in determining fire behaviour (T.M. Everson et al. 1988a).

Burning prescriptions prior to 1981 required biennial burns in spring after the first rains (August to September) or by the end of November in the event of no rain. The ideal burning period is late winter (August to September) while the grasses are still dormant. There was therefore a relatively short period during which managers could implement burns.

Burning once growth has been initiated in the middle of September can have detrimental effects on the grass sward (Everson et al. 1985). For example, summer (December) burning killed 92% of newly emerged tagged tillers of *Themeda triandra*, and caused a marked reduction in growth rate during subsequent recovery. Fires caused by lightning do occur in montane grasslands in summer, but the area burnt is never very large. The only sensible option for managers of large areas is therefore to burn earlier in the season, during early winter. Although fire hazard is high at this time, the grasses are 95% cured and opportunities for safe burns do occur after light rain or snow (T.M. Everson et al. 1988a). In addition, there was a difference between vegetation which had received 30 years of either annual winter or biennial spring burning (Everson and Tainton 1984) in the rate of canopy recovery over the soil surface (Everson et al. 1989). Grasslands burnt in June (winter) green up by October, forming a protective canopy over the soil before the heavy summer rains. The soil is, however, exposed to potential wind erosion before growth starts in September. In contrast, grasslands burnt in October (spring) only recover by the end of November, exposing the soil to the storms of October and November. Differences in season of burn therefore result in exposure to erosive forces at different times of the year. The optimum time for burning is therefore mid-August, when the effects of exposure to rainfall erosion are minimized. However, opportunities for controlled burns during August and early September are severely limited by the hot, dry, and windy conditions common at this time of the year. This is the time of peak fire danger, and an average of only 8 days are suitable for burning at this time (Everson T.M. et al. 1988b). Current burning prescriptions therefore allow burning between May after the first frosts, and mid-September. Previously in dry years (when managers were unable to complete their burning programs), the burning period was extended later into spring. The rationale for this was that growth was delayed in years of drought. However, the initiation of growth is probably not influenced by rainfall, and significant growth often takes place by the end of September (Everson and Everson 1987). To maintain the vegetation in its present condition, burning should not be permitted from October onwards. Extension of the burning season during the period of dormancy of the grass avoids the detrimental effects of burning in the early growing season and has improved the efficiency of the fire management program. The burning season is currently rotated between three periods:

1. May (early winter),
2. June-July (winter), and
3. August-mid-September (early spring).

This seasonal rotation is aimed at encouraging ecological resilience by ensuring that a management compartment is burnt in the same period (season) once every 6 years (Everson 1985). In this way no species is favored at the expense of others. In addition, there is a phased reduction in fuel loads during the dry season, minimizing the risk of wildfire.

11.5.5 Prescribed Burning

Catchments are divided into compartments of about 500 ha in which the appropriate burning prescriptions are applied. Compartments are separated by permanent boundaries along paths and streams or by other landscape features.

Where no natural features occur, fire-breaks are used. Prescribed burns are applied when the prevailing weather conditions are suitable for achieving the aims of the particular burn. This is generally when wind speeds and temperatures are low and atmospheric humidity is high. Since these three atmospheric variables influence the rate of spread of a fire independently, it is difficult to specify their limits for prescribed burning. Fire danger rating models are therefore used to integrate all the atmospheric and plant variables (such as fuel loads and fuel moisture content) into a single burning index (Everson T.M. et al. 1988b). Controled burning can then be carried out safely by applying burns when the burning index is within specified limits. Table 5 shows that most planned burns are carried out, in contrast to fynbos catchment areas. This can be attributed to a longer acceptable burning season (Fig. 8), lower fuel loads, and the lack of an exotic woody weed problem.

Table 5. The actual number of blocks burnt per year in State Forestry areas in the Natal Drakensberg between 1981 and 1986. (Departmental records)

State Forest	Number of blocks burnt				Mean per year
	1983	1984	1985	1986	
Cathedral Peak	20	20	22	33	23.75
Monk's Cowl	13	17	11	13	13.50
Mkhomazi	18	23	25	21	21.75
Garden Castle	33	29	39	28	32.25
Highmoor	13	16	11	18	14.50
Cobham	31	41	30	35	34.25
Total	128	146	138	148	140.00

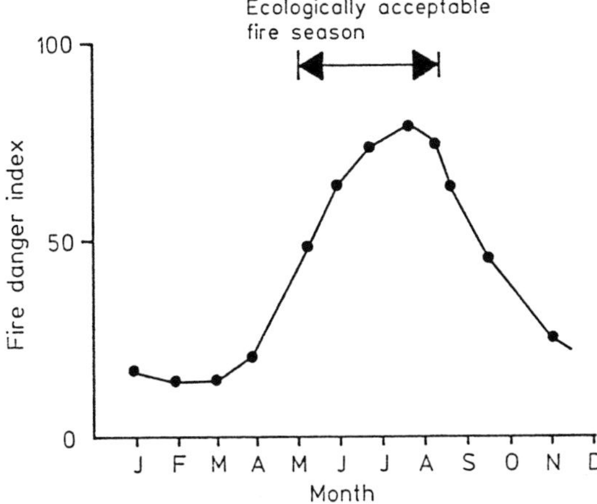

Fig. 8. The annual cycle of fire danger at a typical Natal Drakensberg site. The ecologically acceptable burning season runs from early winter to early spring. Burning operations are feasible for the whole of this period

At present the burning prescriptions defined for the subtropical grasslands (which comprise about two thirds of the grasslands in the Natal Drakensberg) are being used for temperate evergreen (*Festuca costata*) grasslands (which cover the remaining third of the grasslands). Nothing is known of the response of these areas to different fire regimes. Unlike seasonal grasses, temperate grasses are not dormant in early winter (Priday 1989). Early winter burns on low radiation slopes should therefore be discontinued until the implications of these burns on the survival of temperate species have been determined.

11.6 Fire in Savannas: Basic Principles

11.6.1 Natural and Modified Fire Regimes

Fires in savannas burn in the grass layers, and the dynamics of grass biomass that provide the fuel for these fires will determine fire regimes. Grass biomass dynamics are dependent on three major factors: rainfall, grazing pressure, and fire. In many savanna areas, annual rainfall is often highly variable and unpredictable, thus complicating management generalizations.

There is almost no information available on ancient fire regimes. However, the season, frequency, and intensity of natural fires in the savanna were undoubtedly determined by fuel load, fuel moisture, and the incidence of lightning. Savanna in southern Africa is largely confined to the summer rainfall region with a dry season from ca. May to October, at the end of which the herbaceous grass fuel layer is very dry. Africa has a unique fire climate that accentuates the probability of lightning fires, with dry lightning storms that ignite many fires occurring at the end of the dry period (Komarek 1971).

Natural fires occur most frequently at the end of the dry season and just prior to the first spring rains. This is illustrated by data from the Kruger (Gertenbach 1988) and Etosha (Siegfried 1981) National Parks in Fig. 9.

Rainfall is the most important factor affecting grass fuel accumulation, which in turn determines fire frequency. Thus fires are more frequent in moist savanna than in arid savanna due to higher rainfall and the unacceptability of mature grass to herbivores. The natural fire frequency in moist savannas must have ranged between annual and biennial, depending on rainfall and utilization by wild ungulates. For example, in the open savannas (720 mm rainfall yr^{-1}) of Natal, complete protection of the grass sward for 3 years caused it to become moribund and die out (Scott 1971). Conversely, annual and biennial burning maintained a vigorous grass sward with a far greater basal cover. Furthermore, in the Kruger National Park, the most desirable burning frequency under grazing conditions in moist savanna is taken as annual or biennial, depending upon grazing and grass fuel conditions (Gertenbach 1979).

In the arid savannas the frequency of fire is far lower because the rainfall is both less and highly erratic and the grass sward remains acceptable to grazing

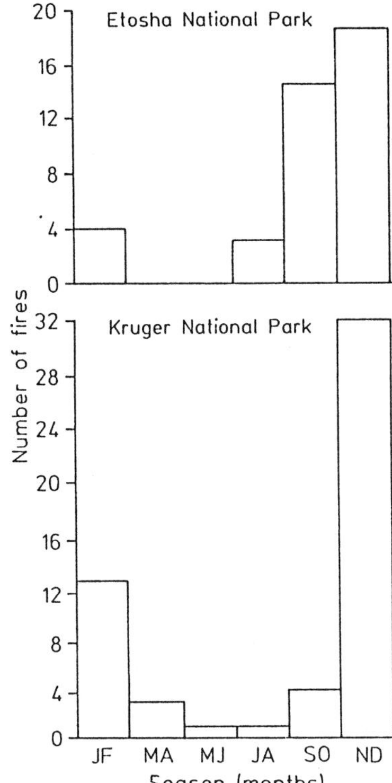

Fig. 9. Frequency of lightning fires in two national parks in southern Africa. Date are from Gertenbach (1988) and Siegfried (1981)

animals even when mature, thus reducing the rate of fuel accumulation. The frequency of fires is determined by the occurrence of exceptionally wet seasons. For example, Kennan (1971) abandoned a burning trial in southwestern Zimbabwe, because it was impossible to apply burning treatments in a regular sequence. The area has very low (450 mm yr^{-1}) and erratic rainfall which causes very marked fluctuations in fuel loads.

Fire intensities in past fires were probably greater than those of present fires. Widely accepted botanical evidence (Acocks 1975) shows that the grass component of the vegetation in South Africa has been drastically altered and reduced in all vegetation types, including savanna, since the advent of settled agricultural practices. A drastic reversal in grassland succession to a pioneer stage can have marked effects on fuel loads. For example, the phytomass of grass produced by pioneer vegetation dominated by species like *Aristida congesta* was only 13% of that produced by climax vegetation dominated by *Themeda triandra* (Danckwerts 1980). Since grass biomass is an important determinant of fire intensity, the fires of the natural fire regime were probably far more intense than those of today. Also, wildfires often are more intense than prescribed burns as presently applied.

11.6.2 Grass/Bush Dynamics

A major problem facing managers of both agricultural and conservation areas in savanna regions is the manipulation of the balance between grass and bush components. Bush encroachment is universally regarded as a problem by managers, as replacement of grass by woody species reduces the capacity for grazing animals. Bush encroachment is encouraged by overgrazing, protection from fire, and lowering of fire intensity. The balance between grass and bush components in African savannas is also influenced by interactions between herbivory and fire. Fire influences the grass/bush balance in arid savannas by maintaining the bush at a height and in a state highly acceptable to browsing animals (Trollope 1974). Under the natural conditions of the past, fierce fires destroyed the aerial growth of encroaching bush species, thus providing coppice growth at an acceptable height and in a highly acceptable state for browsing by wild ungulates. The degree to which fire and browsing influences the balance of grass and bush is determined by fire intensity, the acceptability of the different coppicing bush species and the browsing intensity. This was tested in a *Themeda triandra*-dominated grassland moderately invaded by *A. karroo* and other bush species (1625 plants ha^{-1}). A single intense head fire resulted in an 80.8% topkill of trees and shrubs. Plots in this area were subjected to continuous browsing with goats (1 goat ha^{-1}), no browsing (control), and annual spring burning. After 5 years the browsing treatment reduced the original bush density by 90% and the burning treatment reduced bush density by 32%. On the control plot the bush recovered completely. In subsequent years, the bush density on the burnt plot has steadily increased despite the annual fires, albeit in the form of dwarf coppice bushes, whereas the browsing on the third plot has prevented any reestablishment of the bush (Trollope 1980a). Another example of the interactions between fire and herbivory is provided in Nylsvley in the northern Transvaal, where succession would lead to a closed woodland were it not for regular fires and tree mortality caused by porcupines (Yeaton 1988).

11.6.3 The Importance of Fire Intensity

There has been a growing realization that fire intensity must be considered when managing to achieve certain objectives in savanna, such as the control of woody species. The effect of fire intensity on the topkill of stems and branches of bush of different heights is illustrated by research in the arid savannas of the Eastern Cape (Trollope and Tainton 1986) and the Kruger National Park (Trollope et al. 1988). The results show a greater topkill (Fig. 10) and reduction in height of bush (Fig. 11) with increasing fire intensities. However, the bush became more resistant to fire as the height increased (Figs. 12 and 13). Thus, where fire is applied to reduce woody vegetation, managers will have to aim at producing maximum fire intensity.

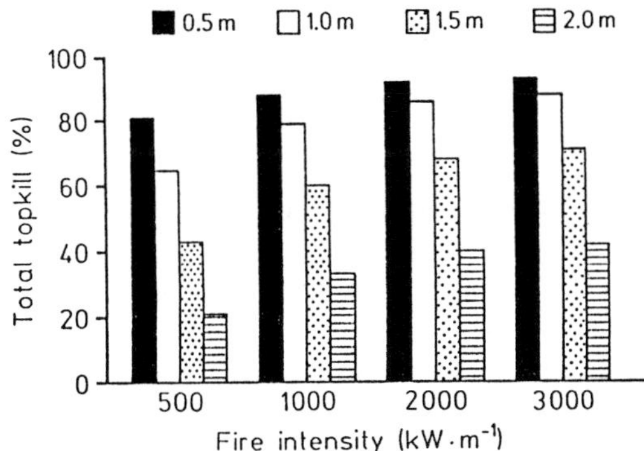

Fig. 10. Effect of fire intensity on the total topkill of bush of different heights in the arid savannas of the Eastern Cape

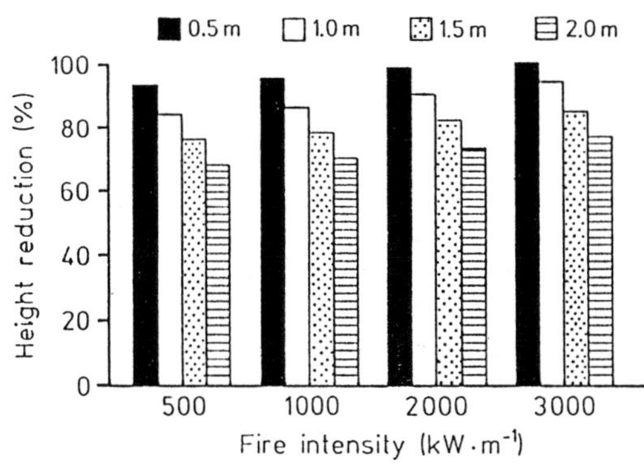

Fig. 11. Effect of fire intensity on the reduction in height of bushes of different sizes in the arid savannas of the Kruger National Park

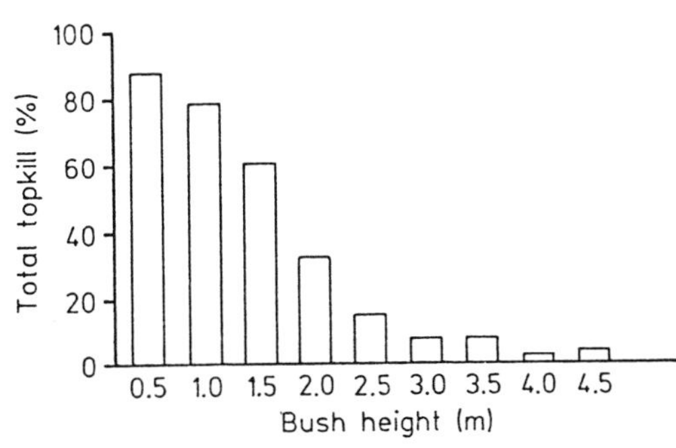

Fig. 12. Effect of height on the total topkill of bush in the arid savannas of the Eastern Cape

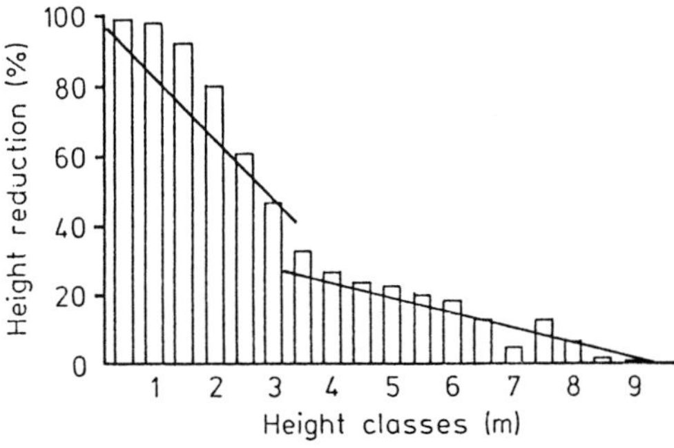

Fig. 13. Effect of height on the reduction in height of bush in the arid savannas of the Kruger National Park

11.7 Agricultural Areas in the Eastern Cape

11.7.1 Aims of Management

The southeastern portions of the Cape Province of South Africa, where both moist and arid savannas occur, are managed primarily for the production of grazing livestock. Bush encroachment has become a serious problem in this region where it has drastically reduced the grazing capacity of the rangeland and poses a serious threat to the important livestock industry of the area. The principal encroaching bush species is *Acacia karroo* but other important associated species are *Scutia myrtia*, *Maytenus heterophylla*, *Rhus lucida*, and *Diospyros lycioides* (Trollope 1984).

11.7.2 Research Background

The eastern Cape has been the center of a program of fire research by pasture scientists based at the University of Fort Hare. Research has concentrated on all major aspects of fire ecology, including the effects of fire season, fire frequency, and type and intensity of fire, as well as interactions between fire and grazing and browsing herbivores. A recent review is given by Trollope (1984). The research has provided the background for the extensive management guidelines available in the eastern Cape.

11.7.3 Current Management

It is difficult to generalize about management in an agricultural area, as each farm is managed individually. What follows are general guidelines that reflect the currently accepted norms for good fire management. Fire is an inexpensive method of controlling bush encroachment which makes it compatible with the inherent low economic potential of rangeland. However, the role of fire differs in the moist and arid savannas. In moist areas it is possible to control bush encroachment with fire alone. Even though the majority of the bush species survive by coppicing after burning, the rainfall is sufficient and reliable enough to enable adequate grass fuel to accumulate under grazing conditions to support frequent fires. These can control coppice growth and bush seedlings. Conversely in the arid savannas the rainfall is too low and erratic to support frequent enough fires under grazing conditions to prevent the regeneration of bush from coppice and seedling growth. Consequently, the role fire can play in controlling bush encroachment in these areas is to maintain bush at an available height and in an acceptable state for browsing animals (Trollope 1980b). This has been successfully achieved by introducing either mutton or mohair goats into a cattle ranching system where the goats control the bush after burning and the cattle utilize the grass.

Fires aimed at controlling bush encroachment can be applied as head fires because they cause least damage to the grass sward but affect the greatest topkill of stems and branches of the bush. High-intensity fires of at least 2500 kW m^{-1} are necessary to achieve a satisfactory topkill of bush up to a height of 2 m. Trollope and Tainton (1986) estimated that such fires will result when the grass fuel load is $>$4000 kg ha^{-1}, grass fuel is fully cured, air temperature $>$25°C, relative humidity 30%, and the wind speed 10–20 km h^{-1}. Fires should be applied before the first spring rains when the grass is dormant. Burning at this time causes least damage to the grass sward and enables the application of high-intensity fires (Trollope 1987).

In the moist savannas the frequency of burning required to control bush encroachment depends upon the rate at which re-encroachment occurs in the form of coppice growth and seedling development. Burning every 3 to 4 years will be adequate to control the re-encroachment of bush. In the arid savannas the frequency of burning is low and is determined by the necessity to reduce the bush to an available height and an acceptable state for browsing by goats. Frequency of burning is also influenced by the occurrence of above-average rainfall conditions necessary to generate adequate grass fuel. Under farming conditions it is both unnecessary and impractical to burn more frequently than once every 10 years in the arid savannas of the Eastern Cape (Trollope 1980b).

The grazing and browsing management after burning is considered very important. Grazing should commence only when the grass has recovered to a height of 100–150 mm. Grazing before this stage causes a loss of vigor in the grass sward and results in excessive levels of soil erosion. In the arid savannas, a high-intensity fire should be followed by browsing with goats when the coppice growth is 100–150 mm long, irrespective of the stage of recovery of the grass sward (Trollope 1980b).

11.8 The Hluhluwe/Umfolozi Game Reserves Complex

11.8.1 Aims of Management

The Hluhluwe/Umfolozi game reserves complex comprises almost 100,000 ha in Northern Natal. The area was originally proclaimed a reserve in 1896, with the primary objective of preserving the white rhinoceros. Today it is also an important reserve for many large mammal and bird species. The management of the Hluhluwe/Umfolozi Reserves has the following objectives:

1. To conserve the diversity of indigenous fauna and flora, and to maintain a desirable diversity of habitats.
2. To perpetuate in a natural state an example of lowland vegetation together with its associated fauna, especially the white and black rhinoceros. To maintain the environment in a condition as close as possible to that which existed before the advent of European influence.

3. To provide for tourism activities consistent with the environment and its conservation.
4. To protect geological, geomorphological, paleontological, and archeological features.
5. To provide facilities for scientific research, education, and recreation.

11.8.2 Background

Since the proclamation of Hluhluwe and Umfolozi Game Reserves in 1896, the vegetation has changed from a predominantly open to a more densely wooded condition. During the past two to three decades, management of the reserves has been aimed largely at retarding this process. The causes of the dramatic change in vegetation are thought to be due to changes in the land use pattern and disturbance regime since 1896, including the impact of fire. Prior to proclamation, the fire regime was largely man-induced. People practicing an iron-age culture burnt primarily to create winter pasture for livestock and also may have used fire for hunting. Lightning does not appear to have been an important source of ignition. Historical fires occurred throughout the winter (May-August), but a greater proportion burnt during late winter (July-August). The frequency of burning was determined by the available fuel (i.e., when the grass was tall and dense, and relatively unpalatable to livestock). This translates into very frequent (annual or biennial) burning during wet periods when fuel production was high, and less frequent burning during dry periods when the impact of indigenous herbivores and livestock limited available fuel. The ignition pattern implemented was usually point source, resulting in much smaller burns than would have been achieved by putting in long fronts of ignition (which may occasionally have been used for hunting). The overall result was a patchwork of smaller burns extending throughout the winter. It created a heterogenous mosaic of burnt/unburnt areas in different stages of regrowth. The different fire seasons resulted in different impacts on the woody component of the vegetation.

Major changes to this regime have taken place since 1896. With proclamation, rural people and grazing livestock were excluded from the area. With the appointment of a game conservator in 1911 control was effectively imposed and burning frequency was probably reduced, although the game conservator himself burnt whenever and wherever it seemed necessary. A program to monitor the tsetse fly was implemented in the 1930's, and burning was actively discouraged due to the danger of burning the Harris fly traps dotted throughout the area. After the removal of the traps, a few winter burns were put in from 1943 to 1946. From 1946 to 1960 no burns were implemented other than peripheral firebreaks. From 1961 to 1982 there was a change in management where large areas within the reserve were burnt during early spring (mid-late August and September). However, burning frequency was very low, averaging out at about one burn every 5–7 years.

11.8.3 Current Fire Management

The primary objective of current fire management is to counter bush encroachment in an attempt to re-create suitable habitat for the large grazing mammals. In addition, tourist game viewing is impeded because of the thick vegetation, leading to numerous complaints. The rationale (based largely on work conducted in the Eastern Cape) has been that fires of high intensity have a greater impact on the woody component of the vegetation than fires of low intensity. In more recent years (1983 to present), cognizance of the possible important impact of frequent fires on woody vegetation has been taken. Furthermore, the burning season has been shifted from early spring to late winter to ensure that the grass is dormant when burnt (a time when woody species are flushing), thus maximizing fire intensity and minimizing damage to the grass layer. Additional means of maximizing fire intensity through the implementation of fire behavior prediction models developed in the U.S.A. (Rothermel 1972; Burgan and Rothermel 1984) have been undertaken in the Hluhluwe/Umfolozi complex (Van Wilgen and Wills 1988). Using appropriate fuel inputs, it is possible to select environmental conditions that will lead to the desired type of fire.

The current fire regime creates very large burnt areas, resulting in a relatively uniform regrowth with a relatively uniform impact (hopefully severe) on the woody component of the vegetation. Early spring rains may prevent the application of fire, as happened in 1981 and 1987. Finally there is a possible problem that by implementing a very uniform burning regime the heterogeneity caused by the pre-proclamation burning regime has been lost. Besides fire management, other measures aimed at combating bush encroachment have been introduced. These include the re-introduction of elephants, and allowing local tribal inhabitants access to collect firewood. Whether any or all of these measures will be effective remains to be seen.

11.9 The Pilanesberg National Park

11.9.1 Aims of Management

The Pilanesberg National Park, covering an area of 55,000 ha, was established in December 1979 in an area that had previously been used for commercial livestock production. The area was extensively restocked with large mammals after proclamation. The vegetation of the Pilanesberg National Park is managed to keep the currently occurring plant species, and to preserve the existing patchwork of communities, such as grassland, scrub, and forest. This is interpreted as meaning that changes in species richness and the dynamic spatio-temporal patterning of flora, which are practically and economically irreversible within 10 years, should not be allowed to occur. Fire is used to help maintain species richness, and spatio-temporal patterning of both plants and animals. In practice, this would be done by ensuring that a range of fire frequencies, areas burnt, and fire intensities are applied, either via provided burns, arson, accidental, or lightning-initiated fires, over a 6-year period. In addition, fire is used to provide high quality forage for grazing mammals during the winter.

11.9.2 Research Background

There has been no formal research into the effects of fire in the Pilanesberg National Park itself, and the prescriptions for season, frequency, and intensity of burns to be applied are based on research done in other parts of the country. This includes important work done on grasses in the Natal Drakensberg, and on the effects of fire in the savannas of the Eastern Cape.

11.9.3 Current Management

Climatic factors are used to guide burning prescriptions. Burning can commence in early winter after the end of the growing season. Depending on the objectives of the burn, curing of the fuels may or may not be required. No burning is allowed after the start of the growing season in spring. Exceptions to this may be in the case of very old fuels. As in many savanna areas, fire in the Pilanesberg National Park is used to control bush encroachment, by applying high intensity fires. This will be done using fire behavior prediction technology developed and applied in the Kruger National Park and Hluhluwe/Umfolozi reserves (Trollope and Potgieter 1985; Van Wilgen and Wills 1988). The accumulation of grass fuel in the Pilanesberg National Park is variable, as a result of erratic and unpredictable rainfall and fluctuating stocking rates of grazing ungulates. As a result, there is no fixed fire frequency.

Besides prescribed burns, wildfires, due to arson or lightning, also occur. Arson fires occur from April to early August, and lightning fires occur from September to November.

The application of fire in the Pilanesberg National Park will be assisted by a locally developed expert system designed specifically for use in the Park (A.W. Bailey and M.T. Mentis pers. commun.). This is the first area in southern Africa to make use of this method of decision-making. The decision of whether or not to burn is arrived at by considering a set of rules, which draw on information on post-fire age, fuel loads, weather conditions, and practical considerations such as the protection of structures from fire. The rules were arrived at by interviewing local managers, as well as ecologists throughout southern Africa. The use of computer-based expert systems in the management of the Pilanesberg National Park should be an interesting test of the acceptability and practicability of the approach.

11.10 The Kruger National Park

11.10.1 Aims of Management

The Kruger National Park is situated in the low-lying savannas of the eastern Transvaal in South Africa adjacent to Mozambique in the east and Zimbabwe in the north. It comprises an area of about 2 million hectares, of which the major portion is arid savanna. Moist savannas occur in the extreme southwest and northwest of the Park. The area is of prime conservation importance. More than 2000 plant species, in a range of communities, occur in the park. In addition, 122 species of mammals (including most large African species), over 400 species of birds, 55 fish species, and 109 reptile species occur. The area is an important tourist destination, and over 6000 visitors can be accommodated per night in 17 camps.

The aims of management are as follows (Joubert 1988):

1. To develop and maintain maximum biotic diversity.
2. To achieve (1) through minimum interference in the natural functioning of the ecosystem.
3. To cater for tourists, and in particular to enable them to observe the maximum biotic diversity.

11.10.2 Research Background

There is very limited published information on the ecological effects of fire on the vegetation in the Kruger National Park. A comprehensive burning experiment, examining the effects of season and frequency of burning, was established in the four major vegetation types in the Park in 1954. Although this trial has been maintained and enumerated periodically, only limited results are available (Gertenbach and Potgieter 1979; Van Wyk 1971) and a comprehensive analysis and evaluation of the effects of the treatments is necessary. A major difficulty, though, in such an evaluation is that the effects are confounded with grazing (Van Wyk 1971). The cumulative effect of grazing after fire, particularly in the annual spring burns and to a lesser extent in the biennial burns applied in April and August, make it very difficult to identify the true effect of season and frequency of burning on the herbaceous grass layer.

Trollope and Potgieter (1985) used fire behavior data (rate of spread, flame height, and intensity) to develop regression models for predicting these parameters from environmental variables. The data were collected during routine burning of the plots in the burning experiment described above. These models were regarded as the initial phase in a project ultimately aimed at determining the relationship between fire behavior and vegetation response. Significant progress has been made in this regard and preliminary results are given in Figs. 11 and 13.

11.10.3 Management

Lightning fires are common in the Kruger National Park (Gertenbach 1979), and prescribed burning has been applied since the Park's establishment in 1926. Up until 1954 burning was used to provide green grazing for the wildlife irrespective of its effect on the vegetation. For example, in the southern moist savannas, annual autumn burning was applied, but this is now regarded as being an unnatural season and also too frequent. Since 1954 several systematic burning programs have been applied. These are dealt with separately below.

Period 1954–1975

A fixed triennial burning program was introduced in 1954 and the Park was subdivided into ± 400 blocks ranging in size from 2500–5000 ha. The blocks were burnt every 3 years in spring after 50 mm of rain had fallen. The reasons for burning the vegetation were: (1) to remove accumulated grass material; (2) to stimulate the production of palatable green grazing and to reduce selective grazing; and (3) to limit the occurrence of wildfires. No mention was made of the role of fire in controling bush encroachment.

During the application of this burning policy, it soon became apparent that a fixed burning program over the whole of the Kruger National Park was impractical. Consequently, the program was adapted in certain areas. In the southern moist savanna the frequency was changed in 1957 to biennial burning during spring and autumn to prevent excessive accumulation of mature grass and to provide short green grazing for a greater portion of the year. Certain arid areas were withdrawn either permanently or temporarily from the burning program because the condition of the vegetation had deteriorated, due to drought and overgrazing.

Period 1975–1980

Arising out of the results of the burning experiment (see above) and field experience, major deficiencies in the triennial burning program were identified. These included the following:

1. Only 60% of the scheduled burns could be applied in practice, while other areas burnt in wildfires.
2. Too large an area was being burnt over a short period of time (spring), resulting in the subsequent poor utilization of the grazing.
3. Too large an area was scheduled to be burnt at any one time, making it difficult to apply the burns at the prescribed times of the year.
4. The frequent absence of adequate rains after the burns had a deleterious effect on the recovery of the grass sward.
5. No areas were being subjected to a natural fire regime.

Accordingly, the following changes were made in 1975 to the burning program:

1. The season of burning was changed to permit burning during late winter, before and after the spring rains, mid-summer, and during autumn (only in the moist savannas). The main reason for these changes was to provide short palatable grazing throughout the growing season for grazing species such as zebra and wildebeest that preferred this type of habitat. The different seasons of burning were applied in rotation to prevent any undesirable accumulative effect of burning.
2. Fire frequency was changed in the arid savannas to permit both triennial and quadrennial burns. For example, quadrennial burns were scheduled in portions of mopane savanna, which is ideal habitat for roan antelope, which prefer grazing long grass.

Period 1980 Onwards

A further adaptation to the burning program was made in 1980 to allow for an apparent 10-year rainfall cycle. A decade of below-average rainfall is usually followed by a decade of above-average rainfall. The rate of accumulation of grass fuel is higher during wet cycles than dry cycles, resulting in high fuel loads during wet cycles and a higher frequency of fires caused by lightning. Consequently, a variable burning frequency based on rainfall and the level of accumulation of grass fuel was introduced. It was felt that such a burning program would approximate a more natural fire regime, where variable climatic conditions are a major driving force.

The season of burning was also adapted to mirror fires caused by lightning which are generally limited to the period immediately before and after the first spring rains during low rainfall cycles. Conversely, during high rainfall cycles fires caused by lightning also occurred during mid-summer. The burning program was changed so that the majority of the blocks are burnt before and after the first spring rains during low rainfall cycles. During high rainfall cycles blocks are also burnt during mid-summer.

The practical application of the current burning program has involved grouping the 400 burning blocks into 31 management units. On the basis of on-site inspections, rainfall data, and the previous fire history, up to 50% of a management unit is burnt during a high rainfall cycle, whereas only up to 20% is burnt during a low rainfall cycle. The season of burning is varied according to the season of the previous fire. All these adaptations have resulted in a more flexible burning program that is believed to simulate a natural fire regime more closely.

11.11 The Etosha National Park

11.11.1 Aims of Management

The Etosha National Park in northern Namibia comprises an area of 2.2 million hectares, most of which is arid savanna. Annual rainfall is low (300–500 mm), variable and erratic. The park is managed for the following objectives:

1. The maintenance, and in some cases the increase, of the diversity of the local biota,
2. To serve the requirements of tourism, even if this means interfering in "natural" processes from time to time.

11.11.2 Fire Management

The fire history of the Etosha area prior to 1979 is summarized by Siegfried (1981). Up until 1980 a fire exclusion policy was applied in the Park and all wildfires, caused either by accident or lightning, were extinguished or limited to the smallest possible area. A burning program was introduced in 1981 as a means of removing moribund herbaceous material and controlling bush encroachment. However, the areas to be burnt were chosen subjectively and this led to inconsistencies. Subsequently, a burning strategy was developed to simulate the expected occurrence of fires caused by lightning. This was based on rainfall data, the period since the last burn, and the natural incidence of fires caused by lightning (Nott and Mentis undated).

For the purposes of fire management, the Park is divided into 24 blocks, making use of existing firebreaks and roads. One hundred and fifty raingauges have been distributed fairly uniformly through the Park. Data from these gauges are used for selecting blocks to be burnt. This is done by comparing the mean seasonal rainfall for a block with the 20-year annual mean. If the rainfall is above the 20-year mean, the block is considered for burning. Of the blocks considered for burning, a further selection is made of the blocks with the longest time since the last burn. If there are blocks with similar times since the last burn then the block that received the highest seasonal rainfall is chosen. This procedure is repeated until not more than 12% of the area of the Park is designated for burning.

Fires caused by lightning are allowed to spread but are retained within the block irrespective of the quantity of rain received during the previous season. Where such fires are prevented from spreading by roads, they are ignited further throughout the block. Conversely, wildfires caused by other agents are confined to the smallest possible area, making use of access roads within the burning blocks (Nott and Mentis undated).

The majority of fires (73%) recorded between 1970 and 1979 were caused by lightning (Siegfried 1981). Most of these occurred during the onset of the rains in spring. Accordingly, prescribed burns are applied in spring during September and October (Nott and Mentis undated; H. Lindeque pers. commun.).

The overall objective of the burning program is to simulate the incidence of fires caused by lightning. Larger areas are burnt either by fires ignited by lightning or through prescribed burning during above-average rainfall seasons. Using the aforementioned procedure for selecting blocks to be burnt, together with permitting fires caused by lightning, the expected burning frequency would be between 5.5. and 9.3 years (Nott and Mentis undated). This is in accordance with the estimate by Siegfried (1981) that the Etosha National Park should burn at least once every 10 years.

11.12 Conclusions

This review has highlighted important differences in the philosophical approach to the use of fire in different areas, brought about largely by differences in the research background, and therefore information available. In fynbos and grassland catchment areas, and in agricultural areas, management prescriptions are based on the biology of key plant species. These include serotinous Proteaceae in fynbos, key grass species in grasslands, and encroaching woody shrubs in agricultural savannas. Fire is applied to induce specific responses from these species, and is based, in many cases, on the results of extensive research. In contrast, many conservation areas in savannas do not have a history of fire research on which to base management prescriptions. Much of the research effort has gone into the ecology and management of the fauna, particularly large mammals. Consequently, fire management prescriptions are based largely on research done elsewhere, and on recreating "natural" fire regimes. The rationale behind this (in conservation areas) is that allowing the fire regime, under which the biota evolved, to continue will ensure their survival. The weakness, of course, is that there is no predictive value in this approach. It is therefore not possible to frame objectives with regard to the desired response from specific elements of the vegetation. We conclude that there is a very real need for autecological studies of key elements of the vegetation in many areas, in order to define acceptable fire regimes for the long-term conservation of the ecosystems involved. For example, there is apparently high mortality in mature specimens of marula (*Sclerocarya birrea*) and leadwood (*Combretum imberbe*) in the Kruger National Park, causing noticeable changes in vegetation structure. In many areas in the eastern Transvaal, there is little or no recruitment under mature stands of kiaat (*Pterocarpus angolensis*), an important timber tree. Despite these problems, no meaningful strategy for their conservation can be formulated, due to the absence of detailed autecological knowledge. Conversely, information on the autecology of key bush species such as scented thorn (*Acacia nilotica*), umbrella thorn (*Acacia tortillis*), and magic guarri (*Euclea divinorum*) in the Hluhluwe/Umfolozi Game Reserves would assist in developing a means of combating bush encroachment in these areas.

The use of procedures, developed for the management of agricultural areas, in conservation areas further highlights deficiencies in our knowledge. For example, the condition of grasslands in the Natal Drakensberg is assessed by comparison to a benchmark site considered to have a desirable composition. However, benchmark sites are all selected on the basis of forage production potential, which may be inappropriate for the conservation of species diversity or streamflow production.

In addition to an expanded knowledge of the effects of fire, there is a need for fire management decision aids (such as fire behavior prediction models and expert systems) to be developed. There has been some work on the development and adaptation of fire behavior models in many areas in southern Africa (Everson T.M. et al. 1988b; Trollope 1978; Trollope and Potgieter 1985; Van Wilgen et al. 1985; VanWilgen and Wills 1988). There has been only limited application of these models to date, but there is increasing interest in the approach, and acceptance of the importance of fire behavior. Researchers should concentrate on developing user-friendly packages to assist managers in selecting conditions that will lead to the desired type of fire. Expert systems offer another useful way of assisting the fire management decision-making process. This has so far only been tried in the Pilanesberg National Park, but the approach deserves attention in other regions. The linking of fire behavior and fire effects models with expert systems would be the best solution, but has yet to be attempted.

Finally, it may seem incongruous that much of the fire management practiced in southern Africa is aimed at the reduction, or even eradication, of trees, while globally, there is concern over the destruction of forests. Alien trees are seen as incompatible with the aims of catchment management as they use more water and reduce biotic diversity. Bush and tree encroachment, at the expense of grass, is not compatible with livestock production or the conservation of large grazing mammals. However, the effects of bush encroachment in savannas on overall biotic diversity are not understood. No studies have compared the overall diversity of encroached areas to unencroached areas. Should they be found to be more diverse, bush encroachment would be desirable in terms of the aim of maximizing species diversity.

Acknowledgments. This review forms part of the conservation research program of the South African Department of Environment Affairs. We thank the following persons for providing information or assisting in the preparation of this chapter: Art Bailey, Bruce Brockett, Terry Everson, Tony Ferrar, Willem Gertenbach, Ann Green, Claire Jones, Fred Kruger, Terry Newby, Andre Potgieter, and Alf Wills.

References

Acocks JPH (1975) Veld types of South Africa. Mem Bot Surv South Africa 40:1–128

Bainbridge WR (1987) Management of mountain catchment grassland with special reference to the Natal Drakensberg. In: Gadow von K (ed) Forestry Handbook. Southern African Inst For, Pretoria

Bainbridge WR, Scott DF, Walker RS (1986) Policy statement for the management of the Drakensberg State Forests. Dept Environ Affairs, South Africa

Bayer AW (1955) The ecology of grasslands. In: Meridith D (ed) The grasses and pastures of South Africa. Central News Agency, Union of South Africa, pp 539–550

Bews JW (1925) Plant forms and their evolution in South Africa. Longmans, London

Bigalke RC (1979) Aspects of vertebrate life in the fynbos, South Africa. In: Specht RL (ed) Heathlands and related shrublands of the world: Descriptive studies. Elsevier, Amsterdam

Bond P, Goldblatt P (1984) Plants of the Cape flora: a descriptive catalogue. J South African Bot (Suppl) Vol 13

Bond WJ (1980) Fire and senescent fynbos in the Swartberg. South African For J 114:68–71

Bond WJ, Vlok J, Viviers M (1984) Variation in seedling recruitment of Cape Proteaceae after fire. J Ecol 72:209–221

Booysen P de V, Tainton NM (1984) Ecological effects of fire in South African Ecosystems. (Ecological Studies 48) Springer, Berlin Heidelberg New York Tokyo

Bosch JM, Van Wilgen BW, Bands DP (1986) A model for comparing water yield from fynbos catchments burnt at different intervals. Water SA 12:191–196

Burgan RE, Rothermel RC (1984) Behave: Fire behaviour prediction and fuel modelling system – fuel subsystem. USDA For Serv, Report INT-167

Clinning CF (1986) Transvaal grasslands. Fauna Flora 44:3–6

Danckwerts JE (1980) (Unpublished data) Dohne Res Stat, Sutterheim

Edwards D (1984) Fire regimes in the biomes of South Africa. In: Booysen P de V, Tainton NM (eds) Ecological effects of fire in South African ecosystems. Springer, Berlin Heidelberg New York Tokyo

Everard DA (1986) The effects of fire on the *Podocarpus latifolius* forest of the Royal Natal National Park, Natal Drakensberg. South African J Bot 52:60–66

Everson CS (1985) Ecological effects of fire in the montane grasslands of Natal. Ph D thesis, Univ Natal, Pietermaritzburg

Everson CS, Everson TM (1987) Factors affecting the timing of grassland regrowth after fire in the montane grasslands of Natal. South African For J 142:47–52

Everson CS, Tainton NM (1984) The effect of thirty years of burning in the Highland Sourveld of Natal. J Grassland Soc Southern Africa 1(3):15–20

Everson CS, Everson TM, Tainton NM (1985) The dynamics of Themeda triandra tillers in relation to burning in the Natal Drakensberg. J Grassland Soc Southern Africa 2(4):18–25

Everson CS, Everson TM, Tainton NM (1988) Effects of intensity and height of shading on the tiller initiation of six grass species from the Highland sourveld of Natal. South African J Bot 54:315–318

Everson CS, George WJ, Schulze RE (1989) Fire regime effects on canopy cover and sediment yield in the montane grasslands of Natal. South African J Sci 85:113–116

Everson TM, Everson CS, Dicks HM, Poulter AG (1988a) Curing rates of the grass sward of the Highland Sourveld at Cathedral Peak in the Natal Drakensberg. South African For J 145:1–8

Everson TM, Van Wilgen BW, Everson CS (1988b) Adaptation of a model for rating fire danger in the Natal Drakensberg. South African J Sci 84:44–49

Foran DB, Tainton NM, Booysen P de V (1978) The development of a method for assessing veld condition in three grassveld types in Natal. Proc Grassland Soc Southern Africa 13:27–33

Gertenbach WPD (1979) Veld burning in the Kruger National Park: history, development, research and present policy. Dept Nature Conser, Kruger National Park

Gertenbach WPD (1988) Fire as a management tool in the Kruger National Park. Fire Protection, Dec 1988, 6–10

Gertenbach WPD, Potgieter ALF (1979) Veldbrandnavorsing in die struikmopanieveld van die Krugerwildtuin. Koedoe 22:1–28

Granger JE (1977) The vegetation changes, some related factors and changes in the water balance following 20 years of fire exclusion in catchment IX: Cathedral Peak For Res Stat. Ph. D. thesis, Univ Natal

Gordon-Gray KD, Wright FB (1969) *Cyrtanthus breviflorus* and *Cyrtanthus lutens* (Amaryllidceae). Observations with particular reference to Natal populations. J South African Bot 35:35–62

Hall M (1984) Man's historical and traditional use of fire in Southern Africa. In: Booysen P de V, Tainton NM (eds) Ecological effects of fire in South African ecosystems. Springer, Berlin Heidelberg New York Tokyo

Hillard OM, Burtt BL (1987) The botany of the southern Natal Drakensberg. Ann Kirstenbosch Bot Gardens Vol 15

Huntley BJ (1982) Southern African Savannas. In: Huntley BJ, Walker BH (eds) Ecology of Tropical savannas. (Ecological Studies 42) Springer, Berlin Heidelberg New York Tokyo

Huntley BJ (1984) Characteristics of South African Biomes. In: Booysen P de V, Tainton NM (eds) Ecological effects of fire in South African Ecosystem. (Ecological Studies 48) Springer, Berlin Heidelberg New York Tokyo

Joubert SGJ (1988) Master plan for the management of the Kruger National Park, Vol 1–6. Unpublished policy document, National Parks Board of South Africa

Kennan TCD (1971) The effects of fire on two vegetation types of Matopos. Proc Tall Timbers Fire Ecol Conf 11:53–98

Komarek EV (1971) Lightning and fire ecology in Africa. Proc Tall Timbers Fire Ecol Conf 11:473–511

Komarek EV (1972) Fire in Africa. Proc Tall Timbers Fire Ecol Conf, Vol 11. Tall Timbers Res Stat, Tallahassee, Florida

Kruger FJ (1977) A preliminary account of aerial plant biomass in fynbos communities of the mediterranean-type climate zone of the Cape Province. Bothalia 12:301–307

Kruger FJ (1981) Use and management of mediterranean ecosystems in South Africa – current problems. In: Conrad CE, Oechel WC (eds) Dynamics and management of mediterranean-type ecosystems. USDA For Serv, General Tech Rep PSW-58

Kruger FJ (1987) Succession after fire in selected fynbos communities of the southwestern Cape. Ph.D. thesis, Univ Witwatersrand

Levyns MR (1966) *Haemanthus canaliculatus*, a new fire-lily from the western Cape Province. J South African Bot 32:73–75

Little RM, Bainbridge WR (1985) Avifaunal conservation in the State Forests of the Natal Drakensberg. Proc Pan-African Ornithol Conf, Gaberone, Botswana

Macdonald IAW (1989) The history and effects of alien plant control in the Cape of Good Hope Nature Reserve, 1941–1987. South African J Bot 55:56–75

Macdonald IAW, Jarman ML, Beeston P (1985) Management of invasive alien plants in the fynbos biome. South African Nat Sci Programmes Rep No 111, CSIR, Pretoria

Macdonald IAW, Kruger FJ, Ferrar AA (1986) The ecology and control of biological invasions in South Africa. Cape Town Oxford

Manders PT (1985) The autecology of Widdringtonia cederbergensis in relation to its conservation management. MSc thesis, Univ Cape Town

Mentis MT, Bigalke RC (1979) Some effects of fire on two grassland francolins in the Natal Drakensberg. South African J Wildlife Res 9:1–8

Noble IR, Slatyer RO (1980) The use of vital attributes to predict successional changes in plant communities subject to recurrent disturbances. Vegetatio 43:5–21

Nott TR, Mentis MT (undated) A proposed burning strategy for the Etosha National Park (unpublished)

Phillips JFV (1974) Effects of fire in forest and savanna ecosystems of subsaharan Africa. In: Kozlowski TT, Alghren CE (eds) Fire and ecosystems. Academic Press, London

Priday AJ (1989) Statistical modelling of fuel loads for development of a fire management strategy in the Natal Drakensberg. MSc thesis, Univ Natal

Richardson DM, Van Wilgen BW (1986) The effects of fire in felled *Hakea sericea* and natural fynbos and the implications for weed control in mountain catchments. South African For J 139:4–14

Rothermel RC (1972) A mathematical model for predicting fire spread in wildland fuels. USDA For Serv, Res Paper INT-115

Rutheford MC, Westfall RH (1986) Biomes of Southern Africa an objective categorization. Mem Bot Surv South Africa, 54. Bot Res Inst, Dep Agric Water Supply-Republic of South Africa

Scott JD (1971) Veld burning in Natal. Proc Tall Timbers Fire Ecol Conf 11:33–51

Siegfried WR (1981) The incidence of veld-fire in the Etosha National Park, 1970–1979. Madoqua 12:225–230

Tainton NM (1981) Veld and pasture management in South Africa. Natal Univ Press, and Shuter and Shooter, Pietermaritzburg

Tainton NM, Mentis MT (1984) Fire in grassland. In: Booysen P de V, Tainton NM (eds) Ecological effects of fire in South African ecosystems. Springer, Berlin Heidelberg New York Tokyo

Trollope WSW (1974) Role of fire in preventing bush encroachment in the eastern Cape. Proc Grassland Soc Southern Africa 9:67–72

Trollope WSW (1978) Fire behaviour – a preliminary study. Proc Grassland Soc Southern Africa 13:123–128

Trollope WSW (1980a) (Unpublished data) Dep Agric, Univ Fort Hare, Alice

Trollope WSW (1980b) The ecological effects of fire in South African savannas. In: Huntley BJ, Walker BH (eds) Ecology of tropical savannas. Springer, Berlin Heidelberg New York Tokyo, pp 292–306

Trollope, WSW (1984) Fire in savanna. In: Booysen P de V, Tainton NM (eds) Ecological effects of fire in South African ecosystems. Springer, Berlin Heidelberg New York Tokyo

Trollope WSW (1987) Effect of season of burning on grass recovery in the False Thornveld of the Eastern Cape. J Grassland Soc Southern Africa 4(2):74–77

Trollope WSW, Potgieter ALF (1985) Fire behaviour in the Kruger National Park. J Grassland Soc Southern Africa 3(4):148–152

Trollope WSW, Tainton NM (1986) Effect of fire intensity on the grass and bush components of the Eastern Cape Thornveld. J Grassland Soc Southern Africa 2:27–42

Trollope WSW, Potgieter ALF, Zambatis N (1988) Assessing veld condition in the Kruger National Park using key grass species. Univ Fort Hare, 30 pp (unpublished)

Van Wilgen BW (1982) Some effects of post-fire age on the above-ground biomass of fynbos (macchia) vegetation in South Africa. J Ecol 70:217–225

Van Wilgen BW (1984) Adaptation of the United States Fire Danger Rating System to fynbos conditions. I. A fuel model for fire danger rating in the fynbos biome. South African For J 129:61–65

Van Wilgen BW, Burgan RE (1984) Adaptation of the United States Fire Danger Rating System to fynbos conditions. II Historic fire danger in the fynbos biome. South African For J 129:66–78

Van Wilgen BW, Kruger FJ (1981) Observations on the effects of fire in mountain fynbos at Zachariashoek, Paarl. J South African Bot 47:195–212

Van Wilgen BW, Richardson DM (1985a) The effect of alien shrub invasions on vegetation structure and fire behaviour in South African fynbos shrublands: a simulation study. J Appl Ecol 22:955–966

Van Wilgen BW, Richardson DM (1985b) Factors influencing burning by prescription in mountain fynbos catchment areas. South African For J 134:22–32

Van Wilgen BW, Viviers M (1985) The effect of season of fire on serotinous Proteaceae in the western Cape and the implications for fynbos management. South African For J 133:49–53

Van Wilgen BW, Wills AJ (1988) Fire behaviour prediction in savanna vegetation. South African J Wildlife Res 18:41–46

Van Wilgen BW, Le Maitre DC, Kruger FJ (1985) Fire behaviour in South African fynbos (macchia) vegetation and predictions from Rothermel's fire model. J Appl Ecol 22:207–216

Van Wyk P (1971) Veld burning in the Kruger National Park, an interim report of some aspects of research. Proc Tall Timbers Fire Ecol Conf 11:9–31

Westfall RH, Everson CS, Everson TM (1983) The vegetation of the protected plots at Thabamhlope Research Station. South African J Bot 2:15–25

White F (1983) The vegetation of Africa: A descriptive memoir to accompany the Unesco/AETFAT/UNSO vegetation map of Africa. Unesco, Paris

Yeaton RI (1988) Porcupines, fires and the dynamics of the tree layer of the *Burkea africana* savanna. J Ecol 76:1017–1029

12 Prescribed Fire in Industrial Pine Plantations

C. DE RONDE[1], J.G. GOLDAMMER[2], D.D. WADE[3], and R.V. SOARES[4]

12.1 Introduction

Industrial plantations of non-indigenous tree species (exotics) can be defined as even-aged stands established outside of their natural habitat. These plantations play a vital economic role in the developing countries of the tropics and subtropics. The ecological benefits of afforestation, however, go far beyond local and regional considerations: the increase in atmospheric CO_2 and its expected negative influence on the global climate may partially be averted through large-scale afforestation with fast-growing species (Maryland 1988). The take-up of carbon in woody matter could potentially balance the discharge of CO_2 from fossil fuel burning and from the vast amount of uncontrolled forest destruction and biomass burning in the tropical and subtropical biota. Although prescribed burning is itself an emission source of CO_2, its main function in plantation management is to increase stability and to protect against destructive wildfires, which are a much larger source. The same is true for particulate matter emissions. Thus, although at first glance it may seem contradictory, prescribed burning is indeed an important link in global fire ecology.

Of the estimated 9,968,000 ha of industrial plantations established in the tropics by the end of 1985, 41% were in softwoods (mainly pines) (Table 1).

Even though almost 60% of the industrial plantations are composed of hardwoods such as eucalypt and teak, most prescribed burning has taken place in pine stands. This chapter will thus be restricted to a discussion of fire in pine plantations. The information needed to plan and safely conduct prescribed fires beneath standing trees (referred to as underburning) and the ecological effects of these fires will be emphasized. The importance of fuel and weather parameters is explained, and techniques of setting fires and monitoring their behavior are described. Some facets of post-harvest burning are also discussed.

The major species used in pine afforestation activities are *Pinus caribaea*, *P. elliottii*, *P. patula*, *P. Pinaster*, *P. radiata*, and *P. taeda*. According to McDonald and Krugman (1986), the leading species are *P. elliottii* and *P. taeda*. They estimate that over 450,000 ha are planted to these two species outside their native

[1]Saasveld Forestry Research Centre, South African Forestry Research Institute, George 6530, South Africa
[2]Department of Forestry, University of Freiburg, 7800 Freiburg, FRG
[3]Southeastern Forest Experiment Station, USDA Forest Service, Dry Branch, Georgia 31020, USA
[4]Department of Forestry, Federal University of Paraná, 80.001 Curitiba, Paraná, Brazil

Table 1. Areas of established industrial plantations (in thousand ha) in 76 tropical countries (estimated/projected at the end of 1985). (FAO 1982)

	Hardwood species				Softwood species			
	Other than fast-growing		Fast-growing		Species		All species	
Region	Total	1981–85	Total	1981–85	Total	1981–85	Total	1981–85
Tropical America (23 countries)	183	54	1393	525	2403	832	3979	1411
Tropical Africa (37 countries)	414	121	232	70	673	132	1319	323
Tropical Asia (16 countries)	2137	324	1560	477	973	367	4670	1168
Total (76 countries)	2734 =27%	499	3185 =32%	1072	4049 =41%	1331	9968	2902

range each year. Fire plays a predominant role in the native habitats of all the above species. Because these species have evolved in close association with fire, they have developed adaptations that make them better able to survive frequent fires than are competing woody species. The dynamic ecological equilibrium maintained by fire thus favors development of pine-dominated stands. Furthermore, in most cases exotic plantations are established in fire-prone environments or even in fire ecosystems (e.g., in savannas, grasslands, fynbos etc.). When pines are planted in such areas, attempting to exclude fire is counterproductive. The resulting unnatural accumulation of litter substantially increases the potential of destructive wildfires and retards formation of a herbaceous/woody understory component necessary for ecological stability.

The information presented in this chapter comes mainly from Australia, South Africa, and the United States, where most of these species are extensively planted and where fairly intensive fire research has been carried out. The intentional use of fire to help manage these plantations is perhaps exemplified in the southern United States, where over 1.5 million hectares of southern pine are treated with prescribed fire each year (Wade and Lunsford 1989). Prescribed burning experiments have been carried out in many tropical and subtropical countries including Bahamas, Belize, Brazil, Costa Rica, Fiji, Honduras, Nicaragua, Panama, Spain, and Venezuela, but there are few instances where the practice is currently operational (e.g., Fahnestock et al. 1987; Munro 1966; Vega et al. 1983). Nonetheless, these studies have in many cases demonstrated that prescribed fire can be safely and effectively used in industrial plantations. Used under the wrong conditions, however, prescription fire can destroy the very resource it was intended to protect, and there are trade-offs associated with every fire that should be recognized and carefully weighed before a decision is made regarding the use of fire. This does not mean, however, that its use should be summarily dismissed.

In many instances prescribed fire can achieve planned benefits at minimum damage and cost in comparison to alternative treatments. Although many specific questions remain to be answered, long-term studies in the southern United States suggest that the benefits of prescribed fire far outweigh any deleterious side effects provided the burns are properly conducted. It is generally agreed that the use of fire greatly facilitates the task of maintaining the overall stability and productivity of fire-climax (pine) forests and in many cases is more environmentally acceptable than are mechanical or chemical means of achieving the same objective.

12.2 Prescribed Burning Objectives

The reasons for using prescribed fire in pine plantation management are fairly universal (see Table 2) although implementation of this practice will vary by species and from country to country. The specific reason(s) for treating an area with fire should be identified and clearly defined in a written fire management plan. This will not only facilitate assessment of the various weather, fuel, and firing technique combinations that will produce the desired fire behavior, but it will allow meaningful comparison of alternative treatments. One of the advantages of prescription fire over many of its alternatives is that a single application can produce multiple benefits. For example, on the coastal plain of the southern United States, the judicious use of fire can enhance range, wildlife, and timber management objectives on a single area to the extent that the net economic return from this combined effort will be greater than if the land was managed exclusively for a single resource (Wade and Lewis 1987).

12.2.1 Wildfire Hazard Reduction

From an economic standpoint, the most important use of prescribed fire in plantation management is to temporarily reduce the hazardous build-up of dead fuels on the forest floor. Prescribed fire is the only practical way to reduce the dangerous accumulation of combustible fuels underneath industrial pine plantations and this, in turn, dramatically reduces the risk of damaging wildfire. These fuels rapidly accumulate, tying up large amounts of nutrients in decomposing organic matter. Fire is nothing more than rapid oxidation of plant matter, the same basic process as decomposition, albeit it takes place at a considerably faster rate. A well-planned low-intensity prescribed fire every 2 to 4 years reduces these fuels and recycles the nutrients into a form useable by the trees. McArthur (1971) states that severe fire suppression difficulties in coniferous plantations should not develop if fuel loads are kept below 11 t ha^{-1}. Studies invariably show that wildfires which burn into areas where fuels have been recently reduced by prescribed burning are much easier to control and cause less damage (e.g., Davis and Cooper 1963; Helms 1979). The interval

Table 2. Potential objectives for the use of prescribed fire in management of industrial pine plantations. (After Goldammer 1983)

Objectives	Target	Desired effects	Undesired effects or potential hazards	Possible Substitution
Wildfire hazard reduction	Thinning or post-harvest slash, forest floor (raw humus), aerial fuels, rank understory	Reduce potential wildfire intensity, remove surface and ladder fuels, Reduce understory stature	Stand/tree damage (crown, bole, or root)	Partial (mechanical treatment/removal by hand, shredding, piling and burning outside of stand, pruning)
Site preparation for natural regeneration or planting	Forest floor, post harvest slash, undesired vegetation	Expose mineral soil (improve germination), increase seedfall	Encroachment, sprouting, or germination of undesired plants	Partial (herbicides to kill undesired vegetation)
Improve accessibility	Thinning of post harvest slash, rank understory	Improve access for silvicultural operations, esthetics (recreation)	Reduction of understory stature	Partial (herbicides to kill understory)
Increase growth/yield	Raw humus layer (forest floor), understory plants	Enhance nutrient availability; reduce competition for moisture, sun and nutrients	Loss of nutrients (leaching), erosion	Fertilization and herbicides
Alter plant species composition	Weeds and other undesirable vegetation	Promote desired species	Increase in weed germination/production of undesirable seeds	Herbicides
Pest management	Pests and diseases and their habitats	Eliminate spores, eggs, individuals, and breeding material	Fire-induced tree stress, increased susceptibility to secondary pests	Pesticides
Silvopastoral land use	Slash; forest floor; mature, unpalatable growth; competing vegetation	Create/improve conditions for desired ground cover		Mechanical removal of dead fuels and vegetation
Improve fire protection	Surrounding buffer zone, fuel breaks and fire breaks	Reduce spread and intensity of wildfires (outside of stands)		

between fuel reduction burns depends on several factors including species, the rate of fuel accumulation, values at risk, and the risk of wildfire. The initial hazard-reduction burn in a young stand should always utilize a backing fire under exacting weather conditions. This initial burn is more difficult to implement in some species, *P. caribaea* in particular, because it retains dead stem needles for a considerable period of time, creating a high potential for "torching out". Luckily, this species can withstand fire intensities of 6900 kW m^{-1} (a factor of 10 above most other pines) without suffering severe mortality (McArthur 1971).

In some instances, prescribed burning by itself will not fireproof a plantation. In Australia, for example, McArthur (1971) reported that fire control was hopeless in pine plantations which were not pruned. He refers to studies which show that fire rate-of-spread is at least double and suppression difficulty is increased by a factor of 4 or 5 in unpruned plantations. He therefore states that pruning is a basic requirement of sound plantation fire management, irrespective of the use of prescription fire.

12.2.2 Prepare Sites For Planting

After harvest, unmerchantable limbs and stems remain scattered across the area, or concentrated at logging decks or delimbing gates, depending upon the method of logging. A high percentage of a site's nutrient capital is locked up in this debris. This material is also an impediment to both people and planting equipment. Moreover, if a wildfire occurs within the next few years, fireline construction can be severely hindered; the result being larger burn acreages and higher regeneration losses. Prescribed fire is often used to reduce this debris; although not all large material will be consumed, what is left is exposed and can be avoided by equipment operators.

12.2.3 Other Objectives

Cattle grazing can be compatible with good plantation management providing cattle numbers are kept in balance with forage yields. The reduction in, and compaction of, fine fuels which help protect the plantation from damaging wildfires is an added benefit. Grazing is profitable from soon after when a site is planted until the herbaceous component is shaded out as the plantation matures. Prescribed fire is often used in conjunction with grazing to increase the quantity, quality, availability, and palatability of grasses and forbs. McArthur (1971) reports that in Australia and the Fiji Islands, grazing results in the replacement of low-quality tussock grasses with a more palatable creeping type of grass.

Even with intensive site preparation, competing woody vegetation often becomes re-established very quickly. Where these woody plants form a dense shade-tolerant understory, periodic burns will topkill the underbrush, limiting competition with the crop trees while at the same time providing succulent browse for wildlife. Keeping the stature of the understory low improves the efficiency of subsequent silvicultural operations and provides greater safety for woods workers. The improved visibility and accessibility can increase stumpage values. Hiking, hunting, and other recreational uses also benefit.

The fact that understory control can improve crop tree productivity is well established (e.g., Haywood 1986; Stewart et al. 1984). However, when using prescribed fire to control competing vegetation, the results are not as clearcut. Intuitively, the increase in available water and the more rapid recycling of nutrients should stimulate growth of the crop trees, but study results documenting reduced growth after low-intensity prescribed fires have been reported (Boyer 1987; Zahner 1989). Other studies show increased crop tree growth even after obvious crown damage (Johansen 1975). A strong possibility is that burning under conditions of low soil moisture results in the death of many feeder roots. Threshold moisture levels are currently unknown, but the problem can be minimized by burning only under high soil moisture conditions and by utilizing weather and firing techniques that will ensure some of the forest floor remains. Crow and Shilling (1980) give an excellent summary of the use of prescribed fire to enhance pine timber production in the southern U.S.

The ecological shortcomings and potential risks associated with extensive areas of pine monoculture are well recognized. Goldammer (1982) suggests that prescribed burning can dramatically reduce some of these risks and shortcomings, thereby helping to stabilize these plantations ecologically. Reducing the heavy accumulation of litter allows an understory of herbaceous and woody vegetation to become established, thereby increasing biodiversity of the area. Establishing a hardwood understory in pine plantations in the Federal Republic of Germany resulted in a reduction of insect pests (Luedge 1971). Craighead (1977) recalled the original forest insect damage survey in California, U.S.A., where much lower bark beetle losses were observed on prescribed burned areas than on unburned lands. Weaver (1959) noted the same situation in Oregon, U.S.A., where bark beetle populations were endemic on an old burn but epidemic in the surrounding unburned forest. Hedden (1978) stated that the only practical way to reduce the risk of southern pine beetle (*Dendroctonus frontalis*) in the southern U.S.A. was to practice intensive forest management, including the use of prescribed fire.

12.3 Fuel Appraisal

Although pine plantation fuels are relatively homogeneous (Fig. 1), they can still vary considerably within a plantation and especially between plantations, depending upon such factors as natural vegetation patterns, species planted, and inherent site productivity, to name a few. Fuels also change over time as a plantation ages and in response to stand-tending measures. Because fuel appraisal is such an important prerequisite to accurate fire behavior prediction,

Fig. 1. A typical fuel load of undecomposed needles in a 9-year-old *P. elliottii* stand. Note the lack of understory and the ladder (aerial) fuels. Fazenda Monte Alegre, Klabin do Paraná, Brazil (April 1981).

a fuel classification and mapping system should be developed for your management area (Fig. 2). Site-specific knowledge of the complexity and spatial distribution of fuels is needed for many other fire management tasks as well, such as in fire-danger rating and to determine initial attack responses to wildfire.

12.3.1 Natural Vegetation

After planting, herbs and other early successional plants tend to rapidly dominate a site. By the second year, woody vegetation generally becomes well established and competes with the pine seedlings for sunlight, water, and other nutrients. Many of these species are light-intolerant and are shaded out as the pine canopy closes with age. Some species, however, persist and eventually form a dense understory. A forest floor begins to form as litter accumulates from the

Fig. 2. Use of the planar intersect technique for assessing the surface fuel load in a 15-year-old *P. taeda* plantation after third thinning. Fazenda Monte Alegre, Klabin do Paraná, Brazil (October 1982)

annual needlefall. This build-up of dead fuels continues until the rate of decomposition approximates that of accretion and a dynamic equilibrium is reached. Cultural practices such as thinning and/or pruning can add significant amounts of flammable debris. The burnable fuels on a site make up the total fuel complex. This total is called the fuel loading and is expressed as an ovendry weight on a per unit area basis. Large-diameter living stems are generally not included in fuel loading estimates, although they can be. That portion of a fuel load that would actually be consumed under specified burning conditions is called available fuel.

12.3.2 Available Fuel

The amount of fuel that is available under a given set of burning conditions depends upon the prevailing weather (mainly wind and relative humidity) fuel parameters (such as size, shape, moisture content, arrangement, and continuity), slope, and the intensity of the fire itself.

To help estimate available fuel, the fuel load is often partitioned into the categories of live and dead because moisture levels in living plants are primarily controlled by life processes while those in dead fuels are dictated by the weather – mainly precipitation, relative humidity, and temperature. A further distinction is often made between surface and aerial fuels.

Underburning

The quantity of readily burnable fuels on the forest floor is one of the most important factors in determining the rate of spread and intensity of a fire. Even though decomposition tends to be fairly rapid in the tropics, growth in industrial pine plantations is also very rapid and a thick litter layer develops (Goldammer 1982). Prior to crown closure, the herbaceous plant cover prevents the sloughed needles from forming a flat compact bed; instead, these needles and twigs remain partially elevated in jack straw fashion producing an extremely flammable fuel bed several decimeters deep. Thickness of the litter layer varies with species and with site parameters. For example, Schutz (1987) found the minimum litter depth was the same (3 cm) under stands of *P. patula*, *P. elliottii*, and *P. taeda* in the Eastern Transvaal, but the average (10 cm) and maximum (35 cm) depths under *P. patula* were about twice that of the other two species. He also found that litter thickness increased with altitude and wet soil conditions and decreased as site index increased. The number of stems per hectare had little effect once crown closure had taken place.

Aerial fuels with the most influence on fire behavior are the canopies of highly flammable understory species, and dead needles, twigs, and small branches suspended on understory vegetation or on the lower branches of unpruned crop trees. These suspended fuels are commonly referred to as ladder fuels because they provide a pathway for fire to carry from surface fuels into the crowns of shrubs and trees. Suspended fuel loads in pine plantations in the tropics tend to be heavy, creating hazardous burning conditions. Although living tree crowns become available fuels only under conditions favorable to the development of crown fires, their volume becomes highly important when timber-cutting operations convert them to logging residues.

Debris Burning

Depending upon arrangement and distribution, all but the larger size classes are generally available when broadcast burning logging debris. It is not generally necessary, nor desirable to consume all fuel on the site (see Sect. 12.2.2). If a burn objective is to consume larger fuels (over 5 or 6 cm in diameter), piling will probably be necessary. Mechanical piling in wet weather should be avoided. Keep the piles small, circular, and free of dirt.

12.3.3 Fuel Moisture

The moisture content of forest fuel is the most important parameter affecting forest fire behavior. This quantity is in a constant state of flux as fuels respond to the ever-changing environmental factors of precipitation, humidity, and temperature. The size and shape of a dead fuel determines how fast it will take on or lose water in response to a change in its surrounding environment. Timelag is a measure of the speed of this response, the shorter the timelag the quicker the response. In the U.S. National Fire Danger Rating System (Deeming et al. 1977; Burgan 1988), fuels are classed by diameter or surface-to-volume ratio to fit 1, 10, 100, or 1000 h timelag classes. Fuels less than about 6 mm (0.25 in) in diameter with a timelag of 1 h or less are called fine fuels or flash fuels and include grass, the uppermost layer of hardwood leaves and/or needles on the forest floor, and draped pine needles.

Before a fuel will ignite, the moisture in it has to be heated to the boiling point and evaporated, a process which takes considerable heat. Moreover, the resulting water vapor dilutes the oxygen and interferes with the gaseous phase of combustion. Thus when moisture content is high, fires are difficult to ignite, and burn poorly, if at all. With little moisture in the fuel, however, fires start easily, and tend to burn intensely and spread rapidly (Brown and Davis 1973).

Underburning

Fine-fuel moisture below 6 or 7% is a good indicator that conditions are too dry to prescribe burn; fires burn hotter, spotting is a problem, containment becomes more difficult, and overstory tree roots are very likely to be damaged. Although the usual objective of hazard reduction burns is to reduce fuel loading on the forest floor, this material should not be completely eliminated. The lower litter should always be checked before burning to make sure it feels damp; this will help ensure that some duff remains to protect the soil surface. A successful burn can be achieved even when the lower layer of the forest floor is soaking wet as long as the moisture content of the uppermost layer is below 30%. As fine-fuel moisture approaches 35%, however, prescribed fires tend to burn slowly and irregularly, often resulting in incomplete burns.

The rate of response of different fuel types can vary, and some types can reach different moisture contents under the same temperature/humidity conditions. For example, areas of cured standing grass can often be burned within hours of a drenching rain if good drying conditions exist. The drying process is slowed down inside a pine stand, depending upon factors such as stand density, height to the base of the overstory canopy, and the presence and stature of an understory. Workers have noted a fairly rapid decrease in fuel moisture during the morning and a much slower moisture recovery during the afternoon and evening, even with a rapid rise in humidity. McArthur (1971) reports that good burning conditions frequently occur in the late afternoon and evening when relative humidities are as high as 80 to 85%. Because of these natural variations, on-the-ground knowledge of fuel moisture is essential. Burgan (1987a) compares three procedures for estimating the moisture content of fine dead fuels. Fuel moisture is so important that virtually all fire prediction schemes include some measure of it.

Debris Burning

After cutting, it takes at least several weeks for the severed tree tops to cure. If branch ends are left attached to larger pieces, the foliage acts as a wick and draws moisture from the larger-diameter fuels so that they dry faster than if cut into shorter lengths. Once the needles turn a greenish yellow the debris is ready to burn. Although heavy concentrations of fuel, such as harvested areas, will burn (once ignited) when fuel moistures are above 30%, logging debris should be burned when dry: fuels will ignite more easily, burn more quickly and completely, require less mop-up and result in less of an impact on air quality. Teh-hour fuel moisture (fuels 0.6 to 2.5 cm in diameter) is a better indicator of burning conditions in logging debris than is fine-fuel moisture.

If logging slash is to be piled, allow this debris to cure for several weeks before piling because drying conditions are exceeding poor in the middle of a pile, especially if it is compacted or contains much dirt. Most of the smoke problems associated with burning are caused by inefficient combustion of damp soil-laden piles. These piles often smolder for weeks.

12.3.4 Evaluating Fuel Inputs

Numerous fuel classification schemes have been devised over the years to predict fire behavior and thus effects (e.g., Byrne 1980; de Ronde 1980; Fahnestock 1970). Such empirical models have many advantages, two major ones being their ease of construction and their ability to utilize accumulated experience. However, the importance of various fuel parameters can significantly change both spatially and temporally, and empirical models are rather inflexible. Models derived using a theoretical approach can accomodate such changes much easier (see Catchpole and de Mestre 1986 for an overview). A good example is the Australian MK III Prescribed Burning Guide developed for eucalypt fuels but also used in pine plantations. This empirical guide initially worked very well, but evolving site preparation, planting, and prescribed burning techniques have changed the fuels to the point where usefulness of the guide has been severely limited (Hunt and Crock 1987). Fuel classification using photographs is a widely used technique in the U.S. and photo series have been developed for most major fuel types (e.g., Anderson 1982). The technology is readily available (Maxwell and Ward 1980; Schmidt 1978).

In the United States, a system of user-friendly computer programs called BEHAVE has been developed to predict behavior of a fire. One of 13 stylized fuel models can be chosen or users can follow instructions to construct their own site-specific fuel models. Fuel data and prescribed weather conditions are combined with slope to produce fire spread and intensity outputs (Burgan and Rothermel 1984; Burgan 1987b). These programs will run on a hand-held calculator so they can be used in the field to make on-the-spot decisions as burning conditions change.

It is difficult to recommend what fuel appraisal method to use because, except for BEHAVE, most existing systems have been based on certain regional requirements. Users will have to evaluate these systems in light of their own requirements to determine whether they can use or modify an existing model, or will have to develop a new one.

Logging residue appraisal is a somewhat unique situation requiring different criteria because the majority of the fuels are severed tree parts. Methods to estimate these fuels have received much atttention in the northwestern U.S. where large volumes are left on site after harvest of old-growth conifer forests. Examples include Brown (1974) and Puckett et al. (1979). Harvesting pine plantations results in much more complete utilization, so relatively little material is left on the site requiring treatment. Moreover, in the past this residue was usually concentrated into windrows or piles, which further reduced the need to develop models.

12.4 Weather and Topographic Considerations

Knowledge of weather is the key to successful prescribed burning! Given adequate fuel, it is past and current weather that determines if and how a fire will burn. Wind, relative humidity, temperature, and precipitation are the more important elements to consider. These factors all influence fuel moisture which, as mentioned previously, is the single most critical factor in determining fire behavior. If smoke management is a concern, airmass stability is also important. Good prescribed burning conditions may exist for only a short period of time (hours or days), so their impending arrival needs to be recognized as soon as possible to fully utilize them. In areas where extensive prescribed burning is done, good burning weather is usually the limiting factor, although aerial ignition techniques have helped tremendously in this area. If you contemplate using prescription fire, it is mandatory for you to become familiar with local weather patterns that are favorable for prescribed burning as well as local "watchout" situations. Before igniting a fire, always obtain the latest weather forecast for the day of the burn and the following night. When possible, get a 2-day weather outlook. Maximum daily fire danger generally occurs between 12.00 and 14.00 h in the tropics, but this is not necessarily when your meteorological agency takes its observations.

Weather observations should always be taken at a prescribed burn site prior to, during, and immediately after the fire. Such observations are important because they serve as a check on the weather forecast and keep the burning crew up to date on any local influences or changes. Precipitation amount in particular

Fig. 3. A typical belt weather kit including additional wind meter

can vary widely between locations. Measurements taken in an open area, on a forest road, and in an adjacent stand are likely to be considerably different. Weather readings during a fire should be taken in a similar fuel type upwind of the burn to avoid heating and drying effects of the fire. Readings should be taken and recorded at 1- to 2-h intervals during a fire and whenever a change in fire behavior is noticed. By taking periodic readings and observing cloud conditions, a competent observer can obtain a fairly complete picture of current and approaching weather. Inexpensive easy-to-use belt weather kits are available from forestry supply mail-order outlets in the U.S. and elsewhere (Fig. 3).

12.4.1 Wind

In level terrain, wind determines the direction of fire spread. On-site wind direction may be influenced by nearby topographic features and on-site windspeed will vary with stand density, understory stature, and height to the bottom of the forest canopy. Windspeed generally increases after daybreak, reaching a maximum in the early afternoon, and then decreases to a minimum after sunset.

Underburning

The preferred prescribed burning windspeed range measured at eye level in a stand is 2 to 5 km h^{-1}. These are the most desirable windspeed ranges for prescribed burning, but specific conditions may dictate other speeds. Windspeed readings in weather forecasts are, however, often taken about 6 m aboveground in the open and are the maximum expected, not the average for the day. The minimum 6 m windspeed for burning is about 10 km h^{-1}. An often used maximum is about 32 km h^{-1}, but higher windspeeds can probably be tolerated. This is because once the tree crowns have closed, eye-level winds in the interior of a stand rarely seem to exceed about 8 km h^{-1} regardless of the strength of the 6-m windspeed.

High windspeeds keep flames bent over and rapidly dissipate heat from a backing fire. The result is less crown scorch than from a fire backing into a low-speed wind. In-stand windspeeds should be in the low to middle range (2 to 3 km h^{-1}) when heading fires are used. With high windspeeds, fires burning with the wind spread too rapidly and become too intense, dramatically increasing the probability of damage to the stand. On the other hand, enough wind should be present to keep the heat from rising directly up into tree crowns. Lulls in the wind allow flames to stand up straight, and heat to rise directly up into tree crowns.

Of equal importance to windspeed is the length of time the wind blows from one direction. Changes in wind direction mean changes in the rate of fire spread. Winds are characterized by continual deviations in speed and direction. Swings in wind direction of plus or minus 10 to 15° are common and are of little consequence providing they do not persist. Changes greater than 45°, however, even if momentary, often result in hot spots (areas of crown scorch) when using a backing fire. Persistent steady winds are associated with certain weather patterns; become familiar with those in your area. For example, near large bodies of water, sea and land breezes are often utilized; but since they are diurnal phenomena, strict attention must be paid to arrival and departure times to avoid dangerous shifts in fire intensity.

Debris Burning

Winds are stronger in open areas than they are in the forest. Since there is no overstory to protect, some increase in fire intensity is acceptable. Moreover, wind is not needed to cool the heated combustion products. When broadcast burning, eye-level winds over 6 or 7 km h^{-1} can create containment problems if a heading fire is used. Always backfire the downwind side first! When piled or windrowed debris is burned, eye-level winds of 14 to 17 km h^{-1} can be tolerated by adjusting the firing pattern. Be aware of smoke-sensitive areas downwind and spotting potential.

With light variable winds, a convection column is often generated. Ignition patterns can be selected to encourage column formation in order to minimize fire control problems along the burn perimeter and to exhaust the smoke high into the atmosphere where it quickly disperses with a minimum impact on ground-level air quality. Wind velocity (speed and direction) can change substantially with height and these upper (transport) winds determine offsite movement of the smoke. Some of these upper wind profiles have dangerous fire behavior characteristics associated with them (e.g., high spotting potential, or a sudden dramatic increase in fire intensity) and are called adverse profiles (Byram 1954). Thus knowledge of existing and forecast upper wind profiles is desirable before igniting a fire that is likely to generate a convection column. This information is, however, generally difficult to obtain. If near a commercial airport, a good source is the weather data used to brief pilots.

12.4.2 Relative Humidity

Relative humidity is an expression of the amount of moisture in the air compared to the total amount the air is capable of holding at that temperature and pressure. Each 11°C rise in temperature (which often occurs during the morning hours on a clear day) reduces the relative humidity by roughly half, and likewise, each 20°C drop in temperature (which often occurs in early evening) causes relative humidity to roughly double.

Underburning

Preferred relative humidity for prescribed burning varies from 30 to 50%. In some locations relative humidities within this range rarely occur and burns have to be conducted under less than ideal conditions. When relative humidity falls below 30%, prescribed burning becomes more dangerous. Fires are more intense under these conditions and spotting is much more likely; proceed only with additional precautions. When the relative humidity exceeds 60%, a fire may leave unburned islands or may not burn hot enough to accomplish the desired objectives. However, when using aerial ignition in heavy fuels or under windy conditions when tall, partially cured grasses are abundant, even higher relative humidities may be desirable.

The moisture content of fine dead fuels (e.g., pine needles, cured grass) responds rapidly to changes in relative humidity but there is a timelag involved. Previous drying and wetting cycles and the amount of moisture in adjacent fuels and soil also influence response time. Therefore, relative humidity and fuel moisture should be assessed independently.

Debris Burning

Relative humidity (along with temperature) controls fuel moisture content up to about 32%. Ignoring the movement of moisture from wet to drier layers of fuel or soil, liquid moisture such as rain or dew must contact a fuel for the moisture content to rise above 32%. The extent of the rise depends upon both the duration and amount of precipitation.

Recently cut pine tops have a drying rate that is somewhat independent of relative humidity, but once this material initially dries to a moisture content below 32%, it then behaves as other dead fuels and becomes more responsive to daily fluctuations in relative humidity. The response to changes in relative humidity is much more rapid in fine dead fuels suspended above the ground than it is in those that have become part of the litter layer. These suspended fuels are not in direct contact with the damp lower litter, and are more exposed to the sun and wind.

When burning piled debris, high humidities have little effect on fire behavior once large-diameter fuels have been ignited. Low humidities (below 30%), however, will promote spotting and increase the likelihood that the fire will spread between piles.

12.4.3 Temperature

Temperature strongly affects moisture changes in forest fuels. High temperatures help dry fuels quickly. Fuels exposed to direct solar radiation become much warmer than the surrounding air. Moisture moves from warmer fuel to the air even if the relative humidity of the air is high. On the other hand, temperatures below freezing retard fire intensity because additional heat is required to convert ice to liquid water before it can be vaporized and driven off as steam. Consequently, it does not take much moisture under these conditions to produce a slow-moving fire that will leave unacceptably large areas unburned.

Underburning

The instantaneous lethal temperature of living plant tissue is about 62 °C. Cool air temperatures are thus recommended for hazard reduction burning because it is less likely that foliage and stem tissue will be heated to lethal temperature levels. When the objective is to control undesirable species, growing season burns with temperatures above 25 °C are desirable. Lethal temperatures are more likely to be reached in understory stems and crowns with high ambient temperatures (see Sect. 12.8.1). Of, course, the overstory pines must be large enough to escape injury. Larger trees have thicker bark and their foliage is higher above the flames, which allows more room for the hot gases to cool before reaching the crowns.

Debris Burning

Cleared areas are often burned when ambient air temperatures are high. There is no overstory to protect and surface heating from direct sunlight usually increases the mixing height which helps disperse the smoke. It is particularly important to use an ignition pattern such as center firing when ambient air temperatures are high. This tactic creates strong convection in the center of the block. Then as the outside perimeter is ignited, the smoke from this ring fire is drawn away from the firelines to the central column thereby preventing heat damage to trees in adjacent stands and reducing the threat of an escape.

12.4.4 Precipitation and Soil Moisture

Rainfall has a pronounced affect on both fuel moisture and soil moisture. It is thus important to have a good estimate of the rain falling on an area to be burned for a few weeks prior to planned ignition. Rainfall amounts are fairly uniform with some large weather systems but they vary widely with others. Shower activity and amount is notoriously difficult to predict, even for nearby sites. The only reliable method to determine the amount of precipitation that actually falls on a site is to place an inexpensive raingauge on the area.

Underburning

The importance of adequate soil moisture cannot be overemphasized. Damp soil protects tree roots and microorganisms. Underburning in pine plantations should cease during periods of prolonged drought and resume only after a soaking rain of at least 2 cm. In older plantations that have been thinned or pruned and previously prescribed-burned, and which have a herbaceous groundcover and good air movement, it is sometimes possible to burn within 4 to 6 h after an early morning shower.

On soils with low infiltration and percolation rates, much rainfall is lost through runoff and duration is more important than amount. For example, on a typical clay soil in the southern U.S., 25 mm of rain occurring in 1/2 h will not produce as large a moisture gain as 13 mm falling over a 2-h period.

Debris Burning

After a rain, fuels in cleared areas dry much faster than those under a tree canopy because of greater solar radiation and higher windspeeds. This differential drying can be used to advantage from a fire-control standpoint. Broadcast-burn the cleared area several days after a hard rain while fuels in the surrounding pine stands are still damp. Burning under these conditions assures good soil moisture. However, when burning harvested areas, soil damage is as much a function of fire intensity and duration as it is of soil moisture. Intense long-duration fires will bake the soil regardless of the moisture present. The biological, chemical, and physical properties of the soil are all likely to be altered (see Sect 12.8.4). Avoid these severe burns which alter the color structure of the mineral soil, particularly on clay soils and steep slopes. These undesirable fire effects are often produced when burning windrowed or piled debris, and are a major reason why piling and especially windrowing logging debris prior to burning is discouraged.

12.4.5 Slope

The slope of the land has an effect on rate of fire spread similar to that of wind. A 5° slope will increase fire spread by 33% and a 10° slope will double the forward rate of spread compared to that on level ground (McArthur 1971). When a fire burns upslope in the absence of wind, the flames are closer to the unburned fuels ahead of the fire than they would be on level ground. Since radiation decreases with the square of the distance, slope steepness has a significant affect on fuel preheating. By the same token, a fire will move downhill at a slower pace because of decreased preheating of fuels, but be on the lookout for rolling debris that can carry fire to the base of the slope where it will then spread rapidly back upslope at a much higher intensity. Because trees grow vertically and not perpendicular to a slope, the uphill sides of their crowns will be much closer to the ground and subject to increased fire damage.

As a slope is heated by the morning sun, an upslope convective flow develops. This breeze will increase to a maximum during the early afternoon and decrease to zero as the slope cools in the evening. As the slope continues to cool, a downslope wind will develop, reaching a reduced maximum after midnight. This breeze ends shortly after sun-up as the slope begins its daily heating cycle. If a fire is ignited at the base of a slope during the day, differential heating will be greatly increased and the fire will rapidly spread uphill. If this burned slope continues to smolder into the night, the smoke will drain downhill and concentrate in low areas.

12.5 Fire Behavior Prediction

Fire behavior is the result of the interaction between weather and fuel conditions, topography, firing technique, and ignition pattern. Measures of fire behavior are useful in comparing fires, in fire suppression planning, and in predicting fire effects. Uniform fire behavior descriptors are necessary for the accurate prediction of prescribed fire effects, but no single, easy-to-measure descriptor has emerged that can be used to assess adequately the full array of fire impacts on a site (Wade 1986). Rate of spread is of prime importance and is thus an integral part of many descriptors, but leaves much to be desired when used by itself. Several descriptors which have proven useful will be briefly discussed below, but each has limitations. Byram's Fireline Intensity is a good descriptor to use for correlating fire behavior with fire effects above the flame zone. Reaction intensity and residence time should both correlate well with fire effects resulting from flame contact. When belowground effects are of interest, either heat per unit area or depth of burn can be used. A number of alignment charts and nomographs have been developed that allow a person to switch from one fire behavior descriptor to another (e.g., Albini 1976; Andrews and Rothermel 1982).

12.5.1 Descriptors

Fireline Intensity

This descriptor is also called frontal fire intensity or Byram's Intensity. Byram (1959) defines this term as the heat energy released per unit length of fire front per unit time irrespective of the depth of the flame zone. It is the product of the low heat of combustion (heat yield), the weight of fuel consumed in the flame zone (available fuel), and the rate of fire spread (Byram 1959). The equation is:

$I = Hwr$; where I = fire intensity in kW m^{-1},
H = heat yield in kJ kg^{-1},
w = weight of available fuel in kg m^{-2},
r = rate of spread in m s^{-1}.

The same output value can result from different values of input variables. The low heat of combustion approximates the heat yield and remains fairly constant so that for a given fireline intensity, a decrease in forward rate of spread implies an increase in the amount of fuel consumed. Slow rates of spread will thus concentrate more heat on the lower portion of a tree bole. Available fuel includes fuel consumed in the flame zone but not that consumed behind the flame zone (which can be considerable). In practice, however, consumption (which also includes that fuel consumed after the fire front passes) is often substituted for available fuel. Alexander (1982) presented an in-depth discussion of the attributes of this descriptor. Wade (1983) described fireline intensity levels and associated headfire behavior to help plan prescription burns in *P. elliottii* stands in the southern U.S.A. (Table 3).

Reaction Intensity

Originally called combustion rate by Byram (1959), this term is defined as the rate of heat release per unit area per unit time in the flame zone. It can be derived by dividing fireline intensity by flame zone depth. Again, the same output value can be produced by varying the values of the input variables.

Residence Time

Residence time is defined as the length of time it takes the flame zone to pass a given point. Numerically, it is the depth of the flame zone divided by rate of spread. As large fuels ignite and burn, residence time will increase, resulting in a stronger heat pulse to the site.

Table 3. Fireline intensity guidelines for prescribed heading fires in *P. elliottii* stands

Fire intensity kW m^{-1}	Description of fire behavior
< 70	Intensity too low. Flame lengths less than 0.3 m. Very patchy burn. Scorch not a problem.
71–250	Optimum range. Flame heights 0.3–0.9 m. Scorch heights generally below 4.5 m. Little chance of fires exceeding control lines.
251–425	Too hot for use in immature stands. Flame heights 0.9–1.1 m. Scorch heights 4.5–9.0 m. Downwind control line should be backfired before setting headfire to prevent fire control difficulties.
426–700	Upper fire intensity limits. Flame lengths generally below 1.7 m. Scorch heights may be excessive even with steady winds. Always backfire downwind side of plot first. Think twice about using a heading fire. Have mechanized fire control equipment standing by.

Heat per Unit Area

The total amount of heat released per unit area during the time period combustion is taking place in that unit area is called the heat per unit area. It can be calculated directly by multiplying the low heat of combustion by the amount of fuel consumed. How this heat energy is distributed vertically within a unit area depends upon the dimensions of the flame envelope. Small flames will concentrate it near ground level, while longer flames will release some higher above the ground surface.

Depth of Burn

This parameter is simply a measure of the vertical depth that the forest floor was reduced during a fire. This descriptor has been expressed as the thickness of the layer removed, as a percentage of the total layer, and as the percentage of the burn upon which mineral soil was exposed. According to Alexander (1982), this depth depends mainly on the moisture gradient of the forest floor.

Even though two fires have the same descriptor value, the effects might be dramatically different. This anomaly can be caused for several reasons including: (1) the output values are only as good as the input values, (2) different combinations of intensity descriptor input variables can yield the same output value, (3) the variability inherent in most fires may make the mean value of a descriptor a poor indicator of fire effects, and (4) pre-fire and post-fire conditions such as plant vigor and the incidence of insect and disease attack may vary between sites.

12.5.2 Fire Behavior Models

In the U.S.A. more than a decade of fire modeling has led to the creation of the BEHAVE computer program that can make use of 13 standard fuel models, revise them, or create new ones, and then combine the output with weather and topographic inputs to produce site-specific fire behavior predictions. (Andrews 1986; Burgan and Rothermel 1984). Rothermel's (1972) fire spread model is used for fire behavior calculations in BEHAVE. Recent work by Nelson and Adkins (1988) using dimensionless correlation has led to a much simpler model that expresses fire spread rate in terms of fuel consumption, ambient windspeed, and residence time. In Australia, a widely used method of predicting fire behavior was developed using fire danger meters to determine a fire danger index and a hazard rating from inputs of a drought index (McArthur 1966, 1967). Crane (1982) added pocket calculator programs to the system.

12.5.3 Predicting Crown Scorch Height

Crown scorch is often used as the primary indicator of burn success. Scorch becomes obvious soon after the burn, is easy to measure, and can result in reduced growth rates or even mortality. Scorch is a function of the amount of heat that reaches the canopy. It is dependent upon fire intensity, understory involvement which is reflected in flame length, the distance to the base of the crown, crown density, ambient temperature, and windspeed. In theory, scorch height should vary with the 2/3 power of fireline intensity. Van Wagner (1973) validated these calculations based on field work in Canada. Interestingly, he found that including ambient temperature and eye-level windspeed did not significantly increase the reliability of his basic model:

$$h_s = 0.1483(I)^{2/3}$$; where h_s is height in m, and I is fireline intensity in kW m^{-1}.

McArthur (1971) graphed the relationship between fireline intensity, scorch height, and flame height for *P. elliottii* and *P. caribaea* stands in Fiji. De Ronde developed a relationship based on a flame height:scorch height ratio of 1:6 with correction factors for windspeed and air temperature. The following tables can be used to arrive at these correction factors (Tables 4 and 5).

To arrive at predicted scorch height use the following procedure:

$$\text{(Predicted scorch height)} = \{(\text{flame height}) \times 6\} \times (\text{wind correction factor}) \times (\text{temperature correction factor}),$$

where both scorch height and flame height are in m.

This height can now be compared with the distance to the bottom of the canopy to determine whether or not it is safe to burn. Increment loss usually does not occur until about 50% of the crown is damaged. Thus a good rule of thumb is to cancel a burn if scorch height is predicted to exceed 33% of the average crown height of the stand. Because *P. caribaea* can withstand

Table 4. Scorch height correction factor for windspeed

Average flame height (m)	Windspeed (km h^{-1}) at ca. 6 m above the forest floor						
	0	2.5	5.0	7.5	10	12.5	15
0.25	–	0.83	0.67	Unreliable data			
0.50	–	0.92	0.83	0.75	0.67	0.58	0.50
0.75	–	0.94	0.89	0.78	0.67	0.58	0.50
1.00	–	0.94	0.89	0.78	0.67	0.58	0.50
1.50	–	0.94	0.89	0.81	0.72	0.61	0.50
2.00	–	0.96	0.92	0.83	0.75	0.67	0.58
2.50	–	0.98	0.97	0.92	0.87	0.81	0.77
3.00	–	0.99	0.98	0.96	0.94	0.88	0.83

Table 5. Scorch height correction factor for air temperature

Air temperature (°C)	Scorch height correction factor
8–11	0.6
12–15	0.7
16–20	0.8
21–24	0.9
25–28	1.0
29–32	1.1
33–36	1.2

more scorch than other pines studied, McArthur (1971) suggested a scorch limit of 50% of the predominant stand height when prescribed burning under this species.

12.6 Prescribed Burning Techniques

Once the desired fire behavior is determined, the proper firing technique can be selected. The technique chosen must be closely correlated with burning objectives, fuels, weather, and topography to facilitate control and prevent damage to the resource. The proper technique to use can change as these factors change.

Fires either move with the wind (heading fire), against the wind (backing fire), at right angles to the wind (flanking fire), or in some combination of the above. The movement of any fire can be described in these terms. For example, a spot fire would exhibit all three types. Heading fire is the most intense because of its faster spread rate, wider flame zone, and longer flames. Backing fire is the least intense, having a slow spread rate regardless of windspeed. This type of fire has a narrow flame zone and short flames. If slight changes in fuels or weather are encountered, consider combining two or more firing techniques to achieve the desired result. A continuous line of fire always spreads faster and thus builds up intensity quicker than does a series of spot ignitions spaced along the same line. Fireline intensity abruptly increases when two fires burn together (called the convergence zone). The magnitude of this increase is greater when fires converge along a line rather than along a moving point. The line of scorch often seen paralleling a downwind control line delineates the zone where a heading and backing fire met.

The residence time of heading and backing prescribed fires is often about the same because the deeper flame zone of a heading fire compensates for its faster movement. Backing fires generally consume more forest floor fuels than do heading fires. The total heat applied to a site may be roughly equal for both heading and backing fires as long as additional fuel strata are not involved, but in a backing fire the released heat energy is concentrated closer to the ground.

Prescription fires are often ignited with driptorch fuel (consisting of 20 to 35% gasoline and 65 to 80% diesel or kerosene) dispensed manually from a driptorch (Fig. 4), or with flamethrowers which use gelled gasoline (made by mixing alumagel with gasoline). Common aerial ignition tools are the helitorch which also uses gelled gasoline and the Delayed Aerial Ignition Device (DAID) which is a round polystyrene capsule (hence the common name "ping-pong ball") containing potassium permanganate that is injected with ethylene glycol (antifreeze) to produce a thermo-chemical reaction. The two basic methods of igniting fires are as a solid line or as a series of point sources but many ignition tools cannot be used for both (at least from a practical standpoint). Ignition equipment must thus be compatible with ignition plans.

12.6.1 Backing Fire

A backing fire must be started along a downwind baseline such as a road or plowline and allowed to back into the wind. Variations in windspeed have little effect on the rate of spread of a fire burning against the wind. Some wind is necessary to give the fire direction but backing fires do not seem to spread much faster than about 1 m min^{-1} no matter what the windspeed. Backing fire is the easiest and safest type of prescribed fire to use provided windspeed and direction are steady. Once the fire backs away from the control line, escape is unlikely, resulting in less worry about damage to adjacent stands. Because crown scorch is minimized, this technique lends itself to use in heavy fuels. The initial fire in pine plantations should be a backfire except in rare instances such as young overgrazed stands where a continuous litter layer has not yet formed.

Major disadvantages are the slow progress of the fire and the increased potential for root damage if the lower litter is not moist enough. When a large area is to be burned, it often must be divided up into smaller blocks with interior plow lines (usually every 100 to 300 m). These blocks must all be ignited at about the same time to complete the burn in a timely manner. Interior plow lines coupled with the extended burning period due to the slow rate of fire spread increase the costs of this technique. Once interior lines are constructed, changes in wind direction are difficult to accommodate. As fine-fuel moisture contents

Fig. 4a,b. Conduction of a prescribed burn for research purposes in the State of Paraná, Brazil (**a**) and in management of pine plantations in the Cape Province, South Africa (**b**)

approach 20% or when litter fuels are discontinuous, a backing fire will not carry well. Backing fires are line-source fires.

12.6.2 Strip-Heading Fire

In strip-head firing, a series of lines are set progressively upwind of a firebreak in such a manner that no individual line of fire can develop a high energy level before it reaches a firebreak or another line of fire. A backing fire should always be set first to secure the downwind base line and then the remainder of the area treated with strip-heading fires. Strips are often set 20 to 40 m apart. The distance between ignition lines is determined by the desired flame length. The spacing between lines should change throughout a burn to adjust for slight changes in topography, stand density, weather, or the type, amount, or distribution of fuel. Changes in wind direction of up to about 45° can be accommodated during a fire by altering the angle of the strips. Strip-heading fires permit quick burnout and provide good smoke dispersion. A short distance between strips keeps an individual line of fire from gaining momentum but is time-consuming and uses a lot of torch fuel. Furthermore, higher intensities occur whenever lines of fire burn together, increasing the likelihood of crown scorch.

An effective method of reducing fire intensity is to use a series of spots or short 1- to 2-foot-long strips instead of a solid line of fire. An added advantage of these short strips or spots is that driptorches will not have to be filled as often. Prescribed burners often start with a backing fire, and after observing its behavior decide to try a few spots, and then a line of headfire if the spots behave as expected. They then increase the distance between successive strips until they reach the balance between speed and intensity that they are comfortable with. This technique is often used under high relative humidities (55 to 65%) and high fine-fuel moistures (20 to 35%), when backing fires will not accomplish the burn objectives. Heading fires are very sensitive to changes in windspeed. Only enough wind to give a heading fire direction (1 to 3 km h^{-1} in-stand) is needed because flame length and thus fireline intensity increase with wind speed. If interior plow lines are present and can be utilized so only a single line of heading fire is used in each small block, the number of zones of increased intensity where each heading and backing fire meet can be reduced or eliminated.

Allowing a heading fire to move across an entire block without stripping should only be done on areas with light fuel loadings. This generally restricts use of this technique to stands burned within the past 2 years. To reduce the potential of canopy scorch, heading fires are usually conducted in the winter when ambient temperatures are cooler.

12.6.3 Point Source (Grid) Ignition

Experienced prescribed burners often switch from one firing technique to another during a burn as conditions change. When properly executed, spot ignitions will produce a fire with an intensity about twice that of a line-backing fire, and roughly 66% that of a line-heading fire. Timing and spacing of the individual ignition spots are the keys to the successful application of this method. First a line fire is ignited across the downwind edge of the block and allowed to back into the block to increase the effective width of the control line. A line of spots is then ignited at some specified distance upwind of the backing fire and the process continued until the whole block has been ignited. The Australian Mk.III fire behavior tables state a grid pattern should be used such that individual fires will link-up in about 2 h. In the USA, a pattern is usually selected such that link-up occurs within the first hour.

To minimize crown scorch, ignition-grid spacing is selected to allow the spots along a line to head into the rear of the spots along the downwind line before the flanks of the individual spots merge to form a continuous flame front. The merger of successive ignition lines thus takes place along a moving point rather than along a whole line at the same time. Merger along a moving point can be ensured by using a square grid. Close spacing between lines helps the individual spots to develop, but ensures that the head of one spot will burn into the rear of the downwind spot before the heading fire's potential flame length and intensity are reached. From a theoretical standpoint, conditions should be such that the individual spots reach a quasi-steady state and can thus be allowed to run as long as burning conditions do not change. However, where the diurnal change in burning conditions is virtually continuous and of considerable magnitude grid spacing is, of necessity, much closer. McArthur (1971) states that the interaction between individual fires can begin at a distance of five times the flame height so that a convergence zone effect can occupy almost 50% of the burnt area. He believes this will trigger a mass fire effect almost instantaneously with even a 20- to 40-m grid pattern. This could certainly happen under severe burning conditions but a 40- by 40-m grid is often used in the southern U.S. without undue worry.

Rectangular grids with wider spacing between lines than within a line should not be used because such a pattern may allow the spots along a line to merge into a line of heading fire before running into the rear of the downwind spots. Once the first few lines have been ignited and their fire behavior assessed, intensity can be regulated to some extent by changing the time between ignition points within a line, the distance between points, and the distance between lines. Thus the balancing act between spacing and timing should be continually adjusted as fire behavior reacts to both temporal and spatial changes in fuel and weather.

Intensity is decreased by widening the interval between ignition points along a line. If fireline intensity is still too high after doubling this interval while maintaining a 40 m distance between lines, firing should be halted. Allow the area to burn with a backing fire or plow it out. Although intensity of an individual spot is increased by widening the distance between lines, the average intensity of the burn as a whole is usually somewhat lower (because that portion of the area burned by heading fires decreases while that burned by flanking fires increases). Check to see that convergence zone flame lengths are within tolerable limits, and that other fire behavior parameters appear satisfactory. If everything is well within prescription, both the between- and within-line distances can be increased. This step will reduce ignition time and the amount of fuel used as well as decrease the number of convergence areas. The time needed to complete a burn can be reduced somewhat by offsetting successive ignition lines by one half of the within-line spacing. The heading fires from one line will then come up between the backing fires on the next line.

When using point source fires, much of the area will be burned by heading and flanking fires and very little by backing fires. Thus if conditions are ideal for line-backing fires, point source fires may get too intense. Preferred burning conditions include low (1 to 2 km h^{-1}) in-stand eye-level windspeeds. Wind direction can be variable. Moisture content of the fine fuels should be above 15% and in many situations it should be above 25%. Guidelines for using aerial ignition in pine plantations have been developed in Australia (Byrne and Just 1982) and preliminary recommendations made in the southern U.S. (Johansen 1987; Wade and Lunsford 1989).

12.6.4 Edge Burning

In those instances where it is not desired to prescribe-burn a plantation, fire is sometimes used to consume fine fuels around the periphery, or to treat the margins so as to create a buffer zone to stop or slow down an approaching wildfire. This practice is commonly called blacklining or edge burning, and can be very effective if done annually, although one must keep in mind that the woody understory will most likely be replaced by herbaceous fuels.

Although edge firing is not a firing technique in the true sense of the term, we include it here. Plantation edges generally burn with higher intensity than plantation interiors. The reasons for this are:

1. Edges are more exposed to the influences of sun and wind, leading to more rapid drying of fuels.
2. Increased sunlight at the edges stimulates the growth of understory species such as blady grass and bracken fern, which tends to cause fires to flare along these transitions zones.
3. Fire behavior at edges is influenced by wind. In plantations with a dense canopy the ratio of windspeed outside the forest to that inside is approximately 5 to 1 (Byrne 1980). Cooper (1965) and Albini and Baughman (1979) have developed some useful tables to predict windspeed inside stands. De Ronde has introduced these relationships in South Africa in a revised form.

Plantation edges are often burned soon after a rain before the plantation interior has dried sufficiently to carry fire. If nothing else is done, this process is called "blacklining", but if the interior is prescribed-burned at a later date, this is simply the first part of a two-step burn designed to minimize damage along the edge of the plantation.

12.6.5 Center and Circular (Ring) Firing

This technique is useful on harvested areas where a hot fire is desired to reduce logging debris and kill any unwanted vegetation prior to planting. This procedure should never be used for underburning because of the likelihood of severe tree damage as the fire fronts converge. As with other burning techniques, the downwind control line is the first line to be ignited. Once the baseline is secured, the entire perimeter of the area is ignited so the flame fronts will all converge toward the center of the plot. One or more spot fires are often lighted in the center of the area and allowed to develop before the perimeter of the block is ignited. The convection generated by these interior fires creates indrafts that help pull the outer circle of fire toward the center, thereby reducing the threat of slop-overs and heat damage to adjacent stands.

Slash disposal burns using this technique can be conducted under a wide range of weather conditions because there is no overstory to worry about. Caution should be exercised, however, particularly when the atmosphere is unstable because this type of fire tends to develop a strong convection column. If the burning materials lifted aloft in this column are not completely consumed before they drop out downwind, a considerable spotting problem can develop. Although the probability of fire whirlwind development is low, they can be produced with this technique, especially where heavy fuel concentrations and uneven terrain exist.

12.6.6 Pile and Windrow Burning

The objective of piling logging debris before burning it is to prolong fire residence time thereby increasing the consumption of large materials. Windrowing should be avoided whenever possible, however, because of the high probability of site degradation. Large amounts of topsoil usually end up in the windrows, preventing an efficient burn. The use of heavy equipment during wet weather exacerbates this situation and can result in soil compaction. Full exposure of the soil to the sun bakes the top layer. Furthermore, the direct force of raindrops will clog soil pores and often results in erosion on slopes. The area beneath the windrows is lost to production because the debris is rarely consumed completely; and what remains makes planting difficult unless the material is repiled and reburned. Even when windrows contain breaks every 25 to 50 m, they are a barrier to firefighting equipment and wildlife.

In many areas, the biggest deterrent to windrow burning is, however, its deleterious effect on air quality. Larger materials that often contain a lot of moisture are consumed, the windrow interior may not have dried, oxygen for good combustion is lacking in wide windrows, and large amounts of dirt are often mixed in. The result is a fire that continues to smolder for days or weeks. The smoke from smoldering combustion tends to stay near the ground, drifting and concentrating in low areas at night, resulting in potentially disastrous reductions in visibility.

Broadcast burning is generally a much better alternative even if more thought has to be given to ignition patterns to prevent formation of a dangerous convection column. Industrial plantations contain little cull material so, even though the area may be covered with branches and tops after harvest, a broadcast burn will usually be sufficient to consume most fuels under 5 cm in diameter. In those instances where an unacceptable amount of large, scattered debris must be concentrated to ensure consumption, round "haystack" piles should be constructed (Johansen 1981). It generally costs a little more to pile than to windrow logging debris, but piles are preferable to windrows because access within the area is no problem, planting is easier, burning is safer, and smoke problems are significantly reduced since piles burn out much more quickly! Keep piles small and minimize the amount of dirt in them so surface water can pass through and the debris can quickly dry. Always machine-pile when the ground surface is dry; less soil compaction will take place, and considerably less soil will end up in the piles. Allow fresh logging slash to cure first and to dry after rain. Then shake the debris while piling to remove as much soil as possible. If material is piled while green or wet, the centers of the piles take an exceedingly long time to dry. Piles that contain little soil and are constructed to permit some air movement will result in a burn that consumes significantly more debris and is over with much quicker. Burning piles can easily be "bumped" to remove any dirt and pushed in to increase consumption.

Techniques used to burn piled debris are somewhat fixed because of the character and placement of the fuel. The upwind side of each pile should be ignited. Under light and variable winds, the whole pile perimeter can be ignited. Burn-out can be speeded up by using gelled gas which results in a quicker build up of fire intensity than with traditional driptorch fuel.

12.7 Prescribed Burning Plans

Planning a prescribed burn involves considerably more than just writing a burning plan. A successful prescribed fire is one that is executed safely and is confined to the planned area, burns with the desired intensity, accomplishes the prescribed treatment, and is compatible with resource management objectives. Such planning should be based on the following factors:

1. Physical and biological characteristics of the site to be treated.
2. Land and resource management objectives for the site to be treated.
3. Known relationships between preburn environmental factors, expected fire behavior, and probable fire effects.
4. The existing art and science of applying fire to a site.
5. Previous experience from similar treatments on similar sites (Fischer 1978).
6. Smoke impact from both an esthetic, and health and safety standpoint.

The first step to a successful prescription is a stand-by-stand evaluation. Determine the needs of each stand and what actions should be taken to meet these needs. Alternatives to prescribed fire should be considered and a decision reached regarding the preferred treatment. Well in advance of the burning season, choose those stands to be burned. Overplan the number of acres to be burned during the coming season by 10 to 20% so substitutions can be made if necessary, and so additional areas can be burned if favorable weather continues. In many locations, the number of suitable burning days varies widely from year to year. Set priorities and specifically designate those burns which require exacting weather conditions. Considerations include heavy fuels, small trees, and potential smoke problems.

12.7.1 The Written Plan

The written burning plan should be prepared by a knowledgeable person prior to the burning season to allow time for interfunctional coordination and any necessary arrangements for manpower, equipment, financing, or other needs. Be ready to burn when the prescribed weather occurs. Some plans will be short and simple while others will be complex. In an area where broad variations in topography and type and amount of fuel exist, developing an effective burning prescription is difficult, if not impossible. Where practical, it is better to divide such areas into several burn units and prepare a separate burning plan for each. Treatment constraints should be included in, or attached to, the plan. Fischer (1978) named the following:

1. Environmental constraints (air quality, water quality, accelerated erosion).
2. Multiple use constraints (protection of other uses, resource management trade-offs).
3. Economic constraints (maximum cost per unit area).
4. Operational constraints (access, terrain, manpower).
5. Administrative constraints (policy, rules, etc.).
6. Legal constraints (fire laws, forest practice acts, etc.).

A prepared form with space for all needed information is best. The form will serve as a checklist to ensure that nothing has been overlooked. Numerous forms exist. A "simple" form that can be used on burns that are well within large landholdings that do not contain public roadways, and a form to use for post-harvest burns, both taken from Wade and Lunsford (1989), are presented (Fig. 5a,b). Some of the more important information that should be found in a plan is:

Required Signatures

Provide space for signature(s) of person who prepared the plan. This identifies the people who know most about the plan.

Simple Understory Prescribed Burning Unit Plan

Landowner ________________________ Permit no. ____________
Address ________________________ Phone No. ____________
S ___ T ___ R___ County ____________ Acres to Burn ____________ Previous burn date ____________
Purpose of burn ________________________
(Draw map on back or attach)

Stand Description

Overstory type & Size ____________ Height to bottom of crown ____________
Understory type & height ________________________
Dead fuels: description and amount ________________________

Preburn Factors

Manpower & equipment needs ________________________
List smoke-sensitive areas & locate on map ________________________
Special precautions ________________________

Estimated no. hours to complete ____________ Passed smoke screening system ____________
Adjacent landowners to notify ________________________

Weather Factors:	**Desired Range**	**Predicted**	**Actual**
Surface winds (speed & dir.)	______	______	______
Transport winds (speed & dir.)	______	______	______
Minimum mixing height	______	______	______
Dispersion/stagnation index	______	______	______
Minimum relative humidity	______	______	______
Maximum temperature	______	______	______
Fine-fuel moisture (%)	______	______	______
Days since rain ______	Amount ______	______	______

Fire Behavior:	**Desired Range**	**Actual**
Type fire	______	______
Best month to burn	______	Date burned ______
Flame length	______	______
Rate of spread	______	______
Inches of litter to leave	______	______

Evalution: **Immediate**	**Future**
Any escapes? ______ Acreage ______	Evaluation by ______
Objective met ______	Date ______
Smoke problems ______	Insect/disease dam. ______
% of area with crown discoloration of	Crop tree mortality ______
5-25% ___ 26-50% ___ 51-75% ___ 76%+ ___	% understory kill ______
Live crown consumption ______	Soil movement ______
% understory veg. consumed ______	Other adverse effects ______
Adverse publicity ______	______
Technique used OK ______	Remarks ______
Remarks ______	______

Prescription made by ________________________
Title ________________________ Date ___ / ___ / ___

Fig. 5a.

Postharvest Prescribed Burning Unit Plan

Prepared by ______ Signature ______ Date ______ Permit no. ______
State ______ County ______ District ______ Comp't ______
Burning Unit No. ______ S ______ T ______ R ______ Gross acres ______ Net acres ______
Landowner ______ Address & phone no. ______
Person responsible & how to contact day & night ______
(Draw map on back or attach)

A. **Description of Area:**

1, Natural stand or plantation ______ Stand age ______ Harvest date ______
2. Clearcut ______ Harvest method ______ Pine basal area removed ______
3. Organic soil ______ Hardwood basal area ______ Hardwoods utilized ______
4. Unmerchantable trees felled ______ Snags felled ______ Debris evenly distributed ______
5. Debris (light, medium or heavy) ______ Brush (light, medium or heavy) ______
6. Herbaceous fuels (light, medium, heavy) ______ Herbaceous fuels continuous ______
7. Herbicide used ______ Date applied ___/___/___
8. Drum chopped ______ Single or Double Pass ______ Date Completed ___/___/___
9. Windrowed and/or piled ______ Date piled ___/___/___ Piled when wet ______
10. Pile or windrow dimensions: Ht. ______ Width (dia.) ______
11. Windrow break interval ______

B. **Preburn Factors and Desired Fire Intensity:**

1. Areas to exclude: ______
2. Chains to plow (see map): Exterior ______ Interior ______ Total ______
3. Chains to fire (see map): Exterior ______ Interior ______ Total ______
4. Equipment needs ______
5. Crew size ______ Type of fire ______ Type of ignition ______
6. Ignition procedure (see map): ______
7. No. of hours to complete ______ Tons/acre to consume ______ Litter to leave (in.) ______
8. Special precautions: ______
9. Notify: ______
10. Regulations that apply ______
11. Passed screening system? ______ List smoke-sensitive areas, critical targets & locate on map: ______

C. **Weather Factors:**

Weather Factors	Desired Range	Predicted	Actual
1. Surface wind (speed & dir.)	______	______	______
2. Transport wind (speed & dir.)	______	______	______
3. Mixing height	______	______	______
4. Dispersion Index (or comparable)	______	______	______
5. Relative humidity (%)	______	______	______
6. Temperature (oF)	______	______	______
7. Fine-Fuel moisture (%)	______	______	______
8. 10-hr. fuel moisture (%)	______	______	______
9. Days since rain ______ Amount	______		______
10. Burning Index ______ Drought Index	______	______	______
11. Best month to burn	______	Dates burned ______	
12. Time of day to start	______	Time set ______	

D. **Summary of Burn:**

1. Type fire & ignition ______
2. All piles, windrows & logging decks ignited ______
3. % of area burned ______ Did area between piles burn? ______
4. Spotting frequency ______ Distance ______ firebrand material ______

E. **Evaluation Immediately After Burn:**

1. Any escapes: Number ______ Adjacent to burn area? ______ Acres involved ______
2. Hours to burnout: Active flaming ______ Smoldering ______ Total hours ______
3. % understory veg. consumed ______ Depth of litter remaining (in.) ______
4. % material < 3″ dia. consumed ______ Did piled debris burn down? ______
5. Objectives met ______
6. Adverse publicity ______
7. Smoke problems ______
8. Remarks ______

F. **Future Evaluation** (Date, signature and remarks) ______

Fig. 5b.

Purpose and Objective

List the reason for prescribing the fire (e.g., reduce hazard). In addition give a specific quantifiable objective. State exactly what the fire is to do – what it should kill or consume, how much litter should be left, etc. Also concisely describe desired fire behavior including rate of spread, flame length, and fireline intensity. In case the prescribed weather conditions do not materialize, this description may provide some latitude so that firing techniques can be adjusted to accommodate the existing weather and still accomplish the objective(s). Such information will also be useful in determining success of the burn.

Map of Burning Unit

A detailed map of each burning unit is an important part of the burning plan. The map should show boundaries of the planned burn, adjacent landowners, topography, control lines (both existing and those to build), anticipated direction of smoke plume, smoke-sensitive areas, holding details, and other essential information. Show any areas that should be excluded or protected such as improvements.

Equipment and Personnel

List equipment and personnel needed on-site and on standby. Assign duties. Chain saws are a useful addition to the equipment list.

Fire Prescription

Include species involved, height to the lower crown, understory type and stature, presence and condition (cured or green) of herbaceous layer, and the continuity and amount of dead fuel by size class. List the desired range of pertinent weather factors including eye-level windspeed and direction, relative humidity, temperature, fine-fuel moisture content, precipitation, and drying days needed. These all determine the maximum intensity that can be tolerated and thus selection of a firing technique. State the type of fire, the desired rate of spread, flame length, fireline intensity, and estimated number of hours to complete the burn. Estimate the fuel amount by size-class to be consumed and/or the amount of litter to leave.

Estimating the Available Number of Burning Days

The correct season of the year when burning should be applied may differ from region to region. In those parts of the tropics and the adjoining regions where the climate is characterized by dry and rainy seasons, it is most suitable to burn during the transition period between the wet and the dry season. Local wind patterns, such as the "bergwind" in southern Africa, may considerably restrict the number of burning days.

Time of Day

As a general rule, plan burning operations so the entire job can be completed within a standard workday. Prescribed fires usually are ignited between 10.00 and 14.00 h. Some burners like to start early, as soon as the sun has evaporated any dew and then switch firing techniques as conditions change during the day. Others like to wait until the maximum burning conditions for a particular day have materialized around 13.00 or 14.00 h before igniting a burn. The decision depends somewhat upon your knowledge of local weather and the reliability of the forecast. Burning conditions are usually better during the day than at night because windspeed is higher and wind direction steadier. Relative humidity also often rises to unacceptable levels at night.

Firing Plan

The firing plan should consist of a narrative section and detailed map. The burning unit map is ideal for this purpose because it already contains much pertinent information. The following should be added: (1) firing technique, ignition method (e.g., driptorch), ignition pattern, and planned ignition time. (2) The planned distribution of manpower and equipment for setting, holding, patroling, and mopping up the fire and managing the smoke. (3) Location and number of reinforcements that can be quickly mobilized if the fire escapes. (4) Instructions for all supervisory personnel, including complete description or illustration of assignment, and forces at their disposal.

Escaped-Fire Plan

Identify potential fire escapes and specify actions to take should such occur. Designate who will be in charge of suppression action and what personnel and equipment will be available.

Control and Mop-Up

List the necessary safeguards to confine the fire to the planned area. Mop-up promptly and completely. Emphasize protection of adjacent lands.

Evaluation

A record of actual weather conditions, behavior of the fire, and its effects on the environment is essential. This information is used to determine the effectiveness of the burn and to set criteria for future burns. Just prior to igniting the burn, record windspeed and direction, relative humidity, temperature, fuel moisture, and dampness of soil and lower litter. Record applicable weather and fire behavior (e.g., flame length and rate of spread) parameters at 2- to 3-h intervals throughout the burn. After the burn, record the amount of crown scorch, consumption of understory, litter, and duff, and any other evidence of fire intensity such as unburned areas, exposed mineral soil, and cracks in bark or cupping on the lower bole due to bark consumption. Also include a short narrative on success of the burn.

12.7.2 Preparing for the Burn

Good preparation is the key to successful burning. It is essential to maximize net benefits at acceptable cost. Preparation consists of all steps necessary to make the area ready for burning and of having all needed tools and equipment in good operating order and ready to go.

Establishing Control Lines

1. Construct lines in advance of burning, preferably after leaf fall to reduce the effect of fallen material on prepared lines.
2. Hold constructed lines to a minimum, keeping them shallow and on the contour as much as possible. Consider igniting from wetlines (a line of water, or water and chemical retardant sprayed along the ground to serve as a temporary control line from which to ignite or stop a low-intensity fire). Use access roads where feasible.
3. Use natural barriers such as streams whenever possible.
4. Keep control lines as straight as possible. Bend them around excluded areas, avoiding abrupt changes in direction.
5. Widen control lines at hazardous places.

After Lines Are Constructed

1. Remove any material above the line that could carry fire across the control line such as vines and overhanging brush.
2. Fell any snags near the line (inside and outside).

Burning-Unit Map

1. Locate all control lines on the map, noting any changes from the original plan.
2. Note on the map any danger spots along control lines having potential for fire escape.

12.7.3 Executing the Burn

There are few days of good prescribed burning weather during the year. When these days arrive, give top priority to burning. With adequate preparation, burning can begin without loss of opportunity. The person in charge should be an experienced prescribed burner with knowledge of the local situation. Radios for communication are useful. The person in charge must make sure the crew has the proper clothing and safety equipment and is in good physical shape. Have plenty of drinking water on hand.

Checklist

1. Make sure all equipment is in working order and safe to use.
2. Notify adjoining property owners and local fire control organizations before starting the fire.

3. Carry burning plans and maps to the job.
4. Check all control lines, clean out needles and leaves, and reinforce as necessary.
5. Check duff and soil for dampness.
6. Post signs on public roads and be prepared to control traffic if visibility becomes dangerously impaired.
7. Check the weather before starting the burn and keep updated throughout the burn.
8. Instruct crew on procedures, including safety precautions and the proper operation of equipment and hand tools.
9. Inform crew of starting point and firing sequence. Give each member a map.
10. Have a means of instant communication with all crew members. Portable radios are very useful.
11. Test burn with a trial fire before firing; check the fire and smoke behavior to make sure the fire is burning as expected. If it is not, decide whether the observed behavior is acceptable. This is the time to cancel the burn if you are not comfortable with the observed behavior.
12. Burn so wind will carry smoke away from sensitive areas.
13. Be alert to changing conditions and be prepared to change burning techniques or put the fire out if an emergency arises.
14. Mop up and patrol perimeters constantly until there is no further danger of fire escape.

12.7.4 Evaluating the Burn

The purposes of a burn evaluation are to determine how well the stated objectives of the burn were met and to gain information to be used in planning future burns. An initial evaluation should be made within a day or two after the burn. A second evaluation should be made during or after the first post-fire growing season.

Points to be considered

1. Were objectives met?
2. Was burning plan adhered to? Were changes documented?
3. Were fuel conditions, weather conditions, and fire behavior parameters all within planned limits?
4. Was burning technique and ignition pattern correct?
5. Was fire confined to intended area?
6. Amount of overstory crown scorch and consumption?
7. Amount of litter remaining?
8. Effects on vegetation, soil, air, water, and wildlife?
9. How can similar burns be improved?

12.8 Fire Effects

Fire effects depend upon the interaction of fire behavior and specific site characteristics such as species and age of vegetation, and soil type. Prescribed burning has direct and indirect effects on the environment. Effects can be short- or long-term and occur both on-and off site. The proper use of this tool requires knowledge of how fire affects vegetation, wildlife, soils water, and air. Burning technique, fire interval, and season of burn can be varied to alter fire effects. Robbins and Myers (1989) present an in-depth review of the seasonal effects of prescribed burning in Florida, U.S.A.

12.8.1 Effects on Trees

As the most valuable components of the plantation ecosystem are its trees, prediction of the occurrence and extent of fire injury should be a prerequisite to the intentional use of fire in these stands. Fire may kill part of a plant or the entire plant depending on how intensely the fire burns and how long a plant is exposed to these elevated temperatures. In addition, anatomical characteristics such as bark thickness and stem diameter influence a plant's susceptibility to fire. For example, small trees of a given species are easier to kill than large ones.

Very high temperatures are produced in the flames of burning forest fuels. Fortunately, the hot combustion gases are usually rapidly cooled in prescribed underburns unless the wind is calm. Wind is needed to dissipate the heat and slow its rise into the overstory canopy as well as to cool tree crowns heated by radiation. Ambient air and fuel temperatures at the time of burning are also important. When the air temperature is 9°C, it takes twice as much heat to kill foliage at a given height above a fire as it does when the temperature is 34°C. An extensive overview of the effects of fire on *P. elliottii* and *P. taeda* (which should also hold for the other species under discussion) is given by Wade and Johansen (1986).

Cambium and Root Damage

Pine bark has good insulating qualities and is what protects the aboveground cambium from injury. Bark thickness varies considerably within and between species. It thickens with age and especially with tree girth. Bark thicker than 12 mm will protect the stem cambium of most pines during prescribed fires (de Ronde 1982; Fahnestock and Hare 1964; Goldammer 1983; Kayll 1963). Susceptibility to cambial damage ranges from *P. radiata* and *P. patula*, the most susceptible, through *P. taeda* and *P. pinaster*, and finally to *P. caribaea* and *P. elliottii*, the most fire-resistant (de Ronde 1982; de Ronde et al. 1986; Langdon 1971; McArthur 1971; Speltz 1968; Van Loon and Love 1973).

Injuries to the cambium are not easy to detect from external signs. In 1966, Curtin stressed the need for a rapid nondestructive method to detect the presence of dead cambial tissue, but this challenge has yet to be met. Several methods of assessing cambial damage have been used (see Wade and Johansen 1986) but in practice, lifting of the bark (where damage is suspected) to detect discoloration is the simplest method (de Ronde et al. 1986; Fig. 6). Cambial damage is usually greatest on the lee side (in respect to fire passage) of a tree just above ground line (Fahnestock and Hare 1964). Except for *P. radiata* and *P. patula*, cambial damage is of little concern in low-intensity fires. Once *P. elliottii* and *P. taeda* reach a groundline diameter of roughly 5 cm, they are pretty much immune to stem damage from prescribed fire (Johansen and Wade 1986).

Even though pine bark is a good insulator, cambial damage can occur from the extended smoldering of deep domelike accumulations of sloughed bark and needles around the root collar. This often results in death of the tree months later

Fig. 6. Cambial damage in *P. pinaster* after a severe wildfire, Cape Province/South Africa. The brown areas on the exposed wood indicate areas of dead cambium

Ferguson et al. (1960) noted this phenomenon in *P. taeda*. Whenever heat penetrates into the soil, feeder roots (which do not have a protective bark covering) are likely to be killed (see comments by Wade regarding *P. elliottii* var. *densa* mortality in Wade and Johansen 1986). Root damage can be virtually eliminated during underburning by proceeding only when the duff is damp.

Crown Damage

Various researchers have observed that mortality is more closely related to crown damage than to cambial damage (Cooper and Altobellis 1969; Crow and Shilling 1980; de Ronde 1983; Jemison 1944). The best indicator of overstory damage (but not death) is percent foliage discoloration.

Pine foliage is quite vulnerable to temperatures above 55 °C. Nelson (1952) determined that the time needed to kill *P. elliottii* and *P. taeda* needles was almost instantaneous at 64 °C, averaged 30 s at 60 °C, and took about 6 min at 54 °C. Death of foliage on a few lower branches is of little consequence, but as crown scorch increases, the probability of reduced growth also increases. De Ronde has observed that the needles of *P. radiata* appear to be more easily heat-killed than those of several other species, perhaps because they are thinner in cross-section. *P. caribaea*, *P. elliottii*, *P. pinaster*, and *P. taeda* can all survive total needle scorch although some loss of height and/or diameter growth is likely. For example, Weise et al. (1989) reported results of a defoliation study in which 4-year-old *P. elliottii* and *P. taeda* pines were subjected to five defoliation levels (100, 95, 66, 33, 0%) at four times (January, April, July, or October) at four different locations in the southeastern U.S. Six out of 30 *P. elliottii* defoliated 100% in October (fall), and all but one *P. taeda* defoliated 100% in October died. No other mortality occurred, but diameter and height growth of trees defoliated at the three most severe levels were significantly reduced (over 50% for some combinations). A similar study in *P. radiata* with similar results was reported by Rook and Whyte 1976).

Needle scorch, however, is not the principal cause of mortality in pines that do not have fully preformed buds. Damage to branch cambium and, especially buds, determines the survival potential of pines that routinely undergo several successive needle flushes during the growing season. The literature shows little mortality in these pine species until scorch approaches 100%; but as consumption begins, dramatic increases in mortality take place (McCormick 1976; McCulley 1950; Tozzini and Soares 1987; Van Loon 1967; Wade 1985; Wade and Ward 1975). But how can an observer on the ground tell whether the buds or branch cambium have been thermally killed? One easy way is to use needle consumption as an analog of bud and cambium death. Temperatures over 200 °C are required to ignite the foliage and these are high enough to kill surrounding meristematic tissue.

Post-fire damage surveys should be conducted within 2 to 3 weeks after a fire before the scorched needles fall. Dead foliage that is retained for much longer periods of time indicates that the branches themselves have also been killed.

Growth and Mortality Prediction

A good rule of thumb is to assume that a tree's chances of survival are poor if more than 20% of the needles are actually consumed by flames even if the remaining foliage is not all scorched. Young vigorous trees are much more likely to survive severe crown damage than are older slower-growing or stressed individuals. Reports of survival after particularly extreme damage can be found (Goldammer 1983; Van Loon 1967; Wade and Ward 1975), but these instances should be considered fortuitous exceptions. Procedures for evaluating wildfire damage have been developed (e.g., Caulfield and Teeter 1988; de Ronde et al. 1986; McCulley 1950; Wade 1985).

The magnitude and duration of growth responses to various levels of fire-caused damage are critical needs that are only now emerging. Both negative and positive responses have been reported, but it is difficult to judge the true merit of many of these observations because fire behavior documentation is often not adequate and sometimes reconstructed at a later date; possible root damage has been almost universally ignored (see also Sect. 12.2.3); and diameter growth assessment techniques are sometimes flawed so that missing rings may be overlooked (see Wade and Johansen 1986). Many studies that tracked diameter and height response of severely damaged trees over time have recorded an inverse relationship between volume growth and damage (e.g., Bourgeois 1985; Weise et al. 1989). Even though increment loss can be substantial (e.g., Johansen and Wade 1987; Villarrubia and Chambers 1978), preburn growth rates are generally reestablished within a few years (e.g., McArthur 1971; Van Loon 1967; Weise et al. 1989). In fact, some studies have recorded an eventual increase in growth rate over the controls (e.g., Johansen 1975; Peet and McCormick 1971). Increased growth rates are a much more likely outcome when crown scorch is nonexistent or very light (e.g., Gruschow 1951, 1952; McCullley 1948, 1950). Several long-term studies that monitored tree growth over multiple burns found no significant differences between burned and unburned plots (e.g., Byrne 1977; Hunt and Simpson 1985; Sackett 1975), although two studies in *P. palustris* have documented retarded growth after several supposedly benign prescribed fires (Boyer 1987; Zahner 1989). De Ronde has unpublished findings that show no growth differences after a maximum of three fires over a 9-year period, and Wade has unpublished results that also show no growth differences after as many as 20 annual winter fires.

Of course, a well-conducted prescribed burn should not result in appreciable scorch, so the question of significant growth loss should be a moot point. However, recognizing that prescribed burns are sometimes hotter than planned, the following rules of thumb adapted from Wade and Lunsford (1989) are presented. They assume no needle consumption (Table 6).

Table 6. Relationship between southern pine growth response and degree of crown scorch

Percent crown scorch	Damage
0 to 33	A slight increase in volume growth may occur the first post fire year unless root damage occurs, in which case a minor growth loss may occur.
34 to 66	Volume growth loss usually less than 40% and confined to the first post fire growing season.
67 to 100	Reduction may be as high as a full years growth spread over 3 years. Some mortality may occur in *P. pinaster* and *P. radiata*.

Secondary Damage

Where trees have been subjected to severe fire-caused damage, bud elongation and some refoliation after the fire may give the impression that the trees are recovering. However, the new needle growth turns brown within a few months and the trees die. If root damage is not the cause, and post fire weather conditions have not resulted in additional stress, secondary pathogens are usually responsible. In industrial pine plantations in the Cape Province/South Africa, the main cause appears to be *Rhizina undulata*, which attacks the remaining live roots and is then followed by bark beetles (*Ips*) and cerambycid beetles which finish the job. This fungus is particularly common where fire has consumed all the humus (Baylis et al. 1986). *Rhizina undulata* has not caused problems to date after underburning, or after burning clearcut logging debris as long as the humus layer is not consumed, which suggests that high-intensity fires are necessary to create conditions favorable for the root rot (Baylis et al. 1986). Observations by De Ronde in the Cape forest regions support this hypothesis. In the Transvaal, Swaziland, and Natal, the abundance of *Rhizina undulata* fruiting bodies after slash burning forced forest managers to abandon this silvicultural practice because of the resulting high mortality in the next crop of pine seedlings (Donald 1979; Lundquist 1984). In many parts of the world, however, fire is used to dispose of logging slash and the associated insect populations that build-up in the freshly cut debris and stumps (e.g., Fellin 1980; Fox and Hill 1973; Hardison 1976; Smith et al. 1983).

There is little doubt that severely damaged pine stands attract and are more susceptible to insects and disease (e.g., Goldammer 1983; Martin and Mitchell 1980; Miller and Patterson 1927). However, some studies have also implicated low-intensity fires. For example, in Spain, De Ana Magan (1981) reported that prescribed fires in *P. pinaster* kill the natural control of *Leptographium gallaeciae*, thereby allowing this fungus to proliferate and attack the stand. On the other hand, Froelich et al. (1978) found that prescribed burning altered the forest floor environment to the detriment of *Heterobasidon annosum*, a particularly damaging fungus in pine plantations in the southern U.S. Investigations in the Philippines have shown that *Ips interstitialis* are attracted to freshly underburnt stands of *P. kesiya* by the monoterpenes released through the heating process in resinous fuels (Goldammer 1987). Schowalter (1983) described the adaptive abilities of insects to disturbances such as fire. The relationship between fire, insects and disease, and their host vegetation is a maze of complex, often subtle interactions that will require a holistic approach to unravel. An example of such an attempt is the work by Schowalter et al. (1981), who looked at *Dendroctonus frontalis* from the standpoint of the overall health of the southern pine ecosystem.

12.8.2 Effects on Woody and Herbaceous Understory Vegetation

The intensity and frequency of prescribed fire, the season during which fire is applied, the number of species present and their abundance, damage to plants and capacity of a species to regenerate after fire, all dictate the extent of fire impact and change in stand composition.

In the southern U.S., prescribed fire is routinely used to manipulate the herbaceous and woody understory for wildlife habitat improvement. Fire increases the amount and palatability of browse by stimulating sprout and sucker growth and keeps plant growth low – within reach of foraging animals. In Australia, Van Loon (1966) reported that higher strata shrubs such as *Synoum*, *Cryptocarya*, and *Acacia* can be drastically reduced by fire. De Ronde observed similar findings in South Africa with *Laurophyllus capensis* under mature *P. elliottii* overstory, while Soares had similar experiences with pine seedlings and small trees in Brazil. Wade et al. (1989) summarized the effects of 20 years of periodic burning on the composition and stature of the understory in a mixed pine/hardwood stand in the southern U.S.

The effects of a low-intensity prescribed burn on lower strata vegetation have such a broad range of possible outcomes that this discussion will be limited to a few examples. Van Loon (1966) recorded an increase in frequency of blady grass (e.g., *Watsonia* spp.) from 4.2 to 17.0% after a single burn. De Ronde observed similar frequency increases in grass communities such as *Ehrharta* spp. (63 to 84%) in *P. elliottii* stands. In some cases, grass species are stimulated in clumps (Vlok and De Ronde 1989; Chippendal and Crook 1976). The improvements in *Ehrharta* frequency and density were always followed up by increased grazing by herbivores. The use of prescribed fire to manage herbaceous species' habitat in Africa has been reviewed by West (1965), and throughout the tropics by Campell (1960) and in various Tall Timbers Fire Ecology Conference Proceedings (see Fischer 1980).

Kellman and others (1987) have developed a scenario regarding underburning in *P. caribaea* stands in Belize that is based on both field data and computer simulations. Annual burning of this savanna results in a graminoid understory that is virtually all consumed in every fire. Less frequent burning cycles allow formation of a relatively nonflammable shrub layer. This shrub understory is capable of storing a much larger share of a site's nutrient capital than is the graminoid layer (except for phosphorus, which is unlikely to be lost from the site because it is rapidly immobilized by soil fixation). Low-intensity fires will topkill but not consume this shrub strata so that rather than being released all at once, the nutrients are mineralized over the ensuing months as the scorched leaves and fire-killed stems decompose. This gradual nutrient release significantly increases the probability that these minerals will again be captured by plants rather than leached from the site.

12.8.3 Effects on Forest Floor Dynamics

Fast-growing pine plantations tend to have a well-developed forest floor with deep, fairly distinct layers of undecomposed litter (AoooL), partly decomposed litter (AooF), and decomposed humus (AoH). Litter decomposition rates are comparatively slow so that heavy accumulations of flammable fuel rapidly build up. Prescribed fire has a considerable effect on the physical, chemical, and biological environment of this forest floor. The amount of fuel removed from the forest floor by a fire depends upon the moisture content of the fuel and the residence time of the fire. The forest floor moisture regime is often characterized by a steep moisture gradient caused by fast drying of the upper litter and considerably slower drying of the F and H layers. In addition, the upper litter has a higher heat value and is less compacted than the lower strata (Hough 1969; Goldammer 1983). Due to the breakdown of all structures in partly decomposed needles, the water-holding capacity (maximum moisture content) increases considerably. Most workers have found the moisture retention capacity of total litter layers to range between 200 and 300% (see Hough 1978) although Goldammer (1983) reported that 3-year-old *P. taeda* needles in the F/H layer may reach a moisture content of more than 600%.

This distinct differentiation in the moisture regime of the forest floor facilitates removal of the most flammable and hazardous litter layer without affecting the lower layers. Consumption of the humus layers should be avoided in order to minimize the overall fire impact on the site and the stand. Exposing the underlying mineral soil predisposes the site to increased erosion and surface runoff.

Needlefall and Crown Scorch Effects on Litter Loading

The natural sloughing of old needles takes place throughout the year in pines although the great majority occurs during the fall months (Fig. 7). Many of the freshly shed needles lodge on lower branches and on the understory plants, where they remain until blown down by the wind. Variables determining the amount of needlefall include species, site productivity, basal area, and age of the stand. Lugo et al. (1980) estimate the production and storage of stemwood and organic matter in tropical plantations that shows the difference in performance between some of these pines. Data by Will (1959) presented in Bazilevic and Rodin (1966) give an annual foliage litter range of 2.9 t ha^{-1} to 10.4 t ha^{-1}

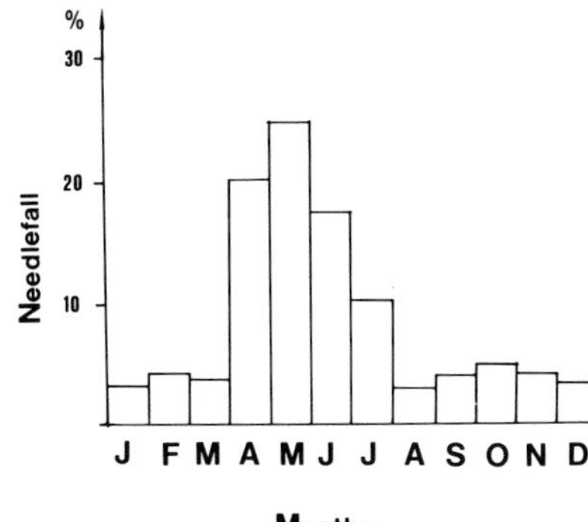

Fig. 7. Average monthly distribution of needlefall (% of annual needlefall) in 5- to 15-year old *P. taeda* plantations during 1981–82. Parana, Brazil. (Goldammer 1983)

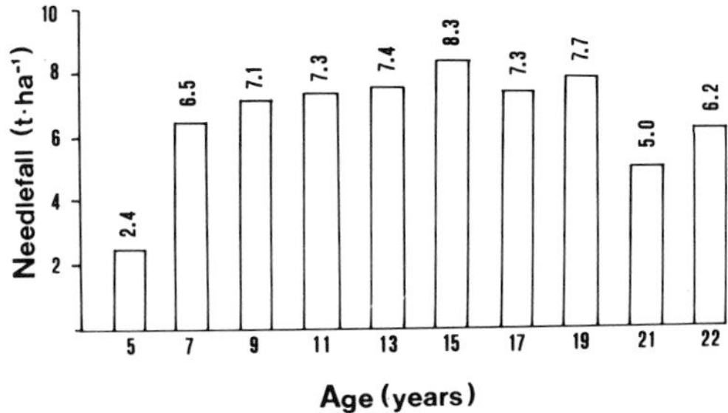

Fig. 8. Annual needlefall (t ha^{-1}) by stand age in 5 to 22-year old *P. taeda* plantations during 1981–82. Parana, Brazil. (Goldammer 1983)

depending upon age for *P. radiata* plantations on the North Island of New Zealand. According to Bazilevic and Rodin (1966), in areas with a distinct dry period (such as dry savannas), the breakdown products are often completely neutralized by the bases due to maceration. The litter thus accumulates much faster than it decomposes. However, the colloidal complex that forms in these tropical savannas helps protect the humus during prescribed fires. Studies in *P. elliottii* and *P. taeda* plantations in southern Brazil showed that annual litterfall was higher there than in their natural range (Goldammer 1983). Figure 8 illustrates the annual needlefall in *P. taeda* plantations in Parana, Brazil in which the total dry weight increased from 2.5 t ha^{-1} in a 5-year-old stand to 8.3 t/ha in a 15-year-old stand and then slightly decreased at older ages.

Crown scorch will result in additional needlefall, the quantity dependent upon the amount of scorch. Sometimes the amount can be large enough to negate the reduction in hazard from the prescribed burn (Haigh 1980). In a prescribed burning experiment at Lottering, South Africa, in a 28-year-old *P. elliottii* stand, needlefall measured by De Ronde ranged from 6.5 tons/ha under unscorched trees (normal yearly needlefall) to 9.8 tons/ha under trees with more than 75% of the crown scorched. The following illustration (Fig. 9) of the results also indicates that as the scorched needles fall to the ground during the first postburn month, they will partially off-set the litter reduction achieved by the fire.

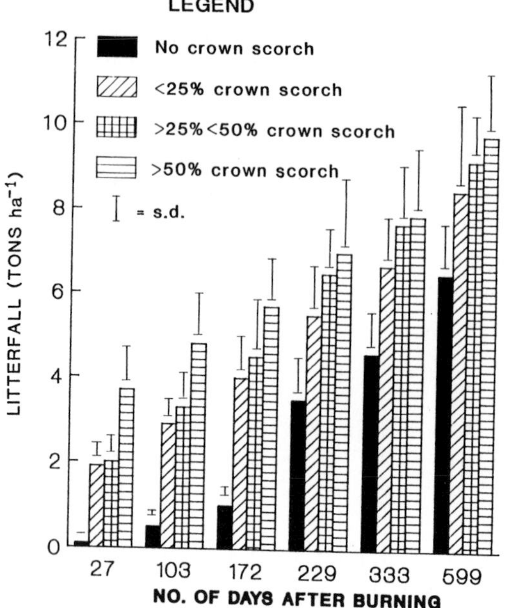

Fig. 9. Histogram of accumulated needlefall in a *P. elliottii* controlled burning trial by four levels of crown scorch. Lottering, South Africa

Effects on Soil Fauna

It is important to understand the effect of prescribed burns and wildfires on soil and litter invertebrates, since these animals have a direct and major role in litter decomposition and therefore nutrient cycling (Abbott et al. 1979; Reichle 1977). They also play an important part in maintaining soil structure (Abbott et al. 1979). Majer (1984) recorded significant long-term effects on some taxa but had little or no effect on other taxa. Abbott (1984) also reported little effect after fire in Jarrah forest of Australia.

All fires influence the biological and chemical processes taking place in the forest floor but intense postharvest fires can drastically alter the biological, physical and chemical characteristics of the upper soil (e.g., Neal et al. 1965). Common changes include increasing soil pH from 0.3 to 1.2 units, widening the C:N ratio, and reducing soil-pore size, aeration, water-holding capacity and water infiltration rates (e.g., Debano 1981; Neal et al. 1965; Ralston and Hatchell 1971). These changes usually reduce microbial activities for varying periods after the burn (Ahlgren and Ahlgren 1960; Wright and Tarrant 1957). Reinoculation occurs from windblown spores or other debris and through invasion from subsurface layers. Because burning changes the soil's physical and chemical properties and eliminates many potential competitors, microbes adapted to the changed soil environment have an advantage in the recolonization process (Harvey et al. 1976). The typical increase in pH after fire favors bacterial population growth over fungal population growth.

When moisture is sufficient, the microbial population quickly recovers, primarily from organisms adapted to the new soil environment. The reconstituted population may be greater and more active than the original one (Ahlgren and Ahlgren 1965), perhaps because of the large quantity of mineral nutrients released from the ash and because of other shifts in soil chemistry. For example, spores of the root pathogen *Rhizina undulata* germinate only after exposure to elevated temperatures (Gremmen 1971). It thereby gains access to soils that are rich in nutrients, that are likely to contain young and susceptible conifer roots, that harbor few competing organisms, and that have low concentrations of growth inhibitors (Watson and Ford 1972). Increases in other potential conifer root pathogens have also been recorded after burning (Tarrant 1956).

Temporary reductions in conifer-fungal mycorrhizal consociations have been found after intense post-harvest burns (Mikola et al. 1964; Tarrant 1956; Wright 1971, 1985; Wright and Tarrant 1957), but this is not a consideration when using low-intensity fires.

12.8.4 Effects on Soil

The impact of fire on soil depends upon such factors as the intensity, frequency, and residence time of the fire; type and moisture content of the soil; the moisture content and amount of protective forest floor; slope; and post-fire weather – particularly precipitation. Wells et al. (1979) found that although both the productivity and stability of soils are adversely affected by excessive heat, low-intensity prescribed fires have little if any effect because they do not raise soil temperatures very much. They concluded further that low-intensity fires facilitate the cycling of some nutrients, may help control some plant pathogens and generally do not increase soil erosion. In contrast, intense fires can volatilize excessive amounts of nitrogen and other nutrients, destroy organic matter, change soil structure and may induce water repellency, thus increasing soil erosion and decreasing productivity potential (Byrne 1980). Good summaries of the effects of fire on soil are given by Brown and Davis (1973) and Wells et al. (1979).

Physical Effects

Prescribed burning generally has negligible effects on soil heating. Moisture in the lower duff and upper soil has to be evaporated before the temperature will increase much above 100°C, which means that some humus remains to protect the soil after a properly conducted prescribed fire. In contrast, the complete consumption of the forest floor has a dramatic effect on soil surface temperatures (Clinnick 1984). Removal of litter is also likely to influence fine root growth by changing post-fire soil properties such as temperature, water, and compaction (Nambiar 1984). During a low-intensity prescribed fire in a 15-year-old *P. elliottii* stand in southern Brazil in which the upper L-layer was burnt and the moist F/H-layer was left unburnt, the topsoil temperature was raised from 28 to 32°C for only a brief period (Goldammer 1983).

As might be expected, the net effect of burning is towards increased surface erosion (Tiedemann et al. 1979). If the forest floor is completely consumed, the underlying soil is exposed and thus vulnerable to raindrop impact. Soil aggregates are dispersed, pores become clogged, and macropore space, infiltration, and aeration are all decreased (Wells et al. 1979). Reduced evapotranspiration causes higher soil water contents that may lead to greater overland flow. Fire may reduce the "hydrologic depth" of a soil by causing formation of water-repellent layers. DeBano (1981) gives a state-of-the-art overview of water-repellent soils.

Virtually all the evidence in the southern U.S. suggests that none of the above deleterious effects will occur after well-planned low-intensity prescribed fires. The single possible exception is on fine-textured soils on steep terrain (Pritchett 1976). A study of the effects of fire during two different seasons in *P. oocarpa* on steep slopes in central Honduras concluded that the small increases in surface runoff and sediment loss were an acceptable alternative to letting the fuels build up, thereby exposing the site and overstory pines to the high risk of damaging wildfires (Hudson et al. 1983a).

Chemical Effects

The role of fire in nutrient cycling is still being unraveled in temperate forests. In the tropics and subtropics, Bazilevic and Rodin (1966) report that knowledge gaps are considerably larger. They go on to show that subtropical coniferous plantations are like their more temperate counterparts in that they take up and return much smaller amounts of chemical elements than do broadleaved forests.

Nutrient losses resulting from fires may occur by volatilization, by removal of particulate matter in smoke, by surface transportation in runoff, and by leaching through the soil (Harwood and Jackson 1975). Volatilization of elements requires high fire intensities so such losses are minimal in low-intensity prescribed fires. However, the consumption of plant materials mineralizes the nutrients stored in them irrespective of fire intensity. These released nutrients are available for immediate uptake by surviving rootstocks and new germinants, but are also subject to loss from surface runoff or leaching if not quickly captured by plants. Christensen (1987) points out that species inhabiting fire-dominated ecosystems in the southeastern U.S. appear to have developed traits that enable them to more efficiently retain and utilize fire-released nutrients. Precipitation affects the amount of ash dissolved into the soil or carried away in overland flow. Hudson et al. (1983b), working in Honduras, determined that increased rates of sediment loss rather that increased losses in soil runoff were responsible for the increase in nutrient losses after prescribed burning.

Recovery after low-intensity fire is generally rapid even though natural mechanisms such as atmospheric deposition add nutrients to a site very slowly. Increases in N-fixing organisms in many soils help offset any nitrogen loss (Metz et al. 1961; Wells 1971). Nitrogen-fixing plants such as legumes are usually promoted by low-intensity fires. Thus, the extent of N fixation and subsequent soil N transformations occurring after fire are critical factors in evaluating the long-term effects on site quality (Harvey et al. 1976). Hudson et al. (1983b) concluded that low-intensity underburns in *P. oocarpa* at least 3 years apart would reduce the rate of nutrient accumulation on the site rather than deplete the nutrient capital. Maggs (1988) took a more conservative approach; even though he found the nutrients in the forest floor had recovered 3 years after underburning a *P. elliottii* plantation in Australia, he cautioned that the repeated loss of nitrogen associated with periodic prescribed burning, when coupled with removal of logs at the end of a rotation, could lead to a depletion of total nitrogen over the long haul.

Close-interval prescribed burns may have the potential to eventually reduce the nutrient capital on some sites if a large portion of the nutrient capital is stored in the plant wastes being consumed. Humphreys and Craig (1981) summarized work done in Australian pine plantations regarding this potential. They reported that Jones and Richards (1977) found a slight but significant decrease in total nitrogen levels in soils that had been subjected to frequent low-intensity fires, but all other studies they refer to found no differences between plantations underburned and those in which fire had been excluded. McKee (1982) looked at soil fertility changes under pine stands in the southern U.S. that had been periodically underburned for 8 to 65 years. He found that these low-intensity fires had no deleterious effects on the soils and, in fact, consistently increased available phosphorus levels and prevented the immobilization of calcium in the forest floor, which could have led to a nutrient imbalance and accelerated soil weathering over time. In an exotic pine plantation in Australia, Hunt and Simpson (1985) even recorded a short-term increase of total phosphorus in both the surface and lower layers of the soil, which contrasted with a decrease in nitrogen. However, no consistent differences were recorded over a period of 9 years.

It is well established that available nitrogen in the soil is usually increased following burning. Davis (1959) concluded that the major reason for this was that fire somehow improves nitrification of the system. Maggs and Hewett (1985) found that the plots with the greatest nitrogen accumulation were the wettest and that this combination, together with the effects of burning, stimulated nitrogen fixation by anaerobic, nonsymbiotic microorganisms.

Davis (1959), Hough (1981), and Bara and Vega (1983) all recorded increased levels of K in the topsoil after prescribed fire, but Wells (1971) and McKee (1982) recorded no change. Similar increases have been recorded in the understory regrowth after fire, where potassium showed the most consistent response of all elements sampled (Harvis and Covington 1983). Because potassium leaches rapidly from ash, there may have been a mass ion movement into the soil, thus providing excess potassium for plant uptake (Grier 1975).

The ash from most plant materials is high in basic ions such as calcium, potassium, and magnesium. This tends to raise the pH of acid soils, especially sandy soils that are poorly buffered (Chandler et al. 1983). The literature also indicates that heat may have the same or an even greater effect on soil pH than oxides in the ashes (Wells et al. 1979). Davis (1959) suggests that the reduced acidity of the surface soil may be enough to stimulate nitrification and growth of subordinate vegetation but that most forest trees do not seem to be significantly affected by the changes in soil acidity brought about by fire.

The effect of fire on pH decreases rapidly with soil depth. Wells (1971) recorded an increase in pH form 3.5 to 4.0 in the F and H horizons after prescribed burning, and an increase in pH from 4.2 to 4.6 in the top 5 cm of the mineral soil. No significant changes were recorded by him in the 5- to 10-cm layer of the mineral soil. McKee (1982) also recorded marginal pH increases after controled burning in coastal plain pine sites in the southern U.S.

The much more intense fires commonly associated with debris disposal after harvest are an entirely different situation. In the past, the use of fire to dispose of logging debris and prepare the site for reforestation was a very common practice. Rennie (1971) reported that prescribed fire was used on more than 90% of all sites in Australia prepared for conifer plantations, but the wisdom of this widespread practice has been questioned. Flinn (1978) showed retention of logging residue after clearcutting pine on infertile sandy soils enhanced early growth of the next crop. He believed that improved soil moisture conditions and the slow sustained release of nutrients over the ensuing years would result in better overall growth of the plantation. In contrast, Turvey and Cameron (1986) demonstrated that all three slash reduction treatments they tested resulted in better survival and growth than the "no treatment" alternative. Turvey and Cameron, do however, go on to warn that burning windrowed logging debris after several rotations could be detrimental to continued productivity because of changes in soil nutrient concentrations. Burning windrows releases the bound-up nutrients, making the soil under the windrows more fertile than the surrounding areas. Trees planted on these ashbeds grow faster than those on the degraded sites between windrows. According to Woods (1981), second-rotation productivity has declined by 37% on South Australian sites that have been subjected to very hot broadcast slash burns, and by 18 to 24% on windrowed sites, primarily due to the loss of nitrogen. In New Zealand, Phillips and Goh (1986) reported that both survival and growth of *P. radiata* seedlings were better on sites where logging debris was broadcast-burned than on those where it was unburned.

State-of-the-art pine plantation harvesting operations leave little logging debris on the site. For example, in the southern United States, fire is almost universally a part of reforestation even though the small amount of slash generated is only a minor hindrance to planting, and air quality concerns make planning and conducting these burns increasingly difficult. Fire alone or in combination with mechanical or chemical treatments temporarily eliminates competing vegetation until the pine seedlings become established. After a burn, equipment operators can more easily see the stumps of just-harvested trees as well as any other hazards. On more mesic sites where the ground surface is restructured into beds before planting, burning first, consumes much of the debris which would otherwise end up in the beds, resulting in a poor rooting medium and thus poor seedling survival. Broadcast burning mineralizes nutrients across the site, making them available for the next timber crop. With the notable exception of *Rhizina undulata*, prescribed fire helps sanitize a site, controling or reducing the potential for many insect and disease problems.

12.9 Conclusions

Fire produces change; whether these changes are deemed beneficial or detrimental depends upon specific burn objectives and overall resource management goals. The pines discussed in this chapter all have a close historic association with frequent fire. They each have evolutionary adaptations that help them predominate in fire-prone environments. In tropical and subtropical countries where prescribed fire is being used operationally or experimentally, the evidence leaves little doubt that its judicious use will result in a net benefit while its indiscriminate use will just as surely have deleterious effects. A total protection policy in industrial pine plantations invariably leads to a dangerous build-up of fuel on the forest floor that dramatically increases fire hazard. It will also result in an accumulation of nutrients in litter layers, and an undesirable habitat for herbaceous and woody plant communities necessary for biodiversity and thus ecosystem stability.

Prescribed fire is no panacea, however. There are trade-offs associated with every fire and in many situations fire may not be the best alternative. If it is the treatment of choice, good judgement in both planning and conducting the burn are mandatory. For instance, the potential of erosion on steep slopes has to be considered with care; but when fuel and weather conditions specified in a written plan of action materialize, this tool can be used to simultaneously accomplish several management and ecological needs. Because these needs are site-dependent, a plan of action that includes firing technique, ignition pattern, and expected fire behavior should be prepared for every burn. As an example, *P. radiata* has the lowest needlefall rate, the fastest litter decomposition rate, and is more susceptible to fire damage than the other species discussed, while *P. caribaea* has the opposite traits.

The evidence presented in this chapter strongly implies that well-planned and conducted prescribed burns should be one of the plantation managers' primary silvicultural tools to ensure sustained timber yields and properly maintained natural resources.

References

Abbott IC (1984) Changes in the abundance and activity of certain soil and litter fauna in the jarrah forest of Western Australia after a moderate intensity fire. Aust J Soil Res 22:463–469

Abbott IC, Parker CA, Sills ID (1979) Changes in the abundance of large soil animals and physical properties of soils following cultivation. Aust J Soil Res 17:343–352 (cited in Majer 1984)

Ahlgren JT, Ahlgren CE (1960) Ecological effects of forest fires. Bot Rev 26:483–533
Albini FA (1976) Estimating wildfire behavior and effects. Gen Tech Rep INT-30. Ogden, UT: US Dept Agric, For Serv, Intermountain For Range Expt Stn, 92 pp
Albini FA, Baughman RG (1979) Estimating windspeeds for predicting wildland fire behavior. Res Pap INT-221. Ogden, UT: US Dept Agric, For Serv, Intermountain For Range Expt Stn, 12 pp
Alexander ME (1982) Calculating and interpreting forest fire intensities. Can J Bot 60(4):349–357
Anderson HE (1982) Aids to determining fuel models for estimating fire behavior. Gen Tech Rep INT-122. Ogden, UT: US Dept Agric, For Serv, Intermountain For Range Expt Stn, 22 pp
Andrews PL (1986) Behave: fire behavior prediction and fuel modeling system – burn subsystem, Part 1. Gen Tech Rep INT-194. Ogden, UT: US Dept Agric, For Serv, Intermountain For Range Expt Stn, 130 pp
Andrews PL, Rothermel RC (1982) Charts for interpreting wildland fire behavior characteristics. Gen Tech Rep INT-131. Ogden, UT: US Dept Agric, For Serv, Intermountain For Range Expt Stn, 21 pp
Bara S, Vega JA (1983) Effects of wildfires on forest soil in the northwest of Spain. In: Goldammer JG (ed) DFG-Symp Fire Ecology. Freiburger Waldschutz Abh 4, Freiburg, pp 181–195
Baylis NT, De Ronde C, James DB (1986) Observations of damage of a secondary nature following a wild fire at the Otterford State Forest. S Afr For J 137:36–37
Bazilevic NI, Rodin LE (1966) The biological cycle of nitrogen and ash elements in plant communities of the tropical and subtropical zones. For Abstr 27(3):357–368
Bourgeois DM (1985) Growth and mortality of dominant and co-dominant trees in prescribed burned pine plantations – first year results. Rep 6–85 Westvaco Tech Dept, Summerville, SC, 5 pp
Boyer WD (1987) Volume growth loss: a hidden cost of periodic prescribed burning in longleaf pine? So J Appl For 11(3):154–157
Brown AA, Davis KP (1973) Forest Fire: Control and Use. 2nd edn. McGraw-Hill, New York, 686 pp
Brown JK (1974) Handbook for inventorying downed woody material. Gen Tech Rep INT-16. Ogden, UT: US Dept Agric, For Serv, Intermountain For Range Expt Stn, 24 pp
Burgan RE (1987a) A comparison of procedures to estimate fine dead fuel moisture for fire behavior predictions. S Afr For J 142:34–40
Burgan RE (1987b) Concepts and interpreted examples in advanced fuel modeling. Gen Tech Rep INT-238. Ogden, UT: US Dept Agric, For Serv, Intermountain For Range Expt Stn, 40 pp
Burgan RE (1988) 1988 revisions to the 1978 national fire-danger rating system. Res Pap SE-273. Asheville, NC: US Dept Agric, For Serv, Southeastern For Expt Stn, 39 pp
Burgan RE, Rothermel RC (1984) Behave: fire behavior prediction and fuel modeling system-fuel subsystem. Gen Tech Rep INT-167. Ogden, UT: US Dept Agric, For Serv, Intermountain For Range Expt Stn, 126 pp
Byram GM (1954) Atmospheric conditions related to blowup fires. Stn Pap 35. Asheville, NC: US Dept Agric, For Serv, Southeastern For Expt Stn 34 pp
Byram GM (1958) Some basic thermal processes controlling the effects of fire on living vegetation. Res Note 114. Asheville, NC: US Dept Agric, For Serv, Southeastern For Expt Stn, 2 pp
Byram GM (1959) Combustion of forest fuels. In: Davis KP (ed) Forest fire control and use. Chapter 3. McGraw-Hill, New York, pp 61–89
Byrne PJ (1977) Prescribed burning in Australia: the state of the art. Paper presented to 5th Meeting of Aust For Council Res Working Group No 6, Melbourne. (cited in Shea SR, Peet GB, Cheney NP 1981)
Byrne PJ (1980) Prescribed burning in Queensland exotic pine plantations. Proc 11th Commonw For Conf, Trinidad, 58 pp
Byrne PJ, Just TE (1982) Exotic pine plantation prescribed burning using a helicopter. Tech Pap 28 Dept For, Queensland, Australia
Campell RS (1960) Use of fire in grassland management. Paper delivered at Working Pary on Pasture and Fodder Development in Tropical America. FAO, Rome, 10 pp
Catchpole T, De Mestre N (1986) Physical models for a spreading line fire. Aust For 49(2):102–111
Caulfield JP, Teeter LD (1988) Using break-even analysis for replanting decisions in damaged pine stands. So J Appl For 12:186–189

Chandler C, Cheney P, Thomas P, Trabaud L, Williams D (1983) Fire in Forestry, vol 1. Forest Fire Behavior and Effects. Wiley, New York, 450 pp
Chippendal LKA, Crook AO (1976) Grasses of Southern Africa, vol I-III. Collins (Pty), Dublin House, Salisbury (Zimbabwe)
Christensen NL (1987) The biogeochemical consequences of fire and their effects on the vegetation of the Coastal Plain of the southeastern United States. In: Trabaud L (ed) The role of fire in ecological systems. SPB Academic Publishing, The Hague, Netherland, pp 1–21
Clinnick PF (1984) A summary-review of the effects of fire on the soil environment. Tech Rep Soil Cons Auth, View, Aust, 24 pp
Cooper RW (1965) Wind movement in pine stands. Georgia For Res Pap 33. Georgia For Res Counc, Macon, Georgia, 4 pp
Cooper RW, Altobellis AT (1969) Fire kill in young loblolly pine. Fire Cont Notes 30(4):14–15
Craighead FC (1977) Control burning will reduce forest fire hazards. Naples, FL: The Naples Star. Fri. May 20, 14A.
Crane WJB (1982) Computing grassland and forest fire behavior, relative humidity and drought index by pocket calculator. Aust For 45(2):89–97
Crow AB, Shilling CL (1980) Use of prescribed burning to enhance southern pine timber production. So J Appl For 4(1):15–18
Curtin RA (1966) The effects of fire on tree health and growth. In: The effects of fire on forest conditions. Tech Pap 13. For Comm NSW, Aust, pp 21–35
Davis KP (1959) Forest fire: control and use. McGraw-Hill, New York, 584 pp
Davis LS, Cooper RW (1963) How prescribed burning affects wildfire occurrence. J For 61(12):915–917
De Ana Magan FJF (1981) Controlled fires in forests induce attack of the fungus leptographium gallaeciae sp. Nov. on *Pinus Pinaster*. Forestal de Zonas Humedas, 37 pp. Dept For Zonas Humedas CRIDA, Pontevedra, Spain
Debano LF (1981) Water repellent soils: a state-of-the-art. Gen Tech Rep PSW-46. Berkeley, CA: US Dept Agric, For Serv, Pacific Southwest For Range Expt Stn, 21 pp
Deeming JE, Burgan RE, Cohen JD (1977) The national fire-danger rating system – 1978. Gen Tech Rep INT-39. Ogden, UT: US Dept Agric, For Serv, Intermountain For Range Expt Stn, 63 pp
de Ronde C (1980) Controlled burning under pines – a preliminary fuel classification system for plantations in the Cape. S Afr For J 113:84–86
de Ronde C (1982) The resistance of *Pinus* species to fire damage. S Afr For J 122:22–27
de Ronde C (1983) Controlled burning in pine stands in the Cape: the influence of crown scorch on tree growth and litterfall. S Afr For J 127:39–41
de Ronde C, Bohmer LH, Droomer AEC (1986) Evaluation of wildfire damage in pine stands. S Afr For J 138:45–50
Donald DGM (1979) Nursery and establishment techniques as factors in productivity of man-man forests in southern Africa. S Afr For J 109:19–25
Fahnestock GR (1970) Two keys for appraising forest fire fuels. Res Pap PNW-99. Portland, OR: US Dept Agric, For Serv, Pacific Northwest For Range Expt Stn, 26 pp
Fahnestock GR, Hare RC (1964) Heating of tree trunks in surface fires. J For 62(11):799–805
Fahnestock GR, Tarbes J, Yegres L (1987) The pines of Venezuela. J For 85(11):42–44
FAO (1982) Tropical forest resources. FAO For Pap 30, FAO, Rome, 106 pp
Fellin DG (1980) A review of some relationships of harvesting, residue management, and fire to forest insects and disease. In: Environmental consequences of timber harvesting in Rocky Mountain coniferous forests. Gen Tech Rep INT-90. Ogden, UT: US Dept Agric, For Serv, Intermountain For Range Expt Stn, pp 335–416
Ferguson ER, Gibbs CB, Thatcher RC (1960) "Cool" burns and pine mortality. Fire Cont Notes 21(1):27–29
Fischer WC (1978) Planning and evaluating prescribed fires. Gen Tech Rep INT-43. Ogden, UT: US Dept Agric, For Serv, Intermountain For Range Expt Stn, 19 pp
Fischer WC (1980) Index to the proceedings: Tall Timbers fire ecology conference, vol 1–15, 1962–1976. Gen Tech Rep INT-87. Ogden, UT: US Dept Agric, For Serv, Intermountain For Range Expt Stn, 140 pp

Flinn DW (1978) Comparison of establishment methods for *Pinus radiata* on a former *P. pinaster* site. Aust For 41(3):167–176

Fox RC, Hill TM (1973) The relative attraction of burned and cutover pine areas to the pine seedling weevils *Hylobius pales* and *Pachylobius picivorus*. Ann Entomol Soc Am 66(1):52–54

Froelich RC, Hodges CS Jr, Sackett SS (1978) Prescribed burning reduces severity of Annosus root rot in the south. For Sci 24(1):93–100

Goldammer JG (1982) Controlled burning for stabilizing pine plantations. In: Proc Int Sem organized by the Timber Committee of the United Nations Economic Commission for Europe, Warsaw, Poland, 20–22 May 1981, Nijhoff/Junk, The Hague, pp 199–207

Goldammer JG (1983) Sicherung des südbrasilianischen Kiefernanbaues durch kontrolliertes Brennen. Hochschulsammlung Wirtschaftswiss, Forstwiss Bd 4, Hochschulverlag, Freiburg, 183 pp

Goldammer JG (1987) TCP Assistance to Forest Fire Management, The Philippines. For Fire Res. FAO, Rome, FO:TCP/PH1/66053(T), Work Pap 1, 38 pp

Gremmen J (1971) *Rhizina undulata*: a review of research in the Netherlands, Eur J For Pathol 1:1–6 (cited in Harvey et al. 1976)

Grier CC (1975) Wildfire effects of nutrient distribution and leaching in a coniferous ecosystem. Can J For Res 5:559–607

Gruschow GF (1951) Effect of winter burning on slash pine growth. Southern Lumberman, December 15 issue, 2 pp

Gruschow GF (1952) Effect of winter burning on growth of slash pine in the flatwoods. J For 50(7):515–517

Haigh H (1980) A preliminary report on controlled burning trials in pine plantations in Natal. S Afr For J 113:53–58

Hardison JR (1976) Fire and flame for plant disease control. Annu Rev Phytopathol 14:355–379

Harvey AE, Jurgensen MF, Larsen MJ (1976) Intensive fiber utilization and prescribed fire: Effects on the microbial ecology of forests. Gen Tech Rep INT-28. Ogden, UT:US Dept Agric, For Serv, Intermountain For Range Expt Stn, 46 pp

Harvis GR, Covington WE (1983) The effect of a prescribed fire on nutrient concentration and standing crop of understory vegetation in Ponderosa pine. Can J For Res 13(3):501–507

Harwood CE, Jackson WD (1975) Atmospheric losses of four plant nutrients during a forest fire. Aust For 38(2):92–99

Haywood JD (1986) Response of planted *Pinus taeda* L. to brush control in northern Louisiana. For Ecol Manage 15(2):129–134

Hedden RL (1978) The need for intensive forest management to reduce southern pine beetle activity in east Texas. So J Appl For 2(1):19–22

Helms JA (1979) Positive effects of prescribed burning on wildfire intensities. Fire Manage Notes 40(3):10–13

Hough WA (1969) Caloric value of some forest fuels of the southern United States. Res Pap SE-187. Asheville, NC:US Dept Agric, For Serv, Southeastern For Expt Stn, 12 pp

Hough WA (1978) Estimating available fuel weight consumed by prescribed fires in the south. Res Pap SE-187. Asheville, NC:US Dept Agric, For Serv, Southeastern For Expt Stn, 12 pp

Hough WA (1981) Impact of prescribed fire on understory and forest floor nutrients. Res Note SE-303. Asheville, NC:US Dept Agric, For Serv, Southeastern For Range Expt Stn, 4 pp

Hudson J, Kellman M, Sanmugadas K, Alvarado C (1983a) Prescribed burning of *pinus oocarpa* in Honduras I. Effects on surface runoff and sediment loss. For Ecol Manage 5(4):269–281

Hudson J, Kellman M, Sanmugadas K, Alvarado C (1983b) Prescribed burning of *pinus oocarpa* in Honduras II. Effects on nutrient cycling. For Ecol Manage 5(4):283–300

Humphreys FR, Craig FG (1981) Effects of fire on soil chemical, structural and hydrological properties. In: Gill AM, Groves RH, Noble IR (eds) Fire and the Australian Biota. Aust Acad Sci Canberra, ACT, pp 203–214

Hunt SM, Crock MJ (1987) Fire behavior modeling in exotic pine plantations: testing the Queensland Department of Forestry "prescribed burning guide mk III". Aust For Res 17(3):179–89

Hunt SM, Simpson JA (1985) Effects of low intensity prescribed fire on the growth and nutrition of slash pine plantations. Aust For Res 15(1):67–77

Jemison GM (1944) The effect of basal wounding by forest fires on the diameter growth of some southern Appalachian hardwoods. Duke Univ Sch For Bull 9

Johansen RW (1975) Prescribed burning may enhance growth of young slash pine. J For 73(3):148–149

Johansen RW (1981) Windrows vs. small piles for forest debris disposal. Fire Manage Notes 42(2):7–9

Johansen RW (1987) Ignition patterns and prescribed fire behavior in southern pine stands. Georgia For Res Pap 72. Georgia For Comm Macon, Georgia, 6 pp

Johansen RW, Wade DD (1986) An insight into thinning young slash pine stands with fire. In: Proc 4th biennial southern silvicultural conference. 1986 November 4–6; Atlanta, GA. Gen Tech Rep SE-42. Asheville, NC:US Dept Agric, For Serv, Southeastern For Expt Stn, pp 103–106

Johansen RW, Wade DD (1987) Effects of crown scorch on survival and diameter growth of slash pine. So J Appl For 11(4):180–184

Jones JM, Richards BN (1977) Changes in the microbiology of eucalypt forest soils following reafforestation with exotic pines. Aust For Res 7(4):229–240

Kayll AJ (1963) A technique for studying the fire tolerance of living tree trunks. Can Dept For Pub 1012, 22 pp

Kellman M, Miyanishi K, Herbert P (1987) Nutrient sequestering by the understory strata of natural *pinus caribaea* stands subject to prescription burning. For Ecol Manage 21(1–2):57–73

Langdon OG (1971) Effects of prescribed burning on timber species in the southeastern coastal plain. In: Prescribed burning Symp Proc, 1971 Apr 14–16. Charleston, SC. Asheville, NC: US Dept Agric, For Serv, Southeastern For Exp Stn, pp 34–44

Lundquist JE (1984) The occurrence and distribution of Rhizina root rot in South Africa and Swaziland. S Afr For J 131:22–24

Luedge W (1971) Der Einfluss von Laubholzunterbau auf die Schädlingsdichte in den Kiefernbeständen der Schwetzinger Hardt. Allg Forst Jagdz 142:173–178

Lugo AE, Schimidt R, Brown S (1980) Preliminary estimates of storage and production of stemwood and organic matter in tropical plantations. In: Proc Int Symp IUFRO S1-07-09 Working Group at the Institute of Tropical Forestry, Rio Piedras, Puerto Rico, Sept 8–12, 1980, pp 8–16

Maggs J (1988) Organic matter and nutrients in the forest floor of a *Pinus elliottii* plantation and some effects of prescribed burning and superphosphate addition. For Ecol Manage 23:105–119

Maggs J, Hewett RK (1985) Nitrogenase activity (C_2H_2 reduction) in the forest floor on a *Pinus elliottii* plantation following superphosphate addition and prescribed burning. For Ecol Manage 14:91–101

Majer JD (1984) Short-term responses of soil and litter invertebrates to a cool autumn burn in jarrah (*Eucalyptus marginata*) forest in Western Australia. Pedobiologia 26(4):229–247

Maryland G (1988) The prospect of solving the CO2 problem through global reforestation. US Dept Energy DDE/NBB-082, 66p

Martin RE, Mitchell RG (1980) Possible, potential, probable and problem fire-insect interactions. In: Land-use allocation: process, people, politics, professionals, Proc 1980 Conv Soc Am For, pp 138–144

Maxwell WG, Ward FR (1980) Guidelines for developing or supplementing natural photo series. Res Note PNW-358. Portland OR: US Dept Agric, For Serv, Pacific Northwest For Range Expt Stn

McArthur AG (1966) Weather and grassland fire behavior. For Timb Bur Leafl 100, Canberra, ACT, 23 pp

McArthur AG (1967) Fire behavior in eucalypt forests. For Timb Bur Leafl 107, Canberra, ACT, 25 pp

McArthur AG (1971) Aspects of fire control in the *P. caribaea* and *P. elliottii* plantations of North Western Viti Levu Fiji Islands. Memeo Rep Canberra, ACT, 45 pp

McCormick J (1976) Recovery of maritime pine (*Pinus pinaster*) after severe crown scorch. For Dept West Aust Res Pap 20, 3 pp

McCulley RD (1948) Progress report on a study of the significant factors affecting fire injury and the effect of such injury on the mortality and growth of slash pine on the Osceola National Forest. Study No 0-113, Lake City, FL: US Dept Agric, For Serv, Southeastern For Expt Stn

McCulley RD (1950) Management of natural slash pine stands in the flatwoods of south Georgia and north Florida. Circular 845. Washington, DC: US Dept Agric, For Serv, 57 pp

McDonald SE, Krugman SL (1986) Worldwide planting of southern pines. J For 84(6):21–24

McKee, WH Jr (1982) Changes in soil fertility following prescribed burning on coastal plain pine sites. Res Pap SE-234. Asheville, NC: US Dept Agric, For Serv, Southeastern For Expt Stn, 23 pp

Metz LJ, Lotti F, Klawitter RA (1961) Some effects on prescribed burning on coastal plain forest soil. Sta Pap 133. Asheville, NC: US Dept Agric, For Serv, Southeastern For Expt Stn, 10 pp

Mikola P, Laiho O, Erikainen J, Kuvaji K (1964) The effect of slash burning on the commencement of mycorrhizal association. Acta For Fenn 77(3), 12 pp

Miller JM, Patterson JE (1927) Preliminary studies on the relation of fire injury to bark-beetle attack in western yellow pine. J Agric Res 34:597–613

Munro N (1966) The fire ecology of caribbean pine in Nicaragua. In: Proc 5th Ann Tall Timbers Fire Ecol Conf, 1966 March 24–25, Tallahassee, FL, pp 67–83

Nambiar EKS (1984) Critical processes in forest nutrition and their importance for management. In: Landsberg JJ, Parsons W (eds) Research for forest management. CSIRO Div For Res Canberra, ACT, pp 52–72

Neal JL, Wright E, Bollen WB (1965) Burning Douglas-fir slash. Physical, chemical and microbial effects in soil. Unnumb Res Pap Corvallis, OR: Oregon State Univ For Res Lab, 32 pp

Nelson RM (1952) Observations on heat tolerance of southern pine needles. Sta Pap 14. Asheville, NC: US Dept Agric, For Serv, Southeastern For Expt Stn, 6 pp

Nelson RM Jr, Adkins CW (1988) A dimensionless correlation for the spread of wind-driven fires. Can J For Res 18:391–397

Peet GB, McCormick J (1971) Short-term responses from controlled burning and intense fires in the forests of western Australia. Bulletin 79. For Dept, Perth West Aust, 24 pp

Phillips MJ, Goh KM (1986) Growth response of pinus radiata to fertilizer and herbicide treatment in a clearfelled logged and a clearfelled, logged and burned northofagus forest. NZJ For Sci 16(1):19–29

Pritchett WL (1976) Considerations in use of fire by prescription for managing soil and water. In: Proc fire by prescription Symp, 1976 October 13–15, Atlanta, GA, Washington, DC: US Dept Agric, For Serv, pp 33–35

Puckett JV, Johnston CM, Albini FA et al. (1979) Users' guide to debris prediction and hazard appraisal. Unnumb Pub Missoula, MT: US Dept Agric, For Serv, Northern Forest Fire Lab, Intermountain For Range Expt Stn Rev 1979, 37 pp

Ralston CW, Hatchell GE (1971) Effects of prescribed burning on physical properties of soil. In: Proc Prescribed burning Symp, 1971 Apr 14–16; Charleston SC. Asheville, NC: US Dept Agric, For Serv, Southeastern For Expt Stn, pp 68–84

Reichle DE (1977) The role of soil invertebrates in nutrient cycling. In: Lohm V, Persson T (eds) Soil organisms as components of ecosystems. Stockholm: Swed Nat Sci Res Counc (cited in Majer 1984), pp 145–156

Rennie PJ (1971) The role of mechanization in forest site preparation. Paper presented at XV IUFRO Congress, Sec 32, Gainesville, FL, 38 pp

Robbins LE, Myers RL (1989) Seasonal effects of prescribed burning in Florida: a review. Tallahassee, FL: The Nature Conservancy, Fire Manage Res Prog, 86 pp

Rook DA, Whyte AGD (1976) Partial defoliation and growth of 5-year-old radiata pine NZJ For Sci 6(1):40–56

Rothermel RC (1972) A mathematical model for predicting fire spread in wildland fuels. Res Pap INT-115. Ogden, UT: US Dept Agric, For Serv, Intermountain For Range Expt Stn, 40 pp

Sackett SS (1975) Scheduling prescribed burns for hazard reduction in the southeast. J For 73(3):143–147

Schmidt RG (1978) An approach to hazard classification. Fire Manage Notes 39(4):9–11

Schowalter TD (1983) Chapter 13: Adaptations of insects to disturbance. In: Pickett STA, White PS (eds) Natural disturbance: an evolutionary perspective. Academic Press, London

Schowalter TD, Coulson RN, Crossley DA Jr (1981) Role of southern pine beetle and fire in maintenance of structure and function of the southeastern coniferous forest. Environ Entomol 10(6):821–825

Schutz CJ (1987) Litter under patula. In: Res Rev, S Afr For Res Inst, pp 35–36

Shea SR, Peet GB, Cheney NP (1981) The role of fire in forest management. In: Gill AM, Groves RH, Noble IR (eds) Fire and the Australian Biota. Aust Acad Sci, Canberra, ACT, pp 443–470

Smith D, Mrowka R, Maupin J (1983) Underburning to reduce fire hazard and control Ips beetles in green thinning slash. Fire Manage Notes 44(2):5–6

Speltz GE (1968) Resistencia ao fogo de diversas especies florestais registrados na Fazenda Monte Algre – Parana Anais Congr. Flor Brazil, Curitiba, 350 pp

Stewart RE, Gross LL, Honkala BH (1984) Effects of competing vegetation on forest trees: a bibliography with abstracts. Gen Tech Rep WO-43. Washington DC: US Dept Agric, For Serv

Tarrant RF (1956) Effects of slash burning on some soils of the Douglas-fir region. Soil Sci Soc Am Proc 20:408–411

Tiedemann AR, Conrad CE, Dieterich JH et al. (1979) Effects of fire on water. A state-of knowledge review. Proceedings of national fire effects workshop, 1978 Apr 10–14; Denver, CO. Gen Tech Rep WO-10. Washington, DC: US Dept Agric, For Serv, 28 pp

Tozzini DA, Soares RV (1987) Relacoes entre compartamento do fogo e danos causados a um povomento de *Pinus taeda*. Floresta 17(1,2):9–13

Turvey ND, Cameron JN (1986) Site preparation for a second rotation of radiata pine: soil and foliage chemistry, and effect on tree growth. Aust For Res 16(1):9–19

Van Loon AP (1966) The effects of fire on understory vegetation. In: The effects of fire on forest conditions. Tech Pap 13, For Comm NSW, Australia, pp 44–52

Van Loon AP (1967) Some effects of a wild fire on a southern pine plantation. Res Note 21, For Comm NSW, Australia, 38 pp

Van Loon AP, Love LA (1973) A prescribed burning experiment in young slash pine. Res Note 25, For Comm NSW, Australia, 53 pp

Van Wagner CE (1973) Height of crown scorch in forest fires. Can J For Res 3(3):373–378

Vega JA, Bara S, Del Carmen G (1983) Prescribed burning in pine stands for fire prevention in the N.W. of Spain: Some results and effects. In: Goldammer JG (ed) DFG- Symp fire ecology. Freiburger Waldschultz Abh 4, Freiburg, pp 49–74

Villarrubia CR, Chambers JL (1978) Fire: its effects on growth and survival of loblolly pine, *Pinus taeda* L. Louisiana Acad Sci 41:85–93

Vlok JHJ, De Ronde C (1989) The effect of low-intensity fires on forest floor vegetation in mature *Pinus elliottii* plantations in the Tsitsikamma. S Afr J Bot 55(1):11–16

Wade DD (1983) Fire management in the slash pine ecosystem. In: The managed slash pine ecosystem Symp Proc, 1981 June 9–11; Gainesville, FL: Univ Florida, Sch Nat Resour, pp 203–227, 290–294

Wade DD (1985) Survival in young loblolly pine plantations following wildfire. In: Proc 8th Nat Conf on fire and forest meteorology, 1985 April 29-May 2; Detroit, MI. Washington DC: Soc Am For, pp 52–57

Wade DD (1986) Linking fire behavior to its effects on living plant tissue In: Proc 1986 Soc Am For Nat Convention, 1986 October 5–8; Birmingham, AL. Washington, DC: Soc Am For, pp 112–116

Wade DD, Johansen RW (1986) Effects of fire on southern pine: observations and recommendations. Gen Tech Rep SE-41. Asheville, NC: US Dept Agric, For Serv, Southeastern For Expt Stn, 14 pp

Wade DD, Lewis CL (1987) Managing southern grazing ecosystems with fire. Rangelands 9(3):115–119

Wade DD, Lunsford JD (1989) A guide for prescribed fire in southern forests. Tech Pub R8-TP 11. Atlanta, GA, US Dept Agric, For Serv, So Region, 56 pp

Wade DD, Ward DE (1975) Management Decisions in severely damaged stands. J For 73(9):573–577

Wade DD, Weise DR, Shell R (1989) Some effects of periodic winter fire on plant communities on the Georgia Piedmont. In: Proc 5th biennial southern silvicultural research conference, 1988 Nov 1–3, Memphis, TN: Gen Tech Rep SO-74. New Orleans, LA: US Dept Agric, For Serv, Southern For Expt Stn, pp 603–611

Watson AG, Ford EJ (1972) Soil fungistais – a reappraisal. Annu Rev Phytopathol 9:327–348 (cited in Harvey et al. 1976)

Weaver H (1959) Ecological changes in the ponderosa pine forest of the Warm Springs Indian Reservation in Oregon. J For 57(1):15–20

West O (1965) Fire in vegetation and its use in pasture management with special reference to tropical and subtropical Africa. Comm Bur Pastures Field Crops Memo Pub 1/1965. Hurley, Berk, England, 53 pp

Weise DR, Wade DD, Johansen RW (1989) Survival and growth of young southern pine after simulated crown scorch. In: Proc 10th Conf on Fire and Forest Meteorology, 1989 April 17–21. Ottawa, Canada (in press)

Wells CG (1971) Effects of prescribed burning on soil chemical properties and nutrient availability. In: Proc prescribed burning Symp, 1971 Apr 14–16; Charleston, SC. Asheville, NC: US Dept Agric, For Serv, Southeastern For Expt Stn, pp 86–97

Wells CG, Campell RE, Debano LF et al. (1979) Effects of fire on soil. A state-of-knowledge review. Proc Nat fire effects workshop, 1978 Apr 10–14; Denver, CO. Gen Tech Rep WO-7. Washington, DC: US Dept Agric For Serv, 34 pp

Will GM (1959) Nutrient return in litter and rainfall under some exotic conifer stands in New Zealand. NZJ Agric Res 2:719–734

Woods RV (1981) The relationship between fire nutrient depletion, and decline in productivity between rotations in South Australia. In: Proc Australian Forest Nutrition Workshop Productivity in Perpetuity, 10–14 Aug 1981. Canberra, ACT (Abstr)

Wright E (1971) Mycorrhiza on ponderosa pine seedlings. Bull 13. Corvallis, OR: Oregon State Univ, 36 pp (cited in Harvey et al. 1976)

Wright E, Tarrant RF (1957) Microbiological soil properties after logging and slash burning. Res Note 157. Portland, OR: US Dept Agric, For Serv, Pacific Northwest For Range Expt Stn, 5 pp

Zahner R (1989) Tree-ring series related to stand and environmental factors in South Alabama longleaf stands. In: Proc 5th biennial southern silvicultural research conference. 1988 Nov 1–3; Memphis, TN: Gen Tech Rep SO-74. New Orleans, LA: US Dept Agric, For Serv, Southern For Expt Stn, pp 193–197

13 Landscapes and Climate in Prehistory: Interactions of Wildlife, Man, and Fire

W. Schüle[1]

13.1 Introduction

Landscapes are more than a simple function of geological, geomorphological, climatic, and botanical parameters. Animals play an important role. Their behavior, especially their trophic habits, is a major force in the forming of landscapes. Herbivores consume the products of the primary biomass production. Fire and man have been doing the same since they appeared on Earth. Moreover, both are not only herbivorous, but also carnivorous, devouring whatever animal wherever they can.

Local and regional climates depend largely on the local and global circulation of air and water, and on the vegetation, as this depends on the general climate. If the vegetation canopy is altered considerably by overgrazing, agriculture, fire, or any other reason over large areas, the albedo as well as the thermal and hydrological conditions are strongly influenced. One can expect global consequences (Shukla and Mintz 1982).

Earth's global climate (Flohn 1985; Wanner and Siegenthaler 1988; Weischet 1988; Schwarzbach 1974; Lamb 1972–1977; Oeschger et al. 1980) is a function of influences from the cosmos, the solar system, Earth's mass and movements within the solar system, of physical and chemical conditions in the lithosphere, hydrosphere, and atmosphere, and of time. Leaving supernovae and meteors aside, the cosmos, solar system, and Earth's mass and movement parameters can change on a large time scale only. We know very little about these factors, and they are not our concern here. Processes in the lithosphere influence the climate on a large time scale through orogenesis and plate tectonics, which in turn affect both other spheres, and on both a large and a small time scale through volcanism which may change the chemical and physical conditions of the atmosphere within years or even days. Volcanic activities have been influencing the atmosphere since the very beginning. That we will leave aside also.

Our concern is life on earth. It conditions the carbon circulation in the three outer spheres of Earth (Fig. 1). CO_2 and some other gases of organic origin are the most important agents in the regulation of the (on a cosmic scale) minimal oscillations of atmospheric temperatures, which in turn are the most important agents in the circulation of water on the planet, and thus of the climate. Primary biomass production by plants is the motor of carbon flux in the biosphere.

[1] Institute of Prehistory, University of Freiburg, 7800 Freiburg, FRG

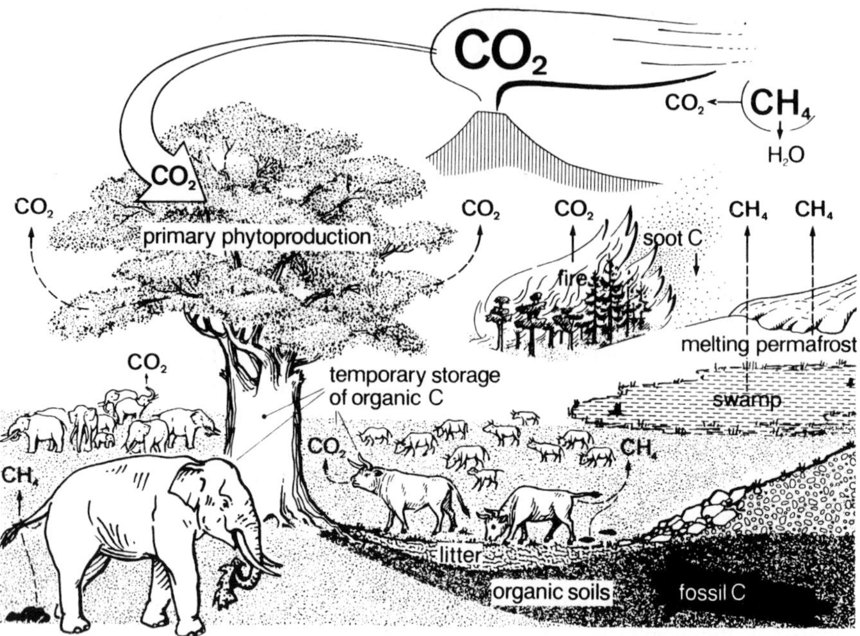

Fig. 1. Main parameters of the terrestrially based part of carbon flux in the Biosphere. Note storage and buffering effect of permafrost before melting. Flux, storage, and buffering by the ocean disregarded. Very simplified. For influences of prehistoric man see legend to Fig. 2

The terrestrial flora has a strong influence not only on the local, but also on Earth's global albedo, as it covers a considerable portion of its surface. The fauna indirectly influences both the albedo and the carbon flux by reducing the vegetation and by the emission of respiration and digestion gases. Carnivorous animals reduce the herbivores and thus their pressure on the vegetation. Man and his ancestors did and do the same thing.

In addition, the flora provides the only fuel for fire, and fire is and has been an important agent in the global carbon flux. Again, man and his ancestors played and play an important part in the global fire regime.

Oceans play a major role in the circulation of CO_2, but they are obviously not directly influenced by fire, prehistoric man, or terrestric herbivores, so we will also leave them aside.

I shall try to analyze some of the interactions and consequences of these processes for biological and cultural evolution and for the forming of landscapes. I will further suggest possible climatic implications, but these, for the time being, have to be seen more as a hypothetical approach than as an exact description of facts.

The importance of fire for human culture has always been recognized, its role as man's indispensable servant. Fire was considered a gift of the gods, or, as the gods are at present a bit unpopular, of man's own spirit. But the gift, when not handled carefully, demonstrates a terrible destructive power. It becomes evident that contact with fire, intentional or unintentional, had a strong influence on the evolution of man's central nervous system.

The positive and negative feedback systems between fire, human evolution, man's influence on the carbon flux in the biosphere (Fig. 1), man's expansion over the Earth (Fig. 2), and the evolution of landscapes and climate, are the main subjects of this chapter. Evidently the problem has to be considered worldwide and on a geological time scale. Evidently again, this can be done only in a very generalizing manner.

13.2 Natural Fires

Up to our days, natural fires were regarded as extremely destructive, devastating grasslands, woods, and forests. Now fire turns out to have been an important agent of biological evolution since the first terrestrial plants began to produce huge masses of organic material 350 or 400 million years ago (dates from Hohl 1985). This accumulating material formed a growing fire potential. Coal seams are accumulated, unburned fire potential, or simply fuel, which was covered and fossilized millions of years ago.

When dry enough, the fire potential can ignite through volcanic activity, atmospheric electricity, or, more seldom, biotic thermo-energy. These natural agents for incineration have existed throughout the history of biological evolution, and one may conclude that vegetation has always been exposed to what is nowadays called a "natural fire frequency" (Goldammer 1978).

Fire frequency depends on the distribution of fuel, its humidity, and its ignition frequency. Leaving the rare incinerations through volcanos or biotic processes aside, fire frequency depends directly or indirectly upon climatic parameters. Given adequate edaphic and thermic values, the primary biomass production, and with it the fire potential, increases with increasing humidity. If precipitation is distributed evenly over the year, opportunities for ignition will be few, and fires will soon be extinguished. If precipitation is seasonal, allowing a high phyto-production during the wet season, and if the dry season is long and dry enough, fire frequency will be high. If thunderstorms without precipitations occur during the dry season, fire frequency will reach its highest value, and fires will burn until they reach natural obstacles like rivers, lakes, or zones with spare vegetation.

Fire intensity depends on the wind and on the quality and quantity of the fire potential. The latter will increase as the fire frequency decreases. Frequency reduces intensity – as fire fighters well know. If the fuel build-up is large and of suitable quality, a wind-driven fire may develop enough heat to burn even wet material. If a drought dries up part of the fuels of a dense, humid forest for instance, even tropical rain forests may burn.

Efficiency of a fire will thus increase with lower frequency, higher primary biomass production, drier fuel, fewer obstacles, and stronger wind.

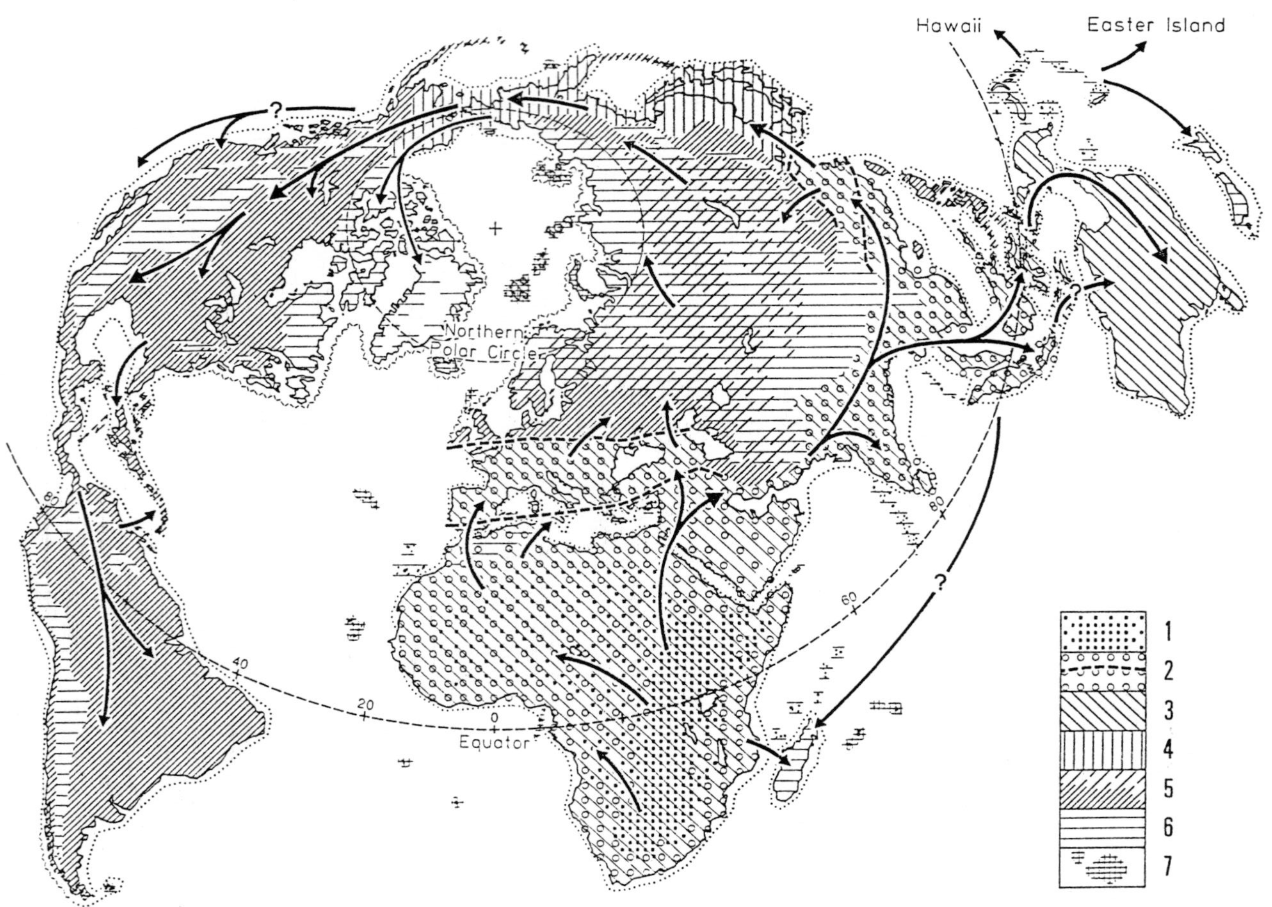
Hawaii
Easter Island
Northern Polar Circle
Equator
0
20
40
60
80
80
1
2
3
4
5
6
7

◄

Fig. 2. The expansion of the Hominidae all over the Earth happened successively by great steps in the scales of biotopic, climatic, and continental parameters. Most values are estimations. Influence on the megaherbivores by use of thrusting spears may begin during stage *1* in Upper Pliocene, about 3×10^6 yr B.P., fire use towards the end of stage *1*, about 1.5×10^6 yr B.P., use of thrown spears and intentional primary production of fire probably during stage *2*. Thus all stages of human expansion were accompanied by considerable changes of the carbon flux in the biosphere (see Fig. 1).

Areals and time scale of the map are derived from the paleo-osteological record for both hominids and megaherbivores, from findings of datable lithic/artifacts, from general ecological parameters, and/or from the shifting climatic zones of the Quaternary, which caused eustatic changes of the sea level in the order of 10^2 m. *Dotted lines* off the actual shores show approximately the rims of the shelfs.

The tolerance of precisation of chronological dates oscillates in the order of 10^6 years in the Upper Tertiary (before 1.8×10^6 yr B.P.); in the Lower Pleistocene ($1.8 \times 10^6 - 7 \times 10^5$ yr B.P.) in the order of $x \times 10^5$ years; in the Middle Pleistocene ($7-1.4 \times 10^5$ yr B.P.) in the order of 10^5 years; in the lower Upper Pleistocene (1.4×10^5-7×10^4 yr B.P.) in the order of 10^4 years; at the Pleistocene/Holocene border (10^4 yr B.P.) in the order of 10^3 years; and for the first settlement of New Zealand and Madagascar in the order of 10^2 years. The first human settlements realized by Europeans rely on historical dates.

Stages:

1 Probable area of evolution of early hominids. *Australopithecus* sp. and *Homo habilis*. Pliocene and lower/middle Lower Pleistocene, since ca. 3.5×10^6 B.P.).

2 First expansion to unspecific biotops in Africa and tropical and subtropical Eurasia. Lower and Middle Pleistocene. In Africa since ca. 10^6 B.P., in Eurasia probably not before Middle Pleistocene, on the shelf islands of southern Asia probably during low sea level of cold phases. *Homo habilis* (known only from Africa) and *H. erectus.* The *dotted line* marks approximately the climatically caused shifting of the Eutrophic Line during up to the end of stage *4* for *H. erectus, H. sapiens praesapiens, H.s. neanderthalensis and earliest H.s. sapiens.* These forms of *H. sapiens* are known only from the western part of the Old World.

3 Expansion to tropical islands off the South-Asian shelf and to Australia, about 4×10^4 B.P. *H.s. sapiens*. Celebes (Sulawesi) might have been settled earlier by *H. erectus*. First indication for passive drifting on rafts or active (?) navigation.

4 Expansion to cooler climates of east Eurasia and the unglaciated North of east Beringia during the last glaciation (Würm/Weichsel/Wisconsin) in later Upper Pleistocene, probably favored by exploitation of marine mammals. The same stage is to be supposed near the European coasts of the Atlantic. Fossil traces are widely destroyed by later glaciation during the last cold phase of Würm/Weichsel in Europe and worldwide by the Holocene marine transgression. Successively with developing technologies of thermo-insulation since about 4×10^4 B.P.; earliest dates in ice-free eastern Beringia about 2.5×10^4 B.P. *H.s. sapiens.*

5 Expansion to very cold and/or arid climates of Eurasia and to the Americas south of the melting North American ice sheet. Near the Pleistocene/Holocene border, since ca. 1.2×10^4 B.P. Earliest settlements on some of the Melanesian (and Polynesian?) islands in Middle and early Upper Holocene. *Homo s. sapiens.*

6 Off-shore islands of the Pacific and India. Upper Holocene, ca. 800 A.D. *H.s. sapiens* (Polynesians, Malayans, and Africans).

7 First settlement by Europeans. Upper Holocene, historical dates. *Homo s. sapiens.*

13.2.1 Vegetation, Herbivores, and Fire

The production of fire potential depends not only on the primary phytoproduction per year and area, but also on the quantity of vegetable material consumed by the fauna, including man (for vertebrate paleontology see Carroll 1988). If there are few or no herbivores, most of the phytoproduction will be stored, and fire potential will accumulate. If a considerable part of the phytoproduction gets eaten, the fuel accumulation in a given area will be low, even if conditions for phytoproduction and its seasonal or occasional drying are favorable.

The more herbivores there are, the more slowly fire potential accumulates. Only a small portion of the annual production remains as burnable material; most of it returns through the animals into the edaphic and atmospheric carbon flux, to be quickly recycled. Too many herbivores will destroy the vegetation, and with it the fire potential. The more efficient the dentitions and digestive tracts of the herbivores are, the quicker the rate of desertification.

For this reason, game preserves and timber production are antagonists. Hypsodont grazers with high-crowned teeth (for general mammal paleontology see Thenius 1980; Thenius and Hofer 1960; Piveteau 1955; Müller 1989; for living mammals see Macdonald 1984; Nowak and Paradiso 1983; Grzimek 1968–1972), like goats, camels or horses, and some rodents, are more efficient destroyers of vegetation than brachyodont browsers with low-crowned teeth, such as moose (*Alces*) and other woodland ungulates. Hypsodont animals chew nearly everything, and they can feed from the ground, not only from fairly near the canopy like real browsers. Tree seedlings have little hope of surviving hungry grazers. Browsers would do them only little damage during their first years. With their long necks, some hypsodont grazers, such as horses and camels, reach branches high above their heads. A ground fire cannot reach the crowns. The fire potential of what escapes hypsodont teeth is too low to damage grown *Pinus* and *Quercus*. That is why horses, donkeys, and even llamas are employed to reduce the undergrowth of Spanish dehesas and regenerating French maquis as an effort in fire prophylaxis. The reverse of the coin is that not one young tree will grow as long as the donkeys are present.

13.2.2 Fire and Evolution

Fire prophylaxis by herbivores is not an invention of modern science. Nature has been employing the same system since the first terrestrial animals began to eat plants in the Middle Paleozoic some 370 million years ago. I shall not bore you with 370 million years of co-evolution of plant self-defense and animals and fire tackling it. To be honest, we know close to nothing about it. It seems clear, though, that herbivores could not prevent the accumulation of fire potential in Middle Paleozoic times, otherwise the coal seams would be less impressive. It is not even clear whether efficient herbivorous vertebrates existed at that time outside the sea and paralic regions and whether the plants were already capable of forming forests outside swamps and/or very humid climate.

Fire frequency may have been low in carboniferous forests because of the high humidity, but new findings have revealed that fires did occur (Goldammer 1988 pers. commun.). During severe droughts, fire intensity must have been high in carboniferous forests some 300 million years ago.

Let us pass over 230 million years of geological and biological evolution and look at the natural fire prevention by herbivores in Late Cretaceous times, about 70 million years ago. At that time, reptiles had developed for about 200 million years and ruled land and sea. You all know what sizes some of them reached in the Upper Mesozoic. Every forest engineer will understand that in Late Cretaceous times few dense woods existed outside swampy or rugged ground or off-shore islands. To know what Late Cretaceous woods were like, we have to imagine a herd of voracious herbivorous dinosaurs the size of a medium house trampling a carefully tended forest. I shall not torture you with detailed descriptions of the result.

Yet dinosaurs were not as ruinous for the vegetation as it might seem. Reptile mastication systems were never as well developed as those of the first hypsodont mammals about 40 million years later. The problem is whether the Mesozoic vegetation was already as resistant to grazing and browsing as recent vegetation. Whether it was or not, donkeys the size of a dinosaur would have been much worse than dinosaurs in Mesozoic woods.

Fire frequency may have been high in semi-arid regions of Late Cretaceous times, but fire intensity was certainly low in most biomes. Fuel was eaten by herbivorous dinosaurs before it could accumulate. Trees needed a lot of good luck to grow beyond the reach of *Diplodocus*, and even afterwards they needed strong trunks to resist the hungry beasts.

No animal survives a fire. The fire-resistant salamander is a myth, as are the fire-spitting dragons. But animals which put their eggs into the water, or dig them deep enough into inorganic soils will survive the fire in their children, as many plants do in their seeds and roots. Living reptiles do so, and they do not care for their offspring, nor do their children need their mother's protection to survive. Thus, even if Mother Reptile dies miserably in the flames, her children happily continue to reproduce. Neither birds nor mammals can do this, and it seems doubtful if the dinosaurs could. At least some of them lived in herds of mixed individual ages.

Certainly, both plants and animals had become more fire-tolerant to survive, some 70 million years ago. Strangely enough, it is not impossible that fire has something to do with the disappearance of the dinosaurs: In Cretaceous-Tertiary boundary sediments a high concentration of graphitic carbon has been found. By some students this has been interpreted as a worldwide layer of soot from wildfires triggered by a giant meteorite (Wolbach et al. 1985). The heat produced by the meteorite (Alvarez et al. 1980) may even have ignited fossilized carbon. The combined physical and chemical effects (Crutzen et al. 1984) of both the impact itself and the fires triggered by it may well have produced the changes of the biosphere on the Cretaceous Tertiary boundary, about 65 million years ago.

Low fire frequency, combined with a corresponding high fire intensity during occasional droughts, will eliminate pyrophobic plants in moist woods, together with all the animals which cannot escape the fire. Burned areas will be repopulated from the unburned surroundings. Moreover, burned areas will seldom be large in these cases. When fire frequency is low and intensity high, there will be little selective value for both plants and animals – there are no survivors to evolve further. High fire frequency, combined with a corresponding low fire intensity makes survival possible. Fire's selective value for biological evolution is certainly high when its frequency is high and its intensity low. The fittest survive.

Most soft-leaved plants are heavily damaged by a high fire frequency regime, while pyrotolerant or pyrophilic plants are favored. Sclerophyllic scrubs, trees with thick bark or other adaptions to fire, and many grasses will spread. Most of these are highly inflammable when dry and cause hot but brief fires which do not heat the soil much. Roots, protected trunks, and resistant seeds are spared. For animals, hope of survival depends on their behavior and speed. Such a fire regime has a high selective value for the evolution of quick legs, perseverance, and the capacity of the central nervous system.

Increasing natural fire frequency, decreasing intensity, evolution of pyrotolerant plants, increasing hypsodonty of ungulates, evolution of adequate digestion, locomotion, and nervous systems, and climatic changes caused by changing vegetation formed a complicated network of positive and negative feedback loops throughout the 65 million years of the Tertiary. The era started with thick woods that covered nearly all of the globe in the Paleocene, and it ended in the widespread savannas and large areas with a more continental climate in the Upper Tertiary and Quaternary.

Besides hard-to-define extraterrestrial parameters, geological processes also had an influence on the Tertiary climate, and thus on the fire regime. The successive merging of India (in the Lower Tertiary) and Africa (at the end of the Oligocene) with Eurasia probably enhanced the above-mentioned changes. These events initiated rapid orogenetic processes, activated volcanism, and altered atmospheric and marine circulations. Something similar happened when the Meso-American Isthmus was formed in the Upper Pliocene. We will not look closer at these problems. It is enough to show that plate tectonics may also influence the fire regime (for literature on plate tectonics see Giese 1986).

13.2.3 Fire and Mammals

At the end of the Mesozoic, the giant reptiles had disappeared. Mammals survived, but in the Paleocene (Lowest Tertiary) they were small, no larger than a fox, their dentitions not yet adapted to hard vegetable food. Woods covered the Earth, humidity increased, precipitations were immediately restored to the atmosphere, causing new precipitations, and so on. In Early Tertiary humid forests, some 65 million years ago, average fire frequency was low, the intensity of occasional fires high.

By the Middle Oligocene, some 35 million years ago, some Eurasian and North-American mammals, most of them ungulates from the order of Perissodactyla, had grown nearly to the size of the dinosaurs and were considerably larger than recent elephants. Some of them survived until the Lower Miocene, when the elephants entered Eurasia. These giant perissodactyles were browsers, not grazers, with brachyodont, low-crowned teeth, and as unable to feed from the ground as modern giraffes are. They certainly did not live in dense forests, but in open woods and savannas. In the shadow of the mega-browsers, smaller ungulates, mainly of the old Perissodactyla and the emerging Artiodactyla orders, with increasingly hypsodont teeth and special ruminant or hindgut-fermenting digestion systems developed, capable of processing the hard grasses and leaves of the spreading savanna vegetation. Thus, some 30 million years ago, fire frequency increased and fire intensity decreased.

13.2.4 Fire and Hominoidea

To avoid terminological confusion, and to give an idea of the time scales concerned, it seems necessary to make some preliminary remarks on plate tectonics and primate systematics. Specialized readers may better skip them and pass to the next section. It may be added that certain details of the plate tectonic background, and some systematic and evolutionary aspects, as well as most chronological data for plate tectonic events and primate evolution, are still discussed. Anyway, these discussions are of little concern here.

Before we consider man's ancestors and their influence on the biosphere, we have to take another look at plate tectonics (Giese 1986; Cocks 1981). During the Upper Paleozoic, about 2.25×10^8 years ago, most of Earth's dry land was united in one giant continent called Pangea. By the Upper Jurassic, about 1.35×10^8 years ago, Pangea had split in two, Laurasia and Gondwana. Laurasia united the three northern continents of Europe, Asia, and North America for most of the Tertiary and Quaternary. During the Upper Mesozoic and Lower Tertiary, since about 10^8 years ago, the smaller southern continent of Gondwana split into Antarctica, Africa, Madagascar, India, Australia, New Zealand and South America. Later, during the Tertiary, between $4–2.5 \times 10^7$ years ago, India and Africa were independently and successively re-united with the northern landmass of Laurasia. The same thing happened to South America when the Meso-American Isthmus formed in the Upper Pliocene, about 3.5×10^6 years ago. Consequently, evolution took different routes on the different continents.

Living man, zoologically seen, is the subspecies: *Homo sapiens sapiens* of the species: *H. sapiens*, genus: *Homo*, family: Hominidae, superfamily: Hominoidea, infraorder: Catarrhini, suborder: Anthropoidea, order: Primates, class: Mammalia, phylum: Vertebrates.

The suborder Anthropoidea includes the fossil and living Old World Catarrhini, the fossil and living New World Platyrrhini and the Old World fossil

Parapithecoidea. The infraorder Catarrhini comprises the fossil and living Hominoidea (tail-less apes) and the fossil and living Cercopithecoidea (Old World monkeys). The superfamily Hominoidea contains the fossil and living Hominidae (man and his next relations), the fossil and living Pongidae (apes and probably the fossil Meganthropus and Giganthropus from the Pleistocene of southern Asia, as well as some other still discussed fossil forms), the living and fossil Hylobatidae (gibbons), and the fossil, still discussed family Ramaspithecidae, which is supposed to include the direct ancestors of all hominids. The originally African family Hominidae includes the fossil African genus *Australopithecus* and the fossil and living genus *Homo*. The originally African genus *Homo* includes the fossil species *H. habilis* and *H. erectus*, and the fossil and living species *H. sapiens*. The subspecies *H.s. praesapiens* and *H.s. neanderthalensis* are only known as fossils. *H.s. neanderthalensis* is regarded by some students as a sideline. The subspecies *H.s. sapiens* is the only living hominid. (Fleagle 1988).

The first Primates appeared in the splitting Pangea during the Paleocene, some 6×10^7 years ago; the first Catarrhini in the Oligocene of Africa, some 3×10^7 years ago; the first Hominoidea in the African Lower Miocene, about 2.5×10^7 years ago. The Hominidae evolved in Africa since about 5×10^6 years ago in the Pliocene, *Homo* in the African savannas of Lower Pleistocene nearby 2×10^6 years ago, *H. erectus* in the same region probably some 10^5 years later, *H. sapiens* around 10^5 years ago and *H.s. sapiens* around 5×10^4 ago, at the best. By some students, all subspecies of *H. sapiens* are regarded to be of African origin, too. (Fleagle 1988; Tobias 1985).

13.2.4.1 Neogene Africa

First, let us look at Africa. There, the history of both mankind and man-made fire began.

Most of Africa was seemingly still covered with the old, Lower Tertiary moist woodlands when it collided with Laurasia at the beginning of the Miocene, about 25 million years ago (Whybrow 1984). Arabia was a part of Africa; the Red Sea and the East African Rift Valley began to form. Many kinds of Laurasian mammals entered Africa. Contrary to the mostly still brachyodont "subungulate" endemits, some of the immigrants had already developed more hypsodont teeth, capable of processing the hard leaves and grasses of spreading Laurasian savannas and steppes. When the newcomers entered Africa, these types of vegetation probably existed only in the extreme south, possibly in some parts of the north, and in the forming rift valley.

Most of the African vegetation was probably not adapted to so ruthless a treatment. The effect must have been similar to what happened 25 million years later, when sailors brought pigs and hypsodont sheep and goats to oceanic islands. Forests retreated, savannas and steppes spread, and a typical savanna fire regime developed. Laurasian ungulates spread and evolved rapidly, most of the old endemic African mammals disappeared. Even the endemic rhino-like subungulate *Arsinotherium*, with already hypsodont teeth, could not stand the Eurasian immigrants (or the rapidly evolving endemic Proboscidea?), if ever it got to meet them. Most living African mammals are neoendemic invaders who came from Laurasia during the Miocene, some of them even later.

Among the few endemic survivors were the ancestors of living elephants (Aguirre 1969), and of living monkeys, apes, and men (Fleagle 1988). They had to struggle hard because of the rapidly changing landscapes – and the increasing fire frequency. Their struggle was successful; otherwise neither we nor the elephants would exist today. Even better: both not only survived in Africa, they colonized the world.

In the beginning, proboscidians were much more successful than primates. Within a geologically very short time, they colonized the united landmasses of Africa, Eurasia, and North America, and, as soon as the Meso-American Isthmus had formed in the Upper Pliocene, they also settled South America. No rain forest was too dense, no tundra too cold, no mountain valley too remote for them. Swimming and drifting, they even reached most of the Wallacean islands between the Asian and the Australian continental shelves, and most Mediterranean islands. Only very steep slopes and waterless regions were off-limits for elephants, as was navigation to Australia, New Zealand, and Madagascar. Wherever they had time enough to adapt, some of them later also survived the high man-made fire frequency and hominid hunting.

African catarrhine primates were not as successful as the proboscidians. In the Lower Miocene, about 20 million years ago, their progress was halted somewhere in the Eurasian subtropics because of climatic and trophic reasons, and because of the sea. Not until 20 million years later did Upper Pleistocene catarrhine primates of the subspecies *Homo sapiens sapiens* finally manage to reach the limits of the proboscidian colonization and, by now being better navigators, to surpass them. But we will come to that later.

In the greater part of Africa, the natural fire frequency was probably high, fire intensity low at the end of the Miocene, about 6 million years ago. Upper Tertiary African savannas were probably not much different from modern undisturbed savannas. Rain forests certainly were. In the Pliocene, 6 to 2 million years B.P., five to ten kinds of megaherbivores and just-about-megaherbivores (Proboscidea, Deinotherioidea, Chalicotheridae, Rhinocerotidae, Giraffidae, Bovidae, Suidae, and Hippopotamidae, most of them in several species) roamed the savannas and woods of the African humid zones. Being so large, most of them had no enemies among carnivores. Saber-toothed cats (*Machairodus*) may have hunted their young. Smaller herbivores were, as always, hunted by their predators. Population densities of megaherbivores were regulated mainly by trophic parameters or occasional catastrophic losses by floods, droughts, fires, or diseases. The total mammal biomass in Plio-/Lower Pleistocene forests must have been considerably higher than in today's African humid forests, and possibly even higher than in today's tropical savannas (for animal biomass see Bourliere in Howell and Bourliere 1963; for living African fauna see Haltenorth and Diller 1977, and Owen-Smith 1988; for fossil African fauna see Maglio and Cooke 1978).

The dentitions of living African forest elephants [*Loxodonta africana cyclotis* and (?) *L.a. pumila*] show that they were originally savanna or open woodland animals that later repopulated the rain forests left empty by extinct Pliocene/Lower Pleistocene megaherbivores. They were not very well off in the Congo rain forests, and consequently they dwarfed. The same can be supposed of dwarfed buffaloes (*Syncerus caffer nanus*), some forest antelopes, and even the okapi (*Okapia johnstoni*). The pygmy hippopotamus (*Choeropsis liberiensis*) is not a dwarfed *Hippopotamus*, but a dwarfed *Hexaprotodon*. The larger forms of that genus disappeared from the fossil record in the Middle Pleistocene. Most recent rain forest ungulates are dwarfed colonizers from out of the savannas, or dwarfed when the forests became increasingly impenetrable after the disappearance (Martin 1984) of the African Pliocene forest megaherbivores during the Lower Pleistocene.

Of today's ungulates from the African rain forests, only the tiny water chevrotain (Tragulidae: *Hyemoschus aquaticus*), some of the smallest woodland antelopes and the Suidae are really adapted to life in thickets. The bush pig (*Potamochoerus porcus*) and the giant forest pig (*Hylochoerus meinertzhageni*) feel at home in the dense rain forests. Their ancestors probably lived in the thickets of Tertiary rain forests from the beginning.

Mammals in the African rain forest are today widely hunted by Pygmies, Africans, and Europeans. Ungulate population densities are low. Their consumption of vegetable material is far from being in a natural equilibrium with the primary biomass production of humid forests in the tropics.

Pliocene megaherbivores were perfectly adapted to forest conditions, as was the vegetation to browsing by mega- and other herbivores. Pliocene megaherbivores were not dwarfed, but gigantic, some of them even larger than recent male savanna elephants. They were hunted by no one. Populations of Pliocene megaherbivores were certainly nearly as dense as the phyto-production permitted. Many kinds of smaller ungulates helped the megaherbivores in their task to keep the forests open. Pliocene humid woods in Africa and elsewhere were not exactly what a modern botanist would call an undisturbed rain forest. They were complicated mosaics of biotopes, ranging from dense rain forest to biotopes we associate with the edges of woods.

Fire frequency in these forests was not much higher than in actual rain forests, fire intensity was probably lower, because huge quantities of stored fuel were lacking. High humidity and high temperatures throughout the year, and a better exposure to light allowed an even higher primary phyto-production than in modern rain forests. Less of it was stored, more was transformed by herbivores into food for carnivores and scavengers, and even more to organic fertilizer and climate-influencing respiration and digestion gases, overlapping with abiotic and marine climate-influencing parameters. A nice puzzle for paleo-bio-chemo-climatologists.

Not only forests are damaged by the lack of suitable communities of herbivores, but also savannas, grasslands, and semi-deserts. Without any grazing, even these cover with dense shrub. Uncleared mine fields of World War II south of El Alamein have transformed into thickets. Fire potential is stored in fuel of excellent quality. If fires occur seldom, their intensity destroys even roots and seeds. Soil erosion accelerates; deserts grow. The main losses of African medium-sized ungulates occurred during the Middle Pleistocene. The Sahara desert extended. When the accumulating fuel is not diminished by frequent fires, undergrazing is just as disastrous as overgrazing.

It goes without saying that plants colonize arid zones before herbivores do. Thus, there should be a correlation between the geologically long tradition of more or less arid climates, as in some parts of Africa, Australia, North and South America, and the Mediterranean, and the evolution of resistance to aridity and fire by plants (see also Sect. 4.3.1). The sooner herbivores colonized these regions, the sooner the stress of stored fuel was removed from the plants. We will leave this evolutionary puzzle aside and return to man's ancestors.

13.2.4.2 End-Tertiary African Primates

Arboreal primates found ideal conditions in the more or less park-like humid Neogene forests. There was a year-round supply of vegetables, fruit, roots, bulbs, eggs, nestlings, and small animals. Primate dentition and digestion was perfectly adapted to these foods, as was their locomotion to life both on and beneath the trees (Ciochon and Corruccini 1983).

The spreading of savanna environments during the Miocene had favored many ungulates, rodents, and birds, but not the primates: The hard savanna plants ruined their delicate teeth, there were seasonal food shortages, and the fire frequency was high. Ungulates adapted to the savanna vegetation by improving their dentitions, which became more and more hypsodont. Most probably their digestion systems became more and more able to process the hard leaves and grasses. In times of food shortages they move to other regions. They are not easily killed by fires. Even if large areas are burned, such as happened in Yellowstone National Park in 1988, the wapiti (*Cervus e. canadiensis* = elk) was hardly affected. The smoke of the fire travels on the wind. Ungulates smell it from afar and simply run away. Even their food recovers soon. That is impossible for primates. They have good eyes, average ears, and bad noses. Primates do not flee from danger head over heels, they flee towards the next tree or rock, and their locomotion is not built for migrations. With the increasing distances between the trees, carnivores became more and more dangerous for savanna primates. In addition, their food takes long to recover. Primates were made for life *on* the trees, not *between* them.

Primates are highly pyrophobic. With increasing fire frequency, their perspectives became dark and smoky in the Miocene, as savannas spread. To solve the problems of a forced savanna life, the primates had to do several things at the same time: They had to adapt their locomotion to the increasing distances between trees and even to long-distance travel on the ground, they had to adapt to hard vegetable food, they had to find means of surviving seasonal food shortages, and they had to live with an increased fire frequency.

The two main lines of African catarrhine primates found different solutions for the savanna problem. The cercopithecoid monkeys adapted their dentitions to harder food. They changed their social system to a more and more hierarchical one and their males developed long canines for their defence against carnivores and competitors. To escape fires, they had to rely on the observation of ungulates, high reproduction rates, and good luck. Recent savanna monkeys (*Theropithecus gelada, Papio anubis, P. hamadryas*) only survive where the accumulation of high quantities of fuel is prevented by ungulates, otherwise they are killed even on high trees. When an intensive fire covers a large area, they have no hope of escape. Even today, they are not fit to travel long distances, much less travel fast.

Hominoids, the line of primates which led to man, found another way: Their teeth lost all defensive value, their capacity to eat hard food was only slightly improved (Pilbeam et al. 1977), their mandibles degenerated. Either they found other ways to process hard food, or they found food that was softer than the plants. Evidence shows that they did both: they processed food with their hands (Kortlandt in Rensch 1968) or with stones and clubs (Leakey 1968; Boech and Boech 1981; Prasad 1982), and they shifted their omnivorous diet in the carnivorous direction. This enabled them to survive the scarcity of vegetables at the end of the dry season. Catastrophic population losses by mass starvation were avoided. Population exploition may have followed. Hominids began to evolve.

Again fire comes into the scene: savanna fires provided terrestrial hominoids (probably of the Ramapithecus group in the Upper Miocene) with toasted termites, cicadas, mice, and snakes. Even today, the smell of roasted meat is appetizing for hominids, but evidently not for apes. The reaction to the smell of roast meat is probably genetically fixed in man, as is the repugnance we have for carrion. That, on the other side, smells lovely to a hyena.

The higher the fire frequency and the lower the fire intensity was, the greater was the probability for savanna-proto-hominids out of the Upper Miocene Ramapithecus group to survive the fires on high trees, and to find tasty toasted meals. Proto-hominids for the first time profited from the high natural fire frequency and the low fire intensity of the spreading savannas of the Upper Miocene, 8 or 10 million years ago. But they badly needed a higher degree of cerebralization to avoid the dangers of fire and to find burned areas. They could not see them from afar, like vultures and storks, but they soon learned to interpret the flight of these birds to the far feast of fried snakes and locusts when little other food was to be found. Yet they had to hurry, lest nothing would be left. Prolonged locomotion on hard ground was needed, but we will get to that later. Once the proto-hominids had learned to avoid its dangers, fire became beneficial. Lately burned areas not only provided food, for a time they were also safe from new fires. If the proto-hominids overcame their genetically fixed pyrophobia and arrived before the large carnivores and scavangers dared to, they had a good chance to get some meat. Not all ungulates escape a fire.

Diminishing fear of fire improved the trophical basis and helped to avoid starvation losses during shortages. With growing experience in its treatment, losses by fire diminished. Populations grew. Selection pressure on the central nervous system of the proto-hominids increased.

Slow-moving megaherbivores were most likely to die in savanna fires. Their big bodies could not get rid of the thermic energy produced by prolonged movement. So they died, even if they were not actually killed by the fire. Yet their carcasses were useless for the proto-hominids. Their teeth could not lacerate the body of an elephant, neither could their hands tear it – until they discovered the cutting edges of stone flakes.

13.2.4.3 Plio-Pleistocene Hominidae and Megaherbivores

Archeologists found thousands upon thousands of so-called "pebble tools" in African Pliocene and Pleistocene sediments, and quite often in connection with smashed bones or among skeletons of extinct megaherbivores (Tobias 1985). Obviously, the hominids needed the cutting edges of the flakes, not the stone cores (Toth 1987).

Most archeologists interpret these finds of stone tools and animal bones as proof for scavenging by earliest hominids (scavenging or hunting see Shipman 1986). I am sure that they are wrong. Primates will not touch carrion. Recent male chimpanzees and baboons hunt small animals up to the size of a baby gazelle or a little monkey, but they never touch carrion, even if it is still warm. Moreover, without weapons, the earliest hominids were near the end of the long list of powerful scavengers. If they had weapons, they had no need to risk collective poisoning by eating carrion. They could hunt their own meat, and they did, probably as early as 2 million years B.P., or even earlier (Schüle 1989a). With sticks and wooden clubs they could kill small antelopes and the like. Once a hominid had discovered that stabbing with a sharp wooden point kills more efficiently than pounding with a club, they certainly made use of the first long pointed splinter that lightning had torn from a tree. If the weapon was to be useful, it had to be carried along all the time. It made no sense if they had to wait for a lightning stroke every time they needed a weapon. Bipedalism was the only possible solution.

Late Tertiary megaherbivores certainly had no genetically fixed patterns of fear and flight from other animals. They had lost them long ago, when their ancestors had become too large to be prey for carnivores. Flight wastes precious energy and always includes a risk of accidents, and this goes even more for attack. Genetically fixed patterns of behavior are lost when they are no longer needed, just like unnecessary organs (Lorenz 1966). Fearless antarctic pinnipeds and penguins and all whales demonstrate these loss of fear patterns. The patterns are lost irrevocably. Whales have a very developed central nervous system and a high intelligence, but they are unable to learn even these reactions.

Even with thrusting spears, it was difficult for early hominids to hunt medium-sized ungulates. These soon learned to flee or to attack their hitherto harmless compatriots. Megaherbivores could not learn to flee or to attack, they were easy prey. The hominids just had to run their spears into the soft skin of the bellies or flanks of fearless deinothers, mastodons, or ancylothers and to follow them until they fell. Their unique combination of an efficient cooling system through nakedness and perspiratory glands, and their perseverance made that easy. *Australopithecus*, the earliest known hominid, probably combined all of these features about 3 million years ago, or even earlier. Most African megaherbivores disappeared from the fossil record during the Upper Pliocene and Lower Pleistocene (Martin and Klein 1984). Only few survived: the ancestors of our elephants and rhinos.

The natural fire regime in Plio-Pleistocene moist forest biomes was much influenced by the hominid's invention of the thrusting spear. Instead of being eaten by megaherbivores and their smaller cousins, the primary phytoproduction was stored in high-quality fuel for wildfires. The resulting high fire intensity damaged the trees much more than before. Conditions became more or less like they are in modern rain forests. The rare fires were catastrophic.

At the borders of more arid zones, invading grazers from the savannas prevented young trees from growing. Savannas and steppes extended where forest once grew. The overall African phytoproduction probably shrank. More biomass was stored in wood. Climatic consequences may have followed.

There were other consequences. Wherever African megaherbivores had disappeared in the Lower Pleistocene, hominids had to fall back on their traditional food of plants and small animals, or they had to explore new resources like the shy and/or aggressive medium-sized animals, mostly artiodactyles and perissodactyles.

Exploiting this resource meant finding a way to neutralize the distances for flight or attack of the game, which had increased considerably when the animals had learned that a pack of tiny hominids walking on their hind legs and carrying long sticks was as dangerous as any big carnivore. That again put a strong selective pressure on the capacity of hominid brains. These animals reproduce much more quickly than megaherbivores. A higher percentage of them can be killed without actual danger of extinction. Probably their populations had grown when their megaherbivorous competitors had disappeared.

Once the hominids had perfectionated the hunt of middle-sized game, their trophic basis improved again. Another possibility for hominid population growth was given.

13.3 *Homo* sp. and Fire

13.3.1 Hominid Use of Fire in Africa

New investigations seem to show that, just as the African megafauna was disappearing about 1.5 or 2 million years ago, hominids began to use fire intentionally. Up to now, about six sites are known in Ethiopia, Kenya, and South Africa (Oakley 1954; Gowlett et al. 1981; Barbetti 1986; Brain and Sillen 1988), where hominids stayed near slow-burning fires, leaving stone tools of the

Olduvan type and smashed animal bones behind. None of these sites proves that hominids really used fire, far less that they produced it (James 1989). We do not even know for sure which of the three different hominids that existed during the Early Lower Pleistocene in Africa (*Australopithecus gracilis*, *A. robustus* and *Homo habilis*) is connected with the traces of fire. It is most likely that it was *H. habilis* or the emerging *H. erectus*. At least one of them left stone tools and smashed bones, some of them with cut marks, near spots where fires were burning for more than a few hours at temperatures of about 400 to 600°C. That is the temperature of a campfire, but also that of a log smoldering after a wildfire. Burnt clay, geomagnetic anomalies, and ashes combined with Olduwan stone artifacts are indications that the fires were fed by hominids. Dates are based on the potassium-argon radiometric analysis in Kenya and paleomagnetic investigations in Ethiopia; paleontological comparisons are employed on all the sites. The most important fact is that the hominids had definitely overcome their fear of fire.

Pyrophobia is genetically well fixed in all terrestrial animals. It is never lost, even if the animals lived in regions with a low or even without any fire frequency for millions of years. Camels are as afraid of fire as African antelopes or South American platyrrhine monkeys. Only some domestic animals stay near fire without visible fear. Lower Pleistocene hominids certainly were afraid of fire, and so were their carnivorous enemies, but the hominids overcame their fear. When they had learned to keep fires alive, they no longer depended on trees or rocks for protection during the night. No hungry *Dinofelis* or *Machairodus* dared to attack tasty hominids when they slept near a fire. Once the hominids knew that, they doubtless took to carrying their protective spirit along, just like they carried their spears. Fire protected our hominid Prometheus and his family as long as he could keep it burning. With the help of spear and fire the hominid population grew; treeless countries could be conquered. Perspectives became even better when the hominids learned to light their own fires, but just when that happened, we do not know.

The ecological advantages of fire were immense for early man, but so was the damage to the ecosystem. To give an idea of the ecological effects of man-made fires, one has to imagine a population of children or chimpanzees playing with matches in an African savanna at the end of the dry season. I think I can see everyone turning pale.

The ecological implications of a rapidly increasing fire frequency on vegetation, slow-moving megaherbivores, arboreal animals, soils, climate, and so on are many; interactions can hardly be calculated (Pyne 1984; Frost 1985; Harris 1980). Again, imagination will be more helpful than detailed descriptions, because a lot of interdisciplinary work which would be necessary for these descriptions is yet to be done.

13.3.2 Fire and the Evolution of the Brain

I tried to show in Section 2.4.2 that the increasing fire frequency put a strong selective pressure on the central nervous system of the Upper Miocene proto-hominids. The intentional use of fire exerted an even stronger selective pressure for more ganglia in the brain. Conserving glowing charcoal after wildfires, carrying it to unburnt areas, feeding it for protection by night or for use in a hunt will, before long, end with the hominid commiting suicide by burning himself. If he only feeds his hot, protective spirit and does not prevent it from taking care of itself, than the Lower Pleistocene Prometheus and his family will soon be lost.

More than any other mammal, the fire-using hominid was forced to choose between several possibilities within his genetically fixed patterns of behavior, and even to act against them. We call this "ethics". Not to do what one's instincts say, or even to act against a genetically fixed fear, demands a lot of additional ganglia in the central nervous system. "Ethical" use of fire became of high selective value for hominids. The situations which demanded "ethic" decisions in fire management were complex, evolutionary pressure was put on the capacity for free decisions, not on the establishment of new genetically fixed patterns of behavior. Selection was harsh when it came to fire: those who did not learn to control it were eliminated at once. To survive, the fittest really had to be fit, and so *Homo erectus* emerged nearly one and a half million years ago.

Additional ganglia in Lower Pleistocene hominids' brains had another advantage: the hominids learned to throw their spears at distant targets. This demanded much preciser movements than just running a spear into a megaherbivore's belly. In throwing their spears, hominids could neutralize the fleeing distances of shy game and put a safe distance between themselves and dangerous animals. Hunting medium-sized ungulates became easier and less dangerous. Defence against carnivores and territorial fights with the Lower Pleistocene cercopithecid biotope rivals of the hominids, some of them the size of a male gorilla and with canine teeth like those of a living gelada or hamadryas, became less adventurous. They disappeared during the Lower Pleistocene, as did the last hominid rival of *Homo*, *Australopithecus robustus* (*boisei*), some time later.

The spear-and fire-bearing hominid evolved into man, ruler of all biotopes, as far as he could stand the climate. In Africa little was left off-limits for early man – high mountains, regions without any open water, and the densened woods wherever the megafauna was destroyed.

13.3.3 Increased Fire Frequency and Climate

We may now suggest that paleobotanical, paleozoological, and sedimentological observations of changing Pleistocene African ecology – traditionally explained mostly as the results of climatic changes – were strongly influenced, if not caused, by ecological interactions of early man's easy-going use of thrusting spear and fire. In the long run, the human use of fire leads to progressive savannization and desertification.

One and a half million years of more or less careless use of fire in Africa seem to be enough to influence the climate, even on a global scale.

It is quite clear that although we have new dates for the use of fire, we are far from knowing when fire use really began. It might have been much earlier. We have to take into consideration that with the successive expansion of humankind (see Sects. 4–5.6, 7; Fig. 2) the fire frequencies all over the world were altered from the natural to the anthropogenic type, and the storage of atmospheric carbon in biomass (Fig. 1) was altered drastically by the human influence on herbivores. The process was not a continuous one, but progressed in leaps whenever man entered new climatic zones, new continents and, in the late Holocene, the last unoccupied large wooded islands, Madagascar and New Zealand. We will come to that later (Sect. 7, No. 11).

13.4 Anthropogenic Fire in Eurasia

In the latest Lower and early Middle Pleistocene, *Homo erectus* began to conquer the Eurasian tropics and subtropics. Low sea levels during a cold period of the Middle Pleistocene let *H. erectus* reach Java, at the southern end of the Eurasian shelf. During warmer periods he temporarily reached central Europe and northern China. As in Africa, only waterless deserts and high mountains were left untouched. Wherever his traces are found in caves, rock shelters, or campsites, they are connected with fire (James 1989). This suggests that he not only carried fire but also knew how to light it. The existence of an untouched and unwary megafauna with all its trophic advantages was reason enough for hungry *Homo erectus* to enter even strange landscapes, as long as he could still survive the alien climate.

By the Middle Pleistocene, all continental biomes in adequate climates of the Old World were influenced by man's spear and fire, either directly or indirectly. Effects on regional climates are certain, effects on the global climate should be considered (Sects. 3.3 and 7, No. 11).

13.4.1 Tropical and Subtropical Eurasia

Before man's arrival, most of southern Asia, abundantly watered by monsoon rains, was covered with seasonally or ever-humid forests. Grassland and dry savanna herbivores lived only in the more arid northwest of southern Asia. The ecological results of the human colonization were similar to those in Africa, but the change was more sudden. *Homo erectus* came as a perfect hunter aided by fire. Flora and fauna had, by geological standards, only very little time to adapt to his habits.

The unwary megaherbivores disappeared, and with them the open rain forests. Even hippopotamids, which had survived hominid persecution in Africa, vanished in the Middle Pleistocene. *Hexaprotodon* was probably less aquatic and nocturnal than *Hippopotamus*, and it therefore became extinct both in Africa and Southern Asia. As in Africa, saber-toothed cats disappeared together with the megaherbivores. There were no savanna cercopithecid biotope rivals for hominids in southern Asia, because open landscapes were less extended, but there was a ground-living hominoid relative. *Meganthropus*, probably a pongid fairly related to the living Asiatic orang-utangs (*Pongo*) and about twice the size of a male gorilla, lived in the bamboo forests of the Pleistocene Far East. *Meganthropus* disappeared from the fossil record in the Middle Pleistocene, surely not for climatic reasons; the bamboo forests are still there.

Occasional fires in humid regions became more intense. The huge fire potential of the new, dense rain forests wrought havoc when ignited. In seasonally dry woods, savannas, and grasslands the fire frequency increased. Due to the different climatic history in the Upper Tertiary and Lower Pleistocene, there were fewer autochthonous fire-tolerant plants present.

Unwary megaherbivores became extinct, humid forests densened, in less humid regions the fire regime changed to the man-made one, savannas and grasslands spread in the Middle Pleistocene south Asiatic tropics and subtropics.

13.4.2 Old World Temperate Zones and the Eutrophic Line

Towards the end of the Lower Pleistocene or early in the Middle Pleistocene, man's expansion was halted by sea shores and climatic limits which he was unable to cross. Naked humans are not suited for cold and/or snowy winters or cool and/or wet summers. North of the Great Eurasian Cordillera, wide steppes and savannas extended. They were kept open by the arid continental climate and hypsodont herbivores like the giant rhinoceros *Elasmotherium*, the mega-camel *Paracamelus* and a number of desert- and steppe-adapted equides and bovides. Boreal and temperate forests covered large areas to the north, east, and west. Further north, the arctic tundra extended. All of these landscapes were populated with herbivores of all sizes (Kahlke 1956, 1981; Kurtén 1968; Vereshagin and Baryshnikov 1984; Kowalski 1986). Following the climatic oscillations of the Pleistocene, both flora and fauna shifted north-south and east-west.

The natural fire frequency in these areas was certainly lower than in the tropics and subtropics, because incineration through electric discharges during the dry season was less likely.

Until well into the last (Würm/Weichsel) glaciation, which means for more than half a million years, human populations, first of *H. erectus*, then *H. sapiens praesapiens*, *H.s. neandertalensis*, and later *H.s. sapiens*, were concentrated along a line which ran from the Atlantic Ocean to the Japanese Archipelago. The line more or less followed the Great Eurasian Cordillera, and shifted with the changing climate during the Pleistocene south-north or north-south when the thermic values were concerned, and east-west or west-east when the hydric values changed.

As long as man had not developed good techniques of thermo-insulation by clothes and artificial shelters, climatic conditions along that line were bad for man, but his trophic situation was very good. Man lived off the surplus of big mammals from beyond the Eutrophic Line. Population pressure of the game which lived off-limits, and seasonal migrations of the herbivores, like the Eurasian mammoth (*Mammuthus trogontherii, M. primigenius*), the woolly rhino (*Coelodonta antiquitatis*) and several medium-sized perissodactyles and artiodactyles, made this surplus available. Different from the old, progressive borderline of *Homo erectus* expansion, this Middle and Upper Pleistocene climatically caused Eutrophic Line was static for long periods and shifted only with the climatic changes (Fig. 2, legend to stage 2).

During the last glaciation of the Upper Pleistocene, the last populations of *H.s. neanderthalensis* and the first of *H.s. sapiens* developed the thermo-insulation which allowed them to stay closer to the border of the ice. In western Europe, the ice sheet extended far to the south. The abundance of caves, rock shelters and flint in Mesozoic limestone, and good grazing conditions for migrating herbivores in the moist Atlantic tundras, provided nearly ideal ecological conditions for man. During the last glaciation, probably about 25,000 years B.P., when the low eustatic sea level left the Bering Strait dry, *Homo s. sapiens* arrived in eastern Beringia, north of the North American ice sheet. The cold-arid zones of eastern Europe and central Asia, although they were free of ice, remained off limits until technologies improved and the climate warmed during the latest phase of the Pleistocene. In eastern Beringia, the way to the south was barred by the huge North American ice sheet until about 11,000 years ago.

Archeologists found indications that the men who lived on the Eutrophic Line throughout the Middle and Upper Pleistocene used fire for hunting. As in the tropics, fire intensity increased considerably because of the hunting-induced reduction of herbivores. The expansion of more fire tolerant species is likely; forests became dense and retreated to more humid zones. North of the Eutrophic Line, the natural fire regime still ruled.

13.4.3 Man and Landscapes in the Hinterland

In the hinterland of the Eutrophic Line, human population density was limited by the equilibrium between the vegetable resources, the behavior and reproductive capacity of the hunted herbivores, and their decimation by man and carnivores. Megaherbivores had disappeared or reactivated their self-defence against predators and high fire frequency. It became dangerous to hunt them without fire. Fire hunting was only possible during the dry season. The less game was left the more dangerous became fire hunting because of the accumulated fuel.

In the warm-temperate and subtropical regions of Europe south of the Eutrophic Line, woodland elephants (*Palaeoloxodon antiquus*) and woodland rhinos (*Dicerorhinus etruscus*/*D. kirchbergensis*) survived until the last interglacial (Riss/Würm) phase (for literature see Sect. 4.2). Smaller herbivores had to carry the burden of sustaining both carnivores and man. Unlike the carnivores, man could turn vegetarian when prey became scarce and use the time to improve his hunting techniques. Thus pressure on the game continued. Population densities of game were certainly low where man lived. The consequences for the vegetation and the fire regime were the same as in Africa except for climate-induced differences. South of the Eutrophic Line, times were hard for both man and game.

13.4.3.1 The Pre-Agricultural Mediterranean Region

The region of the pyrotolerant vegetation of the "mediterranean type" around the Mediterranean Sea requires more detailed consideration. The reasons are the geological history of that area, the long presence of man and anthropogenic fire, and the special Mediterranean climate with long, dry summers, enough precipitations in winter for a high phyto-production in spring, and a tendency to concentrate the precipitations in short events. This climate actually has its parallels in the southwestern regions of Australia, Africa, and both the Americas. In some parts of Australia, Africa, and North America the mediterranean-type climate has a geologically very old unbroken tradition. It is there that the vegetation has achieved the highest degree of adaptation to its special climatic conditions and to its natural fire regime. In the Mediterranean region of mediterranean-type climate, the geological history is quite different.

The present hydrological balance of the Mediterranean Basin, including the hydrological basins of the Nile and the Orontos, of all Turkish rivers west and north of the Euphrates, of all south Russian streams west of the Wolga, and of the Danube, Po, Rhodano, Ebro, and Segura rivers, is negative. Because of that, there is a large influx of water from the Atlantic Ocean through the Strait of Gibraltar. The Mediterranean would be reduced to brackish lakes in its deepest basins if the straits closed. Oceanographical and geological studies have shown that this was exactly what happened when the western outlet of the Mediterranean was closed by tectonic movement in the Upper Miocene, about 5.5 million years ago. For about half a million years, the Mediterranean was reduced to brackish lakes and even brine puddles and salt deserts, about 2000 m below the present sea level (Hsü 1972, 1978, 1983; Hsü et al. 1977; Schreiber 1988). The continental rivers, falling down to the ancient bottom of the sea, cut deep canyons into the shelfs. At Aswan, more than 800 km as the crow flies from the Mediterranean shore, the bottom of the Nile Canyon was 700 m below the present level of the Mediterranean, and below Cairo it was not reached at a depth of 2500 m. The hydrological history was further complicated when the Paratethys, like the Mediterranean a relict of the ancient Tethys Ocean, was partly emptied into the dry Mediterranean Basin.

African, European, and Asian flora and fauna settled the basin as far as salinity permitted. Hsü (1983, p. 177) even suggests that this had an influence on the evolution of our ancestors, the Upper Miocene hominoids. It is perfectly possible that he is right. On the slopes, the flora and fauna that were adapted to semi-arid and arid conditions extended according to their tolerance towards salinity. The natural fire frequency, caused by the volcanoes and by lightning, may have been very high in some biotopes, as was the fire intensity near or even in the canyons when a drought on the continents cut part of the water supply. When the sea returned in the Lowest Pliocene, about 5 million years ago, plants and animals had to escape if they could. What survived gathered in the hottest and less humid locations round the Mediterranean Sea and on its re-established islands (Sect. 5.6).

Man, equipped with spear and fire, arrived maybe one million years ago. For the last one million years or more, the southern Mediterranean region had its man-made fire regime. Vegetation tolerant of fire and aridity spread. During most of the Middle and Upper Pleistocene, the Eutrophic Line shifted near the shores in the northern part of the Mediterranean. What we call "mediterranean vegetation" is the product of a very complex history. It really is a pity that no botanist was present to control the giant experiment. Hsü's discoveries left many problems still to be solved for biologists, paleo-anthropologists, and fire ecologists (Sect. 5.6).

13.4.4 Life and Fire Outside the Eutrophic Line

In the forbidden countries north of the Eutrophic Line and south of the North American ice sheet, as well as in the cold-arid regions of eastern Europe and Central Asia, and in Australia and on off-shore islands nothing changed. Biological evolution and fire regime continued as before, influenced only by the climatic oscillations of the Pleistocene. Laurasian mammal herbivores contitinued to improve their hypsodont dentitions, and thereby altered the vegetation, the fire regime, and perhaps the climate. Animal biomass was as high as ever.

13.5 The Conquest of the Forbidden Countries

South Asian shores had no Eutrophic Line. At best, the shallow bays and reef-protected lagoons provided humans with good fishing grounds. *Homo erectus* had probably learned to use rafts to exploit these resources. Willingly or not, probably he also used the rafts to conquer new lands. Anyhow, he did not get far (Sect. 5.1).

It was not before the last glaciation (Würm/Weichsel/Wisconsin) that *H.s. sapiens* was more successful (Sects. 5.2–6).

For the Europeans of the Age of Discoveries few countries were left to discover. Wherever they arrived, somebody was there (Sect. 5.6).

13.5.1 Celebes-Sulawesi, the Philippines, and the Wallacean Islands

On the island of Sulawesy, bones of dwarfed *Archidiscodon* and/or *Stegodon*, primitive elephants which had reached the island some time before, were found together with stone tools in Middle Pleistocene terraces. The circumstances are uncertain (Bartstra 1977), but after man first appeared on the island, the presence of megaherbivores can no longer be ascertained.

Sulawesy was never connected to the mainland. Stegodonts and other herbivores reached the island drifting in the warm waters of a low eustatic sea level in the Lower or Early Middle Pleistocene. They probably came from Borneo, which was connected to the mainland at that time. Sulawesi, like all humid islands without herbivores, was covered with dense woods. The elephants certainly multiplied rapidly in the trophic paradise, and with their population growth they destroyed it. Their trophic situation worsened, areas were limited, and so they dwarfed, like any other large mammal in an island environment. The fire regime became that of tropical moist savannas.

If the dating of these finds is correct, Sulawesi was the first oceanic island man reached. *Homo sapiens* had not yet emerged, the first human must have been *Homo erectus*. He did not necessarily discover the island on purpose, he might have drifted there on rafts. On Sulawesi, the descendants of the first settlers ate the elephants as fast as they could. Their fires could not keep the woods from recovering when the elephants were gone, neither could the hunting and gathering *Homo sapiens*, who arrived much later. The remaining herbivorous ungulates, anoa (*Bubalus depressicornis*) and babirusa (*Babyrousa babyrussa*), could also not prevent the woods from becoming dense, because they were heavily decimated by human hunting. On the other hand, the dense wood saved anoa and bibarusa from being exterminated like the stegodonts. The fire regime changed to the low frequency and high intensity of the tropical rain forests with only few herbivores.

More or less the same happened on the Philippines and on the islands of the Wallacean zoo region between the South Asian and Australian continental shelves. Their more or less dwarfed megaherbivores, wherever they existed, disappeared when man arrived at an unknown date during Pleistocene or Lower Holocene even there where for lack of archeological evidence we cannot prove the presence of man. I shall not bore you with repeating the story (Sect. 5.6).

13.5.2 Australia, New Guinea, and Tasmania ("Sahul")

The history of the Sahul continental block, which comprises New Guinea, Australia, and Tasmania, is quite different. During the low eustatic sea levels of the Pleistocene the islands were connected. Like India and Africa, Sahul had been part of the Mesozoic southern continent of Gondwana. It split off Gondwana in Middle or Late Cretaceous times, 70 or 80 million years ago, but, unlike India and Africa, it never touched Laurasia.

When *H.s. sapiens* reached Australia, crossing the straits during the low eustatic sea levels of the last glaciation, about 40,000 year B.P. (Gill 1965; Suggate and Cresswell 1975; Gill et al. 1980; Martin and Klein 1984; Singh et al. 1981; Kershan 1984; Christensen and Burrows 1986), large areas were still covered with forests of pyrophobic *Araucaria* and *Casuaria*. New Guinea was covered with tropical forests at that time, Tasmania with temperate woods. In some parts of Australia, mainly in the southwest and center, the long history of semi-arid conditions had led to the development of a "mediterranean type" vegetation with pyrotolerant or even pyrophilic plants, like many species of *Eucalyptus*. Mice, which had arrived drifting in the Late Pliocene or Lower Pleistocene, were the only placental mammals in Sahul. The mammalian fauna was composed of Marsupialia and Monotremata, some of them up to the size of a rhinoceros. A giant lizard, 6 m long and weighing about 2 t, and carnivorous marsupials controlled the populations of marsupialian herbivores and giant flight-less birds, which were about twice the size of emus and casuars.

We can suppose that landscapes were "normal". There were deserts in the arid zones of Central Australia and tropical or subtropical "open rain forests" in the humid zones of the east and north. Dense forests were limited, as elsewhere, to inaccessible ground. The Pleistocene fire regime of Sahul was one of high seasonal fire frequency and low fire intensity in the "mediterranean" areas, and of low frequency and higher fire intensity in the humid zones.

Pollen sequences, taken from lake sediments in Queensland and other semi-humid or humid areas of Australia (Kershaw 1984; Singh et al. 1981), have shown a rapid decline of pyrophobic plants like *Araucaria*, and a parallel increase of pyrotolerant and pyrophilic *Eucalyptus* approximately at the time when radiocarbon dates corroborate man's first presence on the Sahul continent. At the same time, the percentage of coal particles in sediments increased by a factor of about 1000, and then remained at that level. Australia began to cover with the pyrotolerant and pyrophilic vegetation of the semi-arid zones. Australia "sclerophylisized". Personally, I doubt that these phenomena are the effect of a natural climatic oscillation. I believe rather that the changes in both fire frequency and vegetation were caused by man.

Homo sapiens sapiens arrived as a perfect hunter. The megafauna disappeared, probably because of human spears and fire, and vegetation changed. The appetite of the Paleo-Aborigines kept the populations of remaining marsupials low. Whenever the man-made fire frequency decreased for one reason or the other, there were not enough herbivores to prevent the accumulation of fire potential. Fire events became catastrophic, even for pyrotolerant gums.

After a time, the Aborigines got wise. They developed a sophisticated fire management that is much admired by modern ethnologists (Jones 1975, 1979). Shipwrecked James Cook experienced it when Aborigines surrounded him with fire in the afternoon of Thursday, July 19, 1770, on the spot where Cooktown, Queensland, is today. Fire was not only used to hunt or to get rid of uninvited visitors, it was also quite intentionally used to prevent thick undergrowth, to improve the pasture for kangaroos, and to prevent conflagrations. Australians among you know better than I do. When Europeans arrived in the 18th century,

Australia and its islands had a 40,000-year history of man-made fire frequency, with all its effects on flora, fauna, and soils. I am again inclined to see a strong influence on the local climate, and I think that a consideration of influences on the global climate would be worth while.

13.5.3 America

The history of man and fire in both the Americas is similar to that of Australia, only it was settled much later. That means that the pyrotolerant, mediterranean-type vegetation in California and southern Chile, or the grassland type of treeless pampas and prairies, has not had enough time to colonize the huge continents with their mostly more humid and partly much cooler climates.

When during the last (Wisconsin) glaciation, about 25,000 (?) years B.P., Upper Paleolithic arctic hunters crossed the landbridge between east Siberia and Alaska, the way south was blocked by the giant ice sheet which covered North America down to the northern states of the U.S.A. (Kurtén 1966; Kurtén and Anderson 1980; Hopkins 1967; Hopkins et al. 1982; Greenberg et al. 1986; Lewin 1987). The land between the lower courses of the Yukon and Mackenzie rivers was as free of ice as was the landbridge and most of Siberia. The ice-free regions were covered with tundra, cold grassland and subarctic forest. Man-made fires could not have damaged them much. Then the megafauna disappeared, as it always did when human hunters entered a hitherto unoccupied region. Man managed to survive, probably by exploiting the rich marine fauna of the Northern Pacific. Archeological traces of human occupation in the interior of the country are scarce, of the suggested coastal sites none are to be found, because sea levels rose in the Holocene, when the Bering Strait formed.

Man entered the land south of the ice sheet around 10,500 B.P.(Martin 1973; Martin and Wright 1967; Martin and Klein 1984; Lewin 1987; Greenberg et al. 1986). At that time, the Edmonton channel separated the ice sheet east of the Rocky Mountains and opened the way to the south. America had a rich mammalian fauna, adapted to all kinds of landscapes and climates. Even the giant tortoise *Geochelone* had survived all Pleistocene climatic changes in North and South America. As in the Wallaceans and Sahul, no catarrhine primate, far less a hominid, had ever set foot on American ground before *H.s. sapiens* arrived.

This late Pleistocene fauna was a mixture of the old endemic Tertiary Laurasian fauna of North America and the endemic South American fauna of Gondwanian origin, which had existed since the collision of North and South America in the Upper Pliocene, about 3 million years ago (Simpson 1980; Stehli and Webb 1985). One can suppose that, as in Africa during the invasion of Eurasian hypsodont herbivores in the Lower Miocene, the South American vegetation had at that time been considerably damaged by the invading hypsodont Laurasian ungulates, like Equidae, Camelidae, brachyodont Cervidae and, last but not least, the proboscideans of both the *Gomphotherium* and *Mammuthus* lines. The dentitions of endemic South American mammals, including several megaherbivores up to nearly the size of an Indian elephant, were not developed to that degree. They kept the forests open, but the South American vegetation was not adapted to the more resistant dentitions of the Laurasian herbivores and gave way until a new equilibrium was established. Recent investigations have shown that the rain forest of the Amazon was, at least temporarily, less uniform and less extended during the Pleistocene than it is today. This process has been explained as the result of a drier climate during cold stages of the Pleistocene (Flohn 1985). I am certain, however, that the teeth of invading North American ungulates had quite a good part in it. This, of course, should not be misused as an excuse for today's destruction of the rain forest, as has been done with the climatic interpretation.

At the end of the Pleistocene, all American forests were less dense, their primary phytoproduction probably even higher than today. Megaherbivores of both South and North American origin, together with the smaller species, prevented the accumulation of huge quantities of high-quality fuel. This accumulated fire potential makes the destruction of modern rain forests by fire so easy and disastrous. The normally slow shift in the equilibrium of carbon flux is suddenly disturbed.

When *Homo s. sapiens*, hungry and accustomed to the strenuous hunting of shy game, emerged from the frozen tundras of eastern Beringia, he must have thought the flowering prairies and open forests south of the ice sheet were Paradise (Martin and Klein 1984; Kurtén and Anderson 1980). Mighty elephants, mastodons, ground sloths, and glyptodons regarded him without fear, waiting to be killed. Herds of known and unknown game – several kinds of tapirs, horses, camels, bovids, cervids, suids, giant rodents, and many others – surrounded him, taking little notice of the strange, upright figures which did not smell or move like carnivores. Animals biomasses were about the same as in present east African game preserves in most of the American landscapes. If the animals tried to flee, man only had to light a fire encircling them, and he reaped a feast for several days. His only occupation was to eat and to reproduce – and that he did. Within a time that is too short to be measured geologically, nearly all mammals over 50 kg, several big volant and flightless birds, and some large reptiles disappeared, a total of nearly 100 species all over America. The big carnivores and giant birds of prey starved when their prey disappeared. Small animals suffered practically no losses at the end of the Pleistocene. Instead, there had been a rapid change of small species when the climate first began to cool in the Lower Pleistocene. The herbivores that survived the end of the Pleistocene lived in hot or cold semi-deserts, like pronghorn (*Antilocapra*) and South American camels (*Lama*), or they could hide in dense jungles or the canopies of high trees, like peccaries (*Tayassu*), tapirs (*Tapirus*), small capybaras (*Hydrochoerus*), deer (*Odocoileus*), black bear (*Euarctos*), sea cows (*Trichenus* = *Manatus*), tree sloths (*Bradypus* and *Choloepus*), and platyrrhine monkeys. Even of these genera, only the smallest species survived. Large carnivores starved and, in the end, so did most Paleoindians. It certainly was not the climate that killed them.

Vegetation, as far as it had survived the burn-eat-and-reproduce orgies of the Paleoindians, recovered when man destroyed its consumers. Woods became dense and impenetrable, prairies and pampas covered with scrub and trees.

But not for long. Human and animal survivors of the collapse recovered, too. New waves of invaders, both human hunters and hunting-accustomed game, came from west Beringia before the strait formed in the Holocene. Eury-oecious Eurasian bison repopulated the forests and colonized the prairies. Elk (*Cervus*), moose (*Alces*), bighorn (*Ovis*), bear (*Ursus*), and beaver (*Castor*) occupied the empty forests. They reproduced more quickly than the human hunters, who neither dared nor had the chance to forget the environmental ethics they had developed during half a million years of coevolution with their game in the Eurasian hinterland. An equilibrium was again established between plants and animals. This new equilibrium was different from the one that had existed in the biotic paradise of the Pleistocene, because both megaherbivores and hypsolodont Camelidae and Equidae were missing. Forests were denser, prairies probably had fewer trees because population densities of endemic pronghorn and immigrated bison were high for lack of competition with other species. Consequently, the natural fire intensity rose in the forests and declined in the prairies.

The increase of the newly immigrating human population led to the development of a "normal" equilibrium between phyto-production, shy medium-sized herbivores with partly hypsodont dentitions, hungry carnivores, and "ethical" human hunters with a correspondingly elevated fire frequency. No-man's-lands were established between rival tribes – "national parks" which were a more or less reliable source of game for the surrounding humans. All of this, of course, against the background of the oscillating climate of the Holocene.

None of the Eurasian newcomers, except man, passed the Isthmus of Panama. The rain forests between southern Mexico and the Gran Chaco had become impenetrable. South America was repopulated by the surviving endemic Pleistocene fauna. Guanacos, vicugnas (*Lama*) and nandus (*Rhea*) multiplied and prevented scrubs and forests from conquering the outskirts of open landscapes, and the man-made high fire frequency did the same. Tapirs (*Tapirus*), peccaries (*Tayassu*), capybaras (*Hydrochoerus*) and Cervidae were incapable of re-opening the forests. Platyrrhine monkeys and tree sloths profited. The canopy of the forests was more coherent than ever before and protected them from the sharp eyes and poisoned arrows of old and new Amerindians.

The greatest danger for all life forms in the American (and other) rain forests lay in occasional droughts. The gigantic fire potential that piled up in the absence of large herbivores led to a fire intensity that was lethal for all. I doubt that any Indian was ever stupid enough to set fire to the forest during a drought. The rights for this kind of ecological suicide are reserved for modern, civilized man. Amerindians developed their own, adequate system of hunting with fire.

Prairies and pampas not only were kept open by large herds of bison or guanacos and man-made fires, they expanded into the woods. Fire-tolerant and fire-dependent vegetation from the old, semi-arid biotas began to spread, as it had done in Africa one and a half million years before, in Eurasia in the Middle Pleistocene, and in Australia in the Late Upper Pleistocene.

Amerindians went one step further than Australian Aborigines in their ecological management. Five thousand or 6000 years ago, they began to cultivate plants in both the Americas. Wherever necessary, the Indians used fire to clear the land, but we will get to that later (Sect. 6). When Columbus came to America in 1492, all American landscapes had been directly or indirectly changed by man's spear and fire, and in many regions by his hoe, too. American landscapes, vegetation, game, and soils had changed a lot since man first arrived, about 10,500 years ago. And so had the carbon circulation in both the American continents – certainly enough to influence the global climate.

13.5.4 Madagascar

Madagascar split off from the African part of Gondwana in the Upper Mesozoic or very early in the Tertiary. Compared to the living fauna (Walker 1975; Grzimek 1968–1972), the subfossil Holocene fauna is surprisingly rich (Fleagel 1988; Dewar in Martin and Klein 1984). The subfossil Malagasy Aardvark (*Plesiorycteropus*) (Patterson 1975) might have been the only mammalian survivor of the continental fauna, although the giant tortoise *Geochelone* and the endemic crocodiles may also have been paleo-endemic. The origins of the rodent-like primate aye-aye (*Daubentonia* = *Cheiromys*) and the lemurs (Lemuriformes) are controversial. They could be survivors of the old fauna, or they may have arrived drifting during the Lower Tertiary, when Africa was closer. During their long existence on the island, they developed forms that were much larger than today's, like the subfossil terrestric *Archaeoindris fontoynonti*, which was the size of a male gorilla, or the somewhat smaller tree-sloth-like foliovorous *Palaeopropithecus maximus* and *P. ingens*. The subfossil aye-aye was twice the size of the living one.

Hippopotomus certainly arrived by drifting from Africa in the Middle Pleistocene or later, and evolved into an endemic dwarfed species (*H. lemerlei* = *madagascariensis*). How the African bush pig (*Potamochoerus porcus*) managed to reach the island is unknown. Considering its endemization as *P.p. larvatus*, it seems doubtful if it was brought by man or traveled by some mysterious way on its own. Other ungulates never reached Madagascar. Feral pigs (*Sus scrofa*) and cattle (*Bos taurus zebu*) were brought by man.

Ratites (flightless birds) of the genera *Mullerornis* and *Aepyornis* were the most impressive herbivores. The largest had an estimated height of 3 m and the body weight of an ox. Their elephant-like legs show that there were no hungry carnivores from which they had to flee as did their distant African, Australian, and South American relatives. They, too, are of doubtful, possibly Gondwanian, origin.

The largest known endemic carnivores were the subfossil, fox-sized *Cryptoprocta ferox spelea*, and crocodiles. The former certainly did not endanger large or arboreal vertebrates, the latter only when their prey came near the water.

In the absence of any considerable carnivores and competing herbivorous ungulates, the population densities of the paleo-endemic giant herbivorous lemurs and ratites were limited only by trophic conditions, occasional catastrophes, and the like. Consequently, they must have been high before the ungulates hippo and bush pig arrived.

A shifting equilibrium of climate, vegetation, animals, and natural fire regime must have existed throughout the climatic oscillations of the Tertiary and Quaternary in the moist forests of Madagascar's humid north and east, and the more arid savannas of the south, west, and center. Dense populations of herbivorous ratites and lemurs kept landscapes open, although maybe not as efficiently as ungulates would have done.

Some time during the Middle or Upper Pleistocene, the hippos that had stranded on the Malagasy coast must have dealt quite a blow to the unadapted endemic vegetation. The hippos had no natural enemies on the island. Rapid growth of the population was the inevitable consequence. In their search for food they probably ranged further from the open water than African hippos. The same behavior apparently characterized the Pleistocene dwarfed hippos on Mediterranean islands. Dense populations of *Hippopotamus* thinned the vegetation for at least half a day's walk from the shores. The dwarfing suggests an unfavorable trophic situation at the time their bones were buried in subfossil deposits.

If the African bush pig (*Potamochoerus*) arrived prior to man, it assisted the hippos in thinning the vegetation. Dense populations of wild pigs (*Sus scrofa*) are a disaster for any kind of vegetation, and the same can be supposed for *Potamochoerus*. As long as they were not decimated by human hunting, populations of *P.p. lavatus* were certainly dense. Like the hippos, they had no natural enemies on the island. If they arrived with man, they had to share the vegetation with domestic animals and wild pigs and cattle.

When African and Malayan fishermen and farmers settled Madagascar around 800 A.D., Malagasy landscapes must have been the opposite of what is expected of untouched landscapes: zones of open land along rivers and lagoons surrounded by forests and savannas with a relatively low fire potential because of dense populations of endemic herbivores. The first thing the immigrants did was to start exterminating the unwary insular fauna. As in South America, the only survivors were small, arboreal and/or nocturnal. *Potamochoerus*, if it really was neo-endemic, hid in the thickening forests. Fire potential accumulated in forests, savannas, and grasslands, fire intensity increased. The unadapted endemic vegetation was heavily damaged (Humbert 1927; Battistini and Verin 1967). The local climate was certainly influenced, erosion formed the lavakas. The populations of man and domestic animals increased, and so did the damages of the man-made fire frequency and land use. The lavacas expanded, forests retreated. The rapid growth of modern human populations and inadequate technologies are taking care of the rest, as they are everywhere.

The local climate on Madagascar was certainly strongly influenced by these processes, but the Malagasy landmass is probably too small to have a considerable global influence.

13.5.5 New Zealand

New Zealand was also a part of Gondwana, but it was isolated even longer than Madagascar. The islands of the archipelago are all part of the same large shelf. During glaciations, the shelf lay exposed, and thus land loss was compensated. The endemic flora is related to that of Australia, Madagascar, and South America, mainly because of the common Gondwanian origin. Today, all kinds of landscapes, from subtropical rain forests in the north and parts of the west coast to glacial flora in the mountains of the south, are present. None of them ever disappeared completely from New Zealand, otherwise the endemic, adapted plants would have also disappeared.

No mammals, other than seals and bats, ever reached New Zealand. Its Pleistocene glaciations made life difficult for reptiles and amphibians. The tuatara (*Sphenodon punctatus* = *Hattaria punctata*), the last living member of a Lower Mesozoic order of reptiles from Pangea, may be of Gondwanian origin, or it reached the islands by drifting, like all other terrestrial vertebrates probably did. Dominating the endemic terrestrial fauna were about 30 kinds of flightless ratites, the kiwis and moas, ranging in size from a small chicken to a well-fattened ox. As in the case of Malagasy ratites, it is still under discussion whether they are of Gondwanian origin or flew to the islands. All of them were apparently herbivorous and/or fructivorous. No carnivores existed. All of the flightless birds were indubitably fearless towards terrestrial enemies. Even the extinct endemic eagles seem to have fed more on carrion than on prey.

Population density was regulated by reproduction rates, occasional climatic or volcanic catastrophes, and most of all by trophic limits. Herbivorous ratite populations must have been as high as trophic conditions permitted. The bills of some of the moas were perfectly adapted to hard food, like seeds, twigs, grasses, and the like. Consequently, the vegetation was open and the fire potential small in forests, savannas, and grasslands. Only on steep slopes and other inaccessible sites could huge masses of fuel be stored. The volcanic activity led to frequent fires, but fuel was limited. This equilibrium had existed since the Mesozoic. Neither biotic invasions, Pleistocene climatic oscillations, nor occasionally increased volcanic activity and local falls of ashes disturbed it much. For millions of years, it was a land that resembled the Biblical Paradise. Moas did not have to develop quick reactions or sophisticated central nervous systems to escape from enemies or outrun competitors.

The paradise was quickly destroyed when man reached it (Prickett 1982; A. Anderson 1984; Cassels 1984). Polynesian Maoris (probably involuntarily) discovered Aotearoa (Maori name for NZ) around 800 A.D. They were tropical gardeners and fishermen. Their domestic plants, if they had any in their boats, were not adapted to the temperate climate of New Zealand. But they had found the land of milk and honey. Fish and shellfish were plentiful, yet even more plentiful were moas. Moas did not run away or attack – the Maoris harvested them. For years, the Maoris lived on the fat of the land. When there were no moas left near the shore, they were driven with torches and firebrands – moas were afraid of fire, not of man – to the beaches and slaughtered there. The meat was loaded onto the boats and taken to the villages; the bones, thousands upon thousands, were left on the beaches. Years later, the Europeans used these bone deposits as fertilizer for their fields, and as mines for Maori antiquities.

Moas disappeared, the Maoris had no domestic animals to reduce the primary phyto-production. Huge masses of fuel were stored. Fires, kindled by Maoris, lightning, or volcanos became catastrophic, both in intensity and extent. Vegetation was damaged, soils eroded. The shoreline in bays and rivermouths advanced towards the sea because of the erosion in the hinterland. Many moa butchering sites, that were originally near the beaches, are now separated from the sea by broad zones of new land, and piedmont sites are covered with thick layers of eroded soil.

With no moas left, the Maori population, increased during the happy years, had to return to fish and shellfish. Resources were over-exploited. The lower layers of Maori shell mounds contain many big shells; upper layers only small ones. Tribal wars over the diminishing resources were the logical consequence, and the Maoris became warriors, feared and admired by Europeans. No-man's-lands developed between tribal territories. Maori gardening used little land, since their tropical plants did not do well in temperate climate. No domestic animals grazed around the villages, so there were none to escape and run wild. The interior of the islands was not settled. Vegetation recovered and grew rankly, because there were no moas left to keep the landscapes open. The land became impenetrable, protecting the Maori villages from each other. The woods were entered only when necessity demanded. New Zealand environments were neither "natural" nor "untouched" when the Europeans arrived.

The climatic results that I have described for Madagascar also apply to New Zealand.

13.5.6 Small Off-Shore Islands

The only untouched landscapes that modern Europeans ever saw were Antarctica, the highest peaks of bare mountains, and off-shore islands, which had not been discovered by others before. Antarctica is a poor object for studies in fire ecology, and so are the peaks. We will leave them aside.

Most off-shore islands are parts of reefs, submarine volcanos, transoceanic ridges or wedges, or they were lifted by tectonics. Terrestrial life could only come there from the continents by drift in marine or eolic currents, or by active flying. If efficient herbivores do not reach the islands, plants have to struggle only against the hostile environment and other plants. A vegetation cover forms, producing layers of organic soil. As soon as this is done, fuel for natural fires will be stored. Only few such cases are known. The island of Cracatau is the most famous one, but it had little time to develop.

Terrestrial vertebrates are better navigators than one might think. If they can hold on to drifting trees or vegetable islands, and can stand thirst and cold long enough, rodents and primates may travel considerable distances in warm marine currents. Newly arrived mice can diminish the storage of primary phyto-production of colonizing plants quite successfully. Less fuel is stored. Terrestrial tortoises can also travel long distances in warm marine currents. Mauritius, the Galapagos, and the Seychelles are just a few of the islands they reached in that way. Tortoises cannot eat the very hard plants, but they nevertheless keep landscapes open and prevent the accumulation of huge masses of fuel as long as their populations are only limited by losses of eggs and young to skuas and seagulls, and by trophic limits. Soils are fertilized, vegetation profits. When man came to such islands, the tortoises were diminished, if not exterminated. Fuel was stored, natural or anthropogenic fires became catastrophic. Once goats or other hypsodont animals are introduced, most of the vegetation that is edible for tortoises will disappear. The equilibrium between goats and vegetation will leave little fuel for fires. Even with an extremely low fire frequency, fire potential cannot accumulate. If the goats are eliminated to help surviving or newly imported tortoises, fire potential will be stored rapidly. The small surviving populations of giant tortoises are not able to prevent that. Fire intensity increases, the last tortoises are killed. Galapagos Islands are the best, or better, worst example.

Birds, when colonizing off-shore islands, often lose their ability to fly. Flying on an off-shore island implies the danger of being driven off by a storm and never seeing the island again. So, as long as there are no terrestrial carnivores, it is safe and wise not to fly. Thus flightless forms developed on many islands, to become extinct as soon as the first human castaway landed.

Next to the arrival of man, the worst that can happen to the undisturbed vegetation of islands is the arrival of hypsodont mammals. They will eat everything as their populations increase. When trophic limits become decisive, they will correspond with adaptations to harder food, low reproduction rates, and by dwarfing.

The Pleistocene dwarfed elephants, hippos, cervides, and other artiodactyles (*Myotragus*) of the Mediterranean islands are the classic case (Schüle 1990b). Pleistocene vegetation on these islands must have been very open. Regardless of fire frequency, fire intensity was low. Little fuel was stored.

When man arrived (on Corsica-Sardinia in the Middle Pleistocene or earlier, on other Mediterranean islands in the Upper Pleistocene, on Cyprus probably not before the Early Holocene) the unwary island mammals disappeared. Vegetation recovered, fire intensity increased. When there was nothing left to hunt, men starved or left.

Only agriculture and imported domestic ungulates, some of which became feral, gave man the means to survive on the now game-less Mediterranean islands. Some Mediterranean islands were settled by farmers probably as early as the 6th millenium B.C., Cyprus certainly earlier. The domestic animals reduced the vegetation. Some of them – goats, sheep, and pigs – became feral in some of the islands. Wild boars may have reached some islands swimming, but neither the mufflons (*Ovis musimon*) from Cyprus, Corsica, and Sardinia, nor the famous bezoar goats, (*Capra aegagrus*) from Crete and some Egean islands reached the islands on their own. Cyprus, Corsica, and Sardinia were never connected with the mainland after the end of the Miocene (Sect. 4.3.1). Montain ungulates have sharp, slender hoofs and are very bad swimmers. They were brought in boats as game or are feral endemized sheep and goats. Probably already neolithic or later prehistoric people introduced red deer (*Cervus elaphus*) and other ungulates as game or as holy animals for sanctuaries, as the Romans certainly did to some of the Mediterranean islands where they soon dwarfed.

Fire frequency increased by human activities, and its intensity lowered. Goat-resistant and fire-tolerant chaparral and undergrowth spread. This high-quality fuel increased the fire intensity again wherever man did not prevent it by fire management. Anyhow, the pyrophobic components of the vegetation were heavily damaged. Soils, as far as they had outlasted the erosion caused by Pleistocene dwarfed ungulates and giant rodents, were eroded by agriculture, overgrazing, and burning of chaparral for pasture. The soils were sedimented in piedmont situations, in valleys and river mouths, filling up the bays which had formed when the sea rose in the Lower Holocene. Much of the eroded soils necessarily sedimented far away from the islands (and continents) at the bottom of the sea. Mediterranean-type landscapes evolved. On all Mediterranean islands the local climate was strongly influenced, the global climate not at all. The islands are too small.

13.6 Agriculture and Domestic Ungulates

More or less the same happened all over the basin of the Mediterranean and all over the world, wherever man began to cultivate plants and animals. The newly established trophic basis was followed by a corresponding explosion of human populations. Different from the megafauna-founded growth and expansion of the populations during the Pleistocene, this new basis was highly renewable as long as the soils were not spoiled by degradation, erosion, or irrigation-caused salinity. As soon as this happened, or populations became too big, they collapsed or were forced to emigrate, nearly always at the cost of their neighbors (Crosby 1986).

A change in diet, or the invention of new, more efficient technologies (Rindos 1984), could not resolve the problem for long, as long as the growth of the human population continued. It did (and does). Thus the danger of collapse or the necessity to emigrate was (and is) not eliminated, but just delayed, and thus made worse (Cohen 1977).

Regional differences lay in the variations of the abiotic parameters, in the vegetation it concerned, in human behavior and in time scales. Anyhow, compared with geological and even Pleistocene time scales, time scales within the period of human food production are microscopic.

In the semi-arid zones of southwestern Asia, and probably also in the Old World humid tropics, agriculture and domestication of animals began about 10,000 years ago. These were regions where long before a shifting equilibrium had been established between low-density hunting populations, human fire regime, and biological resources. In some parts of Africa, the same seems to have happened soon afterwards.

The fertile soils of climatically favored central European loess regions were occupied by gardeners and farmers about 7000 years ago. They burned the dense woods which had formed when the climate warmed in the latest Pleistocene and there were no longer megaherbivores or herds of horses and bison to keep them open. Successively, less fertile and less favorable regions were occupied. About 6000 years ago, nearly all suitable land was occupied by farmers, gardeners, and herdsmen in southern and part of central Europe. Forests were replaced by cultivated lands and open woodland, heath, or meadows for herds of cattle, sheep, goats, pigs, and, from about 4000 B.P. or earlier, horses.

Large-scale transhumances were probably practised all over Europe and temperate Asia wherever the climate made them necessary and the geographic conditions possible. They are practised up to the present (Hofmeister 1961) wherever they did not disappear because of overpopulation, political reasons, or modern technologies. Up to the last century the flocks from the Upper Jurassic limestone highlands of the Schwäbische Alb in southern Germany, north of the Danube were driven to Burgundy and the French Provence through the narrow passage between the Vosges mountains and the Swiss-French Jura near Belfort. Probably this transhumance was practised already during the European Iron Age (Dehn 1972). Sheep and cattle from northern Spain are still driven from the mountains and highland mesetas in the north to the winter meadows of southern Extremadura and Andalucía (Klein 1920, 1985), a distance of almost 1000 km. It is easy to understand the consequences of this system for the European forests even in moist central Europe if it was practised on a large scale.

Herding certainly was not only a question of economics in prehistoric Europe, but also of social status, as it still is in Africa and elsewhere. Europe had been covered with dense woods when the climate warmed in the Lower Holocene, and neither megaherbivores nor hypsodont game were left to prevent it. Now these woods were kept open again or even disappeared, more because of human burning and overgrazing than by direct land use for agriculture.

Population densities of prehistoric cultures are underestimated most of the time, even by most archeologists, who should know better. Traces of human presence disappear fast with agriculture and the resulting erosion. Others are covered by sediments from just that erosion. The chances that a trace of prehistoric human activities comes to the attention of an archeologist are 10^{-x}.

The value of x is unknown, but it is much higher than most students are inclined to confess. In some areas of Germany, eagerly studied by archeologists for about a century, the number of known archeological sites has increased on a scale of 1:100 with the help of aerial surveys (H. Becker pers. commun.). With this method, only activities which included soil movements can be traced – a shepherd's hut will never be detected. About three quarters of all known prehistoric burial mounds of central Europe are located in forests. Quite naturally, they were not built in the forests, but in open landscapes. These woods are on the bad soils, not on the good ones. On good soils, burial mounds are hardly ever found. They were destroyed by later land use, in some cases even covered by sediments resulting from soil erosion. It makes no sense to suppose that prehistoric farmers deforested first the bad soils and then the good ones if they did so for direct land use by agriculture. Only very heavy soils in the lowlands were unpracticable for them. Thus the deforestation of poor soils could only be useful if the region began to be overpopulated, or if herding was more important for the economy of the tribes concerned than agriculture. Such a hypothesis is not accepted by most archeologists, because it is hardly to be proved by archeological means, but should be considered more seriously. Certainly, most of Middle and Late Holocene Europe was much less wooded than it is supposed to have been.

On the other hand, one can suppose that the agro-silvio-pastoral system of Spanish and Portuguese dehesas is a very old one. It is now limited to poor silicatic soils and soil-eroded limestone mesetas of the Iberian Peninsula, southern France, and Sardinia. In more humid climates and on good soils, it would be very productive for pasture, agriculture, and even a limited timber production. If grazed sufficiently, even in the semi-arid climate of central Spain the probability of dangerous fires is very low, in spite of the long dry season and the high frequency of thunderstorms without rain, which are typical for Castile. The dehesa system produces exactly what mankind needed and needs: shifting fields of cereals, an irrigated garden in or near the village, all kinds of animals, wood for cooking and heating, logs for building, charcoal for mining, and nearly no danger of fires. Combined with vertical or long-distance horizontal transhumances, the carrying capacity for ungulates is very high, even in regions with seasonal shortages of pasture. Without transhumances this is not the case. With the animals near the village all over the year, the vegetation around it will be devastated within a few years (for energy balances see Remmert 1988). The only fire potential left will be the houses of the village. Thus is seems quite reasonable that the Man and Biosphere (MaB) Commission of UNESCO is studying projects to adapt this system to modern necessities. Well organized, it would allow a limited production of timber and solve many of our environmental problems, including the high fire intensity of planted timber-producing forests. Besides, dehesas are ideal biotopes for nearly all kinds of wildlife. The problem is, of course, the organization of transhumances in overpopulated regions.

Pollen diagrams from prehistoric sites in central Europe show exactly what one would expect from the semi-natural open woods or well-stocked savannas which are the basis of the dehesa system: a mixture of local trees, grasses, herbs, and cereals.

In less favored regions of prehistoric Europe, no-man's-lands separated rival tribes. Herds were kept out of these zones for fear of thieving neighbors. The game which managed to survive in these thickets was hunted when it came to feed on the fields and meadows, or by occasional hunting expeditions into the no-man's-lands. Its population density was low. The woods there became ever denser, the trophic conditions for the game became worse and worse, the fire intensity became catastrophic.

Burning was the most comfortable and certainly the most widely employed means to clear the land for agriculture and pasture. When fields were abandoned or regions depopulated by wars, epidemics, or crop failure, secondary forests extended rapidly, temporarily improving conditions for the surviving game, until the lands were again occupied by human populations. The remaining Eurasian hunters prevented the game from recovering even in those regions that were not suitable for herds or agriculture. Forests became dense, fire intensity increased. Far in the north, little changed in spite of the disappearance of the megaherbivores: they were replaced by herds of semi-domesticated reindeer.

In southern North America and in parts of Central and South America something similar happened about 6000 or 7000 years ago. There were fewer species of domesticated plants and animals than in the Old World, because few suitable animals had survived the massacres of the End Pleistocene and Lower Holocene. Economies were based on irrigation systems with maize which did not need the methane-producing swamps of the South Asian rice fields, and on some other plants in the tropical rain forests. The domestication of the two surviving American camelids, guanaco, and vicugna, was only practised in the treeless puna of the Andine region, but not in the lowland pampas. Outside the regions of agriculture and herding, the Amerindians of both continents developed a perfect system to exploit the large herds of camelids and bisons in a kind of semi-domestication. Thus, the pre-agricultural combination of natural and man-made fire regime was hardly modified.

The Australian Aborigines, in spite of having settled their continent long before the Amerindians had reached theirs, did not develop domestication of plants and herbivores until the Europeans arrived. Their perfect fire management improved the trophic conditions for grazing kangaroos and favored for about 40,000 years the expansion of pyrotolerant and pyrophilic plants; their habit of returning again and again to the same camp sites provided the environments of these with seeds of their favorite vegetable food which they carried there in their hands or bellies (Jones 1975).

13.7 Conclusions

1. I have tried to show that fire is, and has been since the beginning of terrestrial life, an important agent in biological evolution. Fires with great intensity have only little selective value – all life is destroyed. Only life forms that are able to escape intensive fires will survive. Animals will run or fly away or dig themselves into the earth. Many plants survive if the soil is not heated much. Strategies for fire tolerance are developed more easily if little fire potential is stored because of high fire frequency or high consumption by herbivores.

2. I further tried to show how Upper Miocene hominoids managed first to escape fire, and later to be its commensals. In the Lower Pleistocene, hominids became fire's symbionts, first by providing fire with additional fuel by killing the unwary megaherbivores and their smaller contemporaries with spears, and then by aiding fire's auto-reproduction, until they finally produced the symbionts themselves. Fire was a strong selective agent in the evolution of the highly developed central nervous systems of the hominids, and later of man. All this happened in Africa more than one and a half million years ago. From there man and his influence on the biosphere (Fig. 1) spread gradually all over the world (Fig. 2). The consequences were numerous. Some of them I wish to recall.

3. Prior to the hominid intervention, even hyper-humid forests were kept open all over the world by high biomasses of herbivores. In these natural, open rain forests, fire frequency was as low as in recent rain forests, and the intensity of fires after droughts less catastrophic because less fuel was stored. When the fearless woodland megaherbivores became extinct, first in Africa and then on the other continents, these rain forests became dense, inhospitable for all herbivorous ungulates. Thus they stored huge, high-quality fire potentials which caused catastrophic intensities of occasional fires.

4. In the less humid zones of Africa, the accumulation of fire potential was prevented by a man-made increase of the fire frequency. Pyrotolerant and pyrophilic types of vegetation spread. In semi-desert zones with their low natural frequency of widespread fires, huge quantities of high-quality fuel were stored as soon as the herbivore biomass was reduced by hominid hunting. Occasional natural or man-made fires became very intense. Deserts spread. The same thing happened all over the world as soon as hominids appeared. Fire frequency increased drastically. This happened in the Old World tropics and subtropics outside of Africa in the Early Middle Pleistocene, prior to 500,000 B.P.

5. The expansion of naked tropical *Homo* was halted at his climatic limit before he invented means of thermo-insulation. Along this Eutrophic Line, climatic conditions were bad for *Homo*, and trophic conditions good because of the game influx from beyond the line. Beyond the Eutrophic Line and overseas, nothing changed until man appeared.

6. In the hinterland of the Eutrophic Line, an equilibrium between vegetation, game, human hunting, and human fire frequency became established. Humid forest became increasingly dense because of megaherbivore extinctions and decimated populations of smaller herbivores; their fire potential was large. Surviving behaviorally adapted savanna megaherbivores and middle-sized ungulates colonized these woods in the Old World continental tropics only slowly and could not prevent the woods from becoming nearly impenetrable. Drier zones were kept open by man-made high fire frequencies.

7. During the last glaciation, about 40,000 years B.P., man reached Australia. He arrived as a perfect hunter in possession of fire. Thus the impact was sudden and compressed. The megafauna became extinct; the fire regime changed. Fire-tolerant vegetation from the southwest and center spread over most of Australia.

8. About 10,500 B.P., man settled America from east Beringia. The process was repeated. Human hunters and hunting-adapted game repopulated North America via the disappearing but already wooded Bering landbridge. Humid forests densened, open landscapes and fire-tolerant vegetation expanded.

9. The same was repeated in Madagascar and New Zealand after 800 A.D. Differences of the ecological consequences lay in a different plate tectonic history and in the different cultural stage of the settlers.

10. The introduction of agriculture brought fire-clearing, the domestication of animals was followed by overgrazing by hypsodont domestic animals and annual burning for improvement of pastures. In the semi-arid zones of southwest Asia, this process began around 10,000 B.P., in some parts of the Old World tropics around the same time or a little later, in the climatically and edaphically favored parts of Europe around 7000 B.P. From about 5000 B.P., much of Europe had open landscapes even in moist regions. Wherever precipitation was high enough, secondary forests extended when human populations and their herds were decimated by wars, epidemics, or climatic catastrophes. No-man's-lands were covered with dense woods.

The consequences of Amerindian agriculture and herding were probably less severe than of the Eurasian types. Australian Aborigines continued with their highly developed spear-and-fire system until Europeans arrived.

11. We are far from understanding the complicated network of biotic and abiotic parameters which regulate Earth's climate, so we cannot make any definite statements about the influence man and his ancestors had on it. Nevertheless, it seems worth while to consider such an influence.

Earth's climate is the result of an exchange of energy between the earth and the universe (Lamb 1972–1977; Schwarzbach 1974; Oeschger et al. 1980; Shukla and Mintz 1982; Weischet 1988; Flohn 1985; Wanner and Siegenthaler 1988).

Life never influenced the energetic input, but it has certainly been influencing the output. The chemical composition of today's atmosphere is, for the most part, a product of biotic activities. It has changed a lot since life first appeared on earth, some 10^9 years ago. Carbon dioxide (CO_2), methane (CH_4), and some other organic gases are the most important agents in the atmospheric "greenhouse effect" (Beauchamp 1989; Pearce 1989; Houghton and Woodwell 1989). Volcanoes seem to be the only primary source of carbonic gases in the atmosphere. Atmospheric oxygen (O_2), the basis of all aerobic life on earth, seems to be the product of biotic activity, as are the low concentrations of sulfur in unpolluted air.

Life regulates the circulation of carbon in the three outer spheres of Earth (Fig. 1). Marine and terrestrial life transforms carbon from atmospheric CO_2 into biomass via the primary phyto-production. Part of this is restored at once to the atmosphere by the respiration of the plants, the rest is stored in living or dead (wood, litter, skeletons etc.) phyto-biomass.

Under favorable geological conditions, part of the biomass will be fossilized in the lithosphere. The formation of limestones, mineral oil, and natural gases is mostly a maritime process. We will not consider the ocean here, but evidently it has a very strong storage, and in this way a buffering, effect. Terrestrial biomass, as far as it is not returned directly or indirectly to the atmosphere, or carried to the sea as detritus, charcoal, ashes, or smoke, may fossilize as coal seams. Plate tectonics and orogenesis play an important role in the sedimentation processes which lead to their fossilization, and thus to the diminishing of the greenhouse effect of atmospheric CO_2. Volcanic activity may moderate this effect. Again, plate tectonics and orogenesis are very important agents in the activation of volcanism. The fossilized part of the primary biomass production had not been touched by man until a very short time ago. How much fossilized carbon has been restored to the atmosphere by erosion and the resulting oxidation during the geological evolution of the lithosphere can hardly be estimated, but these things happen only on a large time scale.

A considerable part of the primary phyto-production is consumed by animals and restored to the atmosphere by respiration and digestion gases with a high greenhouse effect (Pearce 1989), or it is stored in secondary biomass for the lifespan of bacteria, fungi, and animals, or in the organic components of soils and subfossil sediments, before it returns to the atmosphere or fossilizes. Time scales for these processes may vary considerably. On the other hand, the consumption of primary phyto-production by secondary life forms diminishes the short-term storage of carbon in phytic biomass, like leaves, grass and wood, and the long-term fossilization of organic carbon in sediments. Thus, the consumption of phytic biomass by animals has a strong influence on the greenhouse effect. Changes can happen in a very short time scale.

Leaving aside Paleozoic and Mesozoic evolutionary cycles and the sedimentation of organic limestone, the evolution of mammalian digestion systems must have influenced the climate. From the Upper Paleogene (Middle Tertiary) on, several lines of herbivorous mammals not only developed efficient hypsodont teeth, but also integrated methane-producing microbes into their digestive tracts. Neogene (Upper Tertiary and Quaternary) ruminants, equides, some of their perissodactyle relations, and proboscidians produced increasing quantities of methane and other organic gases. In "undisturbed" biotopes, these animals form up to 90% or more of the vertebrate animal biomass. Under consideration of the conditions in east African game preserves (the only ones which can actually be considered similar to really undisturbed biotopes), I tried to estimate the global biomass of methane-producing herbivores in the Upper Tertiary. It may be calculated at $x \times 10^4$ kg herbivore biomass $km^{-2} \times 10^8$ km^2 surface with terrestrial vegetation = $x \times 10^{12}$ kg of Upper Tertiary methane-producing herbivores on Earth. The factor x in the estimation for herbivore biomass km^{-2} may vary between nearly zero and about 3, according to the primary phyto-production of the different landscapes. In most landscapes, the value of x should have been > 1. The high herbivore biomass also produced huge masses of dung that stimulated the primary phyto-production. On the other hand, they diminished the short-term storage of large quantities of carbon in durable substances, and thus their possible long-term storage in fossilized phyto-biomass. It is obvious that the terrestric herbivores always had an influence on the greenhouse effect.

At that point, hominids enter into the calculations. Until 10,000 years ago, their biomass was low, with the possible exceptions of:

a) when they began to hunt the fearless megaherbivores in Africa, 2 or 3 million years ago (Sect. 2.4.3),
b) along the border of their expansion over the Old World tropics and subtropics in the Lower and early Middle Pleistocene (Sect. 4.1),
c) along the Eutrophic Line (Sect. 4.2) of the Middle and Upper Pleistocene, and
d) on the front line-of Martin's blitzkrieg in Australia and both the Americas in the last phase of the Upper Pleistocene (Sect. 5.2–3).

These populations collapsed when the hominids destroyed their trophic basis. Primate digestive tracts do not produce great quantities of gases enhancing the greenhouse effect. But hominid activities considerably altered the production and storage of CO_2 in the biomass: first by the destruction of megaherbivore biomass, and then by decimating the populations of medium-sized herbivores. The production of organic gases by microbes in the digestive tracts of ungulates was diminished.

When early man began to aid fire's auto-reproduction, increasing the natural fire frequency, the circulation of carbon, and thus the greenhouse effect of the atmosphere, was altered even more. Not only the plants and vertebrates were affected by the increased fire frequency, but also the insects, such as the methane-producing termites, and the organic soils which play an important part in the carbon flux.

Things changed again when men began to cultivate plants and to domesticate animals. Agriculture by itself may not have altered the carbon circulation much, but the irrigation of rice fields in South Asia created new swamps with a high methane production in river plains and on hill terraces.

Forests were burned for agricultural land use, and the methane-producing organic components of their soils damaged. Domestic ungulates replaced the vanished game and damaged the vegetation by overgrazing. Now there were only few species of mostly hypsodont and methane-producing herbivores instead of the multi-specied use of the primary phyto-production under undisturbed conditions. Later, man began to drain methane-producing swamps. Mining and ore-melting soon began to have an impact on the circulation of carbon by the production and burning of charcoal, and the input of mineral sulfur and other products into the atmosphere, for about 6000 years. Human impact on the carbon circulation and the greenhouse gases became more and more complicated, the possible feedbacks and the climatic consequences of buffering effects and their sudden breakdown can hardly be estimated.

The consequences of actual human overpopulation and modern inadequate technologies are widely discussed nowadays and not our concern here.

The evolution of mammalian herbivore's hypsodont teeth and microbe-aided digestion systems began in Laurasia about 3×10^7 years ago, in Africa about 10^7 years later (not considering the endemic hypsodent *Arsinotherium* as we do not know anything about its digestion), and in South America about 3.5×10^6 years ago, when Laurasian herbivores entered the continent (again leaving aside the unknown digestion systems of the extinct paleo-endemic herbivores).

The hominid influence on the greenhouse effect began in Africa with the invention of the thrusting spear, probably 3×10^6 years ago, and with the use of fire from around 1.5×10^6 years B.P. In some parts of the Old World tropics, man, spear, and an anthropogenic fire regime arrived nearly 10^6 years ago, in the Old World subtropics and warm-temperate zones a little later, in Australia and north of the Middle and Upper Pleistocene Eutrophic Line around 4×10^4 years ago, in the Americas 10^4 years ago, and in Madagascar and New Zealand 1.2×10^3 years ago. The intervention of man's cultivated and partly irrigated plants and domesticated herbivores began in parts of the Old World tropics and subtropics around 10^4 years ago, in the Old World temperate zones and some parts of the Americas around 7×10^3 years ago, and in the rest of the world not before the European expansion which led to a catastrophic waste of phyto and zoo-biomass and a sudden uncontrolled change of the wisely limited fire management of the aborigines all over the world.

These changes in the biotic carbon flux of the outer spheres of Earth coincide with the at first slow shifts of climate during the Tertiary, and, for the last 3×10^6 years, with the more and more rapid oscillations of Earth's climate in the Upper Pliocene and Quaternary. If it is mere coincidence, it is a strange one indeed. Quite naturally, the changes of climate and biotic parameters do not fit completely. Both the climatic and biotic diagrams are based on poorly known parameters, and both contain an unknown number of internal and mutual, positive and negative feedbacks, as well as a likewise unknown number of widely unknown buffering effects with an unknown time scale. The periods of biotic changes and climatic oscillations coincide, but we still cannot correlate their different stages, as we do not know the causes for the glaciations in former eras of Earth's history.

The effects of the biotic parameters on Earth's climate overlie those of the abiotic ones, much as the wind-made waves overlie the swell, and the swell overlies the tides of the sea. Nevertheless, the breakers of a storm can be higher than the swell or the tide, and they certainly have the higher frequency. If my suggestion is right – and in our parable the climatic consequences of the abiotic parameters produce the tide, the teeth and stomachs of the mammalian herbivores the swell, and man's spear and fire the breakers – then human overpopulation, modern technologies, and the anthropogenic destruction of the vegetation, including the burning of the rain forests, will be a typhoon such as the Earth has not known for about 65,000,000 years, when the dinosaurs died.

I hope that I am a false prophet.

Acknowledgements. This contribution forms part of the investigation program "Cultural Ecology and Prehistoric Landscapes in Southern Spain" sponsored by the *Volkswagen Foundation.*

References

Only modern titles with detailed bibliography are given. Even these are far from being complete.

Aguirre E (1969) Evolutionary history of the elephant. Science 164:1366–1376

Alvarez LW, Alvarez W, Asaro F, Michel HV (1980) Extraterrestrial cause for the Cretaceous-Tertiary extinctions. Science 208:1095–1108

Anderson A (1984) The extinction of moa in southern New Zealand. In: Martin PS, Klein RG (eds) Quaternary extinctions. Univ Arizona Press, Tucson, pp 728–740

Anderson E (1984) Who's who in the Pleistocene: a mammalian bestiary. In: Martin PS, Klein RG (eds) Quaternary extinctions. Univ Arizona Press, Tucson, pp 40–89

Andrews PJ, Franzen JL (eds) (1984) The early evolution of man. Senckenberg, Frankfurt a.M.

Barbetti M (1986) Traces of fire in the archaeological record before one million years ago? J Hum Evol 15:771–781

Bartstra GJ (1977) Walanae formation and Walanae terraces in the stratigraphy of south Sulawesi (Celebes, Indonesia). Quartär 27/28:21–30

Battistini R, Verin P (1967) Ecologic changes in protohistoric Madagascar. In: Martin PS, Wright HE Jr (eds) Pleistocene extinctions. The search for a cause. Proc VII Congr Int Assoc Quat Res 6. Yale, New York, pp 407–424

Beauchamp B, Krouse HR, Harrison JC, Nassichuk WW, Eliuk LS (1989) Cretaceous cold-seep communities and methane-derived carbonates in the Canadian Arctic. Science 244:53–56

Boech C, Boech H (1981) Sex differences in the use of natural hammers by wild chimpanzees: a preliminary report. J Hum Evol 10:585–593

Bourliere F (1963) Observations on the ecology of some large African mammals. In: Howell FC, Bourliere F (eds) African ecology and human evolution. Aldine, Chicago, pp 43–54

Brain CK, Sillen A (1988) Evidence from the Swartkrans cave for the earliest use of fire. Nature (Lond) 336:464–466

Budyko MI, Golitsyn GS, Izrael YA (1988) Global climatic catastrophes. Springer, Berlin Heidelberg New York Tokyo

Carroll RL (1988) Vertebrate paleontology and evolution. Freeman, New York

Cassels R (1984) Faunal extinction and prehistoric man in New Zealand and the Pacific Islands. In Martin PS, Klein RG (eds) Quaternary extinctions. Univ Arizona Press, Tucson, pp 741–767

Christensen PE, Burrows ND (1986) Fire: an old tool with a new use. In: Groves RH, Burdon JJ (eds) Ecology of biological invasions. Cambridge Univ Press, Cambridge, pp 97–112
Ciochon RL, Corruccini RS (eds) (1983) New interpretations of ape and human ancestry. Plenum, New York
Cocks LRM (ed) (1981) The evolving Earth. Cambridge Univ Press, Cambridge
Cohen MN (1977) The food crisis in prehistory. Yale Univ Press, New Haven
Crosby AW (1986) Ecological imperialism. The biological expansion of Europe. Univ Press, Cambridge
Crutzen PJ, Galbally IA, Brühl C (1984) Atmospheric effects from post-nuclear fires. Climatic Change 6:323–364
Dehn W (1972) "Trancehumance" in der westlichen Hallstattkultur? Archäolog Korrespondenzblatt (Mainz) 2:125–127
Dewar RE (1984) Extinctions in Madagascar: the loss of the subfossil fauna. In: Martin PS, Klein RG (eds) Quaternary extinctions. Univ Arizona Press, Tucson, pp 574–593
Elliot DK (ed) (1986) Dynamics of extinction. Wiley, New York
Fleagle JG (1988) Primate adaption & evolution. Academic Press, New York
Flohn H (1985) Das Problem der Klimaänderungen in Vergangenheit und Zukunft. Erträge der Forschung 220, Wiss Buchges, Darmstadt
Frost PGH (1985) The responses of savanna organisms to fire. In: Tothill, Mott 1985, pp 232–237
Giese P (1986) Ozeane und Kontinente. Verständliche Wissenschaft. Spektrum Wiss: Verständliche Forschung (= Sci Am), Heidelberg
Gill AM, Groves RA, Noble IR (eds) (1980) Fire and the Australian biota. Aust Acad Sci, Canberra
Gill ED (1965) The paleography of Australia in relation to the migration of marsupials and men. Trans New York Acad Sci 28:5–14
Goldammer JG (1978) VW-Symp "Feuerökologie", Heft 1: Symposions-Beiträge. Heft 2: Feuerökologie und Feuer-Management. Freiburger Waldschutz-Abhandlungen Bd 1. Forstzool Inst Albert-Ludwigs-Universität Freiburg im Breisgau, Freiburg
Gowlett JAJ et al. (1981) Early archeological sites, hominid remains and traces of fire from Chesowanja, Kenya. Nature (Lond) 294:125–129
Greenberg JH, Turner CG, Zgura SL (1986) The settlement of the Americas: a comparison of the linguistic, dental and genetic evidence. Curr Anthropol 27:477–497
Groves RH, Burdon JJ (1986) Ecology of biological invasions. Cambridge Univ Press, Cambridge
Grzimek B (ed) (1968–1972) Grzimeks Tierleben. Enzyklopädie des Tierreichs. Bd X-XIII, Säugetiere 1–4. Kindler, Zürich
Haltenorth Th, Diller H (1977) Säugetiere Afrikas und Madagaskars. BLV Verlagsges, München
Handtke R (1978–1983) Eiszeitalter. Ott, Thun
Harris DR (ed) (1980) Human ecology in savanna environments. Academic Press, London
Hofmeister B (1961) Wesen und Erscheinungsformen der Transhumance. Erdkunde 15:121–135
Hohl R (ed) (1985) Die Entwicklungsgeschichte der Erde. Dausien, Hanau
Hopkins DM (ed) (1967) The Bering Land Bridge. Stanford Univ Press
Hopkins DM et al. (1982) Paleoecology of Beringia. Academic Press, New York
Houghton RA, Woodwell GM (1989) Globale Veränderungen des Klimas. Spektrum Wiss 1989(6):106–114
Howell FC, Bourliere F (eds) (1963) African ecology and human evolution. Aldine, Chicago
Hsü KJ (1972) When the Mediterranean dried up. Sci Am 227(6):26–36
Hsü KJ (1978) When the Black Sea was drained. Sci Am 238(5):53–63
Hsü KJ (1983) The Mediterranean was a desert. Princeton Univ Press, Princeton
Hsü KJ et al. (1977) History of the Mediterranean Salinity Crisis. Nature (Lond) 267:399 ff
Humbert H (1927) La destruction d'une flore par le feu. Acad Malgache, Mem 5
James SR (1989) Hominid use of fire in the Lower and Middle Pleistocene: a review of the evidence. Curr Anthropol 30(1):1–26
Jones R (1975) The Neolithic, Paleolithic and the hunting gardeners: man and land in the antipodes. In: Suggate RP, Cresswell MM (eds) Quaternary studies. Selected Papers from IX INQUA Congress, Christchurch, NZ. 2–10 Dec 1973. The Royal Society of New Zealand, Bull 13, Wellington NZ, pp 21–34

Jones R (1979) The fifth continent: problem concerning human colonisation of Australia. Ann Rev Anthropol 8:445–466
Kahlke HD (1956) Großsäugetiere im Eiszeitalter. Aulis, Leipzig
Kahlke HD (1981) Das Eiszeitalter. Aulis, Leipzig
Kershaw AP (1984) Late Cenozoic plant extinctions in Australia. In: Martin PS, Klein RG (eds) Quaternary extinctions. Univ Arizona Press, Tucson, pp 691–707
Klein J (1920) The Mesta. Harvard Univ Press, Cambridge
Klein J (1985) La Mesta, 3rd edn. Alianza Univ, Madrid
Kortlandt A (1968) Handgebrauch bei freilebenden Schimpansen. In: Rensch B (ed) Handgebrauch bei Affen und Frühmenschen. Huber, Bern, pp 59–102
Kowalski K (1986) Die Tierwelt des Eiszeitalters. Wiss Buchges: Erträge der Forschung Bd 239, Darmstadt
Kurtén B (1966) Pleistocene mammals and the Bering bridge. Comment Biol Soc Sci Fenn 29 (8): 1–7
Kurtén B (1968) Pleistocene mammals of Europe. Weidenfeld, London
Kurtén B (1988) Before the Indians. Columbia Univ Press, New York
Kurtén B, Anderson E (1980) Pleistocene mammals of North America. Columbia Univ Press, New York
Lamb HH (1972–1977) Climate: present, past and future. I. Fundamentals and climate now. II. Climatic history and the future. Methuen, London
Leakey LSB (1968) Bone smashing by late Miocene Hominidae. Nature (Lond) 218:528–530
Lewin R (1987) The First Americans are getting younger. Science 238:1230–1232
Lorenz K (1966) Evolution and modification of behavior. Methuen, London
MacDonald D (1984) The Encyclopaedia of mammals I-II. Allen and Unwise, London
Maglio VJ, Cooke HBS (eds) (1978) Evolution of African mammals. Harvard Univ Press, Cambridge
Martin PS (1973) The discovery of America. Science 179:969–974
Martin PS (1984) Prehistoric overkill: the global model. In: Martin PS, Klein RG (eds) Quaternary extinctions. Univ Arizona Press,Tucson, pp 354–403
Martin PS (1986) Refuting Late Pleistocene extinction models. In: Elliot DK (ed) Dynamics of extinction. Wiley, New York, pp 107–130
Martin PS (1987) Late Quaternary extinctions: the promise of TAMS ^{14}C Dating. Nucl Instrum Methods Phys Res B 29:179–186, North-Holland, Amsterdam
Martin PS, Klein RG (eds) (1984) Quaternary extinctions. Univ Arizona Press, Tucson
Martin PS, Wright HE Jr (eds) (1967) Pleistocene extinctions. The search for a cause. Proc VII Congr Int Assoc Quat Res 6. Yale, New York
Müller AH (1989) Lehrbuch der Paläozoologie, B III, 3. Fischer, Jena
Nowak RM, Paradiso JL (1983) Walker's Mammals of the World I-II, 4th edn. Hopkins Univ Press, Baltimore
Oakley KP (1954) Evidence of fire in South African cave deposits. Nature (Lond) 174:261–262
Oeschger H, Messerli B, Svilar M (1980) Das Klima. Springer, Berlin Heidelberg New York Tokyo
Owen-Smith RN (1988) Megaherbivores. Cambridge Univ Press, Cambridge
Patterson B (1975) The fossil aardvarks (Mammalia: Tubulidentata). Bull Mus Comp Zool Harvard Univ, 147(5):185–237
Pearce F (1989) Methane: the hidden greenhouse gas. New Sci 6 (May):37–41
Pilbeam D, Meyer GE, Badlgley C (1977) New hominid primates from the Siwaliks of Pakistan and their bearing on hominoid evolution. Nature (Lond) 270:689–695
Piveteau J (1955) Traité de Paléontologie. Paris (t. V pp 1044)
Pool R (1989)Ecologists flirt with chaos. Science 243:310–313
Prasad KN (1982) Was Ramapithecus a tool-user? J Hum Evol 11:101–104
Prickett N (ed) (1982) The first thousand years. Regional perspectives in New Zealand archeology. Palmerston North (NZ)
Pyne SJ (1984) Introduction to wildland fire. Wiley, New York
Remmert H (1988) Energiebilanzen in kleinräumigen Siedlungsarealen. SAECULUM XXXIX (2):110–118
Rensch B (ed) (1968) Handgebrauch bei Affen und Frühmenschen. Huber, Bern

Rindos D (1984) The origins of agriculture. An evolutionary perspective. Academic Press, Orlando
Roberts L (1989) How fast can trees migrate? Science 243:735–737
Schüle W (1990a) Prähistorischer Faunenschwund. Ursachen und Wirkung. (in preparation)
Schüle W (1990b) Mammals and the initial human settlement of the Mediterranean islands. Int Conf Early Man in Island Environments, Oliena (Sardegna), Sept 25th – Oct 2nd 1988 (in press)
Schüle W (1990c) Martin's blitzkrieg, animal behaviour and human evolution. (in preparation)
Schwarzbach M (1974) Das Klima der Vorzeit. Eine Einführung in die Paläoklimatologie, 3. Aufl. Enke, Stuttgart
Schreiber BC (ed) (1988) Evaporites and hydrocarbons. Columbia Univ Press, New York
Shipman P (1986) Scavanging or hunting in early hominids. Theoretical framework and tests. Am Anthropol 88:27–43
Shukla J, Mintz Y (1982) Influence of landsurface evapotranspiration on the Earth climate. Science 215:1498–1501
Simpson GG (1980) Splendid isolation, the curious history of South American mammals. Yale Univ Press, New Haven
Singh G, Kershaw AP, Clark R (1980) Quaternary vegetation and fire history in Australia. In: Gill AM, Groves RA, Noble IR (eds) Fire and the Australian biota. Aust Acad Sci, Canberra, pp 23–54
Stehli FG, Webb SD (eds) (1985) The Great American Biotic Interchange. Plenum Press, New York
Suggate RP, Cresswell MM (eds) (1975) Quaternary Studies. Selected Papers from IX INQUA Congress, Christchurch, NZ. 2–10 Dec 1973. The Royal Society of New Zealand, Bull 13, Wellington NZ
Thenius E (1980) Grundzüge der Faunen- und Verbreitungsgeschichte der Säugetiere, 2. Aufl. Fischer, Stuttgart
Thenius E, Hofer H (1960) Stammesgeschichte der Säugetiere. Springer, Berlin Göttingen Heidelberg
Tobias Ph V (ed) (1985) Hominid evolution. Past, present and future. Proc Taung Diamond Jubilee Int Symp Johannesburg 1985. Liss, New York
Toth N (1987) The first technology. Sci Am 256:112–121
Tothill JC, Mott JC (1985) Ecology and management of the world's savannas. Int Savanna Symp 1984. Commonwealth Agricultural Bureau. Aust Acad Sci Canberra
Vereshchagin NK, Barishnikov GF (1984) Quaternary mammalian extinctions in northern Eurasia. In: Martin PS, Klein RG (eds) Quaternary extinctions. Univ Arizona Press, Tucson, pp 483–616
Walker EP (1975) See Nowak and Paradiso (1975)
Wanner H, Siegenthaler U (eds) (1988) Long and short term variability of climate. Lecture Notes in Earth Sciences 16. Springer, Berlin Heidelberg New York Tokyo
Whybrow PJ (1984) Geological and faunal evidence from Arabia for mammal "migrations" between Asia and Africa during the Miocene. In: Andrews PJ, Franzen JL (eds) The early evolution of man. Senkkenberg, Frankfurt/M, pp 189–198
Weischet W (1988) Einführung in die allgemeine Klimatologie. Physikalische und meteorologische Grundlagen, 4. Aufl. Teubner, Stuttgart
Wolbach WS, Lewis RS, Anders E (1985) Createaceous Extinctions: Evidence for Wildfires and Search for Meteoric Material. Science 230:167–170

14 Fire Conservancy: The Origins of Wildland Fire Protection in British India, America, and Australia

S.J. Pyne[1]

"Of the many problems which have to be dealt with by the forester, there is none which is so constantly with him as that of fire. Its shadow is always over him: the dread possibilities are ever present in his mind."

– C.E. Lane-Poole (1920)

"He [the Forester] is a soldier of the State and something more."

– Sir D.E. Hutchins (1916)

14.1 Introduction

The advent of modern wildland fire protection was almost everywhere associated with the advent of modern forestry. Professional forestry itself evolved beyond folk practices when the Enlightenment applied its rationalizing impulses to the peculiar environmental and social circumstances of central Europe. It became a vital export to overseas colonies as the industrial revolution and imperialism established a global economy and a global ecology, and as Western science promulgated a global scholarship. Foresters joined other transnational cadres of European engineers and administrators. But everywhere that European foresters ventured they encountered fire practices vastly different from those of central Europe in the 19th century. Everywhere their precepts conflicted with local lore, their practices with local custom. Everywhere they found themselves immersed in a conflict over fire practices that was virtually instantaneous, often violent, and unavoidable.

The story is complex, yet there are remarkable parallels among different places and peoples. In particular, the controversy over fire practices – what became known as "fire conservancy" – often took on the character of a formal debate not only between foresters and locals but within the forestry community itself. To most foresters with academic training, the value of fire exclusion seemed self-evident, and the primary vehicle was system – organized, aggressive fire prevention and suppression. To others, and to many intellectuals outside forestry, some compromise with local burning practices seemed advisable. Regardless, foresters absorbed the question of fire as their own, they laid out the infrastructure for industrial fire management, and they brought to the subject both the conceptual rigor of science and the evangelical fervor of a moral mission. The shock encounter between European and non-European and between traditional and industrial practices defined the heroic age of wildland fire protection.

[1] History Department, Arizona State University (West Campus), Phoenix, Arizona 85306, USA

What grants the story added urgency is the promise of a certain parallelism between the kind of landclearing and unconstrained burning that occurred in the 19th century and what is occurring today in the tropics and subtropics. Historical analogies are inherently treacherous, but there are enough similarities to tease out some fundamental insights, if only to emphasize the ease with which prior experience may be misapplied in new social and geographical environments. As a historical cameo, consider the experiences of Great Britain and three English-speaking former colonies – British India, the United States, and Australia. Many of the problems experienced during the heroic age stemmed from trying to translate European experiences onto other lands. Accommodations resulted that became the basis for distinctive national styles of fire protection. To translate the experiences of those nations onto the tropics may be less likely to recreate their eventual successes than to repeat their old errors.

14.2 Home Fires: A Synoptic Fire History of Britain

While the British Isles merit scant mention when considered relative to the notorious fire climates of the world, nearly every fire practice that British explorers, settlers, or foresters encountered had its cognate in British history. There were natural fires, including a kind of spontaneous swidden in which windstorms and lightning burned out patches of conifer forest. Pleistocene hunters contributed anthropogenic fires of hearth and field. Waves of new migrants from the continent introduced livestock and crops, and fire practices were adjusted accordingly. It was easy to adapt burning for hunting to burning for herding, somewhat more complex to adapt foraging fires to the cultivation of alien crops, for the problem was not simply to extract resources from the indigenous flora but to replace it with an alien flora. Fire was essential for both: there was no slashing without burning. Sites once felled might be maintained in grass or browse through broadcast fire. "Even though Mesolithic societies were technologically simple," a group of British archeologists has concluded, "their usage of fire seems to have conferred on them the ability to alter their surroundings in a purposeful way" (Simmons et al. 1981). Some of those environmental changes became more or less permanent.

The full revolution came with Neolithic agriculture. The appearance of landnam and livestock multiplied human fire practices and enlarged the geography of anthropogenic fire. Fire made possible the clearance of a felled deciduous forest; it expanded the domain of forage, within the woods and beyond them; it cleared birch-pine uplands; it extended heathlands, and probably set into motion the lengthy process by which upland sites degenerated into moor and peat. Once established, the mossy mats, during drought, became vast fuses to carry fire through the wastelands. Fire prepared ground for cereal crops of wheat, flax, and barley, and for pulse crops like the bean, lentil, and domestic pea. It assisted in the gathering of wild foodstuffs and in the hunting of wildlife. On many marginal sites, anthropogenic fire helped tip the biotic scale from forest to moor, and moor to mire.

Versions of shifting agriculture persisted long after the initial bout of forest landclearing. Successive waves of farmers and pastoralists steadily brought the landscape under a different regimen of burning, one tied to the rhythms of crops and domesticated beasts. Farmers burned straw after harvest and fired their fallow to fertilize or to purge pests. Moor burning became endemic where herders exploited the native heath or gorse. Welsh graziers fired the moors as part of a transhumant migration between summer and winter pastures. The English word swidden (from an Old Norse svithinn, "singed") referred to the burning of the Yorkshire moors in which decadent heath was fired on a roughly 10-year cycle. Additional lands were broken to cultivation by ax and fire. As organic soils replaced forests, a new cycle of swidden agriculture became common in which the sod was slashed (pared), burned, planted to a scenario of crops, then left fallow. During the agricultural revolution of the 18th century, several observers concluded that "sodburning was essential to success" (Marshall 1796). It was, in effect, a second wave of agricultural reclamation in which analogous techniques played out against a different environment.

There were wildfires, of course. Droughts allowed for occasional outbreaks of fire into wildland or field. Moor fires were a frequent nuisance, and often a threat. The Earl of Fife all but despaired of eliminating fire in his woods, whose danger was "scarce to be conceived" (Anderson 1963). Endemic unrest, chronic warfare between England and the continent, and later between England and Scotland, and then civil war, left large stretches of the countryside subject to episodic burning. Later, hunters replaced soldiers, and they and other migrants left fires in moor and woods that, during spring and fall droughts, ravaged the local landscape. But lacking routine lightning fire, savage droughts, or desert winds, the amount of fire prevalent in Britain was almost wholly determined by humans. Accordingly, fire seemed remote, far from the normal consciousness of Britons.

There was fire aplenty, but it was increasingly confined, sublimated, restricted in place and season, disciplined into agricultural cycles and strictly prescribed sites. Daniel Defoe noted with approval the observations of two "Foreign Gentlemen" in the 1720's, who asserted that "England was not like other Country's, but it was all a planted Garden" (East 1951). By the time of the American Revolution, Britain had become a land of minute manipulation. Agriculture had expanded, intensified, and harnessed Enlightenment science. Enclosure changed the crop but not the intensity of manipulation. Improved breeds of sheep demanded improved pasture grasses which demanded fertilizers more sophisticated than ash beds. Woods were confined to ceremonial groves, or were otherwise coppiced or pollarded. The underbrush was repressed into hedges or driven into remote moors. Grassland become manured paddocks or manicured lawns. Everywhere the human hand shaped, pruned, replaced. To cite Defoe again, "In a Word, it was all Nature, and yet look'd all Art" (East 1951).

If Garden replaced Nature, then the hearth replaced free-burning fire. Biological agents – sheep, cattle, domesticated plants – supplanted generic treatments like broadcast burning. Potential fuel became fodder for livestock or was carted into ovens. Fires escaping during droughts had little land in which to roam. Peat was cut and burned as fuel; the woods were coppiced for fuelwood or converted to charcoal; debris was burned in piles. Underwood, furze, and bracken were harvested for fuelwood or rendered into charcoal; small branches were gathered into faggots for hearth fires or to kindle the heartier fires that burned lime and brick. Fire fertilized lands indirectly by preparing lime. Coal replaced wood as critical fuel, reducing even further the value of wooded lands. Human pruning and harvesting, rigorous browsing and grazing, the planting of new species of pasture grasses and crops and ornamental trees – all substituted in specific detail for the generic power of free-burning fire. Fire became an implement of gardening, like rakes and spades. Even forestry emphasized plantations, not the management of natural or quasi-natural woods.

Something similar happened with Britain's human firebrands. Fire and sword went elsewhere; English society became as disciplined as the English landscape. Authorities eager to protect property strengthened the laws against arson and careless fire. Even where new ignitions appeared courtesy of the railroads, there was little opportunity for fires to burn freely. By the onset of the 19th century, Britons had largely strangled fire from the Home Islands.

In fact, they had gone too far. Without fire, heath degenerated, and hunting plantations in Scottish moors suffered massive losses of game birds (Kayll 1967). When, after the Great War, Britain attempted reforestation, it experienced serious losses in conifer plantations and discovered that woodland fire had vanished from Britain because British woodlands had vanished (e.g., Forestry Commission 1934). The vigor of inbred sheep required periodic rejuvenation from "hill sheep," who thrived on the heath of open moors, and such moors senesced without fire. Gradually fire returned, like sparkling artifacts from some lost treasure. As control over the rural landscape slackened, as surplus straw and furze were no longer marketed for fuel or fodder, as lands were planted to conifer forests, as heath became valued for its natural beauty and habitat, as new sources of ignition like railroads and tourists proliferated, fire rekindled the newly liberated fuels and Britain rediscovered a sampling of its ancient fire past.

Even among European nations Britain was exceptional, most closely approximated by the Low Countries. Around the perimeter of Western Europe free-burning fire continued – with pastoral burning endemic throughout the Mediterranean Basin, swidden cultivation in Scandinavia and Russia, agricultural fires in field and forest, wildfires in the steppes, the taiga, and other remote lands. But by meticulous control over fuels and peoples, by intensive cultivation and land use, fires became increasingly alien in the newly industrialized heartland. Only along the frontier was organized fire protection essential.

Those frontiers were by and large located in overseas colonies. Certainly by the 19th century British fire history had moved to distant dominions. Britain's own expulsion and containment of fire perhaps made all the more intense the shock of rediscovering free-burning fire as an endemic and universal phenomenon. To travel to India, America, or Australia was to step back into British history to the millenia of big game hunters and the early centuries of landnam and to do so in environments where nature could escalate fire into conflagrations. So ill-equipped were the British that they lacked even foresters. They had to import German professionals – a remarkable company of intellectuals and adventurers – to spearhead the administration and industrialization of their colonial wildlands.

14.3 An Empire Strategy: Fire Protection in British India

There were varied outposts of European forestry in which fire protection was part and parcel of the drama of expansionism – the Dutch in Java, the French in Algeria, the British in Cyprus, Canada, and a score of other African, Asian, and Australian colonies. Liberated colonies like the United States added other lands in need of reformation. In all these cases foresters discovered circumstances vastly different from those in temperate Europe with its reclaimed landscape. Beginning in the 11th century, driven by the Cistercians and other monastic orders, a kind of agricultural crusade had swept Europe, mopping up uncleared woodlands or converting swidden sites to sedentary agriculture; by the 18th century, peasants even carried landclearing and mixed farming to the high slopes of the Alps. As in Britain, the surge pushed free-burning fire to the margins. But in the newly colonized interiors of Asia, Australia, and America fire was endemic; there were no formal institutions to oversee fire practices or to organize fire lore; and foresters confronted natural forests of vast dimensions for which they had to seek natural regeneration. For every point of apparent similarity with European experience there was another of disparity.

It was in British India that European forestry really learned what it meant to establish forestry outside the environmental and social context of Europe. Forests fueled the industrialization of India. From its forests came the "sleepers" and trestles upon which railway construction across the subcontinent depended, and from those same forests came the firewood to power the locomotives. Soon after the Mutiny, however, after Great Britain supplanted the East India Company as Raj, the Government of India began to set aside forest reserves to assure a perpetual source of timber and water, and it appointed Dietrich Brandis as Conservator of Forests. Originally trained as a botanist, then formally educated as a forester in the grand European manner, Brandis was the archetype of the transnational forester, an indefatigable agent of Empire, a Clive of natural resource conservation in Greater India, a Humboldt of forestry who managed to combine precise technical knowledge with the kind of journey to exotic lands so fundamental to the literature of German Romanticism. Brandis quickly established the institutional and conceptual framework for forestry in Burma, India, and elsewhere in the Empire (Stebbing 1923).

Almost immediately, forest conservation was overwhelmed by fire. "Every forest that would burn," recalled E.O. Shebbeare, "was burnt almost every year." Fire littered the landscape almost year-round. Natives burned to drive game, improve grazing, render pathways safe for travel, ward off predators (notably tigers), and "to placate the local deities." As various portions of the biota cured, they carried fire; and where tallgrass or other annual fuel dominated the environment, fierce fires annually swept the "jungle," a term which served generically for India as "bush" did for Australia. It was natural then, Shebbeare observed, for the "pioneers" to see in "fire their chief, almost their only enemy" (Shebbeare 1928).

Inspector-General Ribbentrop declared that the nearly annual "conflagrations are the chief reason of the barren character of so many of our Indian hill ranges, and are more closely connected with distress and famine than is usually supposed." This was matched by "a most marvellous, now almost incredible, apathy and disbelief in the destructiveness of forest fires" (Ribbentrop 1900). Stebbing noted matter-of-factly that "in every Province . . . the officers of the Department had to commence the work of introducing fire conservancy for the protection of the forests in the face of an actively hostile population more or less supported by the district officials, and especially by the Indian officials, who quite frankly regarded the new policy of fire conservancy as an oppression of the people" (Stebbing 1923). Even the Forest Officers, Stebbing recalled, "were openly sceptical" of the possibility of fire protection. The prevailing sentiment was, among the native populace, that fire was useful, and among British administrators, that it was damaging and probably ineradicable.

Thus when Brandis instructed Colonel Pearson of the Central Provinces to try to stop the burning in 1863, he did so more out of hope than of conviction. "Most Foresters and every Civil Officer in the country", Pearson affirmed, "scouted the idea." He added that had the attempt failed, "any progress in fire protection elsewhere would have been rendered immeasurably more difficult," if not impossible in any systematic way (Shebbeare 1928). Pearson laid out fuelbreaks, sent out patrols, exhorted locals to give up burning, and enjoyed a couple of exceptionally wet seasons. The Bori Forest became a show-case of fire conservancy. The experiment succeeded.

Not completely, not without considerable debate and second-guessing, but thanks to militant enthusiasm, patience, and some relatively wet years, the experiment evolved into a demonstration program, and then into a prototype suitable for dissemination through Greater India. As with the native principalities, so with the native forests: more and more were reduced to British rule by fire protection. By 1880–81, some 11,000 square miles were under formal protection; by 1885–86, some 16,000 square miles; by 1900–01, an astonishing 32,000 square miles that spanned the spectrum of Indian fire regimes, from semi-arid savanna to wet-dry tropical forest to montane conifers (Ribbentrop 1900). Protection forests targeted, in particular, the great timber trees of the subcontinent – sal (*Shorea*), teak (*Tectona*), and chir (*Pinus*).

Within 20 years, however, doubts were voiced about the wisdom of "too much fire protection." There had always been dissenters, but now discontent bubbled publicly to the surface. In wet forests fire protection seemed to retard natural regeneration and it allowed fuels to accumulate that, once dried, exploded into all-consuming conflagrations. In drier forests, years of seemingly successful protection would be wiped out in massive fires during drought years.

Exhortations and bribes with goats had not extinguished all the native firebrands. Protected forests were fortified enclaves in a landscape that simmered annually with fire. Critics argued for a compromise program, with controlled burning to supplement fire suppression. The protests were most vigorous at the margins – most spectacularly in Burma, which wanted to reintroduce fire to support teak regeneration. In 1897 Inspector-General Ribbentrop had to intervene in favor of fire exclusion. But the debate continued across the pages of the *Indian Forester* and the annual reports of Conservators. In 1905 a compromise was proposed under the name of "early burning" that hybridized techniques and goals. Sub rosa burning in Bengal, Burma, Assam, and elsewhere scorched the landscape like an insurgency.

The revolution finally boiled over in Burma in 1907. Faced with a choice between excluding fire and excluding fire protection, the Inspector-General began withdrawing fire protection from prime teak forests. By 1914 Conservators of sal forests recognized that regeneration "had ceased throughout the fire-protected forests of Assam and Bengal and that no amount of cleanings and weedings would put matters right" (Shebbeare 1928). They tried to reintroduce fire, but fuels had so changed that it was no longer possible to run benign light fires through the understory. The taungya system by which swidden fields were restocked with planted timber trees evolved as a partial compromise. Nearly everywhere some form of early – that is, spring – burning of grassy understories became the norm. Whatever the causes for the failure of natural regeneration, Shebbeare concluded for an audience of Empire foresters, "fire appears to be the only real cure" (Shebbeare 1928). By 1926 the cycle of fire practices had come full circle. A Conservator's Conference amended the rules of the *Forest Manual* to make early burning the general practice and to extend complete protection only for special temporary sites. That included the Central Provinces.

The legacy of British India was felt throughout the British Empire. Nearly all British foresters (or foresters in British employ) progressed through a combination of formal schooling in European forestry followed by a field apprenticeship in Greater India (Fernow 1907). It is possible to speak of an "Empire strategy" of fire protection that combined European and Indian models, one later modified by other experiences, particularly in Cape Colony. The emphasis was on organization and system, on structural design with lookouts and fuelbreaks, on the integration of fire protection with working plans, and on the critical role of the Forester as fire guard. Its rigor and moral intensity derived from the attempt to impose a set of expectations developed in one social and ecological environment onto another very different environment.

The Empire strategy showed sufficient flexibility to claim status as a kind of universal model. In an early avatar it was brought to America and to Australia around the onset of the 20th century. In both nations the concept came too soon in its life cycle to carry with it the lessons of fire use, and by the time early burning became an integral component of the Empire strategy, the transplants had sprouted their own peculiar identities. The learned the lesson of system for fire control, but they did not learn – or chose to ignore – the lessons of fire use. They reached those conclusions on their own.

14.4 An American Strategy: Systematic Fire Protection

The American story unfolded in remarkable counterpoint. There was no disputing the ubiquity of fire. Lightning was a prominent source of ignition in the arid West, and burning was integral to American agriculture, pastoralism, landclearing, and other frontier economies; it was regarded as a folk right. Bernhard Fernow – a Prussian forester and naturalized American citizen – thought fire was the "bane of American forests" and dismissed its causes as a case of "bad habits and loose morals" (Rodgers 1968).

Along the borders of settlement, fire was endemic. Where associated with landclearing, it frequently got out of hand – a fatal eruption of frontier violence that incinerated forests, savaged fields and towns, and claimed an unconscionable number of lives. In 1904 a surveyor laying out the Plumas Forest Reserve in northern California wrote that:

"The people of the region regard forest fires with careless indifference .... To the casual observer, and even to shrewd men ... the fires seem to do little damage. The Indians were accustomed to burning the forest over long before the white man came, the object being to improve the hunting by keeping down the undergrowth, which would otherwise shelter the game. The white man has come to think that fire is a part of the forest, and a beneficial part at that. All classes share in this view, and all set fires, sheepmen and cattlemen on the open range, miners, lumbermen, ranchmen, sportsmen, and campers. Only when other property is likely to be endangered does the resident of or the visitor to the mountains become careful about fires, and seldom even then" (Barrett 1935).

And this was only a part of the fury. In the Lake States railroads opened up the big woods to markets, and settlers and loggers poured in. Forests cleared by logging were burned by farmers. During droughts and exceptional winds, fires raged on the order of hundreds of thousands of hectares, wiped out whole villages, and stripped forest soils of all organic matter, rendering them unusable to farming. Most of the devastated forests were marginal to farming anyway; sites spared the worst holocausts quickly exhausted soil nutrients.

On public lands and the larger private forests of America models of fire protection groped towards definition during the 1880's. In 1885 the State of New York adapted the administrative machinery of its rural fire warden network to accommodate the Adirondacks Preserve, and a year later the U.S. Cavalry assumed responsibility for protecting Yellowstone National Park; in Canada, Ontario instigated a system of fire-ranging. In 1897 the General Land Office received a mandate for the administration of the burgeoning national forests. By the end of the century private timber protective associations staffed lookouts and paid for patrols on high-value lands. A National Academy of Sciences committee on forest policy alluded to British India as a possible model and recommended that, for the present, the forest reserves be turned over to the Army and forestry be taught at West Point (Pyne 1982).

Civilian foresters successfully protested, established forestry schools at major universities, and began the exhilarating, maddening task of transferring European models to American landscapes. It was a century of Germanic influence – Germans were the largest immigrant group to America in the 19th century – and forestry joined kindergartens, graduate education, and journals of Hegelian philosophy in reshaping American society. German foresters like Carl Schenk and Bernhard Fernow journeyed to America as their colleagues did to India. The charismatic Gifford Pinchot studied at Nancy under the tutelage of Brandis, and after the national forests were transferred to his control in 1905, he corresponded with Sir Dietrich about how to establish a national corps of foresters. His successor as Chief Forester, Henry Graves, formerly Dean of Forestry at Yale University (a school endowed by the Pinchot family), also passed through the classic curriculum of British colonial forestry. Clearly, forestry was a case of technology transfer. From that time onward American foresters interested in foreign models looked almost exclusively to Europe or, in more typically American fashion, let the world swirl into America. A contingent of forestry engineers sent to France during the Great War left with convictions that there was much to admire in European forestry. And after France, American foresters looked obsessively to Sweden rather than to India or Africa.

This pioneering generation shared a crusading ethos, a militant commitment to conservation, that brought a moral energy to the question of "forest devastation," alarms over "timber famines," and to wanton burning. Elers Koch marveled at what a wonderful thing it was to belong to an organization composed almost wholly of young men (Koch 1944). Inman Eldridge wrote of those devotees that they believed Gifford Pinchot was a prophet, that he was all-knowing and far-seeing (Eldredge 1977). Coert duBois described his entry into forestry as head down and tail up, animated by the cry "*Gifford Pinchot le vult!*" (duBois 1957). They grappled with fire as foresters in new lands everywhere did. As Americans they shared in that age of aggressive political and economic reform known as the Progressive Era.

Then came 1910. That summer two crises simultaneously flared into public controversy. Parched by a vicious drought, the northern and western landscapes burned; an estimated five million acres burned throughout the national forest system, more than three million in the Northern Rockies. The great fires traumatized the Forest Service, still reeling from Pinchot's dismissal earlier in the year. Some 79 firefighters died in the line of duty. The Service sank into debt until Congress agreed to honor a deficit-funding statute enacted 2 years earlier. It was the first great crisis for Henry Graves, Pinchot's hand-picked successor and another Brandis protégé. To add to the embarrassment, the same month (August) that the fires blew up, Graves finished publication in serial fashion of a treatise on fire control. Two future chiefs, William Greeley and Ferdinand Silcox, weathered the fires as regional officers. The fires were the Valley Forge – the Long March – of the Forest Service. Soon afterwards Congress passed the Weeks Act (1911) which allowed for national forests to expand by purchase, for Federal-State cooperative programs in fire control, and for interstate compacts for firefighting.

Yet that same August *Sunset* magazine placed into public view a smoldering debate about fire policy. While professional foresters argued for fire protection by means of aggressive fire suppression, others advocated a program of controlled, light burning on the front-country and let-burning in the backcountry. The light-burners were an unwieldy amalgam – the State engineer of California, sentimentalist poet Joaquin Miller, timber owners and stockmen, the Southern Pacific Railroad and novelist Stewart Edward White, the San Francisco newspapers, and Secretary of the Interior Richard Ballinger. They argued that fire protection only aggravated fuels, stoking uncontrollable burns; that absolute fire control was technically impossible and fiscally irresponsible; that fire exclusion damaged forests, starving regeneration and encouraging insect plagues. They proposed instead a rationalization of frontier fire practices organized around controlled burning (Pyne 1982; Clar 1959).

Thus the events of 1910–1911 challenged the Forest Service as a fire agency across the board – the 1910 fires, its technical ability to control fire; the light-burning controversy, its capacity to formulate policy; the Weeks Acts, its ability to establish a national policy, not just a national forest policy. The Forest Service responded with vigor and discipline. Former Chief Pinchot was already on the record as saying that "like slavery forest fires may be shelved for a time, at enormous cost in the end, but sooner or later they must be faced" (Pinchot 1898). Chief Forester Graves declared that fire protection was 90% of American forestry (Graves 1914). Future Chief Greeley made "smoke in the woods" his yardstick of progress (Greeley 1972). The Service exploited the provisions of the Weeks Act to establish a Federal-State cooperative fire program that allowed formal fire protection to extend well beyond the lands of the national forest system and the Federal lands overall. Surveying the wreckage of 1910, German forestry professor Dr. Deckert noted that "devastating conflagrations of an extent elsewhere unheard of have always been the order of the day in the United States" – that, by contrast, its success with fire prevention constituted "a brilliant vindication of the forestry system of middle Europe" – and he expressed confidence that American foresters would follow their example (Deckert 1911). Within a decade after the 1910 holocaust, the U.S. Forest Service was doing exactly that.

The light-burning controversy flared up throughout the Western states and across the South. In Florida and Arkansas new national forests were almost consumed by woods burning. But the ideological battleground was northern California. Here the Forest Service engaged light burning in what was less a dialog than a dialectic. Against light burning, district forester Coert duBois proposed "systematic fire protection." In 1914, after "retiring to his cell like John Bunyan," he wrote a classic treatise on fire suppression that provided over the next two decades the basis for fire planning throughout the national forests (duBois 1914a,b). Essentially duBois applied the industrial engineering of his day – Taylorism – to the problem of fire control. Each component was broken into its atomic parts, measured, timed, assessed, then reassembled into a machined whole. Its rigor made light burners appear like woolly-headed folk philosophers; the intensity of the controversy forced proponents to sharpen their concepts and lessened the opportunities for compromise. Except for allusions to Europe, there were no other counterexamples for critics to propose. Examples of fire protection from elsewhere in the world, with the exception of Canada, were ignored. America accepted what came to it but did not search, on its own, for new exemplars.

Behind the logic, moreover, lay a fervor that had been burned into the Service by its trial by fire in 1910. No one who had weathered the 1910 holocaust could seriously argue for more fire in the woods. It was a generation that was prepared for action, not contemplation – an era whose President, Theodore Roosevelt, advocated the life of "strenuous endeavor," whose greatest philosopher, William James, argued for a "moral equivalent of war" to be waged against the threatening forces of nature, whose most popular author, Jack London, wrote novels and stories about struggle in the wilderness, and whose Eastern Establishment was enamored of the Western Experience as a place to test manhood. The peculiar circumstances that combined Harvard gentility with Oklahoma cowboys into Roosevelt's Rough Rider regiment now sent Yale foresters to battle forest fires in the Far West.

But like a fire in a damp log, light burning rekindled every few years. In the early 1920's a panel of foresters convened to compare the two philosophies and judge between them. They ruled for systematic fire protection. Light burning became heresy; controlled burning, anathema. The Service even suppressed research findings that contradicted what had now become an ex cathedra declaration. Here again, as in India, chance intervened. The year designated for field experimentation was abnormally wet and the burns inconclusive; then, following the condemnation, drought returned and the 1924 fire season broke all records. Only the means at hand prevented a universal extension of systematic fire protection. Subsequently, when, in the 1930's, the Administration of Franklin Roosevelt, a gentleman forester, released enormous resources to the Forest Service, the Chief Forester announced a 10 A.M. Policy, which ordered all fires to be controlled by 10 A.M. the day following their report, or failing that, by 10 A.M. the subsequent day until extinguished. The means at hand determined the ends.

What had been possible now became mandatory. What was announced as an "experiment on a continental scale" became dogma. Controlled burning continued in scattered enclaves, without official sanction, condemned as the personal obsession of eccentric wildlife biologists, Indians on remote western reservations, and deranged southern arsonists. In America fire protection meant fire suppression. Fire protection steadily assimilated new lands. The one exception was the South, where controlled burning to reduce fuels became accepted practice in the 1940's; but this was considered another manifestation of southern exceptionalism.

The American strategy intensified after the Korean War, catching a rising tide of national affluence, a commitment to scientific research, and a cornucopia of war-surplus material that allowed forest firefighting to mechanize virtually overnight. The means remained sufficient to sustain the artificial ends to which they were put. By the 1960's, however, cracks appeared in the system. After formal fire protection came to Alaska in the late 1950's, there were no further regions to subject to a first-order fire protection system. All the public lands and every state had at least a minimum level of wildland fire protection; firefighting was no longer a frontier institution. The larger liabilities of the strategy became apparent. There were social and economic costs, a point of diminishing returns. Pouring more and more money into fire suppression did not lead to a corresponding diminution in burned area; expenditures rose regardless. Then there were the environmental costs. Mechanized firefighting frequently inflicted scars on the land; the effectiveness of fire suppression often allowed fuels to build up, stoking larger fires; and researchers recognized that fires were a natural phenomenon, such that it became difficult philosophically and legally to justify the expulsion of lightning-caused fires from wilderness. In the 1970's, the singular American strategy fractured into a pluralism of fire practices that sought a better balance between fire control and fire use.

14.5 An Australian Strategy: Bringing System to Burning Off

Australia was slower to develop bushfire protection. There was no denying the need. Fire – most likely anthropogenic fire – had helped a stage a biotic revolution that tipped the entire continent toward pyrophilia (Gill et al. 1981; Pyne in press). For at least 40,000 years Aborigines had torched nearly every environment into a matrix of burned corridors and various-sized patches within which lightning fire had to operate. On this, the hottest and driest of the vegetated continents, nomadic bands habitually walked around the countryside clutching firesticks that dribbled embers and inflamed whatever was dry enough to burn. "Fire-stick farmers," Rhys Jones labeled them – prehorticulturalists, who cultivated the landscape with fire (Jones 1969). To these uses Europeans added others. They burned to improve pasturage, to clear land, to clean fields of stubble, and not least of all to protect themselves from the bushfires that otherwise threatened to engulf them. By the time professional foresters arrived in the early 20th century, fire was considered inevitable, endemic, and useful, and formal fire protection misguided and impossible. "Burning off" appeared to be an inextricable part of the Australian physical and social environment.

Australian foresters had exemplars from Europe, North America, and elsewhere in the Empire. Both India and South Africa offered climatic cognates, and there were haunting similarities among their shared Gondwana forests; teak and sal, for example, resembled eucalypts in some important properties. The mobility of Empire foresters and the transnational character of forestry conferences put the experiences of India and Africa squarely before their Australian colleagues. So did exchanges with North America. California, too, had a mediterreanean climate, ferocious winds, a biota of grasses and woody scleromorphs, and a resident population intent on burning wherever and whenever it could. During his inspection tour in 1914, Sir D.E. Hutchins dismissed with scorn Australian protests that theirs was a uniquely fire environment. All Australia needed, he insisted, was what had reduced wildland fire in France, Germany, India, South and East Africa – system, purpose, commitment (Hutchins 1916).

The Empire strategy worked marvellously on protected forests. But the critical problem everywhere was with unprotected lands subject to folk fire practices and with exceptionally bad years that could readily incinerate the good effects of years, if not decades, of labor. Australian foresters struggled to apply the tenets of the Empire strategy to their reserved lands and to resolve the questions of bushfire protection on Crown or private lands. They were keen students of foreign experiences, but after the Great War they tended to look to North America (particularly Canada, another dominion of the Empire) rather than to Europe or India. They also tolerated a higher level of burning – deliberate and sub rosa – than did European or North American foresters. Lacking the resources to expunge fire, they accommodated it in practice even as they typically condemned it in theory.

Australia drifted. From the island continent there emerged nothing original, and the apology that Australian circumstances were unique lost much of its bite. India and America also experienced routine burning and frequent conflagrations, and they also sheltered populations accustomed to free access to fire and hostile to fire suppression. Foresters there also polarized around the proportional commitment to fire control and fire use. India's early burning debate was virtually indistinguishable from America's light-burning controversy, and both strongly resembled Australia's debate about burning-off. The light-burning controversy, one Australian noted with satisfaction, was so "practically similar" to what he confronted that he felt convinced that Australian circumstances "are not vital to the issues involved, nor do they affect the conclusions arrived at" (Kessell 1924).

The peculiar attribute about Australia was not its choice one way or the other, but its absence of choice. So, while India and British Africa moved towards early burning and America hammered out the tenets of systematic fire protection, Australia muddled through without a coherent, autonomous philosophy or system of its own. The federation of the Australian colonies in 1901 made not the slightest difference. By the time Australia established schools of forestry in South Australia and Victoria (1910), India had fought through the early burning controversy and America was poised to begin its light-burning equivalent. By the time Australia established its national forestry school (1927), early burning had entered the Indian *Forest Manual* and light burning in America had suffered a humiliating condemnation.

Each Australian State adopted variants on fire protection suitable to its circumstances. The critical battlegrounds were two. The southeast quadrant, where the vast majority of white Australians lived, lay in a kind of fire flume where drought and desert winds could combine to propel incredible holocausts on a roughly 10–13-year rhythm. The specter of the holocaust fire exaggerated arguments on both sides and inspired, after a fashion, a kind of fire fatalism. The southwest, by contrast, suffered less acute outbreaks and under the long tenure of Stephen Kessell, an Oxford-trained forester, adapted the classic principles of

Empire fire protection.The experiment was first attempted in jarrah forests, appeared successful, then disseminated throughout the forest reserves. So powerful was the Westralian *system* that foresters increasingly tolerated forms of fuel reduction burning, both for fuelbreaks and on surrounding unprotected lands. The difference between the two regions can be illustrated by their attention to California. Where Victoria (in the southeast) emphasized firefighting, Western Australia (in the southwest) studied the light-burning controversy and quietly adopted aspects of controlled fire for fuel reduction.

The pivotal events came in 1939. The Black Friday fires devastated the forests of Victoria, and the Second World War plunged isolated Australia into a global maelstrom (Stretton 1939). That double shock led, after the war, to a major revolution in bushfire protection. The postwar years introduced, too, a new generation of foresters – Australia-born, sympathetic to rural Australia, eager to reconcile rural fire practices with the purposes of forestry, distrustful of North American and European models, and haunted by the specter of the holocaust fire. They sought to rationalize and put into systemic form those traditional fire practices loosely known as "burning-off." Within two decades they evolved a distinctive Australian style of bushfire protection.

The revolution encompassed all aspects of bushfire protection (Luke and McArthur 1978). It inspired an institutional reformation that centered on cooperative fire protection through the use of volunteer bushfire brigades under the aegis of oversight councils. The Hume-Snowy Scheme in New South Wales, developed in 1951, became a prototype by which some form of protection could be extended from State forests to the rural countryside. The unprotected Crown lands, in particular, were notorious for breeding damaging fires. Soon afterwards, the Forestry and Timber Bureau (Commonwealth) hired A.G. McArthur, formerly fire officer on the Hume-Snowy Scheme, to inaugurate a program of bushfire research; later, the Commonwealth Scientific and Industrial Research Organization (CSIRO) established a complementary bushfire research unit. The architects of the revolution made a strategic commitment to controlled fire as a method of fuel reduction; the research agenda sanctioned that commitment. Meanwhile, Western Australia, faced with a massive backlog of burning, adopted controlled fire *within* protected blocks. A Royal Commission called upon to investigate the 1961 Dwellingup fires confirmed the logic of that extension (Rodger 1961). The Westralian system had, in effect, expanded to incorporate new fire practices, another prototype for a national system. Throughout, skepticism over "North American" firefighting – with its aerial hardware and unblinking addiction to suppression – served as a useful counterexample, and bestowed on the revolutionaries a glow of nationalism. In 1965 researchers devised incendiary capsules capable of igniting vast expanses of wildlands during favorable conditions. The means were now at hand to encompass all of Australia under a universal system. Aerial ignition stood symbolically for the Australian strategy's emphasis on fire use, as aerial suppression did for the North American strategy of fire control.

In retrospect it is increasingly apparent that the Australian strategy was the invention of a specific generation of foresters who sought to adapt rural fire practices to the ends of forestry and to use the political and scientific resources of public forestry and who sought, in return, to extend formal fire protection back over rural Australia. But rural Australia eroded economically and fragmented into a mosaic of new lands – the urban bush, Aboriginal reserves, parks and nature preserves. A generic method for extensive management was of much less value than specific methods that targeted particular practices for particular outcomes. As Australia promoted an environmental pluralism to match its new social pluralism, it demanded a spectrum of fire practices. The Australian strategy came under attack from an increasingly urban population unfamiliar with the rituals of rural burning and from scientists disenchanted with the single-minded broadcast burning of valued nature preserves for fuel reduction. The reformation of American fire policy and programs gutted the Australian strategy of a vital negative exemplar, against which Australians had defined their own invention. By the late 1970's the Australian strategy was in an advanced state of dieback.

14.6 Stirring the Ashes: Concluding Thoughts

The advent of modern wildland fire protection went hand in glove with the advent of European imperialism and industrialization. The essence of the forest reservation was to substitute one kind of use for another, and in the process forest officials often denied traditional folk usage in the name of forest protection. Perhaps no element of this conflict was more volatile or immediate than fire. Traditional fire practices were mandatory for the traditional usage of woods and pasture. To deny access to fire was often to deny access to many natural resources, which were otherwise in biologically useless forms. The shock encounter between traditional and industrial defined the heroic age of wildland fire protection.

Everywhere European foresters, particularly French and Germans, pioneered the strategies, the techniques, and the philosophy behind the reformation. Foresters became of necessity – and often to their surprise and consternation – firefighters and fire strategists. They encountered scenes in overseas colonies that had not been played out in central Europe for decades, if not centuries. From indigenous materials and inherited concepts, they compounded new fire practices. Foresters were, as Hutchins declared of them, "soldiers of the State." They belong with other engineers of the global economy, and with other empire-builders of European colonialism. They spearheaded that rationalization of resource use that became known as conservation.

A militant forestry rationalized fire practices and endowed them with a rigor of purpose and a moral intensity that had never before existed. Foresters found laissez-faire pandemonium and created system; where folkways had guided practice, they imposed science; on land ruled by faction, they brought uniformity. They built empires not by fire and sword but by overthrowing fire and axe. Like the ancient Romans, however, each victory – every new territory aggregated to the old – compelled them into further expansionism. They could not guarantee protection of their civilized enclaves so long as wild tribes roamed the surrounding lands free to launch fiery raids against them. To defend forest reserves it was necessary to defend the surrounding lands, and foresters increasingly laid claim to a fire protectorate of vast, almost limitless dimensions. The thrust of their program was to shatter traditional patterns of fire use and to install a fire protection system where none had previously existed.

But fire exclusion was unworkable over any length of time unless it was accompanied by the kind of intensive silviculture and the large-scale landscaping that had emerged out of centuries in central Europe. Some accommodation with wildland fire was essential for wildlands managed as wildlands. In all three examples – British India, America, and Australia – fire control had to compromise with fire use. In both America and Australia new patterns of fire use have become prominent in the form of prescribed burning. Fire use has gone beyond fuel reduction and become a valued instrument to advance biological goals. Ecologists have replaced foresters; a new environmentalism has challenged the moral fervor of the old conservationism; habitat, wildlife, and wilderness have supplanted commercial forests as the nuclear core of public fire protection; what had been a singular system of fire protection has evolved into a pluralism of fire practices.

The heroic age bestowed important legacies. It created institutions for fire management where none had existed before, none at least within the kind of context promulgated by industrialization. It formalized fire lore and bound it to modern science. It established national traditions in wildland fire, a style that would continue to shape fire practices long after the formative years had passed away. When the reformation came, it had an infrastructure within which to function, there was knowledge on which to build, and there existed practitioners skilled in field operations who it could redirect to new purposes. There were also inherent biases, often the result of chance events and personalities, that became the basis for national styles of fire management. The assumptions they embodied were revealed, in the end, as logical truths, not operational truths. It was inherently true that a reduction in fuels could alter fire behavior and effects, as the Australians argued, or that rapid detection and initial attack could suppress virtually any new ignition, as the Americans insisted, but the real test of these precepts was whether they could be implemented in the field, under the actual environmental, political, and social circumstances they confronted.

If there were limitations inherent in each national strategy, so there are limitations in their application to contemporary analogs. It would be as inappropriate to transfer American fire management in toto to a developing country in the tropics as it was to transfer German fire management unaltered to North America. There are some universal principles that European, American, Australian, and British Indian fire history illustrate, and they almost certainly apply to any environment in which, as in the developing world today, traditional fire practices have broken out of their confinement with disastrous consequences – the conflict between industrial and traditional fire practices; the volatility of the transitional era, when mixed land usage and mixed fire practices combine in unstable compounds; the way in which fire practices interact with other anthropogenic practices and exotic flora and fauna to disturb ecosystems; the special role of foresters; and the pertinence of chance and personality in shaping ideas and institutions. The history of fire conservancy is not that of a template stamped relentlessly onto new materials, or of a life cycle of events tirelessly repeated, but of shared problems acted out idiographically. Their common denominator was a special generation of foresters.

But much is different, too. Industrializing nations are no longer political colonies, and foresters do not possess the clout – political, intellectual, moral – they could once claim. Concerns with biodiversity, global climate change, air quality, and acid rain have compounded old obsessions with timber and water and made fire an ecological phenomenon, not merely an exercise in forest protection. There are remarkable similarities between the fires that swept the Lake States in the late 19th century and those that have infested East Kalimantan and Rondonia in the 1980's, but if North American expertise is to be transferred, it is important to understand the whole – not just the machinery of fire suppression, but the philosophy of fire management. It is critical to see the "fire problem" as one dimension in a complex of human disturbances. Fire per se is not the cause of devastation; and fire exclusion will not, by itself, restore a damaged biota. Wildland fires, moreover, are only one manifestation of a dramatic reconstitution in the Earth's combustion regime that has witnessed important changes in chemistry, geography, seasonal timing, and biology of fire. Fire is a multiple, not a singular, phenomenon. When exploited by humans, it rarely acts alone but rather in concert with other human manipulations – an often irreversible synergism.

There is irony in the saga of fire conservancy. Contrary to the expectations of many foresters, fire did not vanish once traditional fire practices were routed. In too many of the critical environments fire was too integral to be dismissed, and traditional fire use too valuable to ignore. Intensive management in the classic European mode was impossible. Fire protection was not a one-time act, like surveying a forest boundary, but as much an on-going, inexpungable facet of wildland administration as silviculture. Foresters could not shed their experience with fire; nor did they wish to. Particularly in America and Australia firefighting became the romance of forestry, and foresters became the principal depositories of fire lore and practice in their societies. Instead of conserving wildlands from fire, fire conservancy as practiced by modern forestry conserved fire within wildlands. This preservation of wildland fire has turned out to be an accomplishment every bit as significant as its selective extinction.

References

Anderson ML (1963) A general history Scottish forestry, Vol 1. Nelson, London

Barrett L (1935) A record of forest and field fires in California from the days of the early explorers to the creation of forest reserves. US Forest Service

Carron LT (1985) A history of forestry in Australia. Aust Nat Univ Press, Canberra

Clar CR (1959) California government and forestry, Vol 1. Division For, Sacramento
Deckert E (May 1911) Forest fires in North America: a German view. Am For 17:275–279
duBois C (1914a) Organization of forest fire control forces. Soc Am For Proc 9:512–521
duBois C (1914b) Systematic fire protection in the California forests. US Forest Service
duBois C (1957) Trailblazers. Stonington, Connecticut
East WG (1951) England in the eighteenth century. In: Darby HC (ed) An historical geography of England before A.D. 1800. Univ Press, Cambridge
Eldredge F (1977) Interview. In: Maunder ER (ed) Voices from the south: recollections of four foresters. For Hist Soc, Santa Cruz, California
Fernow BA (1907) A brief history of forestry. Univ Toronto Press, Toronto
Forestry Commission (1934) Journal of the Forestry Commission 13:9–29
Gill AM et al. (eds) (1981) Fire and the Australian biota. Aust Acad Sci, Canberra
Graves Henry (1914) Report of the forester for 1913. Gov Print Office, Washington, DC
Greely W (1972) Forests and men. Arno, New York (reprint)
Hutchins DF (1916) A discussion of Australian forestry. Dept For, Perth
Jones R (1969) Fire-stick farming. Aust Nat Hist 16:224–228
Kayll AY (1967) Moor burning in Scotland. In: Proc 6th Ann Tall Timbers Fire Ecol Conf. Tall Timbers Res Stn, Tallahassee, Florida, pp 29–40
Kessell SL (1924) The damage caused by creeping fires in the forest. For Dept Bull 33, Perth
Koch E (1944) Region One in the pre-regional office days. In: Early days in the forest service, vol 1. US For Serv, Missoula, Montana
Lane-Poole CE (1920) Notes on the forests and forest products and industries of Western Australia. For Dept, Perth, Western Australia
Luke RH, McArthur AG (1978) Bushfires in Australia. Aust Gov Print Service, Canberra
Marshall W (1970) Rural economy of the West of England, 2 vols reprint of 1796 edn, New York
Pinchot G (1898) Study of forest fires and wood protection in southern New Jersey. Ann Rep Geol Surv New Jersey (Trenton NJ), Appendix
Pyne S (1982) Fire in America: a cultural history of wildland and rural fire. Princeton Univ Press
Pyne S (in press) Burning bush: a fire history of Australia. Holt
Ribbentrop B (1900) Forestry in British India. Office of the Superintendent of Govt Print, Calcutta
Rodger GJ (1961) Report of the Royal Commission appointed to enquire into and report upon the bush fires of December, 1960 and January, February and March, 1961 in Western Australia. (For Dept, Perth
Rodgers A (1968) Bernhard Eduard Fernow. Hafner, New York
Shebbeare EO (1928) Fire protection and fire control in India. 3rd British Empire For Conf, Canberra
Simmons IG et al. (1981) The Mesolithic. In: Simmons I, Tooley M (eds) The environment in British prehistory. Duckworth, London
Stebbing EP (1923) The forests of India, 3 vols. J Lane, The Bodley Head, London
Stretton LEB (1939) Report of the Royal Commission to inquire into the causes of and measures taken to prevent the bush fires of January, 1939. . . Govt Print, Melbourne

15 The Contribution of Remote Sensing to the Global Monitoring of Fires in Tropical and Subtropical Ecosystems

J.-P. MALINGREAU[1]

15.1 Introduction

Recent developments in the use of satellite observation technology have led to new insights into the dynamics of vegetation communities and of large-scale disturbances occurring therein. In particular, data derived from thermal sensors carried on board space platforms have provided regular information on the occurrence and distribution of fires in tropical and subtropical ecosystems. Whilst the role of fire in vegetation communities has been extensively studied by plant ecologists, it is only recently that global views of the phenomena have been directly acquired. Satellite data shed additional light on fire distribution and on ecological interactions leading to vegetation burning. In some cases, they may force adjustments in current views on worldwide fire regimes.

Natural fires, as well as those associated with human occupation, have been a dominant factor in the establishment and maintenance of savanna ecosystems, where burning has long been commonly associated with traditional land clearing, hunting, and grass regeneration for grazing (Daubenmire 1968; Mueller-Dombois 1981; Menaut 1983). In the forest, biome fire is linked to shifting cultivation practices which use fire as a tool for land clearing and biomass mineralization. The recent boom in deforestation activities linked to agriculture and logging has led to the rapid expansion of fires in forested areas where burning was hitherto a relatively rare occurrence. Events of the last few years have shown that fires can radically affect large areas of semi-seasonal or even evergreen broadleaf forests of the tropics. The spectacular developments of these last years in the Amazon or in South East Asia have been well reported. They should, however, not hide the fact that most of the burning in the tropics remains unreported. Because of the size of the area increasingly involved, biomass burning in the tropics is recognized as a major agent of change in atmospheric chemistry (CO_2 and trace gas release). The phenomena has received growing attention in the framework of what is now commonly called the process of "Global Change" (Woodwell 1984; Seiler and Crutzen 1980; Crutzen et al. 1979). As noted by Robinson (1988), "among the way in which humans affect biosphere-atmosphere interactions fire is outstanding for the extensiveness of the transformation caused in proportion to the amount of effort applied". All these recent developments have led to a renewed interest in the role of fire in tropical ecosystems and in the analysis, in a new light, of emerging

[1]Institute for Remote Sensing Applications, Joint Research Center EEC, 21020 Ispra (Varese), Italy

interactions between man, climate, and burning. Despite some limitations, remote sensing techniques can open new vistas on the spatial and temporal dimensions of the mechanisms at play.

It is the objective of this chapter to review the current state of the art in remote sensing of fire in the intertropical belt. The occurrence and distribution of fires as seen on satellite data are related to vegetation conditions and land use practices. The review is followed by a survey of satellite data covering various parts of the tropics where important fire-related events have been recently taking place. This provides an opportunity for a comparison between fire patterns as found in different parts of the tropical belt. The contribution of biomass burning to atmospheric chemistry is reviewed in detail in a companion chapter by Kaufman et al., who describe how the relevant information can be derived from satellite data. Proposals related to the structuring of a global tropical fire accounting approach by satellite are summarized in the concluding section.

15.2 Satellite Monitoring of Vegetation Dynamics

Remote sensing analysis of fire in the tropical biomes cannot be separated from an examination of the vegetation conditions prior to or at the time of burning. Much of the research on fire in natural ecosystems is actually concerned with such an analysis. In this perspective, burning is considered as one factor in a particular vegetation disturbance regime. In other instances, burning can be considered as an anomaly imposed by external determinants such as the conjunction of exceptional climatic conditions and human activities. In both cases an understanding of ecosystem linkages is a requisite for the analysis. It has now been well demonstrated that satellite remote sensing techniques can provide important elements of information on vegetation dynamics over large areas (Justice et al. 1985). Current approaches to continental and global vegetation monitoring rely on the collection of radiometric data at regular intervals of time, on the derivation of parameters of interest, on the processing of the data in consistent time series and on the analysis of spatio-temporal patterns revealed by the data sets. The success of the approach relies on a right combination of the system characteristics. These are related to the dimension of the coverage, the frequency of acquisition, the ground resolution, and the nature of the measured parameter.

The "global" dimension of fire and burning in ecosystems dictates that, above all, the satellite coverage must be exhaustive. While, because of the nature of the satellite's orbits, most of the earth's surface is systematically covered, there are still portions of the world which cannot be recurrently covered by satellite observations because of the lack of a local receiving station. These are necessary to capture at the time of overpass the data which cannot all be recorded on board. Systematic gaps may thus exist in the earth coverage by a particular instrument.

The second requirement is that data be acquired at intervals of time compatible with changes in the vegetation canopy. When related to plant distribution (i.e., succession, colonization, range shift, etc.) a long-term time frame can be considered (i.e., annual, decennal). A higher frequency of data is, however, necessary for observations related to plant phenology or to processes such as photosynthetic activity, evapotranspiration or in the event, burning episodes. Constraints associated with weather conditions over the area of observation must also be taken into account. The probability of obtaining satellite images not obscured by clouds is partly determined by the frequency of data acquisition passes. Maximizing this latter is thus one way to reduce the cloud problems. Microwave sensors which can see the surface regardless of the presence of clouds are not yet in the operational domain.

The resolving power of the instrument (ground resolution) must be compatible with the level of detail desired. Current technology allows the acquisition of earth surface data at a resolution of up to 10 m, for the SPOT Panchromatic Instrument for example. Other satellites provide a ground resolution of 1 km. The choice of resolution is crucial for vegetation monitoring and, given all the constraints, the maximum ground resolving power is not necessarily retained. Scene variability will usually increase with ground resolution and thus may unnecessarily complicate the interpretation (Irons et al. 1986). By integrating the spectral response of a variety of features present in the field of view, large "pixels" (the image resolution element) present the advantage of providing a level of information compatible to the needs of a global analysis. Furthermore, the volume of data is directly proportional to the size of the area, to the frequency of data acquisition and to the resolving power of the instrument. This latter parameter is the only one which, in the limits of instrument availability, can be easily controlled, since the area to be covered and the frequency of acquisition are dictated by the application itself. Reducing the ground resolution to an acceptable level is thus of primary concern in remote sensing for global vegetation monitoring.

Finally, the parameter measured by the sensor must be linked to a relevant characteristic of vegetation. The radiometric characteristics of the canopy must be interpreted in terms of physical parameters such as biomass, spatial distribution, composition, etc. Two such measurements which are currently used are the vegetation index (a linear combination of red and near infrared reflectance) and the surface brightness temperature (emittance in the thermal infrared). The first measurement is related to the green biomass and canopy geometry, the second to surface canopy temperature and thus thermal radiation balance.

A compromise between the above requirements must be sought for global vegetation monitoring. At present the optimum, yet not perfect formula is to be found in the use of the data derived from the AVHRR sensor carried on board the NOAA series of satellites. The Advanced Very High Resolution Radiometer is an instrument designed for meteorological applications, yet it has been found to be particularly relevant to our concern since it leads to information related both to the condition of the vegetation canopy and to particular events such as fires. The technical characteristics of this sensor are summarized in Table 1 (see also Kidwell 1984). It is worthwhile to note here

Table 1. Characteristics and status of the NOAA/AVHRR systems

TIROS-N	launched October 1978, NASA funded protoflight				
NOAA-6	launched June 1979, NOAA funded				
NOAA-7	launched June 1981, NOAA funded				
NOAA-8	launched March 1983, NOAA funded				
NOAA-9	launched December 1984, NOAA funded				
NOAA-10	launched September 1986, NOAA funded				
NOAA-11	launched 88				
Coverage cycle	9 days				
Scan angle range	±55.4°				
Ground coverage	2700 km				
Orbit inclination	98.8°				
Orbital height	833 km				
Orbital period	102 min.				
IFOV	1.39–1.51 mrad				
Ground resolution	1.1 km (nadir); 2.4 km (max. off-angle along track); 6.9 km (max. off-angle cross track)				
Quantization	10 bit				
Equatorial crossing	Des.	Asc.			
(at launch).	07.30	19.30	(NOAA-6, NOAA-8 and NOAA-10)		
	14.30	02.30	(NOAA-7 and NOAA-9), NOAA-11		
Spectral channels	1	2	3	4	5
Spectral range (μm)	0.58–0.68[a]	0.725–1.1	3.55–3.93	10.3–11.3	11.5–12.5[b]
NOAA-6	Deactivated 5 March 1983 Reinstated 22 June 1984, Deactivated 31 March 1987				
NOAA-7	Deactivated 6 July 1987				
NOAA-8	Deactivated 12 June 1984				
NOAA-9					
NOAA-10	Operational status				
NOAA-11	Operational Status				

[a] Channel 1 range on TIROS-N: 0.55–0.90.
[b] Not on NOAA-6, NOAA-9 or NOAA-10.

those characteristics most suited to fire and vegetation monitoring in the tropics.

1. The high frequency of coverage (near-daily) is an advantage for overcoming poor atmospheric conditions. Good coverage of areas otherwise difficult to observe can be obtained. Daily observations can be compiled in weekly or decadal summaries for monitoring changes in the plant canopies. Consistent data sets of AVHRR data now exist for the entire globe since the early 1980's and their systematic collection is on-going. Their current availability is summarized in Table 2.
2. The visible and near infrared bands of the instrument can be combined in vegetation indices related to the photosynthetic activity of the green biomass (Sellers 1985). Time series of such indices can be studied with respect to phenology and climatic events (Justice et al. 1985; Goward et al. 1985; Malingreau 1986). These wavelength regions are, however, particularly sensitive to smoke, water vapor, and aerosols in the atmosphere.

Table 2. Type and availability of AVHRR data

	Source	Resolution	Repeat period	Area covered	Channel
LAC	Request at NOAA	1.1 km	Daily	Selected	All
HRPT	Local	1.1 km	Daily	In range	All
GAC	NOAA	4 km	Daily since 1981	World	All
GVI	NOAA	15 km	Weekly since 1982	World	NDVI

3. The thermal sensors of the instrument record changes associated with physical transformations in the vegetation canopy. They can, for example, detect the changes in surface temperature associated with the replacement of an evergreen tree cover by a more seasonal vegetation type (Tucker et al. 1984; Malingreau and Tucker 1987). The same parameter is also intimately linked to water stress in the vegetation (Becker and Seguin 1985).
4. The thermal channel centered on 3.8 microns is particularly suited for detecting elevated heat sources which indicate the presence of active fires (Matson and Dozier 1981). This characteristic is described in more detailed in Section 4. Channel 3 of the AVHRR also has a good transmittance through water vapor and haze or smoke, and is thus particularly suited for observations in tropical areas.
5. The original resolution of the AVHRR data (1.1 km) is adequate for regional observations (HRPT or LAC – Local Area Coverage formats – see Table 2). The sampling at 4 km resolution (GAC – Global Area Coverage format) can form the basis for worldwide monitoring of vegetation. Detailed vegetation mapping should obviously not be attempted at such resolutions.
6. A growing number of receiving stations are being operated to receive the AVHRR-HRPT data. Currently, the following stations provide the coverage of most of the intertropical belt: New Orleans (USA), Cachoeira Paulista (Brazil), Maspalomas-Canary Islands (Spain), Niamey (Niger), Nairobi (Kenya), La Reunion (Fr.), Pretoria (Sth Africa), Dakha (Bangladesh), Bangkok (Thailand), Kuala Lumpur (Malaysia), Jakarta (Indonesia), several in the P.R. of China, and Alice Springs (Australia).

15.3 Fires in Vegetation – Data Needs

An exhaustive review of the data needs related to the study of fire in tropical vegetation is beyond the scope of this chapter (for a review of fires in tropical ecosystems see Mueller-Dombois 1981). It is proposed instead to examine here how a basic information set can be extracted from remote sensing observations in support of a global fire ecology assessment. In such a framework, a wide range of parameters on vegetation, biosphere-climate interactions and land use practices are to be considered. Linkages between biomass burning and those other system components are illustrated in the flow chart included in Fig. 1. The drawing also emphasizes the growing relevance of studying fire in the broad context of global change. Synoptic and repetitive views offered by satellite sensors are particularly suited for such a task. Remote sensing data can in addition attract attention to potential feedback mechanisms associated with expanded burning in the tropics.

This framework is used to list the various fire-related parameters of interest which one could hope to derive from remote sensing. Many of these parameters are discussed in the text.

1. Active fires
   – detection and count
   – temperature (e.g., smoldering – 400–500° K, flaming – 800–1200° K)
   – smoke characteristics.
2. Geographical distribution
   – location of fires

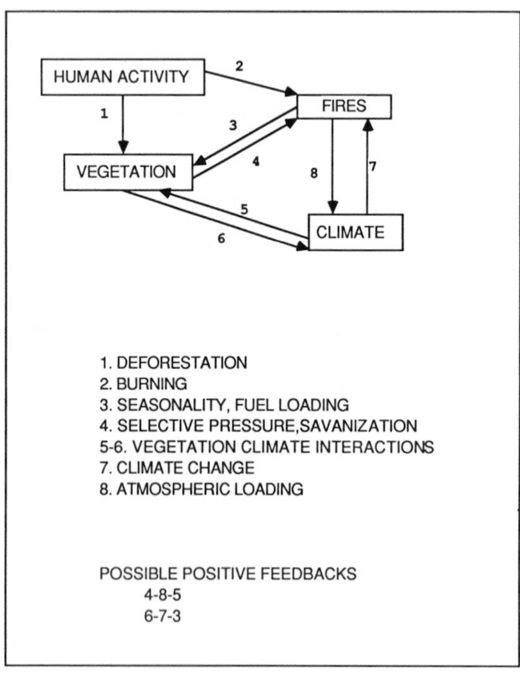

Fig. 1. A framework for fire analysis in tropical ecosystems

   - density
   - distribution patterns.
3. Temporal dimension
   - spatial dynamics (movement, expansion)
   - seasonal evolution and relationships with climate parameters
   - return period.
4. Target vegetation
   - type and biomass
   - seasonality, phenology
   - regrowth, regeneration after fire.
5. Land use
   - type of land use practices, trends.
6. Impact
   - area affected by burn
   - severity of burn.

In the perspective adopted here the need to carry out these measurements in a systematic fashion throughout the intertropical belt must again be emphasized. Such a scale requirement has important implications in the selection and development of the appropriate approach.

15.4 Fire Detection Using the AVHRR Instrument

Measurements provided by the thermal channels of the AVHRR sensor can be used to detect fire in vegetation through the effect of combustion on radiative temperature. Depending upon its temperature and size relative to the ground resolution of the instrument, an active fire will increase the pixel brightness temperature above that of the background. The term radiant or brightness temperature is used here because radiance measurements are not transformed into actual target temperature; this would indeed necessitate a knowledge of ground emissivity and of atmospheric path absorption/emission. The AVHRR channel centered around the 3.7 microns (Ch 3) is more sensitive to heat sources than the other channels in the 11-micron range (the wavelength of peak radiant power decreases with temperature according to the Wien's Displacement Law) and is thus most commonly used for fire detection (Matson and Dozien 1981). Recent applications to biomass burning have for the major part made use of this unique, and certainly unforeseen, characteristic of the AVHRR sensor.

There are, however, difficulties associated with the application of the approach. These have to be properly evaluated lest errors of interpretation occur. Since the instrument saturates at 320 K (Matson and Holben 1987) it cannot provide information on the fire temperature itself which is commonly higher. Bare soil surfaces, especially in the tropics and in early afternoon (at least for the NOAA 7,9,11), can also have a radiative temperature higher than the saturation level (Cass et al. 1984); confusion can then occur with active fires. In addition, at 3.5–3.9 microns some solar radiation is still reflected and measured by the sensor. While the contribution of this reflected component is limited for vegetation (5% according to Wong and Blevin 1967) it can lead to pixel saturation over highly reflective surfaces. This is the case for some types of clouds and is particularly apparent in areas peppered with small and individualized cumulus clouds (Malingreau et al. 1985). In this case, the sunward edge of the cloud can show a saturated reading due to high reflectance in contrast to the other edge which will show a low temperature. Algorithms based on a combination of AVHRR channels can be used to somewhat eliminate such problem and exclude cloud pixel from fire counts (Kaufman, pers. commun.).

The separation of fire from hot soil background areas is, however, not always straightforward. An obvious way to proceed is to compare successive passes which should show the transient nature of the fire versus the permanence of the warm soil. In some cases warm bare ground also tends to occupy several contiguous pixels as compared to fires, which will form a dotted pattern. The presence of smoke is, of course, an indicator of burning, but smoke is not always visible. Comparison with the radiant temperature in other thermal infrared wavelengths may also be used (Flaningan 1985; Kaufman 1988; Laporte and Malingreau 1989) although thresholding techniques clearly have to be adapted to the background radiance emitted by surrounding land surfaces. While in a relatively "cool" forest background fire detection is straightforward, it cannot be done in a xeric savanna environment where in the dry season surface temperatures can easily reach 40–50 °C. This lack of contrast is also observed when fires are found in or at the edge of large burned-over areas which are still very warm (Laporte and Malingreau 1989). This limitation restricts the application of the AVHRR data when working in the dry savanna environment, where fire is a major regulating feature. Night observations, also provided by the NOAA satellite, could offer an alternative approach if it were not for the fact that controlled burning associated with agricultural activities is likely to be put out at night. The measurement of post-burn scars is another alternative to biomass burning assessment (Vickos 1986; Langaas and Muirhead 1988 and Langaas this Vol.).

The spatial resolution (1.1 km at nadir) and the particular calibration (saturation at 320 K) of the AVHRR instrument have additional consequences for the interpretation. First, it is not possible to distinguish what portion of the pixel is occupied by the fire. It is believed that the radiometric temperature recorded by the instrument is proportional to the product "fire fraction – fire temperature". In other words, subpixel resolution but intense fires will be detected (see for examples gas flares in Matson and Dozier 1981, or straw burning in Muirhead and Cracknell 1985). Inversely, extensive but low temperature fires will also mark the pixel. Finally, let us note that because emission increases in a steep nonlinear fashion with temperature (Planck's curve) the "average" temperature measurement of a pixel will be strongly weighted by the maxima present in the field of view. The result is that a wide range of fire conditions will be similarly expressed at the pixel level. This can obviously lead to a "fire pixel" count, but not to a measure of the burning area. Indeed, in the absence of spatial information it is not correct to multiply the number of

AVHRR fire pixels by an area coefficient to obtain the surface of the burn. Recent field observations in the savanna woodland of Guinea showed for example that a large number of small fires lit by farmers (mainly by farmer's children, apparently) are indeed expressed in an aggregated fashion on the AVHRR scene (27.2.89). It is found, in this case, that the satellite measurement underestimates the number of individual fires but overestimates the area burned (Malingreau and Gregoire 1990). Possible approaches to subpixel fire analysis have been proposed (Robinson 1988; Kaufman, this Vol.). Since most fires in the tropical belt are related to controlled agricultural burning, it is essential to realize that fire counts in those conditions will be especially sensitive to lower spatial detection thresholds.

Finally, another sampling aspect is introduced by the time of satellite overpass. The 14.30 h equatorial crossing probably ensures that for most ecosystems observations are made at the peak of burning (high wind activity, low atmospheric moisture, dry fuel, and peak of field work). However, given the fact that some fire episodes are short and can be started at any time of the day, the afternoon pass must be considered as a temporal sample. Until further information is available on specific ecosystems, it is difficult to assess how representative it is. Other satellite passes (NOAA-11 at 02.30, NOAA-10 at 07.30 and 19.30) can be exploited but this approach has not been used extensively. A comprehensive fire monitoring system could also be based on a geostationary satellite providing data at regular time intervals throughout the day. Currently, the existing meteorological satellites (GOES, METEOSAT, GMS) do not have the appropriate ground resolution in the thermal bands.

The above shows that an overriding concern with the AVHRR approach is the need to evaluate and correct for biases related to fire size, duration, diurnal distribution, and to environmental effects. The understanding of the relationships between AVHRR measurements and fire behavior still present serious gaps which should now be filled by systematic testing in a range of ecosystems. The task is considerable, given the possible variables to take into account and the difficulty in studying fire in field conditions. Instrument-related limitations will be reduced in future earth observation sensors such as the MODIS which will have a 500-m resolution, a temperature saturation of 700 K at 3.75 microns and of 1100 K at 11 microns (NASA-EOS 1986). These are, however, experimental instruments. Efforts are currently made to promote improvements in the sensitivity of future AVHRR instruments (AVHRR K,L,M – Holben pers. commun.).

15.5 Fires and Environmental Conditions

The occurrence of fire events is related to a conjunction of factors among which the climate and the vegetation characteristics are predominant. Land use practices are to be increasingly considered in the tropics as, in most instances, it is man who provides the ignition trigger. Climate (mainly rainfall in the tropical belt) is an overriding control factor in fire occurrence and frequency since it determines the succession of wetness and dryness period and thus of green biomass accumulation and fuel loading. It also influences soil microorganism activity and thus litter decomposition. The nature of the plant canopy (plant density, green biomass, combination of life forms) determines the fire potential. The condition of the canopy in terms of water stress or dryness at certain periods during the season integrates in a sense the climatic conditions leading to fire situations. Insights into such variations can be provided at biome levels by time series of satellite-derived parameters such as the vegetation index and surface temperature.

A vegetation index is a combination of radiometric values measured over a portion of the earth's surface. Among the many indices developed for vegetation analysis, the normalized difference vegetation index (NDVI) is the most commonly used. Literature references on its interpretation are abundant. Suffice to say here that by combining the near infrared (NIR) and the red reflectances [NDVI = (NIR-RED/NIR + RED)] the NDVI carries information related to the absorption of the red wavelength by the green plant and to the reflection of the near infrared wavelength by the plant canopy. Recent theoretical work on the significance of the normalized difference vegetation index has established that this parameter is closely related to the photosynthetically active radiation intercepted by the canopy (Hatfield 1983; Sellers 1985). Light absorption mechanisms and radiative transfer in the plant canopy are thus involved. It follows that such spectral measurements are related to plant processes as much as to the state (nature) of the canopy. Vegetation index data must thus be examined in a dynamic perspective. For the sake of discussion we will assume here that time series of vegetation indices are linked to green biomass photosynthetic activity and thus to phenological development. One data set used in the establishment of NDVI time series over selected areas of the tropics is derived from the Global Vegetation Index (GVI – at 15 km resolution) maximized over 3-week periods. Other time series are derived from the GAC 1981–88 data set over Africa (daily vegetation index maximized over a month). The maximization referred to above is a procedure whereby daily recordings are summarized over selected periods (i.e., a week or a month) to avoid the noise associated with poor atmospheric conditions (for a description of the methodology, see Holben 1986). The tropical rain forests of various continents being currently the site of much disturbance by land clearing and fire, we now examine series of the vegetation index data for that particular biome. The analysis of time series of vegetation index data is to be considered as an important component of the identification of long-term changes in ecosystems.

Typically, the satellite-derived vegetation index of a tropical rain forest tends to remain within a narrow range of values throughout the year, thereby indicating the evergreen nature of the plant community. The values found in the AVHRR data set are in the 0.4 to 0.5 range. Slight variations reflecting a seasonal pattern are, however, discernible. This seasonality is observed to various degrees in all the temporal series obtained over the tropical rain forest. Peaks of vegetation indices (usually early in the "wetter period") may be related to the phenology of individual species (i.e., leafing) although it is difficult to imagine how irregular changes in the physiognomy of isolated and often widely spaced individuals could affect the response over a pixel area of the order of a 1 km^2. It must be considered indeed that, at such resolutions, measurements are related to the vegetation canopy and not to individual species; however, phenological changes at the community level are not easily identified in the rain forest. Variations in the day time short wave radiations (0.3–3.0 μm) with the cloudiness level may be of significance for photosynthetic process (Pinker et al. 1980). One may expect the reduction in the in PAR to be translated in a lower vegetation index during the rainy season. This is, however, the season during which a slight but significant increase in vegetation index is noticed for many samples in the tropical belt. Temporal variations in evapotranspiration and the development of stress in the plant canopy may partly explain some of the variations in the NDVI. Facultatively deciduous species, which shed leaves only during periods of reduced soil water potential may then account for differences in the NDVI signals between years (Malingreau 1986). This latter hypothesis is reinforced by the fact that as one goes away from the ombrophilous rain forest per se toward the limits of the biome, the amplitude of the seasonal NDVI signal increases (see Fig. 2c). The satellite data for the Amazon Basin, for example, show that the transition between the evergreen and the seasonal forest (drought deciduous) follows a continuum along lines of enhanced seasonality in rainfall patterns. This continuum can also be assumed to be one of increased susceptibility to fire events especially in the presence of intensive human clearing activities and during exceptionally dry years. It is this last point which is of particular relevance to the present analysis.

Several cases are presented in Fig. 2. The first one relates to the dipterocarp forest of central Kalimantan. The GVI curve showing the monthly maximum vegetation index for the 1982–85 period is, as often in the tropical belt, affected by noise (high frequency variations) related to poor cloud conditions. A smoothing of the curve (low pass filter) somewhat reduces the amplitude of the series but at the same time allows the determination of consistent periods of changes in the vegetation index; these are identified by calculating the slope of the smoothed curve (Malingreau 1986). The main "growing" seasons identified here are consistently found around November to January for the 3 years on record. A sharp decrease of this vegetation index over East Kalimantan was observed in relation to the El Nino drought of 1982–83 (Malingreau et al. 1985). It indicated a progressive dessication of the forest and increasing tree mortality, which provided fuel for one of the largest tropical forest fires recently recorded (see Sect. 6).

The time series of vegetation index for the rain forest of the Central Amazon Basin present similar characteristics (Fig. 2b) but here, two "growing seasons" of different amplitude are observed, the main one in December and a smaller one in June-July (the curve is averaged for a 40 × 40 pixel area or 360,000 km^2). The area is characterized by a short dry season (less than 2 months). As one goes away from the central part of the basin, the seasonal character of the vegetation increases, as shown in Fig. 2c. Of particular interest is the larger amplitude of the

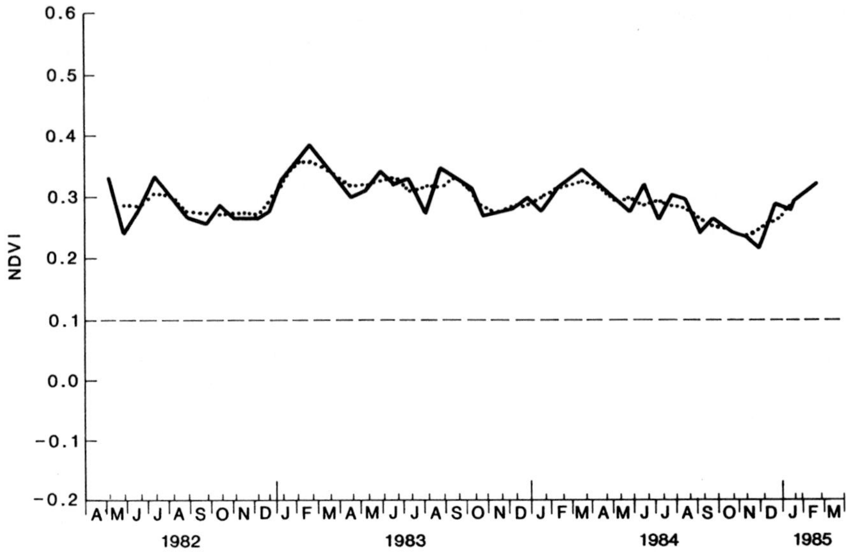

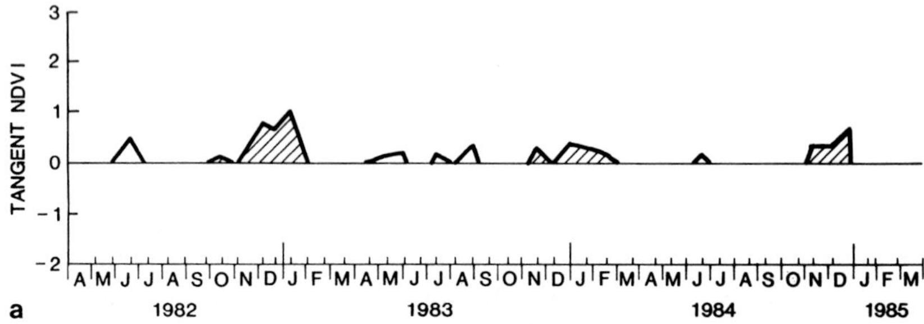

Fig. 2a-f. Time series of vegetation index data derived from NOAA-AVHRR for selected tropical ecosystems (Global Vegetation Index data are shown in tri-weekly maxima in **a**, **b**, and **c**. The data derived from the Global Area Coverage are shown in monthly maxima in **d**). **a** Dipterocarp forest of Central Kalimantan (Indonesia) 1982 to 1985 (n = 4 pixels of 15 km resolution). Normalized Difference Vegetation Index at the top (*dotted line* is moving average) and slope of the curve at the bottom (positive values only). (After Malingreau 1986). **b** Central Amazon Basin (n = 144 pixels at 15-km resolution). NDVI at the top and slope of the curve at the bottom. Note the bimodal pattern in "growing seasons". **c** Two forest sample points across the Amazon Basin from the moist evergreen (Xingu) to the semi-seasonal forest (Rondonia). A vegetation index curve for the cerrado is added for comparison. Note the impact of the 1983 drought (August). **d** Seasonal forest-woodland transition area in Central African Republic. (month 1 is Jan. 1982). **e** Rain forest of the Central Congo Basin (Zaire). The seasonality in the signal can be related to the bimodal rainfall distribution pattern. **f** Woodland-tree savanna transition in southern Guinea. The area is submitted to intensive burns associated with high fuel loading, contrasted seasonality, and land use practices

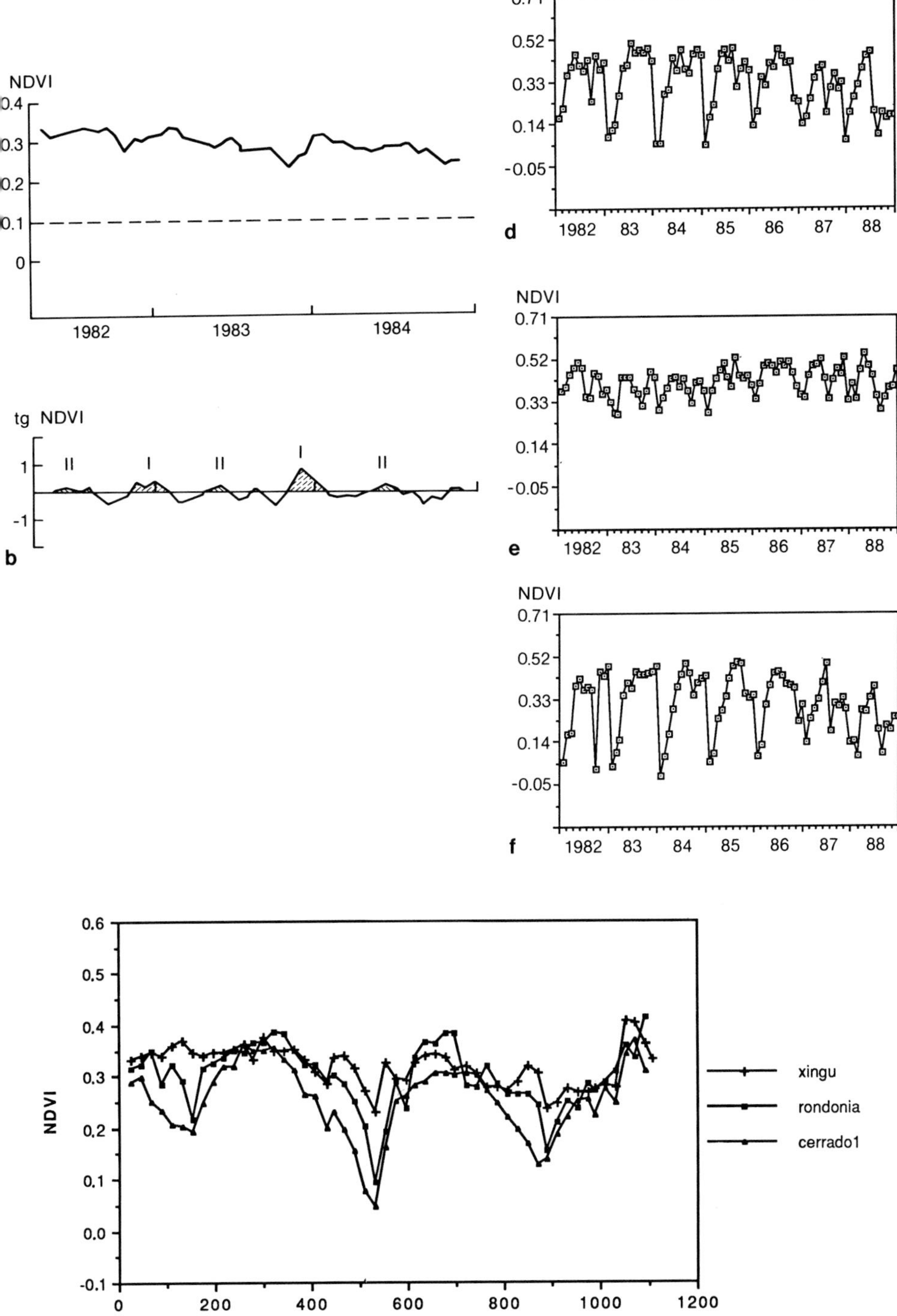

NDVI
0.4
0.3
0.2
0.1
0
1982
1983
1984
tg NDVI
1
-1
II
I
II
I
II
b
NDVI
0.71
0.52
0.33
0.14
-0.05
1982
83
84
85
86
87
88
d
e
f
0.6
0.5
0.4
0.3
0.2
0.1
0.0
-0.1
0
200
400
600
800
1000
1200
DAY (1=April 15, 1982)
xingu
rondonia
cerrado1
c

curve for 1983, a year which experienced a strong drought in the Brazilian northeast and north (Gasques and Magalhes 1987). In the southern part of the Amazon Basin, August marks the start of the main burning period and 1983 saw extensive burning in the state of Rondonia and in northern Mato Grosso. Given the sensitivity of the vegetation index to atmospheric loading by smoke and aerosols, it is likely that the NDVI values are somewhat artificially depressed during those burning weeks.

Time series of GAC data at 4 km resolution allow a better selection of sample points. Such data are now becoming increasingly available and examples are given in Fig. 2d,e. The sample points were selected at the boundary between the rain forest of the Central African Basin and the woodland savanna of the Central African Republic to the north. The two curves illustrate the seasonality of vegetation in two adjacent but ecologically different environments. North of the forest limit where fire activity is intense, (Fig. 2d) the seasonality is more pronounced than in the rain forest canopy itself, (Fig. 2e) indicating a higher fuel loading potential. It is to be noted, however, that the Congo Basin forest shows an amplitude of vegetation index variations higher than for the other rain forest samples reviewed above (the smoothing effect of the GVI sampling may partly explain this fact). For comparison, the case of the woodland savanna of West Africa has also been added. (Fig. 2f)

The progressive drying out of seasonal vegetation is also marked by an increase in surface temperature. This is due to the progressive warming of the plant canopy itself as evapotranspiration stress develops. Maximum temperature contrasts between evergreen and seasonal vegetation are observed at the peak of the dry season. This characteristic has been usefully exploited in tropical areas for discriminating on satellite data between forest and nonforest areas (Tucker et al. 1984; Malingreau and Tucker 1987). The watersheds of western Africa exhibit a similar thermal behavior when the dry season progresses, as shown by Grégoire (1989), who has attempted to exploit such characteristics for landscape stratification. It has been demonstrated that the thermal channels in the 11-micron range can also be used for such discrimination (Kerber and Schutt 1986). The possibility of using this thermal approach in conjunction with the vegetation index to determine a threshold of "flammability" is examined in more detail in the case studies dealing with the Amazon Basin.

15.6 Post-Fire Landscapes

As common experience will attest, burning imprints a strong radiometric signal on vegetation. Black carbon left on the ground has a high absorption coefficient which will cause a sharp decrease in the near infrared wavelength, as compared to area covered with green or drying vegetation. This characteristic has been exploited for measuring burned area in savanna environment using LANDSAT or SPOT data (Vickos 1986): AVHRR data at 1 km resolution have been used for measuring the impact of bush fires in Australia (Hick et al. 1986) and in

Senegal as presented by Langaas in this Volume. Burn scars will usually be characterized by a mosaic of patches of vegetation burned to various degrees and this makes the accuracy of fire impact estimation very much dependent upon the resolution of the observation instrument. In addition, many factors can influence the appearance of the burns; hot burns may leave whitish ash of relatively high albedo, moisture will darken the ashes and the charcoal, wind activity will change ash deposition patterns.

On low resolution satellite data burn scars present continental features which have been little investigated. A feature of such a kind is, for example, clearly seen across the African continent. A large band of reduced near infrared albedo extends almost uninterrupted at the end of the dry season from West Africa to the footslope of the Eastern African Ridge. This belt is located between the forest domain to the south and the tree savanna to the north. Its characteristics are further described in the section dealing with Africa. Regional burn patterns are also seen on AVHRR data of South East Asia.

Of particular interest for remote sensing approaches to the measurement of fire impact is the fact that the burned areas will show higher daily average temperature and amplitude because black surfaces are highly absorptive (Cass et al. 1984). This differential regional warming is clearly observed in the case of large-scale burns like those of West Africa and Kalimantan (Malingreau and Laporte 1988; Malingreau et al. 1985).

15.7 Fire in Tropical Ecosystems

15.7.1 The Amazon Basin

Frontier expansion in the Amazon Basin has experienced a rapid growth during the last decade. Increased economic hardship in the south of Brazil, policies of active colonization, opening of new roads, and financial incentives have combined to cause rapid deforestation in sections of the Amazon Basin. The situation in the Brazilian States of Rondonia, Acre, and Mato Grosso has received extensive coverage in the last few years, essentially because of the exceptional pace of deforestation activities which have taken place in those areas. For selected periods (i.e., after the opening or paving of new access roads) the clearing can even show an exponential rate of increase. Here, most of the deforestation can be linked to the expansion of agriculture done at the expense of the terra firme forest (high ground); while the amount and rate of deforestation vary greatly among regions, it is the drier transitional forest at the southern edge of the hylea which has received the brunt of the recent colonization movement. Pioneer farming is of the slash-and-burn type, and fire has become a normal feature throughout the settlement areas. Ranching also makes use of burning for shrub control. The burning of felled vegetation is greatly variable; because of weather uncertainties poor burns are common, creating serious problems for the planting of crops. The amount of aboveground biomass being burned in such land clearing operations is estimated by Fearnside at 30% (Fearnside 1985). The variability in burning conditions leads to many different fires with different temperatures and areas. Field observations indicate, however, that the controlled burning will usually take place over a few hectares of holding whether in a clustered fashion or along cutting lines.

AVHRR data were obtained for the 1982 and 84 to 87 dry seasons over the southern portion of the Amazon Basin (7 to 13 °S and 50 to 70 °W). This exercise led to the compilation of a large data set which had to be screened for cloud cover, image quality, sensor problems etc. The process allowed the elimination of many unsuitable day passes. The statistics of this data compilation exercise indicate that the temporal sampling is rather poor in tropical areas even with a daily pass throughout the dry season. The number of images acquired in July, for example (the most cloud- and smoke-free period of the year over the southern Amazon) is 5 in 1982, 10 in 1984, 15 in 1985, and 11 in 1986. Over the whole season the screening procedure eliminates more than 75% of the data. For fire detection, however, image quality must not necessarily be of the level required for forest nonforest discrimination because of the high contrast between the fires and the background. A fire count was done in 1987 by Setzer et al. throughout the dry season, per state in the Legal Amazon. The two bar charts derived from that study (Fig. 3) illustrate the constrasting burning calendars for two states occupied by different ecosystems. Given all the caveats associated with the interpretation of AVHRR thermal data, it is difficult at this stage to estimate the accuracy of the count.

Given the size of the territory to be covered and the diversity of deforestation features observed across the Amazon basin, a stratification was first carried out. It was based upon the detection and delineation on the AVHRR images of areas where activities associated with forest conversion were taking place (Malingreau and Tucker 1987). Activities such as road opening, mining, clearing for ranching or field crop planting identify the active deforestation fronts. The so-called "disturbance area" thus defined comprises a mix of marginal modifications and radical conversions of the forest canopy. The "deforestation" per se becomes a measure of the magnitude of these transformations. In this framework, fires are used as indicators of human activities, since their spatial distribution follows very closely the patterns of population movement and installation. The case of Rondonia is used to illustrate this last point (Fig. 4).

In the Rondonia settlement area, burning is associated with three main types of activities related to colonization: clearing of forest for establishment of new plots, clearing of old fields (bush, shrub burning), and ranch burning (bush control and grass regeneration). The spatial distribution of fires as revealed by the thermal channel of the AVHRR sensor is closely linked to such features. Along and beyond the feeder roads opened for facilitating the access to the colonization areas, fires are generally organized in the following manner. The first zone includes the old settlements, where most of the burning is done on agricultural fields and ranches. Some extension of agriculture in the forest remnants is also possible and the operation then reduces the forested area within

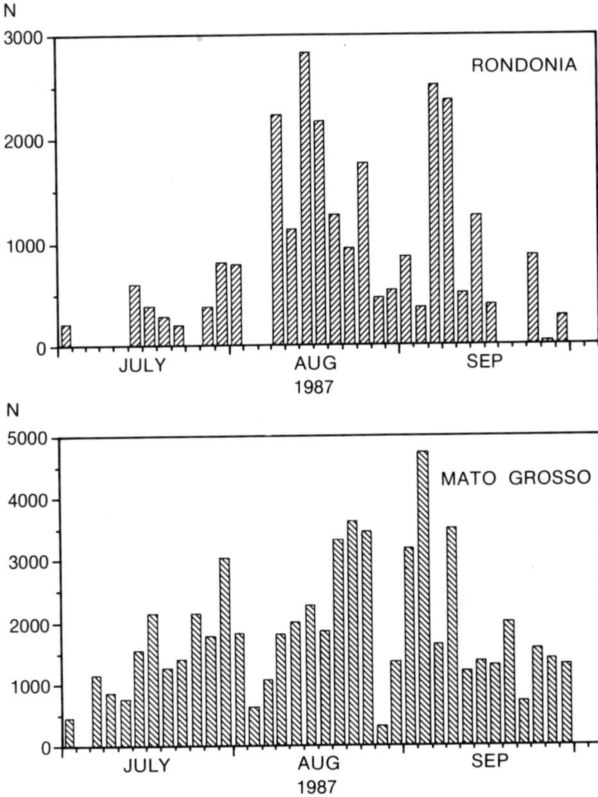

Fig. 3. 1987 Fire statistics derived from the AVHRR data over the state of Mato Grosso and the State of Rondonia in Brazil. (After data provided by Setzer et al. 1988). *N* in ordinate is the number of "fire pixels"

the settlement themselves. Fire pixels are very close together and may coalesce, background temperature increases (this complicates the satellite discrimination of fire pixels). The whole thermal pattern reflects a high land use intensity. In his study on settlement concentration, Fearnside indicates that at the level of the individual exploitation the upper bound to the density of land occupation may be reached after 5 to 6 years of occupation (Fearnside 1984). Away from the centers, feeder roads which facilitate the spread of deforestation to more remote areas are underlined by isolated fire pixels (Zone II), often close to the roads. They indicate the clearing of new plots in the original forest cover. The highly controversial highway 429, for example (lower left on Fig. 4), is underlined by a spread of isolated fires and openings in the forest canopy, which indicate the progression of the colonists. Further away, the activity of front runners off the feeder roads (probably along surveying lines) is revealed by isolated fires dispersed across the forest without apparent order. This was, for example, the case for the Machadin and Cujubin project areas where, before 1986, a fairly

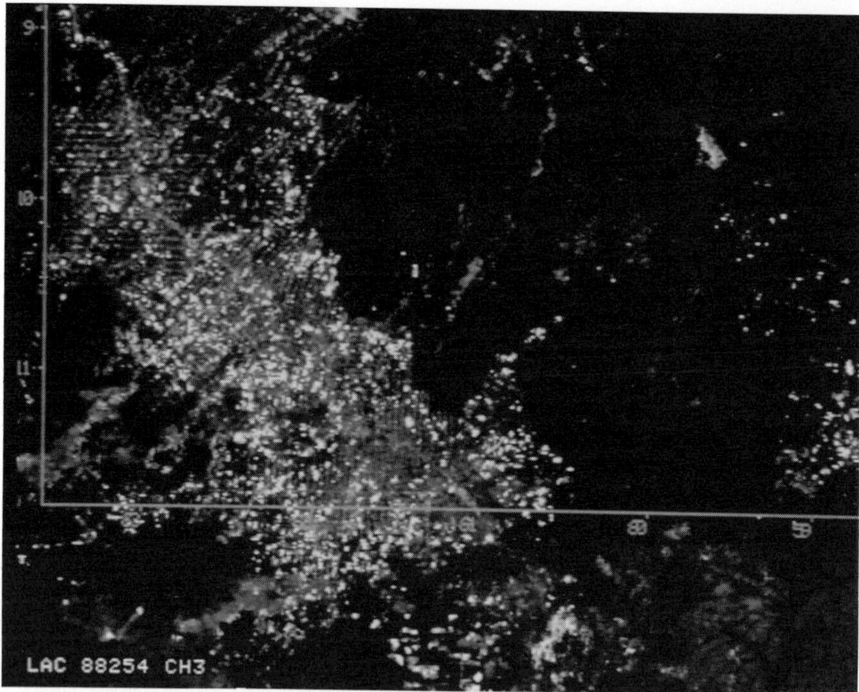

Fig. 4. Thermal AVHRR image taken in August 1985 showing the distribution of fire points (*white dots*) in and around the main colonization area of the State of Rondonia. Most of the fires are seen to be closely associated with the agricultural settlement pattern (*light areas*). Isolated fires and smaller forest openings are found away from the main access roads. The image is 500 km across

random spread of hot spots was observed. Once those projects were well established, a more organized spatial arrangement of fires and clearings took shape. This simplified model of fire distribution and expansion pattern can lead to inferences regarding the type of biomass burning (mainly bush, shrub, and grass in Zone I, original forest in Zones II and III) and regarding the risk of fire extension beyond the clearings in drier years (Uhl and Buschbacher 1985) (high risk in Zone I, lower risk in Zone II, and very limited risk in the sparse pattern of Zone III). Such stratification can also guide the spatial sampling for elaborating vegetation index time series, which could add additional information on seasonality and fuel loading.

Finally, an additional fire pattern observed in the Amazon Basin must be mentioned. It is directly related to the opening of large ranches like in northern Mato Grosso and Para (between the river Xingu and Araguia). Here, fires are used for ranch establishment or enlargement and for shrub control. On satellite data the drying prior the burning is marked by a progressive warming. The histogram of brightness temperature values associated with such landscape typically moves toward higher values with the season. At a given time, when vegetation is ripe for burning, the same histogram suddenly shows the appear-

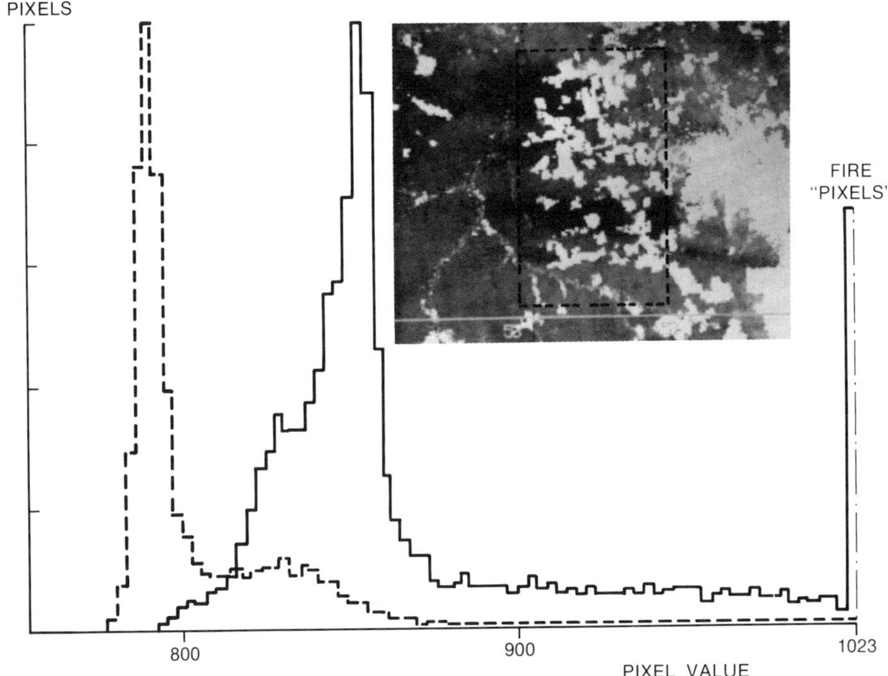

Fig. 5. The overall surface brightness temperature of a seasonal vegetation canopy increases as drought stress develops. Fires (saturated pixels) appear as dessication has reached an adequate level for burning. This is seen on the histogram of AVHRR Ch3 values for an area of large ranches carved in the seasonal forest of Para (Brazil). The histogram covers the area *inside the dotted line on the image*

ance of saturated pixel values in the middle infrared, indicating that burning has started (Fig. 5).

At the basin level, it can thus be seen that fire distribution is not haphazard but follows some organized pattern which can to some extent be predicted. However, with growing pressure on land and a possible evolution to uncontrolled burning situations, a constant monitoring of the whole area will become increasingly necessary. The drastic increase in the occurrence of fire in the southern Amazon Basin, clearly shown by the time series of AVHRR data obtained since 1982, raises several issues linked to the ecology of that part of the tropics. First, is the recent appearance of a fire belt along the southern periphery of the Amazon a precursor of the extension of a more xeric environment towards the rain forest domain? To what extent is this process controlled by climatic, anthropogenic, or edaphic parameters? At the meso-climatic level, are there feedback mechanisms leading to changes in the structure and seasonality of the transition forest making this latter more susceptible to fires? Finally, what is the significance of the pattern of forest opening (i.e., large blocks with forest islands versus smaller gaps in a forest cover) observed across the area in terms of hydrology, biological conservation, fire dynamics and, ultimately, climate impacts?

15.7.2 South East Asia

The 1982–83 events in the tropical forest of Kalimantan and North Borneo have been instrumental in bringing a new realization of the emerging importance of fire in tropical ecosystems. The intense El Nino episode of 1982–83, which caused the well-reported significant warming of the eastern Pacific (Philander 1983) and anomalous wet conditions over its western rim is also at the origin of the less publicized drought in Australia and South East Asia (Nicholls 1987). The drought, which had an impact upon the agriculture of that region (Malingreau 1987), was so pronounced in mid 1982 and especially in early 1983 that it caused serious damage to the tropical evergreen forest of the eastern and northern part of the Island of Borneo. Drought stress led to the drying of vines and the shedding of leaves by evergreen species. In the low atmospheric humidity conditions prevailing during the ENSO episode, the accumulated organic material provided abundant fuel for the spreading of fires. These were mainly triggered by agricultural practices of slash-and-burn and to an unknown extent by permanent coal seam fires. Accelerated settlement programs along the edge of the forest and logging operations of the last 10 years have stimulated the occupation of large tracts of land along the coast as well as inland along the main access rivers. The drought of 1982–83 was seen by the many migrants occupying that area as a unique opportunity to engage in a drastic land clearing campaign. The burning rapidly went out of control. Field observations, confirmed by satellite data, indicate that it is the selectively logged forest (a common practice in Indonesia) which suffered the greatest damage. Road opening, dry woody material accumulation, and maybe fuel depots all contributed to the expansion of fires. In the dryland dipterocarp forest the mosaic of dried, burned, and unburned patches was found to be mainly site-related (soil moisture and dominance of drought-resistant species as shown by Leighton and Wirawan 1985). In addition, another ecosystem which was surprisingly severely damaged by the 1983 fires is the peat swamp forest of the large inland freshwater marshes of the Mahakam River (Lennertz and Panzer 1984). This edaphic variant of the tropical rain forest, usually considered immune to fire (Mueller-Dombois 1981), saw a lowering of the water table, a progressive dessication of the turflike accumulation of organic matter, and a toppling over of butressed trees. After burning the surface woody material, fire continued underground for a long period of time.

By all accounts, the Borneo fires were major ecological events which underscored once more the fragility of the tropical rain forest when exposed to a new conjunction of environmental stress factors. What is, however, truly extraordinary in this case is that fires raged completely unmonitored and unabated for close to 3 months in 1983 and that outside a small group of forestry officials and foreign scientists, the extent and severity of the Kalimantan fires were little noticed until the end of 1983. Satellite observations were made daily over the area. Their poor use failed to alert scientists and officials alike to the importance of the events. Unfortunately, none of those satellite data were archived by the local stations and we had to resort to an analysis of the 4-km data archives of the resampled GAC (4 km) format to reconstruct the history of the burning (Malingreau et al. 1985).

The AVHRR GAC data show that on March 13, 1983, large amounts of smoke generated by the burning in East Kalimantan was extending westward over most of the island, forming a mass of more than 350,000 km^2. The fire count, which included at that date ten fire areas (zones of clear thermal anomalies), culminated in mid-April with more than 110 fire areas. By mid-May, most of the burning in East Kalimantan was extinguished by heavy rain, but a large "warm area" remained visible in the hinterland swamps. This is believed to have been caused by the persisting peat fires (Fig. 6). At this time, however, numerous fires had appeared in North Borneo, where it is estimated that 1.5 million hectares of forest were damaged during this extreme El Nino episode (Beaman et al. 1983).

An estimate of the damages caused to the tropical forest biome of East Kalimantan was made by German scientists using small airplane transects (Lennertz and Panzer 1984); they estimated that over 3.5 million hectares of forested area were affected to one degree or another by the drought and fire. The AVHRR image of October 1983 showed that an increase in brightness tem-

Fig. 6. AVHRR image of the island of Borneo taken on April 19, 1983. The data are extracted from the GAC (Global Area Coverage) archives. Fires (*red dots*) are visible in North Borneo. The still-burning hinterland peat swamp of the Mahakam River (East Kalimantan) is seen as a warm area

perature values closely follows the boundary of this area (Malingreau et al. 1985). Some details on the mosaic of burned-unburned patches was derived from LANDSAT data obtained more than 10 months after the burning period; unfortunately, only a small portion of East Kalimantan was covered by appropriate high resolution images, and a comprehensive assessment was not feasible.

The Borneo-Kalimantan fires were caused by a unique conjunction of climatic and human factors which led to the development of a high fire risk situation. Prior to the occurrence of the fires themselves, the vegetation index of the eastern part of the island showed an anomalous drop which could have alerted the foresters to the seriousness of the situation and possible fuel loading conditions. It was the presence of many migrants engaged in traditional agricultural activities which provided the trigger to what has been dubbed the major fire event of the century. Once again, one can ask whether catastrophic fires of this kind are not dynamically amplified through feedback mechanisms leading to increases in fire susceptibility. Fires in the forest invite more fires (Mueller-Dombois 1981).

AVHRR data obtained in early 1983 at the Bangkok Station did not cover much of the Island of Borneo, but provided striking views of the fire situation over most of continental South East Asia in a year of exceptional drought. These data, which are currently under analysis for an assessment of tropical deforestation in the region (Malingreau and Tucker 1989), indicate three areas standing out for their burning intensity (February to April 1983):

– Northern Thailand with traditional shifting cultivation in the seasonal monsoon forest of the Chiang Mai and Chiang Rai regions.

– Northern Laos where the intensity of burning in March 1983 is very high. The fire points are evenly spread across the mountainous area covered with seasonal forest without showing much organization in deforestation fronts. The density of fire pixels seen in this forest biome is high and equalled only by that observed in the African woodland savanna or in the Brazilian cerrado. Fires extend to southern Yunnan in China.

– Northern Burma. The higher reaches of the Mt. Arakan chain (east of Manipur-India) are characterized by a very high density of well-individualized fire points evenly distributed across the range.

With only 1 year of data available, it is, however, not possible to ascertain if these observations are exceptional and related to an extreme El Nino year or if they reflect recurrent events in these areas of intense land occupation. Additional satellite data sets are unfortunately not yet available on a regular basis over South East Asia.

15.7.3 Africa

The African continent represents by all account a major contribution to biomass burning in the tropics (Hao et al. this Vol.). It is occupied by large areas of savanna and transition seasonal forests; the dominant land use is agro-pastoral and population growth is among the highest in the world. All these factors combine to create a burning situation of extensive dimensions on that continent. This was well illustrated in the mid-1970's already by the stunning night time low illumination visible imagery showing a large fire belt south of the Sahara Desert (Croft 1978).

The savanna ecosystems, which encompass a wide range of facies from the grass steppe to the savanna woodland (the Yangambi classification recognizes ten major categories) are those mainly affected and indeed most often maintained by recurrent fires. The variety of climatic and edaphic conditions in the savanna biome types gives rise, therefore, to a large range of burning situations. The bioclimatic stratification of the African continent would thus be an obvious way to start a fire pattern analysis insofar as such stratification would take into account controlling parameters such as biomass accumulation, fuel loading, and fire susceptibility. Many such maps are available to support this approach (Keay 1959; White 1983). We have here elected to concentrate upon the satellite data themselves and to relate the burning information derived from a series of such data to known vegetation patterns. This more empirical approach is justified in our mind by the current availability on an extensive and repeated coverage of some areas of interest and by the ready identification of fire features of such data set.

15.7.4 West Africa

The Maspalomas Station receives daily AVHRR readings for the afternoon pass over West Africa since 1986. Dry season images of the subcontinent show a very clear thermal stratification which follows from south to north the broad vegetation types. This horizontal stratification which is readily noted on the now familiar vegetation index images of the continent (Tucker et al. 1984) is especially marked in channel 3 of the AVHRR (Malingreau and Laporte 1988). A transect from the Gulf of Guinea in the South to the Sudanian zone in the north cuts across three distinct regions: the "cooler" moist forest area of the atlantic side, the mixed woodland savanna transition belt and the tree-shrub savanna in the north. The main characteristics of the three zones are summarized in Table 3. Boundaries between thermal regions are more or less set by mid-December (Grégoire et al. 1989) and are stable from one year to the other, but their contrast is enhanced by increasing water deficit. Distinct fire patterns and periodicity are associated with this stratification; they are related to the ecoclimatic characteristics as well as to the dominant land use practices.

Limitations related to sensor saturation prevent a consistent analysis of "hot spots" in the warm subsaharian savanna where monitoring must rely on a post-fire scar analysis. Early in the season (October) some fire points are, however, visible on the AVHRR images ("feux précoces", early fires). The following will concentrate on fire patterns in the woodland savanna and moist forest regions of the southern part of West Africa with particular attention to the

Table 3. Radiometric and ecobiological characteristics of the three regions identified in the brightness temperature profile of West Africa (14° N–5° N). AVHRR data 12/7/86

Thermal region	1	2	3
CH3 mean pixel value (transect)	165	124	110
NDVI mean value (transect)	0.16	0.16	0.21
Mean annual precip. (mm)	750 to 1500	1500 to 2000	>2000
Dry months (P ≤ 2 T°)	5 to 6	3 to 5	0 to 2
Climate type	"Sahelo-Soudanian"	"Soudano-Guinean"	"Guinean forest"
Natural vegetation:			
→ Climax	Dense bush to dense dry deciduous forest	Dense moist semi-deciduous forest	Dense moist evergreen forest
→ Present	Shrub savanna	Tree savanna	Woodland savanna and dense moist evergreen forest
Fire density (estimation)	Low	High	Very low

transition between the two units. Intensive burning is observed in the woodland savanna from Sierra Leone to Ghana. Fire points appear for the first time around mid-November in a dotty pattern throughout the region. Changes in this pattern are then observed with the advancement of the dry season. By mid-January a strong differentiation between regions is visible. In some parts, the individual fire points have extended into large but still isolated patches of high-temperature pixels. In others, burning has progressed almost uninterrupted across extensive areas. Fire lines are seen at the edge of these large burns. The first pattern is seen mainly in northern Sierra Leone and northern Ivory Coast, while the large-scale burns are seen mainly in Guinea. By the end of the dry season, however, a more or less continuous burned belt 100–200 km wide is found in the southern part of the savanna range. At that time, the boundary between the woodland savanna and the forest domain is marked by an abrupt drop in the thermal channel (Fig. 7). The southern edge of the fire belt is thus very well delineated. The northern one is more difficult to identify with precision in the thermal channel but is more clearly marked in the near infrared data due to the presence of dark burn scars.

The differences in the evolution of the well-individualized fire points observed early in the season are most likely due to agricultural practices and population density. In the Ivory Coast, for example, a very large number of individual fire points indicate intensive agricultural clearing activities. Fires are fairly well controlled and the burned-over area remains patchy. In Guinea, where hunting and grazing are the dominant forms of land use, fire can continue completely unfettered over very large areas. In dry or high fuel loading situations burning can take on massive proportions, as shown by an AVHRR image of early 1987, where long fire lines are seen at the edge of a burn of more than 10,000 km^2 in Guinea (Malingreau et al. 1989). It is not uncommon during a field survey to drive for several hundred kilometers in a landscape where the trace of fire is always present. The interpretation of the satellite data must, however, take

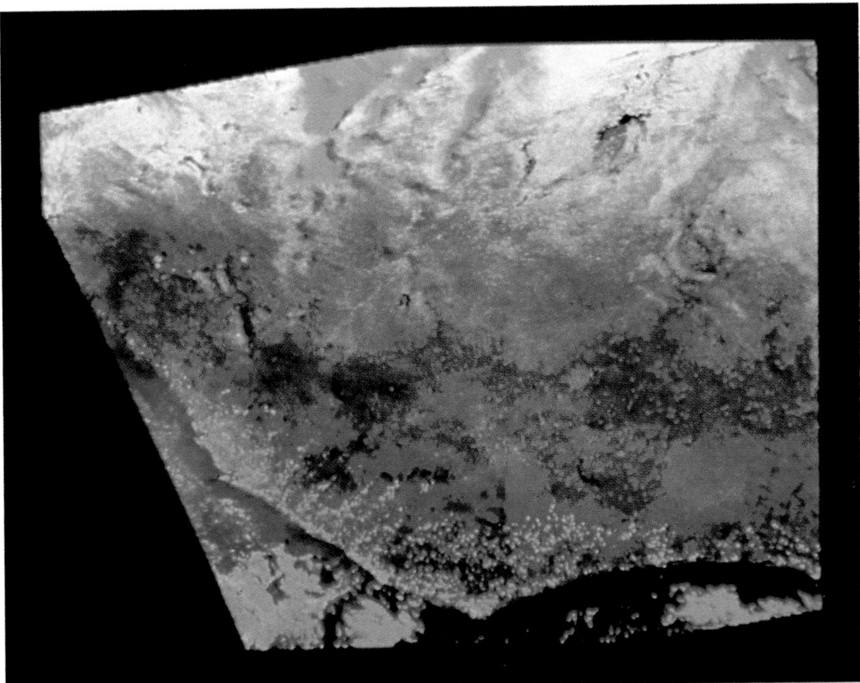

Fig.7. A color composite AVHRR image of West Africa received by the Maspalomas Station (Feb. 5.87). The thermal infrared channel 3 data are printed here in *red*, Ch2, in *green* and Ch1 in *blue*. The thermal stratification of the subcontinent is well marked from the cooler forest domain in the south to the warmer woodland savanna and shrub savanna to the north. The belt of large-scale burns immediately north of the forest area is visible (*dark red*)

into account the characteristics of the sensor. As already mentioned, we have verified that many small fires (a few tens of meters long and a few meters wide) can "accumulate" in a given saturated pixel. Inversely, small "cool" fires can be indicated by a nonsaturation increase in the pixel value which will stand out by contrast with the background, but for which threshold values are not easily determined. In large areas of continuous burning, very high thermal values form the background and it is equally not easy to identify which parts are still burning at the time of satellite overpass. Preliminary data appear to indicate that in such case the difference between the readings in the thermal channel at 11 and 3 microns (Ch4–Ch3) is negative and larger for active fires pixels than for burned-over area (Laporte and Malingreau 1989). The interpretation is often confused by the presence of smoke or haze.

These difficulties encountered in resolving all the details included in a thermal image of such a complex area as the savanna woodland of West Africa indicate the necessity of better understanding the local mosaic of vegetation and land use practices. It remains, however, that on a regional scale it is entirely feasible to use the AVHRR satellite data to detect the appearance of the thousands of fire points appearing early in the dry season, and to monitor the evolution of the burned-over area as the season progresses. The next step will be to use time series of low resolution GAC data to assess if the patterns observed in 1987 and 1988 are normal or if interannual variations are important. The feasibility of this approach rests, however, upon the capability of detecting the relevant information in the highly sampled product represent by the Global Area Coverage data. This point is discussed in the concluding section.

The "fire belt" of West Africa ends abruptly at the edge of the moist forest domain. This boundary follows the line formed by the very contrasting floristic and physiognomic features of the woodland savanna versus the forest biome. In the southern part of Western Africa, however, remnants of the primary rain forest are few and located mainly in less accessible areas or very poor soils. The major transition facies is thus found between the woodland savanna (north) and a mix of secondary regrowth interspersed with plantations and dry agriculture to the south. This secondary regrowth is semi-seasonal but evergreen species are dominant (see NDVI temporal development curve in Fig. 2d). These formations are essentially nonflammable in their natural state and it is only after cutting and drying that they can be burned. The pattern of fires in this region is understandably different than in the woodland-savanna. The AVHRR data for the 2 years of record show mainly isolated fires which do not coalesce as the dry season progresses. As in other wet tropical areas, the contrast between the "hot spots" and the surrounding evergreen vegetation is good. In the forest part of Guinea the number of fires increases abruptly in April. It is not understood at this stage why fires peak at the time of the first, albeit localized rainfall. Of interest in the framework of large-scale changes in transition ecosystems is the fact that this area, which has been essentially deforested since colonial time and consequently regained by a secondary tree cover, is currently submitted to renewed pressure from population growth and migration from the subsahelian zone. Change is also stimulated by economic development (e.g., opening and paving of roads, industrial plantations). It is likely that new clearings and thus new fires will soon appear in the region. The AVHRR image of early 1988 over the so-called Guinée Forestière shows a much higher density of "hot spots" than expected. Given the fact that many of the primary forest remnants are located in difficult terrain (this may account for their survival), it is likely that the next "deforestation" wave will take place in these more accessible secondary formations. Studying changes in fire patterns of those ecosystems is thus most relevant.

15.7.5 Central Africa

The "fire belt" already noted continues to the central portion of the continent where recent observations also show well contrasted thermal patterns related to vegetation type and burning. A series of 1988 LAC images acquired over the Central Congo Basin has provided preliminary information of those areas. The fire analysis was included here as part of a tropical deforestation assessment. As in other cases described in this chapter, the presence or absence of fire is interpreted in terms of forest disturbance by human occupation. The preliminary observations presented herewith are still of a tentative nature. The processing of Central African AVHRR data set (10° S to 10° N, daily passes from Dec. 1987 to May 1988) has been undertaken at NASA GSFC (Justice et al. 1985) is described by Kaufman in this Volume. The preliminary analysis proposed here refers to two different settings. First the rain forest-seasonal forest-woodland transition in the north of the Central Basin and second, the rain forest biome itself.

15.7.5.1 Transition Rain Forest – Seasonal Forest-Woodland Savanna

1. The transition between the rain forest of the Central Congo Basin and the more seasonal forest-woodland mosaic to the north is marked by a sharp thermal transition which is most visible at the end of the dry season (see Fig. 12 in Kaufman this Vol.). This differential increase in surface brightness temperature underlines the contrasted seasonality between the evergreen forest and the dry deciduous formations.
2. The northern edge of the rain forest is also underlined by a 100–150-km-wide belt characterized by an increase in background temperature and by the presence of a large number of active fires dotting the area. The spatial variability of the AVHRR thermal channel 3 data is high and reflects complex land cover conditions. The density of the "hot spots" and their high contrast with the background suggest a high biomass-fuel loading situation (Fig. 8a). Surprisingly, however, little smoke is visible.
3. Further to the north, into the tree-shrub savanna, the fire patterns disappear because of the loss of contrast with the warm background temperature. It is likely that the magnitude of burning is smaller in these low biomass ecosystems.
4. The thermal belt referred to above is also underlined by a drastic drop in the near infrared albedo (Ch. 2 in AVHRR – see Fig. 8a). This strip of seasonally low albedo coinciding with the area of intense burning is also noted on the continental albedo maps derived from Meteosat data (Dedieu at al. 1987). The authors of these maps indicate that in those areas the albedo can vary seasonally between 0.20 (rainy season) and 0.12 (dry season).
5. The situation found just beyond the northern edge of the rain forest biome can be explained by the conjunction of the following factors: high biomass situation in the rainy season, a marked seasonality and thus high fuel loading and an intense land occupation. Vegetation maps of the area show that a complex woodland-secondary forest mosaic is present here with dominant closed tree formations. This may account for the more patchy nature of the burning and the relatively low amplitude of the vegetation index time series (Fig. 2c). The continuity of this line of high burning conditions-low albedo across the whole African continent is also clearly visible on the AVHRR GAC images.
6. The strong influence of burning on the radiation balance of such a large area located at the edge of the primary forest biome is an important feature of

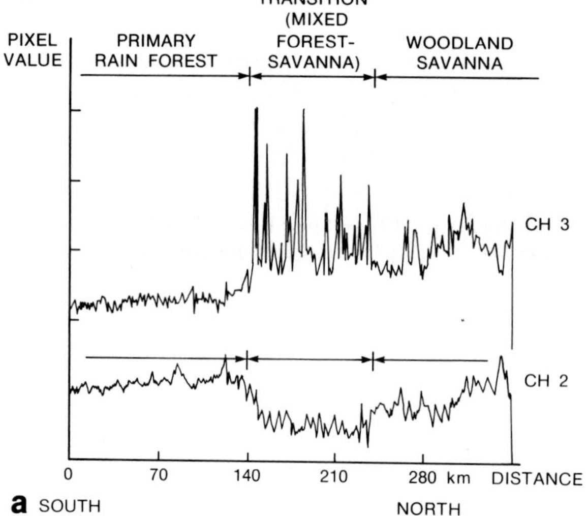

PIXEL VALUE
PRIMARY RAIN FOREST
TRANSITION (MIXED FOREST-SAVANNA)
WOODLAND SAVANNA
CH 3
CH 2
0
70
140
210
280 km
DISTANCE
a
SOUTH
NORTH

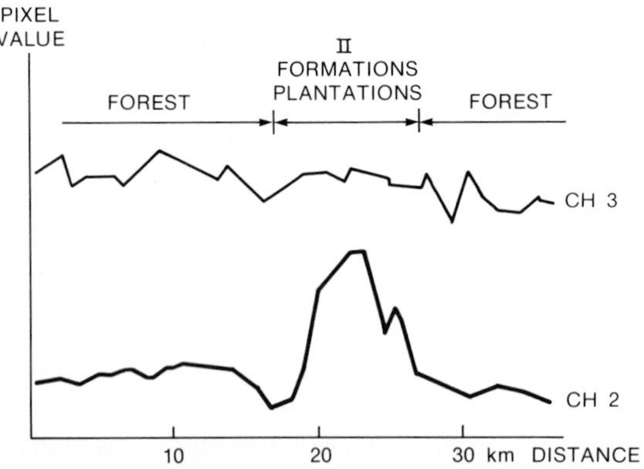

PIXEL VALUE
FOREST
II FORMATIONS PLANTATIONS
FOREST
CH 3
CH 2
10
20
30 km
DISTANCE

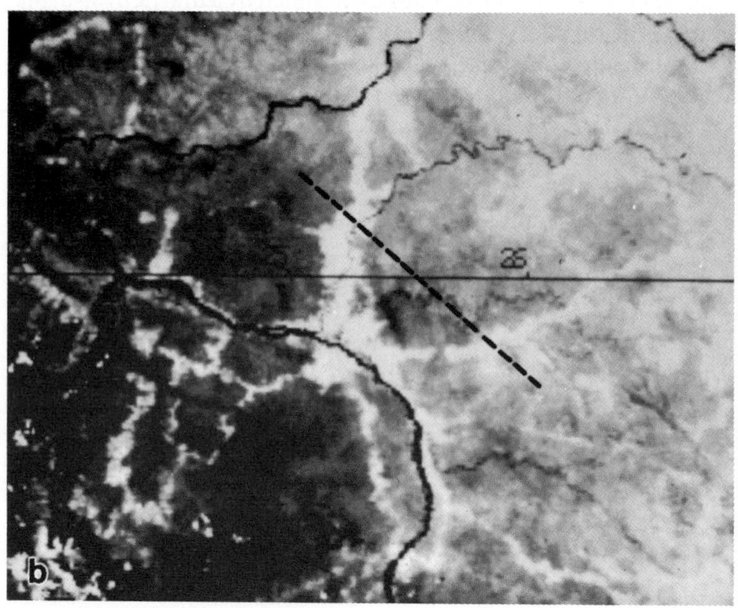

26
b

large-scale terrestrial ecosystem dynamics. The size and geographic position of the feature warrants further study in terms of biosphere-atmosphere interactions. Judging from the density of fires points seen on the AVHRR images, land use intensity is high and in all likelihood most dynamic. It is not known if the southern part of this fire belt represents an actual deforestation front.

15.7.5.2 The Central Congo Basin

Further to the south, the presence of "hot spots" is hardly detected on the available AVHRR data. The rain forest biome presents a very uniform thermal landscape. No clearings or sign of human activities are seen during the dry season of 1987–88. A few alignments of warmer points close to the northern border are the only exception. What is however new compared to other tropical forest scenes analyzed so far is that the near infrared data unmistakably underline vegetation formations of higher reflectance within the darker rain forest biome (Fig. 8b). A reticulated network appears on the few AVHRR scenes unencumbered by clouds; they correspond closely, in part at least, to the road network of Zaire (ref. Michelin map). We believe that what is revealed here is a pattern of secondary formations and plantations associated with human settlements. The secondary formations typically have a higher degree of heterogeneity, a more open architecture, and more seasonality character than the rain forest (Lebrun and Gilbert 1954). These physiognomic characters could explain a higher near infrared reflectance than the closer and highly structured rain forest canopy. What is surprising, however, is the complete absence of fire underlining the areas of human occupation. This may be due to sampling effects (wrong time of satellite overpass, small "cool" fires because of low fuel loading etc.) or may properly reflect the land use conditions.

Beyond the features described above, no signs of human activity which could indicate deforestation on a major scale are found in the Congo Basin itself. Fire is not a common feature in the forest, as one would in such a low population density situation. As mentioned, the most dynamic situations are found at the edge of the basin, towards the more seasonal vegetation formations. The northern edge in particular seems to warrant a close monitoring.

The observations made over West and Central Africa are summarized in Fig. 9.

◄

Fig. 8. **a** Transect of radiometric values (AVHRR LAC data) for the transition area between the rain forest of the Central Congo Basin and the seasonal forest-woodland savanna to the north. The thermal contrast between the two formations is evident in the Channel 3 data. The decrease in the near infrared reflectance in the mixed vegetation area indicates a dense distribution of fire and burns (Jan. 1988). This feature forms a temporal low albedo belt across the northern part of the African continent (see Fig. 9). **b** Transect of radiometric values (AVHRR LAC data) in the forest biome of the Central Congo Basin showing a complete absence of thermal features as one crosses from the rain forest to a secondary formation (including plantations). This formation is however well indicated by an increase in the near infrared reflectance (Ch 2). See black and white image

Fig. 9. The main observations related to fire and burning in Africa are summarized on this map. The following zones are identified: *Zone 1.* Shrub-savanna, low biomass. Early fires (October) are detectable using AVHRR thermal data. Later, the high ground temperature saturates the sensor signal. Burn scars can then be measured using near infrared data (low albedo). *Zone 2.* Savanna-woodland mosaic. Numerous fires related to intensifying agricultural activities but also to grazing and hunting appear in November. Peak of burning is in February-March. High biomass and high fuel loading situations favor extensive burning in some areas. A large belt of low albedo burn is observed at the end of the dry season. *Zone 3a.* Moist forest domain. Covered mainly by secondary formations (including plantations, agriculture). Fires are isolated and do not expand. Growing number of fires may be observed at the northern edge with opening of new agricultural-cattle raising land. *Zone 3b.* Rain forest domain. No or few fires. Low human occupation. No active deforestation front is visible

15.8 A Global Fire Monitoring System – Conclusions

Despite the limitations inherent to the existing sensors, satellite remote sensing can today provide, at regular intervals of time, elements of a global fire assessment for the intertropical belt. The experience gained so far in the use of the thermal data provided by the NOAA-AVHRR system supports a better evaluation of the problems related to the resolution and sensitivity of the sensor. These limitations appear to have a greater impact upon the correct determination of fire size and fire temperature than upon the ability to assess the overall distribution and frequency of fire points in the ecosystems. More important in this respect are the shortcomings introduced by the temporal sampling (time of overpass) and cloudiness conditions which still prevent the formation of a fully reliable count.

The AVHRR instrument has so far provided a large number of data on biomass burning in a wide range of tropical ecosytems. Approaches developed for their analysis seem to work equally well in evergreen and semi-seasonal formations with the notable exception, however, of the warmer shrub savanna-steppe ecosystems, where instrument saturation prevents consistent readings. In this case, the analysis of post-burn scars is the only viable approach.

Notwithstanding the already large volume of data collected, the AVHRR HRPT or LAC full resolution archives acquired so far represent only a limited data set for a truly global fire monitoring. As discussed at the beginning of the chapter, a wall-to-wall coverage of the tropical areas, acquired at appropriate intervals of time and in a uniform mode, can only be made using the AVHRR system in the GAC (Global Area Coverage 4 km resolution) format. These data have existed since 1981 daily for the whole world. Before embarking upon their systematic analysis, it would, however, be of paramount importance to carry out a comprehensive evaluation of the effect of the LAC-GAC sampling transform upon fire detection capabilities. Here, a quantification of the effect of the pixel averaging and line skipping operation used to produce the GAC data must be made. Currently, most of the research on scale effects in remote sensing refers to the visible and near infrared characteristics of landscape features (Townshend and Justice 1988; Woodcock and Strahler 1987) rather than on fire-related features themselves. On an empirical basis, it has been shown that the low resolution GAC data can reveal anomalies associated with biomass burning (Malingreau et al. 1985; Matson et al. 1987) and that a first stratification of such a kind could support a global fire survey. The use of Meteosat data for similar purposes should also be investigated, since the thermal information provided by this instrument several times a day appears to contain information related to vegetation burning (see Citeau et al. 1989 on savanna fires in Senegal).

The systematic comparison of a GAC and HRPT data set obtained on the same date over West Africa shows that the ability to pick out fire points is retained in the low resolution GAC data set but that the results are much conditioned by the lay-out of the fires (Belward and Malingreau 1989). Saturated GAC pixels will thus only occur if all the original 1 km pixels in a line are saturated; smaller values will be obtained if this condition is not respected. Whole lines of fire points can be missed if, in the original data set, they happen to be located in a pixel line skipped entirely by the sampling process. More work is needed to quantify these effects before assessing the true representativity of a fire count carried out on a global product of the GAC type.

To cope with the very large amount of AVHRR data necessary for global monitoring, procedures will have to be devised to automatically extract the most likely candidates for fire points. This will require that fairly standard cloud screening approaches can be implemented from one region to another. The resulting data can be arranged into a global fire picture showing the distribution and dynamics of burning. Such information will then have to be interpreted in the context of relevant surface data such as meteorological parameters, biomass estimation, and land use practices. These surface data can partly be extracted from the satellite products. It is obvious at this stage that the success of a global fire monitoring exercise depends upon the capability to spatially integrate satellite data with other data sets and to interpret their temporal variations.

Their inclusion into vegetation burning models should also be a promising avenue of investigation (Antonovski and Ter-Mikhaelian 1987). The experience gained with the AVHRR approach described in this chapter is encouraging. For parts of the world well covered by receiving stations, a semi-operational fire monitoring system could already be set up. The extension to global scales implies, of course, the monopolization of large data processing and interpretation means. Fire in tropical ecosystems being a growing agent of global change, its monitoring now warrants such effort.

Acknowledgments. The exciting recent developments in fire analysis using satellite data were shared by many enthusiastic colleagues. M. Matson, C (Jim) Tucker, B.N. Holben, C. Justice, T. Goff, J.M. Grégoire, A. Belward and many others have been instrumental in pressing ahead with those activities. Results reported here were only possible with a strong data processing support; the tireless efforts of B. Rank and W. Newcomb at NASA GSFC, S. Flasse and M. Righetti at JRC are warmly acknowledged. Through her constant probing of the West African data set, N. Laporte did much to progress in the understanding of fire-related features in that part of the tropics; her willingness to share is much appreciated. The GAC data referred to in this report have been analyzed in the framework of the EEC-JRC NASA-GSFC scientific agreement.

References

Antonovski MY, Ter-Mikhaelian MT (1987) On spatial modeling of long-term forest fire dynamics. IIASA Working Paper 87–105, Laxenburg

Beaman RS, Beaman JH, Marsh CW, Woods PV (1983) Drought and forest fires in Sabah in 1983. The Sabah Foundation, Kota Kinabalu (Malaysia)

Becker F, Seguin B (1985) Determination of surface parameters and fluxes for climate studies from space observations: methods, results and problems. Adv Space Res 5(6):299–317, COSPAR 84

Belward A, Malingreau JP (1989) A comparison of AVHRR GAC and HRPT data for regional environmental monitoring. Presented at IGARSS 89, Vancouver, Canada

Cass A, Savage MJ, Wallace FM (1984) The effects of fire on soil and microclimate: Chapter 14. In: Booysen P de V, Taintor NM (eds) Ecological effects of fire in South African ecosystems. Springer, Berlin Heidelberg New York Tokyo

Citeau J, Demarq H, Mahé G, Franc J (1989) Une nouvelle station est née. Veille Climatique Satellitaire. 25.23–29. Lannion, France

Croft TA (1978) Nighttime images of the earth from space. Sci Am 239:86–98

Crutzen PJ, Heidt LE, Krasnec JP, Pollock WH, Seiler W (1979) Biomass burning as a source of atmospheric gases CO, H_2 N_2O, NO, CH_3Cl, and COS. Nature (Lond) 282:253–256

Daubenmire R (1968) The ecology of fire in grasslands. Adv Ecol Res 5:209–266

Dedieu G, Deschamps PY, Kerr YH, Raberanto P (1987) A global survey of surface climate parameters from satellite observations: preliminary results over Africa. Cospar Adv Space Res Vol 7(119):129–137

Deshler W (1975) An examination of the extent of fire in grasslands and Savanna of Africa along the southern side of the Sahara. Proc 9th Int Symp Remote Sensing of the Environ. ERIM, Ann Arbor, MI

Dozier J (1981) A method for satellite identification of surface temperature fields of subpixel resolution. Remote Sensing of Environ 11:221–229

Fearnside PM (1984) Land clearing behavior in small farmer settlement schemes in the Brazilian Amazon and its relation to human carrying capacity. In: Sutton SL (ed) Tropical rain-forest: the Leeds Symposium. Leeds Philos Literary Soc, Leeds, UK, pp 255–271

Fearnside PM (1985) Burn quality prediction for simulation of the agricultural system of Brazil's Transamazon Highway colonists for estimating human carrying capacity. In: Govil GV (ed) Ecology and Resource Management in the Tropics. Int Soc Tropic Ecol, Varanasi, India

Flannigan MD (1985) Forest fire monitoring using the NOAA satellite series. M Sc Thesis, Colorado State Univ, Fort Collins Colorado

Gasques JG, Magalhaes AR (1987) Climate anomalies and their impacts in Brazil during the 1982–83 ENSO Event. In: Glantz M et al. (eds) The societal impacts associated with the 1982–83 worldwide climate anomalies. Report NCAR, Boulder, pp 30–36

Goward SA, Tucker CJ, Dye D (1985) North American vegetation patterns observed with the NOAA-7 AVHRR. Vegetatio 64:3

Grégoire JM (1989) Effect of the dry season on the vegetation canopy of some river basins of West Africa as deduced from NOAA-AVHRR data. 3rd Sci Assembly Int Assoc Hydrol Sci, Baltimore. 10 May 1989

Grégoire JM, Flasse S, Malingreau JP (1988) Evaluation de l'action des feux de brousse de novembre 1987 à février 1988 dans la région frontalière Guinée-Sierra Leone. Exploitation des images NOAA-AVHRR. SPI 88:39. Joint Res Cent Ispra, Italy

Hatfield JL (1983) Remote sensing estimators of potential and actual crop yield. Rem Sens Environ 13:301–311

Hick PT, Prata AJ, Spencer G, Campbell N (1986) NOAA AVHRR satellite data evaluation for areal extent of bushfire damage in Western Australia. In: Proc 1st Aust AVHRR Conf, Perth

Holben BN (1986) Characteristics of maximum value composite images from temporal AVHRR data. Int J Remote Sensing 14:65–76

Irons JR, Markham BL, Nelson RF, Toll DL, Williams DL, Latty RS, Stauffer ML (1986) The effects of spatial resolution on the classification of Thematic Mapper data. Int J Remote Sensing 6(8):1385–1403

Justice CO, Townshend JRG, Holben BN, Tucker CJ (1985) Analysis of the phenology of global vegetation using meteorological satellite data. Int J Remote Sensing 6:1271–1318

Kaufman YJ, Tucker CJ, Fung I (1988) Remote sensing of biomass burning in the tropics. Cospar, Helsinki

Keay RWJ (1959) Vegetation map of Africa. Oxford Univ Press, Oxford

Kerber AG, Schutt JB (1986) Utility of AVHRR channels 3 and 4 in Land-Cover mapping. Photogr Eng Remote Sensing Vol 52(12):1877–83

Langaas S, Muirhead K (1988) Monitoring bush fires in West Africa by weather satellites. Presented at XXIII Int Symp Remote Sensing Environ, Abidjan

Laporte N, Malingreau JP (1990) Fire patterns in the savanna-forest transition of Guinea using AVHRR data. (in preparation)

Lebrun J, Gilbert G (1954) Une classification écologique des forêts du Congo. Série Scientifique N° 63. INEAC, Yangambi

Leighton M, Wirawan N (1984) Catastrophic drought and fire in Bornean tropical rain forest associated with the 1982–83 ENSO event. In: Prance GT (ed) Tropical rainforest and world atmosphere. AAAS, New York

Lennertz R, Panzer KF (1984) Preliminary assessment of the drought and forest damage in Kalimantan Timur. Transmigration Area 8 Development Project (TAD) Report of the Fact Finding Mission, Samarinda, Indonesia

Malingreau JP (1986) Global vegetation dynamics: satellite observations over Asia. Int J Remote Sensing 7(9):1121–1146

Malingreau JP (1987) The 1982–83 drought in Indonesia. Assessment and monitoring. In: Glantz M et al. (eds) The Societal Impacts Associated with the 1982–83 Worldwide Climate Anomalies. Report, NCAR Colorado, Report, pp 11–18

Malingreau JP, Laporte N (1988) Global monitoring of tropical deforestation. AVHRR observations over the Amazon Basin and West Africa. Forest Signatures Workshop. Joint Res Cent Ispra, Italy

Malingreau JP, Tucker CJ (1987) The contribution of AVHRR data for measuring and understanding global processes: large-scale deforestation in the Amazon basin. In: Proc IGARRS' 87, Ann Arbor, MI, 18–21 May 1987, pp 443–448

Malingreau JP, Tucker CJ (1988) Large-scale deforestation in the Southeastern Amazon basin of Brazil. Ambio 17(1):49–55

Malingreau JP, Tucker CJ (1990) Tropical deforestation assessment. Preliminary evaluation of the AVHRR data available over Southeast Asia. (in preparation)

Malingreau JP, Stephens G, Fellows L (1985) Remote sensing of forest fires: Kalimantan and North Borneo in 1982–3. Ambio 14(6):314–315

Malingreau JP, Laporte N, Grégoire JM (1990) Exceptional fire events in the tropics. Southern Guinée. January 87. Int Remote Sensing (in press)

Malingreau JP, Grégoire JM (1990) Groping in the smoke. Field verification of AVHRR fire analysis in a woodland savanna environment. (in preparation)

Matson M, Dozier J (1981) Identification of subresolution high temperature sources using a thermal IR sensor. Photogr Eng Remote Sensing 47(9):1311–1318

Matson M, Holben B (1987) Satellite detection of tropical burning in Brazil. Int J Remote Sensing 8(3):509–516

Matson M, Stephens G, Robinson JM (1987) Fire detection using data from the NOAA-N satellites. Int J Remote Sensing

Menaut JC (1983) The vegetation of African savannas. In: Bourlière F (ed) Tropical Savannas. Elsevier, Amsterdam, pp 109–149

Mueller-Dombois D (1981) Fire in tropical ecosystems. In: Fires, regimes and ecosystem properties. Proc Conf General Tech Report WO-26. US Dept of Agric For Serv, pp 137–176

Muirhead K, Cracknell (1985) Straw burning over Great Britain detected by AVHRR. Int J Remote Sensing 6(5):827–833

NASA-EOS (1986) Earth Observing System Science and Mission Requirements Working Group Report, Volume 1. Tech Mem 86129, NASA Goddard Space Flight center, Greenbelt, MD

Nichols N (1987) The El Nino southern oscillation phenomenon. In: Glantz M et al. (eds) The Societal Impacts Associated with the 1982–83 Worldwide Climate Anomalies. Report NCAR, Boulder, Colorado, pp 2–10

Philander SGH (1983) Anomalous El Nino of 1982–83. Nature (Lond) 305:16

Pinker RT, Thompson OE, Eck TF (1980) The albedo of a tropical evergreen forest. QJR Meteorol Soc 106:551–558

Robinson JM (1988) The role of fire on Earth: A review of the state of knowledge and a systems framework for satellite and ground-based observations. NCAR/CT-112. Cooperative Thesis No 112. NCAR, Boulder, Colorado

Seiler W, Crutzen PJ (1980) Estimation of gross and net fluxes of carbon between the biosphere and the atmosphere from biomass burning. Climatic Change 2:207–247

Sellers PJ (1985) Canopy reflectance, photosynthesis and transpiration. Int J Remote Sensing 6:1335–1372

Setzer AW, Da Costa Perreira M, Da Costa Perreira Jr A, De Oliveira Almeida SA (1988) Relatorio de Atividades de Projeto IBDF-INPE "SEQE". Ano 1987. INPE 4534 RPE/565. Sao Jose dos Campos, Brazil

Strahler AH, Woodcock CE, Smith JA (1986) On the nature of models in remote sensing. Remote Sensing Environ 20:121–139

Townshend JRG, Justice CO (1988) Selecting the spatial resolution of satellite sensors required for global monitoring of land transformations. Int J Remote Sensing 9(2):187–236

Tucker CJ, Townshend JRG, Goff TE, Holben BN (1986) Continental and global scale remote sensing of land cover. In: Trabalka JR, Reichle DE (eds) The changing carbon cycle: a global analysis. Springer, Berlin Heidelberg New York Tokyo, pp 222–241

Tucker CJ, Holben BN, Goff TE (1984) Intensive forest clearing in Rondonia, Brazil as detected by satellite remote sensing. Remote Sensing Environ 15:225

Uhl C, Buschbacher R (1985) A disturbing synergism between cattle ranch burning practices and selective tree harvesting in Eastern Amazon. Biotropica 17:265–268

Vickos JB (1986) Etude des feux de savanes. Thèse, DEA. Univ Paul Sabatier, Toulouse, France

White F (1983) The vegetation of Africa. UNESCO, Paris 356 pp + maps

Wong CL, Blevin WR (1967) Infrared reflectance of plant leaves. Aust J Biol Sci 20:501–508

Woodcock CE, Strahler AH (1987) The factor of scale in remote sensing. Remote Sensing Env 21:311–332

Woodwell GM (1984) The carbon dioxide problem. In: Woodwell GM (ed) The role of terrestrial vegetation in the global carbon cycle. Measurement by remote sensing. Scope Vol 24. Wiley, New York

16 Remote Sensing of Biomass Burning in the Tropics

Y.J. KAUFMAN[1], A. SETZER[2], C. JUSTICE[3], C.J. TUCKER[4], M.C. PEREIRA[2] and I. FUNG[5]

16.1 Introduction

Biomass burning in the tropics, a large source of trace gases, has expanded drastically in the last decade due to increase in the controlled and uncontrolled deforestation in South America (Setzer et al. 1988; Malingreau and Tucker 1988), and due to an increase in the area of cultivated land with the expansion of population in Africa and South America (Seiler and Crutzen 1980; Houghton et al. 1987). In the burning process trace gases and particulates are emitted to the atmosphere, and the ability of the earth to fix CO_2 is substantially reduced (17% of the primary productivity occurs in the humid tropical forests – Atjay et al. 1979; Mooney et al. 1987), and as a result has a strong contribution to the anticipated climate change (Houghton and Woodwell 1989).

In the Amazon, as an example, a common practice associated with forest clearing and land management is the burning of the existing vegetation cover. Although these burnings occur in large numbers every year and have adverse effects on the environment, no estimates have been made to evaluate the magnitude of the problem. No official statistics exist for the extent of burning, and legal procedures to restrict or obtain authorization to burn the vegetation are seldom followed. Considering that the burning occurs in an area of many million square kilometers with poor road access, and that federal and state forest services are not suited to survey the forest, orbital sensing becomes the only practical technique to monitor the Amazon forest on a regular basis (Malingreau and Tucker 1988; Setzer and Pereira 1989; Kaufman et al. 1990).

An example of satellite-derived rate of deforestation is given in Fig. 1 for the state of Rondonia, Brazil (this figure is an expansion of the work of Malingreau and Tucker 1988). The deforestation in 1982 (see Fig. 1a) was much smaller than in 1987 (also in Fig. 1a). The rate of deforestation (per year) is shown in Fig. 1b. Although delayed in Rondonia, similar deforestation trends are found in other parts of the Amazon Basin (Setzer et al. 1988).

[1]NASA Goddard Space Flight Center, Greenbelt, Maryland 20771, also at University of Maryland, College Park, Maryland 20742, USA
[2]National Institute of Space Research (INPE), 12.201 Sao José dos Campos, Sao Paulo, Brazil
[3]NASA Goddard Space Flight Center, Greenbelt, Maryland 20771, also at University of Maryland, College Park, Maryland 20771, USA
[4]NASA Goddard Space Flight Center, Greenbelt, Maryland 20771, USA
[5]NASA Goddard Space Flight Center, Goddard Institute for Space Studies, New York, N.Y. 10025, USA

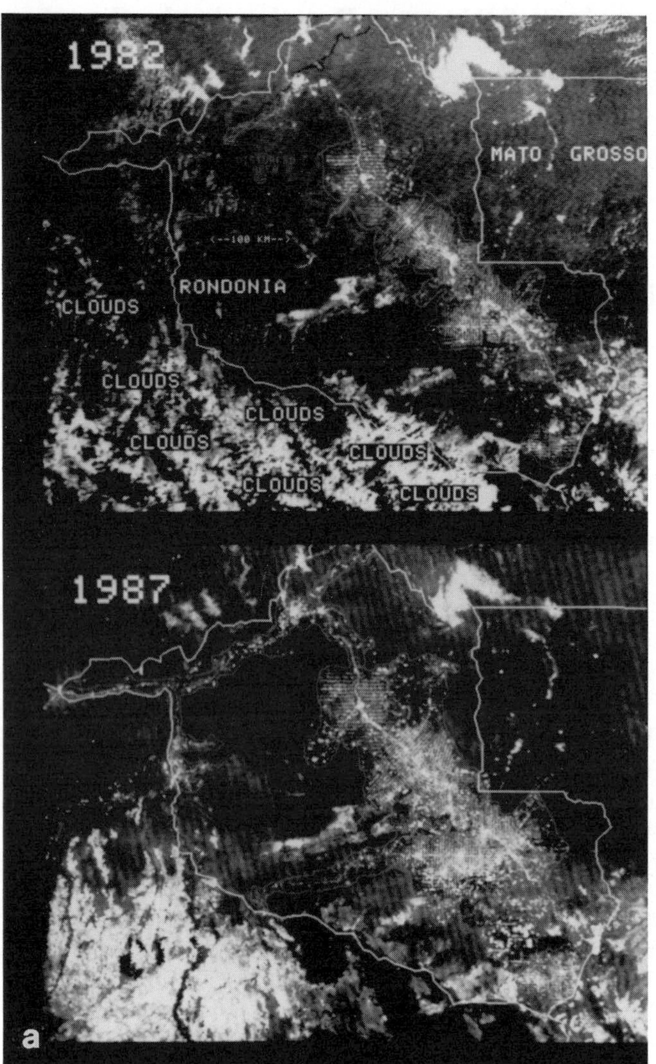

Fig. 1. Rate of deforestation (per year) in the state of Rondonia, Brazil. **a** Deforestation in 1982 and 1987, shown by the response in AVHRR channel 3 (3.7 μm). **b** Accumulated deforestation

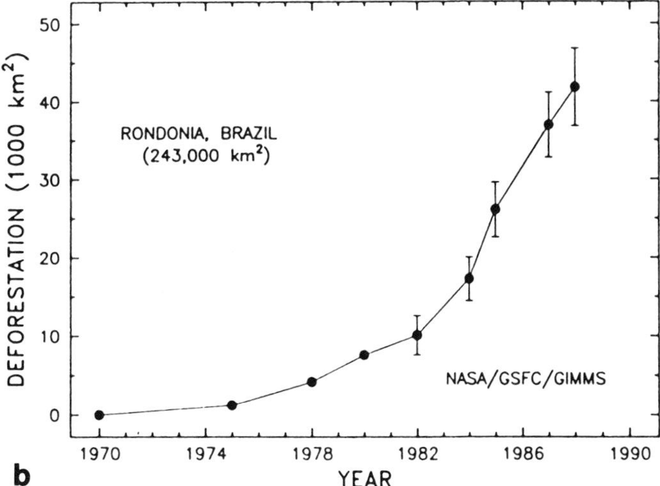

Fig. 1b.

The rationale for the study of biomass burning is linked to the need of the Global Tropospheric Chemistry program (GTC 1986) to understand the sources and sinks of the atmospheric trace gases and their variations in the atmosphere. The importance of the trace gases other than CO_2 (e.g., CH_4 and CO) is in their greenhouse effect, which is comparable to that of CO_2 (Ramanathan et al. 1985), and in their effect on tropospheric chemistry (Crutzen 1988). We shall concentrate in this chapter mainly on remote sensing of trace gases other than CO_2, since biomass burning is generally an inefficient combustion and the relative abundance of trace gases is high. Trace gases also play an important role in atmospheric chemistry, whereas CO_2 is completely oxidized.

Removal of the trace gases from the atmosphere is mainly by oxidation processes. Measurements of the emission rates of trace gases can show the potential of changes in the atmospheric balance. Comparison of the changes of the emission rates with the variations in the background concentration can teach us about the ability of the atmosphere to take care of extra emission, i.e., the cleansing process. For example, methane (CH_4) concentration is increasing by 1% a year (GTC 1986; Blake and Rowland 1986) and has doubled in concentration over the last 100–300 years after being constant for thousands of years (Stauffer et al. 1985). It is not clear whether this increase is due to an increase in the emission of CH_4, a reduction in the oxidation of methane (GTC 1986), and/or an increase in the conversion of other organic gases to methane. The present increase in the tropospheric concentration of ozone (an oxidant), not only in strongly polluted areas (Fishman and Larsen 1987), but also at background stations (Oltmans and Komhyr 1986), suggest that there cannot be a reduction in the oxidation process, and the increase in the methane concentration has to be explained by an increase in the methane emission. The larger seasonal cycles in the CH_4 concentrations in the Northern Hemisphere than in the Southern Hemisphere also suggest the dependence of CH_4 concentration on source variation.

During prescribed and natural fires large amounts of trace gases (e.g., CO_2, CO, CH_4, NO_x) and particles (organic matter, graphitic carbon, and other constituents) are released to the atmosphere. It is estimated that biomass burning contributes 5–30% of the global amounts of CH_3Cl, CH_4, NO_2 (GTC 1986; Crutzen and Graedel 1988) thus having a similar contribution to the atmospheric trace gases as fossil fuel burning (Crutzen et al. 1985; Fishman et al. 1986) and more than half of the contribution of fossil fuel in CO_2 (Seiler and Crutzen 1980; Crutzen et al. 1985; Mooney et al. 1987). It increases the atmospheric greenhouse effect as well as affecting tropospheric chemistry (Crutzen et al. 1979). Trace gases such as N_2O and CH_4, which are stable in the troposphere, can penetrate to the stratosphere, be activated by UV radiation, and affect the chemistry of the stratosphere (Crutzen et al. 1985). There is also evidence of an increased production of N_2O from soil following fires (Anderson et al. 1988). Note that most of the stratospheric NO, that serves as a catalyst in the ozone destruction, originates from tropospheric N_2O (Wayne 1985). NO is responsible for about 70% of the annual global destruction of stratospheric ozone via the nitrogen oxide catalytic cycle (Turco 1985; Cofer et al. 1988). One radiatively active trace species, ozone, is not emitted directly by burning, but despite its secondary nature, it may be affected as strongly by tropical burning as methane or aerosol. Observations made in situ (Crutzen et al. 1985; Andreae et al. 1988) and from space (Fishman and Larsen 1987) suggest that high ozone in the continental tropics, forested or agricultural ecosystems, is most likely associated with burning emissions of pollutants. The fires emit NO, and extrapolation of studies of mid-latitude, predominantly urban pollution (Seinfeld 1986) suggests that it is the NO emission that is most important in the tropics for ozone production (Chatfield and Delany 1988).

The particles emitted from biomass burning may affect the radiation budget and boundary layer meteorology by reflecting sunlight to space and absorbing solar radiation (Coakley et al. 1983). The particles emitted from biomass burning are a major source of cloud condensation nuclei (Radke 1989). Therefore an increase in the aerosol concentration may cause an increase in the reflectance of thin to moderate clouds (Twomey 1977; Coakley et al. 1987) and a decrease in the reflectance of thick clouds (Twomey et al. 1984). Since for most clouds the aerosol effect dominates through an increase in the cloud albedo, Twomey et al. (1984) suggested that the net cooling effect due to the increase in the aerosol concentration can be as strong as the heating effect due to the increase of global CO_2 and other trace gases, but act in the opposite direction (Coakley et al. 1987). Therefore, the net effect of increased concentration of aerosol and trace gases on the earth's energy budget is not obvious.

Biomass burning in the tropics is of particular interest, due to the large extent of forest clearing and agricultural burning. Estimates based on satellite imagery are of 350,000 fires corresponding possibly to 200,000 km^2 of burned vegetation in Brazil's Amazon alone (Setzer et al. 1988; Setzer and Pereira 1989). Biomass burning is also important due to the high solar irradiation in the tropics, that enhances atmospheric chemical reactions (due to the U.V. radiation) and the aerosol climatic effects (more solar radiation to be reflected to space). In addition to the study of the emission of trace gases, it is also important to monitor the emission of the particles, due to their possibly strong, but poorly understood effect on climate (through cloud modification), and their ability to serve as the surface for heterogeneous atmospheric chemistry.

Present methods for the estimation of trace gas emissions from biomass burning are based on measurements of the ratio of the trace gas concentration to the concentration of CO_2 (Andreae et al. 1988), and on a crude global estimate of the rate of emission of CO_2 during biomass burning (Seiler and Crutzen 1980). This estimate may suffer from uncertainty in the overall extent of biomass burning, e.g., rate of deforestation, or agricultural burning practices, and in the density of the burned material. As a result, it is very important to obtain direct information about the fires and the emission products. Satellite imagery provides such an observational tool. It can show the spatial and temporal distribution of fires (at least the ones that are observable during the satellite pass) and provide a measure of the rate of deforestation (Tucker et al. 1984; Malingreau and Tucker 1988), as well as measure one emission product – the emission of particles from the fires (Kaufman et al. 1989). Although satellite imagery cannot be used directly to sense the emitted trace gases, both the mass of emitted particles and the mass of emitted trace gases are proportional to the mass of the burned biomass. In Figs. 2 and 3, the relation between the emitted trace gases and particulates is shown, based on the measurements of Ward (1986) of the mass of the emitted species per unit area of the fire. Due to the variability of the density of the consumed fuel in each case, a strong relation is observed between the trace gases and CO_2 (Fig. 2) or particulates (Fig. 3). Ward (1986) also normalized the emission measurements for the mass of fuel (see Figs. 4 and 5). The normalized emission ratios show the relations between the different emission species, due to variability of the fire efficiency. The correlation between methane emission (CH_4) and particulate emission (see Figs. 3 and 5) shows that estimate of methane emission can be more accurately done from particulate measurements than from fuel consumption measurements.

Once the relationship among the emitted particles, trace gases, and the burned biomass is established (for a given land cover) measurement of the emitted particles can be used to determine the mass of the emitted trace gases, and to some extent, the mass of the emitted CO_2. For example, CH_4 and CO can be computed directly from the measured mass of particles using the relationship between them (Figs. 3 or 5), while N_2O can be computed from the mass of emitted CO_2, using a conversion ratio measured by Cofer et al. (1988) of 0.0002 between the mass of emitted N_2O and the mass of emitted CO_2.

Among satellites that can be used to monitor large fires, the meteorological NOAA series of polar orbiters present a unique capability: daily coverage currently by two satellites with at least four daytime and four nighttime passes for any area; multispectral coverage with the same 1.1 km resolution in visible, near, and thermal infrared bands; and a relatively easy capability of near-real time receiving and processing of images covering areas of continental dimen-

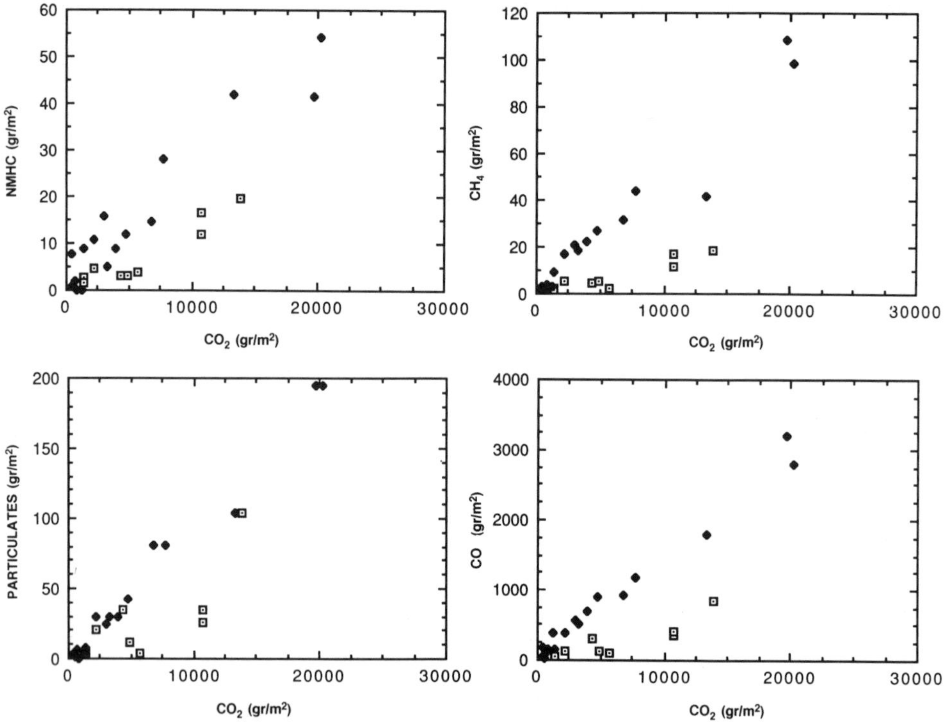

Fig. 2. The relation between emission of CH_4, NMHC, particles and CO and the emission of CO_2 in g of the emitted substance to m² of the burned area. Data are taken from Ward (1986) for flaming (□) and smoldering (♦) conditions

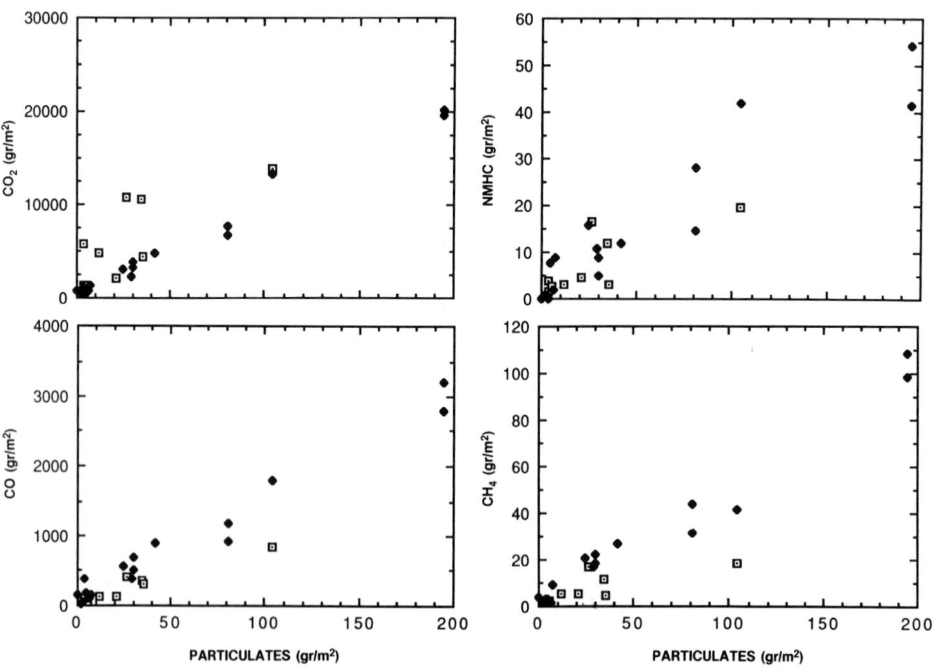

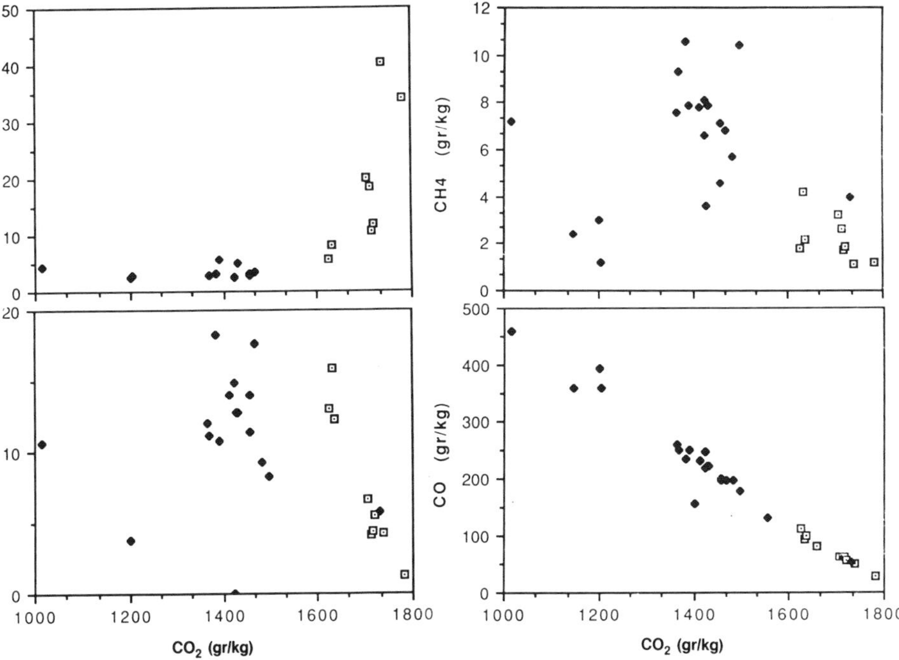

Fig. 4. The relation between emission of elemental carbon, CH_4, particles and CO and the emission of CO_2 in units of the ratio of mass of the emitted substance to the mass of burned fuel. Data are taken from Ward (1986) for flaming (□) and smoldering (♦) conditions

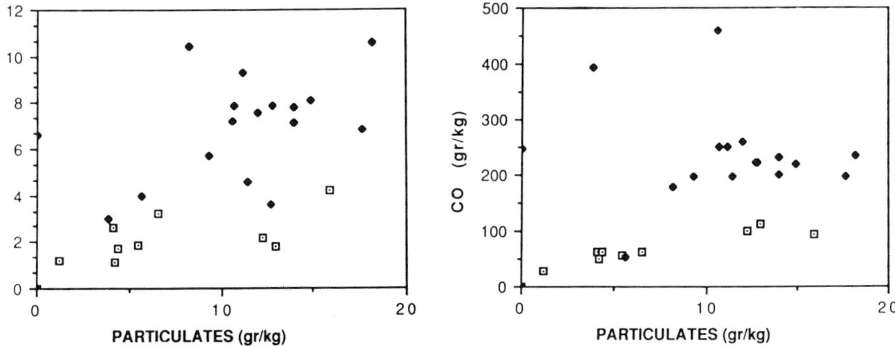

Fig. 5. The relation between emission of CO and particles, and CH_4 and particles (in units of the ratio of mass of the emitted substance to the mass of burned fuel). Data are taken from Ward (1986) for flaming (□) and smoldering (♦) conditions

Fig. 3. The relation between emission of CO, CO_2, CH_4, and NMHC and the emission of particles (in gram of the emitted substance to m^2 of the burned area). Data are taken from Ward (1986) for flaming (□) and smoldering (♦) conditions

sions. LANDSAT TM and SPOT (30 and 10 m resolution, respectively) can also be used and provide much finer details of the fires and burning areas, but such systems are of limited use for regional monitoring due to the high data volumes associated with the high spatial resolution and low temporal repeat frequency of many days.

16.2 The NOAA-AVHRR Series

The NOAA-AVHRR series provides a set of images in five spectral bands (four AVHRR passes are available daily). Its visible (0.63 μm) and near-IR (0.85 μm) channels can provide information (during the day) on the surface characteristics and the smoke loading and characteristics (Kaufman et al. 1990; Ferrare et al. 1990). The 3.7-and 11-μm channels can be used, during day and night, to monitor the number and spatial distribution of the fires. Therefore it should be an adequate tool for regional and global remote sensing of fires and their emissions. The analysis of AVHRR bands 1 and 2 requires accurate satellite calibration (less than 10% absolute error, less than 1% change during a 20-day period and less than 5% error in relative calibration between the channels (Kaufman et al. 1990). Presently, the AVHRR first two channels are calibrated using radiances measured over remote ocean areas and deserts (Kaufman and Holben 1990) and by an aircraft-calibrated scanning radiometer above White Sands National Monument in New Mexico (B. Guenther pers. commun.). It is expected that with these calibrations such requirements can be met.

A preliminary operational scheme for monitoring biomass burning using 1-km data from the AVHRR system was devised by the Brazilian Space Institute (INPE/MCT) for the 1987 Amazon burning season. The afternoon pass of the NOAA-9 was recorded by INPE over central South America. The images were processed on the day of recording to detect large forest burnings, and by the following day the geographical location of the main fires was available to state representatives of the Brazilian Institute of Forest Development (IBDF/MA).

16.3 Remote Sensing of Fires, Smoke, and Trace Gases

The radiative temperature of the pixel from AVHRR channels at 3.7 and 11 μm can be used to identify fires (Dozier 1980; Matson and Dozier 1981; Matson et al. 1987). Since the fire line cannot fill a pixel, we may consider the pixels to be only partially filled with fires. Although for some hot objects the two channels can be used to identify the hot object temperature and its subpixel size, for fires (temperature > 500 K), even if 0.01–0.1% of the pixel is covered by a fire it is enough to saturate the 3.7 μm channel of the AVHRR, and have a small effect in the 11 μm channel (see Fig. 6). The AVHRR sensor was not designed for fire monitoring and both AVHRR channels saturate around 320 K (Matson et al.

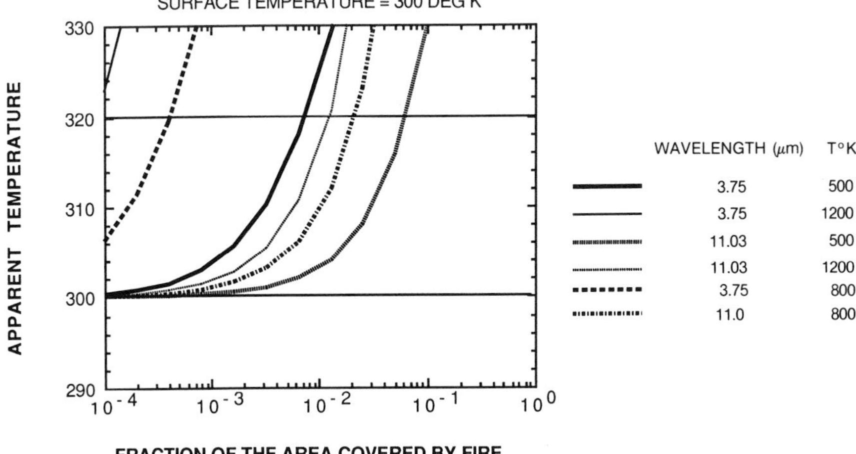

Fig. 6. The apparent radiative temperature of a satellite pixel composed of a fire of temperature of 500, 800 or 1200 K and a nonfire area (300 K) as a function of the fraction of the area covered by the fire, at 3.7 μm and 11 μm. The apparent temperature was computed from the Planck black body equation, as the temperature of a black body that emits radiance equivalent to the weighted average radiance of the fire and the background

1987). As a result, AVHRR digital imagery can be used to count the number of fires, and check the magnitude of the fire size, but not to estimate the fire size. Note that flaming fires (temperature > 800 K) that occupy 10^{-4} of the pixel area can be observed, while smoldering (temperature > 500 K) areas have to occupy more than 10^{-3} of the pixel area to be observed.

There are two major fire phases in which particles and trace gases are emitted. The flaming phase ranges from a condition of very high rates of heat release (1 to 10 mw/m^{-2}) to fires of only a few cm of flame length with very low rates of heat release (less than 0.1 mw/m^{-2}-Ward pers. commun.). During a low intensity flaming phase the combustion is relatively efficient, releasing only small amounts of particulate matter and trace gases. High intensity flaming combustion results in the release of a higher concentration of graphitic carbon but only a moderate amount of trace gases (Ward and Hardy 1984 and Fig. 4). Smoldering results in an inefficient burning and a release of large amounts of trace gases and particles that contain a higher percent composition of organic matter (Ward and Hardy 1984; Ward 1986), but a much smaller amount of graphitic carbon (see Fig. 4). From satellite imagery it may be possible to estimate the relative contribution of smoldering and flaming processes to the mixture of combustion products, by deriving the single scattering albedo (ω_0) of the smoke (Kaufman 1987). ω_0 is the ratio of light scattering to light extinction by the smoke and is related to the amount of graphitic carbon in the smoke (Ackerman and Toon 1981). Laboratory measurements (Patterson and McMahon 1984) show that under flaming conditions 25 times more graphitic carbon is released than under smoldering conditions. Graphitic carbon is a very strong absorber of light in the solar spectrum, and therefore ω_0 is around 0.97 for smoldering and as low as 0.66 for flaming conditions (Patterson and McMahon 1984). Measurements of the aerosol absorption in field fires (Patterson et al. 1986) showed a much smaller range of variation that corresponds to ω_0 in the range 0.90–0.96. Radke et al. (1988) used aircraft sampling to find a variation in ω_0 from 0.75 for a strong fire of dry wood to 0.82–0.91 for typical large prescribed fires (Radke et al. 1988). Measurements of total emitted particulates mass and graphitic carbon (Andreae et al. 1988) in the Amazon basin show that 7% of the particulates is graphitic carbon which corresponds to $\omega_0 \sim 0.96$ and indicates a large portion of smoldering in the emission.

Given current sensor capability, satellites cannot observe directly the emitted trace gases in the lower troposphere. Only concentration of tropospheric ozone (Fishman et al. 1986; Fishman and Larsen 1987) was derived indirectly by subtracting from the total ozone estimated from the TOMS (Total Ozone Mapping Spectrometer) data and stratospheric ozone derived from the SAGE (Stratospheric Aerosol and Gas Experiment) experiment. Therefore, in order to obtain information about the emission products from fires, a remote sensing method is proposed that concentrates on detection of fires and remote sensing of the emitted particles, and translation of these observations into trace gas emission using measured ratio of particle concentrations and trace gas concentrations in the smoke for several fire conditions (see Figs. 3 and 5). Alternatively, the number of fires can be used to estimate the trace gases emission by using measurements of the burned biomass in an "average" fire and average ratios between the emitted trace gases and the emitted CO_2.

The data of Ward and Hardy (1984) and Ward (1986) indicates that there is a strong relationship between particulate matter emission (M_{part}) and CH_4 emission (M_{CH4}), see Fig. 3. The relation is somewhat different for flaming or smoldering:

$$M_{CH4} = 0.6 M_{part} \quad \text{for smoldering conditions, and} \qquad (1)$$
$$M_{CH4} = 0.3\ M_{part} \quad \text{for flaming conditions.} \qquad (2)$$

M_{CH4} can be derived from M_{part} with uncertainty of 20% if the stage of the fire is determined, and 50% if it is not (see Fig. 3). Similar relations were found for NMHC:

$$M_{NMHC} = 0.3 M_{part} \quad \text{for smoldering conditions, and} \qquad (3)$$
$$M_{NMHC} = 0.2 M_{part} \quad \text{for flaming conditions.} \qquad (4)$$

For CO similar relations were found, except that the emission of CO is better correlated with the emission of CO_2 (see Fig. 4), than with the emission of particles for the smoldering phase. Nevertheless, CO can also be determined from the derived particulate mass with similar uncertainties (see Figs. 3 and 5). The measurements of Ward (1986, see Fig. 5) indicate that the relations between CO and particles emission are:

$$M_{CO} = (17 \pm 5)M_{part} \quad \text{for smoldering conditions, and} \quad (5)$$
$$M_{CO} = (10 \pm 5)M_{part} \quad \text{for flaming conditions.} \quad (6)$$

Since particulate matter concentrations in the atmosphere can be derived from satellite imagery (Fraser et al. 1984; Kaufman and Sendra 1988a,b; Kaufman 1987), therefore the mass of CH_4, CO and other trace gases can be derived as well. The relationship between the emission of particulates and the emission of CO_2 can also be found from Fig. 4 (data of Ward 1986) as:

$$M_{CO2} = (100 \pm 30)M_{part} \quad \text{for smoldering conditions, and} \quad (7)$$
$$M_{CO2} = (170 \pm 80)M_{part} \quad \text{for flaming conditions.} \quad (8)$$

It is not completely clear at this point how well these relationships represent the conditions found in the tropics. Measurements of emitted trace gases were performed in Brazil (Greenberg et al. 1984; Crutzen et al. 1985; Andreae et al. 1988) but the concentration of submicron aerosols was not measured. Greenberg et al. (1984) measured the relation between the emitted CH_4 and CO and the emitted CO_2. They found that the volume ratio of CO and CO_2: V_{CO}/V_{CO2} is 0.12 (range between 0.05 and 0.30) which corresponds to mass ratio M_{CO}/M_{CO2} of 0.08 (range between 0.03 and 0.20). It is assumed here that the low part of that range corresponds to flaming conditions, while the upper part of the range to smoldering conditions, similar to the measurements of Ward (1986) shown in Fig. 4). This range of mass ratio results in production rate for CO of 50 (g kg^{-1} fuel) for flaming and 280 (g kg^{-1} fuel) for smoldering. These production rates are similar (within 30%) to the production rates measured by Ward (1986) and presented in Fig. 4. Table 1 summarizes these results. For comparison, the derived rates from the volume ratios measured by Andreae et al. (1988) for mixed smoldering/flaming conditions result in emission rates smaller than the above (see Table 1). A similar comparison is carried for CH_4 in Table 1, resulting in a similar fit. The variation between the measurements of Greenberg et al. (1984), Ward (1986), and Andreae et al. (1988) results from different fuel and fire characteristics. By using the ratio between CH_4 and aerosol and between CO and aerosol, this variation can be decreased, since all these three biomass burning products are released in larger quantities with the reduction of the fire efficiency (which is not true for CO_2).

Water vapor condensation on the aerosol particles may result in an uncertain estimate of the nonwater part of the emitted particles. The moisture contained in the biomass and water generated through combustion can make only a small increase in the relative humidity of the smoke (Ward et al. 1979). Most of the water vapor in the smoke was in the ambient air before the fire. Therefore, the effect of water vapor on aerosol scattering characteristics depends on the atmospheric profile of the relative humidity and the altitude of the smoke. Andreae et al. (1988) found that in the Amazon Basin the smoke was emitted in thin layers located 1–4 km from the ground. In the relatively dry conditions in the fire season, the relative humidity in the lowest 4 km is between 40 and 70% for savannah and 50 and 85% for tropical forest regions (Crutzen et al. 1985). Hänel

Table 1. Comparison between production rates for CO and CH_4

Quantity and source	Smoldering	Flaming	Mixed
M_{CO}/M_{CO2} [1]	0.20	0.03	0.08
CO_2 production rate ($g\ kg^{-1}$ fuel)[2]	1400	1700	1550
CO resultant production rate ($g\ kg^{-1}$ fuel)	280	50	120
M_{CO}/M_{CO2} [3]			0.056
CO resultant production rate ($g\ kg^{-1}$ fuel)			85
M_{CO}/M_{CO2} [4]			0.09
CO resultant production rate ($g\ kg^{-1}$ fuel)			140
CO production rate ($g\ kg^{-1}$ fuel)[2]	220	70	150
M_{CH4}/M_{CO2} [1]	0.008	0.002	0.004
CO_2 production rate ($g\ kg^{-1}$ fuel)[2]	1400	1700	1550
CH_4 resultant production rate ($g\ kg^{-1}$ fuel)	11	3.5	6
CH_4 production rate ($g\ kg^{-1}$ fuel)[2]	7	2	5

[1] Greenberg et al. (1984), [2] Ward (1986) and Fig. 4, [3] Andreae et al. (1988), [4] Crutzen et al. (1985).

(1981) developed a model of the relationship between relative humidity and particle size. According to this model, for a relative humidity of 70%, it is expected that roughly 50% of the particulate mass will be water, as long as there is no saturation (see also Kaufman et al. 1986). Saturation can be detected as cloud formation resulting in high spatial nonuniformity in the smoke. We estimate that the error in the derivation of the dry aerosol mass due to the error in the humidity effect is $\pm$ 30%.

16.4 Remote Sensing of Aerosol Characteristics

Satellite imagery of the earth's surface in the visible and near-IR part of the spectrum were used to derive the aerosol optical thickness (a measure of the aerosol mass loading in the atmosphere), its single scattering albedo (ratio between scattering and total extinction – a measure of the presence of graphitic carbon in the aerosol) and the particle mass median size. The remote sensing technique is based on the difference in the upward radiance, reaching the satellite sensor, between a hazy day and a clear day (Fraser et al. 1984; Kaufman 1987; Kaufman et al. 1990; Ferrare et al. 1990). The aerosol optical thickness can be detected from the radiances above water or land for most land covers (except bright soil or sand). The single scattering albedo can be detected over a sharp contrast in the surface reflectance, e.g., a sea shore as well as over vegetated areas (here the contrast is between the low reflectance in the visible and the high reflectance in the near IR). The particle size is derived from the wavelength dependence of the optical thickness. The aerosol optical thickness can be detected in AVHRR imagery in more than one wavelength only over water. Therefore, the particle size can be derived only from images acquired over water. Derivation of the aerosol optical thickness using a single image was applied by Kaufman and Sendra (1988a,b) over forests. The methods were applied to GOES (Fraser et al. 1984), AVHRR (Kaufman et al. 1990; Ferrare et al. 1989) and LANDSAT MSS imagery (Kaufman and Sendra 1988a,b). The derived aerosol optical thickness was verified by sunphotometer measurements from the ground (Fraser et al. 1984; Kaufman and Sendra 1988a,b). The derivation of the single scattering albedo was tested in a laboratory simulation (Mekler et al. 1984; Kaufman 1987).

16.5 Satellite Estimation of Gaseous Emission from Biomass Burning

The remote sensing techniques are based on the analysis of NOAA-AVHRR 1 km resolution imagery. The 3.7-μm and the 11-μm channels are used to detect fires and distinguish them from clouds and hot soil (that may also saturate the 3.7-μm channel). The mass of particulates in the smoke and their single scattering albedo (ω_0) are determined from the visible and near-IR channels (ω_o is used to distinguish between the relative contribution of smoldering and flaming to the smoke). In the following sections two methods for the remote sensing of the emitted trace gases are applied. In the next section, the average emission of particulates per fire is determined from the satellite imagery, and converted to the emission of trace gases by Eqs (1)–(8). The total emission is found by multiplying by the total number of fires. In Section 16.5.2, the total number of fires is used to compute the total burned biomass using the amount of biomass burned in an "average fire". The amount of burned biomass is converted into the emitted trace gases using emission rates from the literature.

16.5.1 Estimate Based on the Average Emission of Particulates per Fire

Since fires are better spatially defined in the image than smoke, first the average emission of particulates per fire (in a given area and season) is estimated and the corresponding total amount of fires (in the same season and area) is computed. The total mass of the emitted particulates is computed as a product of the emitted mass per fire and the total number of fires. This indirect procedure has two advantages. Statistically, it surveys the area by counting fires, that due to their distinct spatial nature are more easily defined than smoke. On the other hand, the disadvantages in counting fires (we cannot distinguish between large and small fires, we may miss very small fires or fires that were extinguished before the satellite pass) are eliminated by the fact that in the procedure we first divide by the number of fires in a given area and day to find the emission per fire, and then multiply by the total number of fires to find the total emission. In this way the subjectivity of defining fires by the satellite capability is canceled in the process. The mass of the emitted particulates is converted to the mass of emitted trace gases and CO_2.

16.5.1.1 Basic Assumptions

The basic assumptions in the remote sensing analysis are the following:

- For each land management zone and season it is possible to define an "average fire" so that if the characteristics of a large sample of fires are established, the total emitted mass of particles and trace gases can be found from the total number of fires.
- The relations between the emission rate of particles, trace gases, CO_2, and the mass of burned biomass, that are found from prescribed fires (Ward and Hardy 1984; Ward 1986) can be applied to the fires in the tropics and other regions of the world. In this regard it is helpful that the fires monitored by Ward and others consumed different types of wood and were carried in varying meteorological conditions (Ward 1986).
- There is a direct relationship between the particulate characteristics measured by Ward (1986) and the particulate characteristics measured from space. The relation between the mass of particles measured from the ground and derived from satellite radiances were discussed by Fraser et al. (1984). Past measurements of particles and gases made using sampling devices suspended from towers have compared favorably with measurements made using airborne sampling techniques (Ward et al. 1979).
- It is possible to separate the contribution of liquid water and the emitted material from the fire to the light scattered by the smoke. The contribution of the liquid water is found from climatology of the area as well as from local meteorological information.
- It is possible to assume a value of the refractive index and size distribution of the particles.

16.5.1.2 Estimation of the Emission Rates per Fire

For a given land cover category and season, it is assumed that the range of fires and resulting emissions can be represented by an average fire. To find the emissions from an average fire, the emitted smoke has to be related to the fires from which it is emitted. For this purpose, an area in the satellite image is determined for which there is a direct relation between the smoke and the fires that emitted the smoke (see Fig. 7). The average rate of emission and characteristics of the aerosol part of the smoke is estimated from AVHRR channels 1 and 2. These two reflective channels are calibrated using the calibration of Kaufman and Holben (1990). The aerosol optical thickness is computed from the difference in the radiances in channel 1 between the area that contains the smoke and a nearby background radiance. The integrated mass of smoke is computed from the derived optical thickness, integrated on the area covered by the smoke. A conversion factor is used to convert the optical thickness into the mass of aerosol per unit area of the surface. For the assumed average particle radius (averaged on the mass distribution) of 0.3 μm, the conversion factor is 0.22

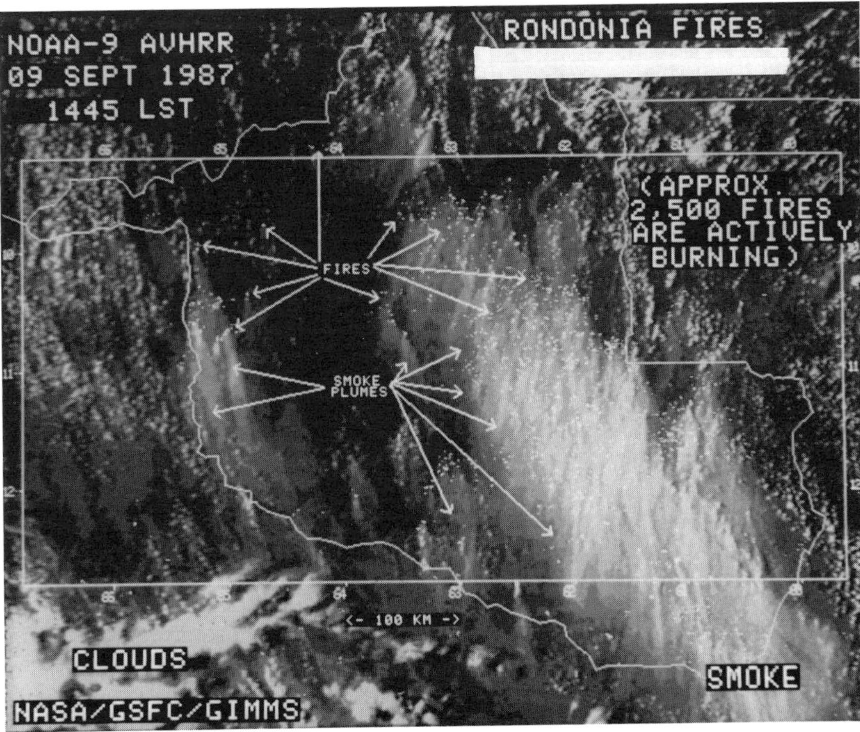

Fig. 7. The fires and smoke of biomass burning due to deforestation in the state of Rondonia, Brazil. The image was generated from AVHRR channels 1–4. Fires are shown by *white dots* (based mainly on the radiance in channel 3–3.7 μm), vegetation is *dark* (based on the radiance in channels 1–0.63 μm and 2–0.85 μm), smoke is *gray* (channels 1 and 2) and thick clouds are *white* (channels 1, 2, and 4–11 μm). The latitude and longitude lines are indicated. In the image there are 1500 fires and as a result, heavy smoke covers 50,000 km^2

g m^{-2} (assuming density of 1 g cm^{-3}). This particle radius of the emitted smoke is taken as the average of the radius measured in situ (Radke et al. 1988) and by remote sensing (Puschel et al. 1988; Kaufman et al. 1990; Ferrare et al. 1990). For comparison Tangren, (1982) derived a value of 0.28 g m^{-2} from a comparison of particulate mass and a nephelometer scattering coefficient. Since our remote sensing method is sensitive only to small particulates (less than 2 μm) it is reasonable to use a somewhat smaller value. If half of the aerosol mass is liquid water (see discussion in Sect. 16.3), then the conversion factor from the aerosol optical thickness to the dry aerosol mass is 0.11 g m^{-2}.

The value of ω_0 (the single scattering albedo) is determined from the relation between the increase in the upward radiance, due to the presence of the smoke, in channel 1 and channel 2 (Fraser and Kaufman 1985; Kaufman 1987; Kaufman et al. 1990; Ferrare et al. 1989) and is used to estimate the relative contribution of smoldering and flaming phases to the total emissions. Since the AVHRR can measure the smoke release only once or twice a day (morning and afternoon passes), there is a need to estimate the relation between the smoke seen from the AVHRR and the actual daily generation of smoke, from more frequent platforms – GOES (Fraser et al. 1984) and/or ground observations (Kaufman and Fraser 1983). Cloud analysis from the AVHRR is used to estimate the effect of clouds on fire detection and the effect of condensation and cloud generation on the aerosol. The following is the algorithm to estimate the emission of particulates per fire:

1. From GOES and/or ground observations estimate the ratio t_{fire}/t_{avhrr} where t_{fire} is the average daily duration of most of the emission and t_{avhrr} is the duration of substantial emission till the AVHRR image is taken.
2. From the AVHRR find the total dry mass of the smoke M_a in a given area, the average value of ω_o and the number of fires that contribute to this smoke.
3. Estimate the fractional contribution of flaming and smoldering to the smoke from ω_o, then estimate the emission ratios $C_{CH4}(\omega_o)$, $C_{CO}(\omega_o)$, etc.
4. Compute the emitted mass M_{CH4}, M_{CO}, M_{af} (total dry aerosol mass) etc: $M_i = (t_{fire}/t_{avhrr}) * M_a * C_i(\omega_o)$, where i is CH_4, CO etc.
5. Compute the average emission of particles and gases per fire (by dividing the emitted mass by the number of fires).

In 1987 the AVHRR afternoon pass was around 15.30 h LST. Assuming that the fires start around 11.00–12.00 h LST and produce most of the emission in 7 h, then the ratio is t_{fire}/t_{avhrr} and 0.5 ± 0.2.

16.5.1.3 Remote Sensing of Fires and Total Emitted Mass

The radiative temperature of the pixel from AVHRR channels at 3.7 and 11 μm is used to identify fires. The 11 μm channel is used to distinguish between fires, reflective clouds, and hot surface areas. A hot surface ($320 < T < 400$ K) will saturate both the 11 μm and the 3.7 μm channels. For pixels that include fires, the radiance in the 3.7 μm channel will be much larger than in the 11 μm channel and usually saturated (see Fig. 6). Water clouds that saturate the 3.7 μm channel, will usually result in low radiances in the 11 μm channel, and have an irregular shape. A computer program determines the amount of fires, based on these criteria (see Sect. 16.5.1.4 for details).

The total emission during the fire season is computed by multiplying the emission per fire by the number of fires. The emission is adjusted for the cloudiness (that covers some of the fires) by the cloud fraction. The cloud fraction is found from the satellite images, based on a threshold reflectance in the visible band.

16.5.1.4 Accuracy Estimates

The process of computation of the particles and trace gases emission involves several steps (as discussed in the previous paragraphs). In the following the errors in each step are estimated:

- Computation of optical thickness from the detected radiances – 30% error (Fraser et al. 1984).
- Computation of the total aerosol mass from the optical thickness – 30% error (Fraser et al. 1984).
- Computation of the dry aerosol mass from the total aerosol mass – 30% error (Kaufman et al. 1986).
- Conversion to trace gases from the dry aerosol mass – 40% error (see Figs. 3 and 5).
- Integration on a whole area and emission time – 60% error (Ferrare et al. 1990).

These errors are errors in multiplicative factors (e.g., the relation between wet aerosol mass and optical thickness or trace gases and dry aerosol mass), as a result the final product "P" can be described as: $P = \Pi P_i$, where P_i are the multiplicative factors. The multiplicative error in P (e) can be described by:

$$eP = \Pi e_i P_i, \text{ and therefore } \quad e = \Pi e_i.$$

The accumulated r.m.s. error, is found from the logarithm of e ($\log e = \Sigma \log e_i$) and the r.m.s. error is:

$$\text{r.m.s.}(\log e) = \sqrt{[\Sigma(\log e_i)^2]} = 0.76. \rightarrow e \approx 2.$$

Therefore the estimate of the emission rate is within a factor of 2.

16.5.1.5 Application of the Techniques

Satellite measurements of the aerosol emission and the presence of fires was used to detect the rate of emission of particulates and trace gases from biomass burning. In the following an example is given of analysis of emission from biomass burning due to deforestation during the dry season in Brazil (a limited area 6.5–15.5° south and 55–67° west is analyzed currently). The analysis included the following steps:

Calibration of the AVHRR Visible and Near-IR Channels. These channels were calibrated using the calibration derived for 1987 by Kaufman and Holben (1990). The satellite data were first calibrated by the preflight calibration supplied by NOAA together with the data. The radiances in the visible and near-IR channels were further divided by 0.81, in order to compensate for the deterioration in the calibration of the two channels.

Smoke Analysis. The surface reflectance was determined from radiances recorded over areas where local smoke was not observed (i.e., uniform areas) and assuming that the aerosol optical thickness is zero. The meaning of this assumption is that the background smoke is considered as part of the background radiance that includes the surface reflectance. Therefore, the determination of the amount of smoke emitted per fire is based on the freshly emitted smoke, and on the observed fires. This analysis is applied whenever the fresh smoke is easily observed and related to the fires from which it was emitted. In the State of Rondonia the range of surface reflectance in channel 1 was found to be 0.05–0.13, and in channel 2: 0.22–0.34. Similar reflectances were determined in other areas (e.g., Bolivia, Para, and Matto Grosso). The fresh smoke increased the radiance (normalized to reflectance units: $L' = \pi L/F_o\mu_o$ where F_o is the extraterrestrial solar flux and μ_o is cosine of the solar zenith angle) in channel 1 from 0.06–0.12 to 0.15–0.25; and in channel 2 from 0.16–0.26 to 0.18–0.32. The increase in the radiance in channel 1, corresponds to an increase in the aerosol optical thickness of 0.5–3.0, and the increase in channel 2 implies single scattering albedo values approaching 1 (see Tables 2 and 3). Therefore it is concluded that smoldering rather than flaming is the major contributor to the emission products that we see in Brazil.

Table 2. Examples of the results for the smoke area (A), the number of fires (N), the aerosol single scattering albedo (ω_o), and the aerosol mass per fire (M_{af}) for the state of Rondonia, Brazil, as analyzed from the AVHRR imagery

Date	A smoke area (km^2)	N number of fires	ω_o single scat. albedo	M_{af} aerosol mass /fire (t)
7.31.87			0.99	60
8.10.87	54,000	1361	0.99	15
8.11.87	75,000	2302	0.99	22
8.12.87	70,000	3597	0.99	11
8.20.87	130,000	2010	1.00	23
8.21.87	153,000	2036	0.99	40
8.22.87	43,000	1570	0.98	21
8.23.87	54,000	720	0.96	78
8.27.87	197,000	1929	0.96	58
9.01.87	36,000	760	0.98	52
9.08.87	27,000	1830	0.97	26
9.09.87	150,000	4709	0.97	21
Average quantity			0.98 ± 0.01	35 ± 20

Table 3. Average emission rates from different fire regions (the emission represents the fire emissions before the satellite pass, per fire)

Fire region	Central location		ω_o single scat. albedo	Aerosol mass/ fire (t)	CH_4 mass/ fire (t)	CO mass/ fire (t)	CO_2 mass/ fire (t)
	Latitude	Longitude					
Rondonia	12°S	62°W	0.98	35 ± 20	20	600	3500
Bolivia	14°S	64°W	0.97	55 ± 20	33	930	5500
Mato Grosso	11°S	57°W	0.99	43 ± 15	26	730	4300
Average			0.98	45 ± 20	26	750	4500

The uncertainty in the estimation of the trace gases is 70% for CH_4 and CO and 80% for CO_2.

Analysis of Emission Rates per Fire. Due to the derived high value of the single scattering albedo (0.98) the emission rates per fire summarized in Table 3 are based on the relationship between the emission of particulates and trace gases for smoldering conditions [Eqs. (1), (3), (5), and (7)]. Remote sensing of the emitted aerosol and its conversion to the emitted trace gases (CO, CH_4, CO_2, etc.) shows (see Tables 2 and 3) that on average each fire in the tropics contributes 45 t of particulates, 25 t of CH_4, 750 t of CO, and 4500 t of CO_2. These emission rates per fire are for the total emission (before and after the satellite pass during 1 day), based on the ratio of the emission before the satellite pass and the total emission of 0.5 (see Sect. 16.5.1.2). If the fire continues more than 1 day, it will be counted again in the second day, but also its emitted smoke will be measured in each day separately. Therefore, in this analysis, in each day a fire can be considered as an independent fire.

Fire Count. Fires were counted based on the radiative temperatures in channel 3 – T_3 and 4 – T_4 of the AVHRR. A fire was identified if the radiative temperatures T_3 and T_4 fulfilled the following criteria:

$$T_3 \geq 316\ \mathrm{K},\ T_3 \geq T_4 + 10\ \mathrm{K}\ \text{and}\ T_4 > 250\ \mathrm{K}. \tag{7}$$

The first criterion requires that in order that a pixel is considered to contain a fire (or fires), it has to be very hot (close to the 320 K saturation level). The second criterion requires that the radiative temperature has to be much larger in channel 3 than in channel 4 if the pixel contains a fire rather than a warm surface (bare soil or dry grass in the tropics may be as hot). The third criterion requires that the pixel does not include strongly reflective clouds in the 3.7-μm channel, which may saturate the 3.7-μm channel.

Subject to these criteria, fires were counted during 49 out of 77 days (July 4 till Sept 18) in the dry season in Brazil, for which data were available. The fire count for the area between 6.5 and 15.5° south and 55 and 67° west is shown in Fig. 8 as a function of the Julian date. In order to display the sensitivity of the fire count to the apparent temperature of the fire in the 3.7-μm channel, three thresholds for the first criterion in Eq. (7) were used: 316 K, 318 K, and 320 K (shown in accumulative form in Fig. 8). The high correlation between these three temperature ranges shows that the criteria identify fires, that there is a good correlation between small (or less hot) fires and larger (or hotter) fires. Decreasing the threshold further resulted in contamination of the fire count from warm surface areas. A total of 115,000 fires were counted. The cloudiness was also estimated from the satellite imagery for the same period. Clouds may hide part of the fires from the sensor, but they also may indicate the presence of rain and thus less fires (see Fig. 9). Due to the small correlation between clouds and fire, it was assumed in this analysis that the presence of fires is independent of the presence of clouds, and that clouds with reflectance larger than 0.4 in the visible do hide fires. Accounting for the cloudiness in each day for which fires were counted increased the total number of fires to 132,000. In Fig. 10 the fires counted in each individual day are plotted together with the number that is corrected for the effect of clouds. Extrapolating these results to the whole fire

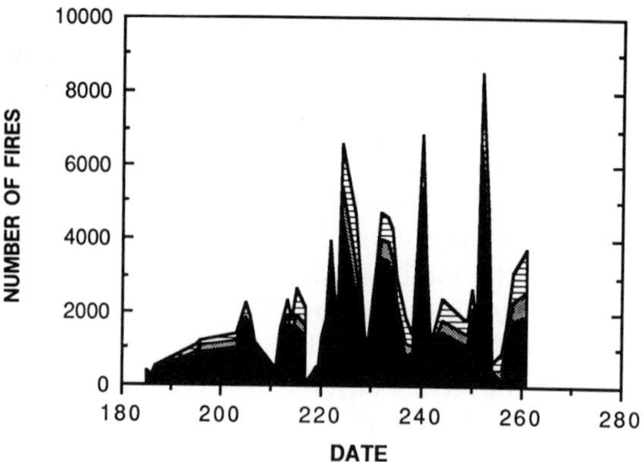

Fig. 8. Number of fires as a function of the day of the year for 1987. The fires were counted, based on the radiance in the AVHRR 1 km resolution imagery, in channels 3 and 4, over the area between 55 and 67° west, and 5.5 and 15.5° south. The *black area* represent fires that saturate channel 3 (3.7 μm), the *gray area* are additional fires that are 2 K colder, and the *dotted area* fires that are 4 K colder

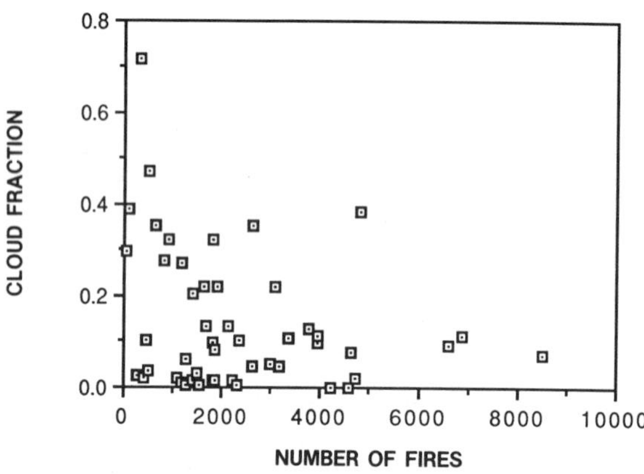

Fig. 9. Total cloud fraction as a function of the number of fires during the dry season in 1987, for the area between 55 and 67° west, and 5.5 and 15.5° south

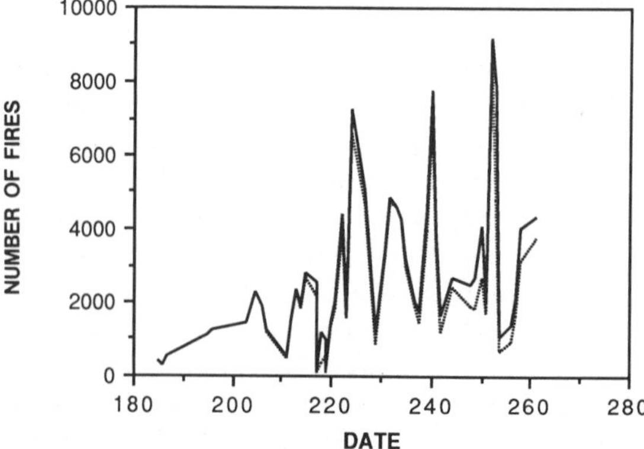

Fig. 10. Number of fires counted as a function of the day of the year for 1987 (*dotted line*) and corrected for the effect of clouds (*solid line*)

Table 4. Emission from biomass burning

Source	Particulates ($\times 10^{12}$ g)	CH_4 ($\times 10^{12}$ g)	CO ($\times 10^{12}$ g)	CO_2 ($\times 10^{12}$ g)
1. Estimate based on fire count and remote sensing of the aerosol (limited area: 6.5–15.5° south and 55–67° west)	11	7.3	180	1,100
2. Extrapolation of results in (1) to global deforestation in the tropics		73	1800	11,000
3. Estimate based on fire count and "average fire size" (area covering most of Brazil)		6	90	1,800
4. Global estimate (Crutzen et al. 1985)		40	800	10,000

Comparison to global budgets

Source	CH_4 ($\times 10^{12}$ g)	CO ($\times 10^{12}$ g)
Sources (Seinfeld 1986)	570–825	650–2200
Sinks (Seinfeld 1986)	605–665	1500–2500

season of 90 days, between July 1 and Sept. 31, we get 242,000 fires. Using the average emission rates per fire (in its present definition) from Table 3 we get total emission from the study area during the fire season shown in Table 4a.

16.5.2 Estimate Based on Average Biomass Burned per Fire

Forty six images from July 15 to October 2, 1987, were analyzed and pixels in band 3 corresponding to an AVHRR temperature reading above 315 K were selected as burning sites by a digital nonsupervised clustering algorithm. These pixels were next examined in the visible channel to verify if a smoke plume was associated with each of them; the image was also visually screened to ascertain if any significant plumes existed without a "fire pixel". Figure 7 corresponds to such an image for the area of the State of Randonia. The number of pixels thus classified was obtained for the Brazilian Amazon as well as for the individual states of the region. The most northern areas of the states of the Amazons and Para, and the Territory of Rorãima and the State of Amapa were not examined because they were out of the satellite range as viewed by the direct receiving station, which is located at about 23°S. Figure 11 shows thousands of fires burning along a belt from Belem, at the mouth of the Amazon River, to Porto Velho, capital of the State of Rondônia, stressing the current geographical limits of forest conversion in Brazil.

Table 5 presents a summary of the data obtained through 1987 and estimates of areas burned. The states of Acre and Maranhão had less frequent

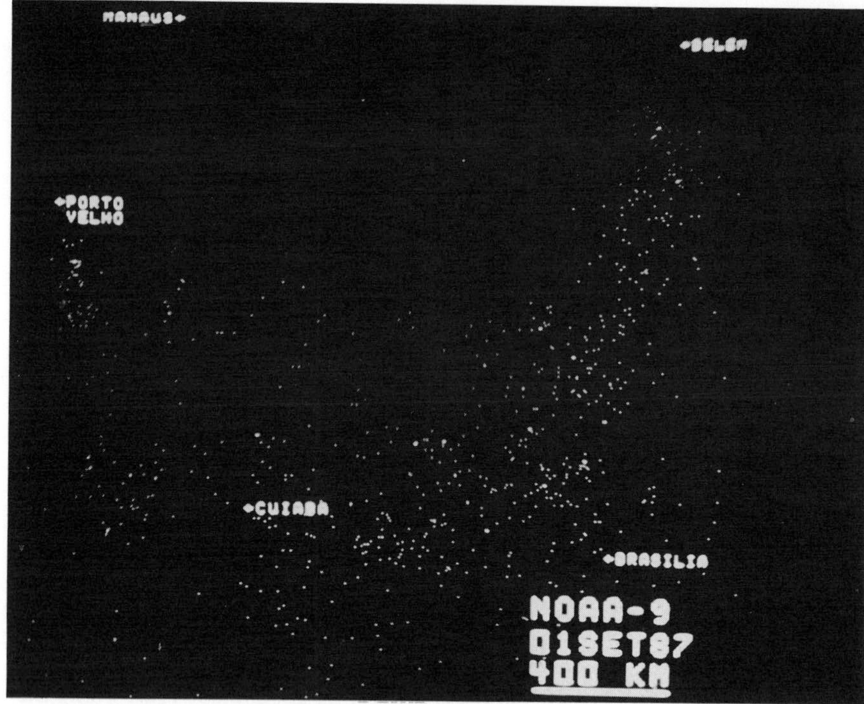

Fig. 11. Thousands of fires burning along a belt from Belem, at the mouth of the Amazon River, to Porto Velho, capital of the State of Rondônia, stressing the current geographical limits of forest conversion in Brazil

Table 5. Data summary of the 1987 burning/dry season

State	Acre	Mazonas	Para	Rondonia	M. Grosso	Goias	Maranhao	Total
Number of analyzed images	10	29	35	27	37	34	11	
Detected number of fire pixels	1557	679	14507	26267	62341	28338	3241	
Total number of fires	12456	1873	33159	77828	134791	66678	23571	350,000
Burned area (km^2)	7274	1094	19365	45452	78718	38940	13765	204,608

coverage because they appear at the far sides of the area of interest. "Average number of fire pixels" is the number of fire pixels registered in the images analyzed divided by the number of images; "total estimate of fire pixels" is the average number of fire pixels multiplied by 80, the number of days that the burning season is assumed to have lasted. In order to obtain the actual burned area some adjustments had to be made. First, some large fires lasted more than 1 day and were possibly registered twice by the satellite; field work information obtained in North Mato Grosso indicated 1.5 days to be a reasonable average value for areas in the 100-ha range, and the total number of fire pixels were adjusted by this factor. Second, in many cases the area actually burned, although smaller than a pixel, was depicted as a full pixel because of its high temperature. By comparing LANDSAT TM and NOAA-9 images for selected areas where burnings occurred also in North Mato Grosso, it was found that an overestimate of about 37% occurred in the NOAA-9 images, and therefore another adjustment factor was used. And finally, to obtain the corresponding burned areas, the adjusted pixel number was multiplied by 1.2, the area of a pixel in km^2 at nadir. These factors can certainly be elaborated to produce more precise figures, and the values issued represent a conservative estimate. As an example, the duration of fires is smaller for areas like the state of Rondônia where properties are usually in the 1–10-ha range. In this region the burned area overestimate is thus much larger, but many fires occur in the same pixel, and some small ones may be ended or not yet started when the image was obtained. Also, very dense smoke clouds from some fires may prevent the AVHRR from detecting other fires. As shown in Table 5, the total estimate of areas burned in the Brazilian Amazon basin in 1987 alone is about 200,000 km^2, close to five times the area occupied by the country of Switzerland. Of great importance is the knowledge of the percentage of this area that corresponds to recent forest conversion. This number also varies regionally, from 100% in regions of new development in the forest to a few percent for regions developed a few years ago or with predominant cerrado type vegetation (Setzer et al. 1988). As no statistics exist also in this case, the authors, based on limited field work activities and examination of LANDSAT TM images, suggest about 40%, or 80,000 km^2 as a conservative first guess. For the whole Amazon tropical forest this value must be greater, since other countries, such as Bolivia, are also pursuing deforestation policies.

Emissions of the 1987 burnings into the atmosphere were by no means small. Press reports relate that in the Amazon region and Central Brazil low visibility due to burnings that occurred hundreds of km away closed airports for many days, caused boat accidents in rivers, and also significant increases of respiratory diseases (Setzer et al. 1988). Carbon monoxide concentrations in remote areas of the Pantanal, thousands of km downwind from the main fires, increased by a factor of 10 (Kirchhoff et al. 1988). The smoke clouds were so large that they could be easily detected on the visible channel of images of the meteorological geostationary GOES satellite, and covered areas of millions of km^2 (Setzer et al. 1988).

An estimate of the fire emissions was made as follows. Based on Seiler and Crutzen (1980), the amount of dry matter burned can be estimated from

$M = A_i \times B_i \times ai \times b_i$, for i = 1 and 2, and where
M = Mass of dry matter burned, in g;
i = type of basic vegetation burned, where 1 is dense forest, and 2 cerrado;
A_i = Area of vegetation type i burned, in km^2
B_i = Average biomass of vegetation i, in g km^{-2};
a_i = Fraction of biomass over the soil in vegetation i; and,
b_i = Combustion factor for vegetation i.

If A1 = 40% and A2 = 60% of the estimated above, B_1 = 22.6 kg m^{-2} (Fearnside 1985), B_2 = 4.3 kg m^{-2} (Santos 1987), a_1 = 0.8 (Seiler and Crutzen 1980), a_2 = 0.65 (Santos 1987), b_1 = 0.6 (Setzer et al. 1988) and b_2 = 0.75 (Seiler and Crutzen 1980), resulting in $M = 1.15 \times 10^{15}$ g of dry matter burned in 1987.

Also following Seiler and Crutzen (1980), the associated emissions of carbon dioxide should be 45% of dry matter burned, or 0.52×10^{15} g, and based on the methodology presented by Andreae et al. (1988), the emissions of trace gases can be estimated, see Table 4a.

16.6 Discussion

The main advantage of the remote sensing method suggested in this chapter to sense the emission products from biomass burning in the tropics is in the availability of inexpensive and reliable daily satellite data (the NOAA-AVHRR) that provides information from every km^2 of the tropical forest on the number of fires and the smoke emitted by them. Since the 3.7-μm channel of the AVHRR is sensitive to fires as small as 10^{-4}–10^{-3} km^2 (10×10m of flaming or 30×30m of smoldering), it is possible to count all significant fires (unless there is more than one fire in a km^2 pixel), and detect new deforestation frontiers that could be gone undetected otherwise. The disadvantage of the relatively low saturation level of the AVHRR (320 K) that prevents one from distinguishing between small or large fires, and smoldering or flaming, is overcome by the ability of the satellite to detect the particulates emitted in the smoke from the fires, and to distinguish between smoke emitted from flaming fires and smoldering fires by the grayness of the smoke in the visible and near IR parts of the spectrum. Note that remote sensing of smoke by itself is not good to estimate the emission from biomass burning, due to the diffusiveness of the smoke and difficulties in smoke tracking (Ferrare et al. 1989). The main weaknesses in the present technique are in the integration on the emission time and in the transformation from particulates to trace gases. The time integration can be helped by analysis of GOES satellite imagery that can provide hourly information about the smoke (from its visible channel). The transformation from particulates to trace gases has to be studied by field measurement programs. Other uncertainties, such as the relationship between the measured radiance and the aerosol optical thickness, as well as the relationship of the optical thickness and the total and dry particulate matter can be reduced by in situ measurements of the aerosol characteristics (e.g., size distribution) and meteorological conditions.

One unique characteristic of satellite-based techniques for monitoring biomass burning is the synoptic coverage of the data. The AVHRR polar orbiting sensors provide the basis for a global monitoring system. The techniques previously described in this chapter have been developed for the Amazon Basin where deforestation rates are high and forest burning is extensive. Other regions of tropical forest are undergoing various degrees of forest clearance through an increased demand for agricultural land. The techniques developed for the Amazon are currently being applied to other tropical regions of the World to test their general applicability for monitoring global biomass burning. Figure 12a is an AVHRR image of the northern boundary of the tropical forest of Equatorial Africa, including parts of Cameroun, Congo, Central African Republic, and Zaire. The major river systems are included in the image for geographic location. The image was derived from AVHRR channels 1, 3, and 4 to highlight the forest boundary and to reduce the effects of high cloud reflectance and warm land surfaces. The image shows a sharp contrast in brightness temperature between the tropical forest and the regrowth forest and

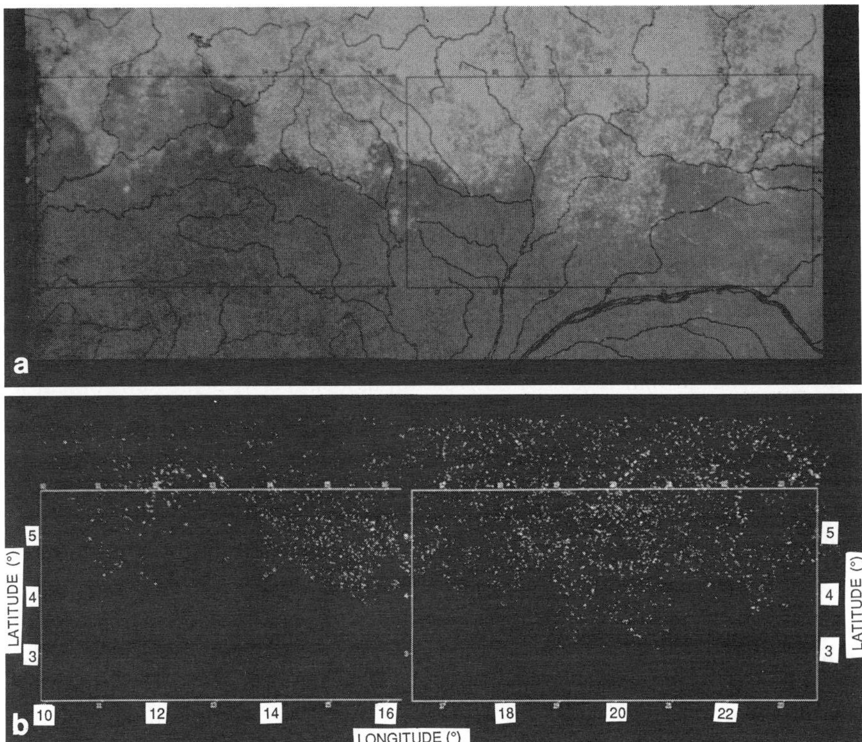

Fig. 12a,b. One week composite of the fire in Equatorial Africa, from the 3.7 μm channel of the 1 km

savannah mosaic. The image is a composite of five dates from January 1988 during the short dry season. Figure 12b is an enhanced image of the fire distribution for the same period. In this image 12,000 fires are depicted, the majority of which occur to the north of the forest margin within the mosaic of shifting agriculture and regrowth. The identification of fires within the savannah zone where there is extensive burning annually is a subject for further research and is complicated by the low saturation level of the AVHRR. Areas of active burning at the forest margin and within the forest domain can be identified and may provide evidence of active forest clearance. Preliminary analyses using these data are currently being undertaken to estimate the extent of biomass burning and to estimate the resulting gaseous emission within the entire equatorial forest region of Africa.

16.7 Conclusions

Deforestation and accompanying burnings in the Amazon are occurring at exponential rates which started in the early 1970's with the official policy to develop the region. Current yearly deforestation increases of 30% are expected for the states of Amazonas, Mato Grosso, Para, and Rondônia, where most deforestation takes place. If compared to a tentative evaluation of emissions for 1985 (Pereira 1988), the present estimate shows an increase of about 100% in the number of fires.

The burning emissions are in the lower troposphere, although some smoke clouds reach up to 4 km before starting to disperse horizontally. Considering that almost no rain occurs during the burning/dry season, the emissions enter the circulation of the anticyclonic cell that prevails in central Brazil during this period, and are transported to latitudes of 20 to 30°S over the Atlantic Ocean. In that region, increases in ozone tropospheric concentrations have been reported in a possible link with biomass burnings (Fishman and Larsen 1987). Also in this region, as seen in meteorological satellite images, tropospheric air can be lifted to high levels by convection associated with the subtropical jet stream, and which will eventually carry the emissions to Antarctic latitudes as corroborated by wind data. Biomass burning emissions from the Amazon basin present a substantial source of carbon dioxide and many atmospheric reactive gaseous compounds, and aerosols that provide catalytic surfaces for photochemical reactions, which could cause changes in the planet's climate and atmospheric chemistry.

Comparison with the global estimates of biomass burning (Crutzen et al. 1985; Table 4), shows that the Amazon region presently analyzed emits 1/10 of the global estimate of biomass burning (Crutzen et al. 1985). Extrapolation of the present results to global emission from deforestation can be performed by assuming that the ratio between the regional emissions reported by Houghton et al. (1987) for 1980, are also valid for 1987. According to Houghton et al. (1987) the presently studied area emitted carbon in the rate of 160×10^{12} g year^{-1}, while the global emission from tropical deforestation was 1650×10^{12} g year^{-1}. If the same ratios were applied for 1987, then the emission from tropical deforestation in 1987 are similar to estimates of total global emission from all types of biomass burning (deforestation as well as others) in 1980. This is mainly due to the expansion of deforestation and biomass burning due to population expansion, as found in the Amazon.

Acknowledgements. We would like to thank: Robert Chatfield from NCAR, Robert Fraser from NASA/GSFC, Jennifer Robinson from Penn State Univeristy, A.C. Pereira, F. Siquerira, G. Rodriqeus, J.C. Moreira, M.M. Cordeiro, P.C.D. Santos, and V.A.S. Oliveira, at INPE/Brazil, C. Paive and S. Almeida at IBDF/Brazil, J.P. Malingreau at EEC/Italy, and B.N. Holben at GIMMS/NASA. Darold Ward from the USDA Forest Service, for fruitful discussions throughout this study, as well as Jackline Kendall, Shana Matto, and Robert Rank for the software development and the image processing involved in this work.

References

Ackerman TP, Toon OB (1981) Absorption of visible radiation in atmosphere containing mixtures of absorbing and nonabsorbing particles. Appl Optics 20:3661–3668

Anderson IC, Levine JS, Poth M, Riggan PJ (1988) Enhanced emission of biogenic nitric oxide and nitrous oxide from semi-arid soils following surface biomass burning. J Geophys Res 93:3893–3898

Andreae MO, Browell EV, Garstang M, Gregory GL, Harris RC, Hill GF, Jaqcob DJ, Pereira MC, Sachse GW, Setzer AW, Silva Dias PL, Tablot RW, Torres AL, Wofsy SC (1988) Biomass burning emissions and associated haze layers over Amazonia. J Geophys Res 93:1509–1527

Atjay GL, Ketner P, Duvigneaud P (1979) In: Bolin B, Degens ET, Kempe S, Ketner P (eds) The global carbon cycle. Wiley, New York, pp 129–182

Blake DR, Rowland FS (1986) World-wide increase in tropospheric methane, 1978–1983. J Atmos Chem 4:43–62

Chatfield RB, Delany AC (1988) Efficiency of tropospheric ozone production: a tropical example of NO_x dilution and cloud transport. Presented at American Geophysical Union Fall Meeting, San Francisco, December 5–9, 1988

Coakley JA Jr, Cess RD, Yurevich FB (1983) The effect of tropospheric aerosol on the earth's radiation budget: a parametrization for climate models. J Atmos Sci 40:116

Coakley JA Jr, Borstein RL, Durkee PA (1987) Effect of ship stack effluents on cloud reflectance. Science 237:953–1084

Cofer WR, Levine JS, Riggan PH, Sebacher DI, Winstead EL, Shaw EF, Brass JA, Ambrosia VG (1988) Trace gas emissions from a mid-latitude prescribed chaparral fire. J Geophys Res 93:1653–1658

Crutzen PJ (1988) Tropospheric ozone: an overview. In: Isaksen ISA (ed) Tropospheric ozone. Reidel, Dordrecht, pp 3–32

Crutzen PJ, Graedel TE (1988) The role of atmospheric chemistry in environment-development interactions, Chapter 8. In: Clark WC, Munn RE (eds) Sustainable development of the biosphere. Cambridge Univ Press, Cambridge

Crutzen PJ, Heidt LE, Krasnec JP, Pollock WH, Seiler W (1979) Biomass burning as a source of atmospheric gases. Nature (Lond) 282:253–256

Crutzen PJ, Delany AC, Greenberg J, Haagenson P, Heidt L, Lueb R, Pollock W, Seiler W, Wartburg A, Zimerman P (1985) Tropospheric chemical composition measurements in Brazil during the dry season. J Atmos Chem 2:233–256

Dozier J (1980) Satellite identification of surface radiant temperature fields of subpixel resolution. NOAA Tech Mem NESS113, Washington DC

Fearnside PM (1985) Summary of progress in quantifying the potential contribution of Amazonian deforestation to the global carbon problem. In: Workshop on biogeographic of tropical rain forest, Piracicaba, SP, Brazil, Sep/30-Oct/04. Proceedings, CENA/USP, 1985, pp 75–82

Ferrare RA, Fraser RS, Kaufman YJ (1990) Satellite remote sensing of large scale air pollution – application to a forest fire. J Geophys Res (in press)

Fishman J, Larsen JC (1987) The distribution of total ozone and stratospheric ozone in the tropics: Implications for the distribution of tropospheric ozone. J Geophys Res 92:6627–6634

Fishman J, Minnis OP, Reichle HG Jr (1986) Use of satellite data to study ozone in the tropics. J Geophys Res 91:14451–14465

Fraser RS, Kaufman YJ (1985) The relative importance of scattering and absorption in remote sensing. IEEE Trans Geos Rem Sens 23:625–633

Fraser RS, Kaufman YJ, Mahoney RL (1984) Satellite measurements of aerosol mass and transport. J Atmos Environ 18:2577–2584

Greenberg JP, Zimmerman PR, Heidt L, Pollock W (1984) Hydrocarbon and carbon monoxide emissions from biomass burning in Brazil. J Geophys Res 89:1350–1354

GTC – Global Tropospheric Chemistry (1986) UCAR, PO Box 3000, Boulder CO 80307

Hänel G (1981) An attempt to interpret the humidity dependencies of aerosol extinction and scattering coefficients. Atmos Environ 15:403–406

Houghton RA, Woodwell GM (1989) Global climatic change. Sci Am 260:36–44

Houghton RA, Boone RD, Fruci JR, Hobbie JE, Melillo JM, Palm CA, Peterson BJ, Shaver GR, Woodwell GM (1987) The flux of carbon from terrestrial ecosystems to the atmosphere in 1980 due to changes in land use: geographic distribution of the global flux. Tellus 39b:122–139

Kaufman YJ (1987) Satellite sensing of aerosol absorption. J Geophys Res 92:4307–4317

Kaufman YJ, Fraser RS (1983) Light extinction by aerosols during summer air pollution. J Appl Meteor 22:1694–1706

Kaufman YJ, Holben BN (1990) Calibration of the AVHRR visible and near IR sensors using molecular scattering, ocean glint and desert reflection. J Appl Meteor (in press)

Kaufman YJ, Sendra C (1988a) Satellite remote sensing of aerosol loading over land. In: Hobbs PV, McCormick MP (eds) Aerosols and climate. Deepak, Hampton, VA

Kaufman YJ, Sendra C (1988b) Algorithm for atmospheric corrections. Int J Rem Sens 9:1357–1381

Kaufman YJ, Brakke TW, Eloranta E (1986) Field experiment to measure the radiative characteristics of a hazy atmosphere. J Atmos Sci 43:1136–1151

Kaufman YJ, Tucker CJ, Fung I (1989) Remote sensing of biomass burning in the tropics. Adv Space Res 9:265–268

Kaufman YJ, Fraser RS, Ferrare RA (1990) Satellite remote sensing of large scale air pollution-method. J Geophys Res (in press)

Kirchhoff V, Setzer AW, Marinho E, Pereira MC (1988) Effects of forest burnings in the increase of CO over remote areas of Brazil. Brazilian Soc Adv Sci Congr, Oct 88

Malingreau J, Tucker CJ (1988) Large scale deforestation in the Southeastern Amazon Basin of Brazil. Ambio 17:49–55

Matson M, Dozier J (1981) Identification of subresolution high temperatures sources using thermal IR sensors. Photogr Engineer Rem Sens 47(9):1311–1318

Matson M, Schneider SR, Aldridge B, Satchwell B (1984) Fire detection using NOAA-series satellite. NOAA Technical Report NESDIS-7. Washington DC, Jan/84, 34 pp

Matson M, Stephens G, Robinson J (1987) Fire detection using data from the NOAA-N satellites. Int J Rem Sens 8:961–970

Mekler Yu, Kaufman YJ, Fraser RS (1984) Reflectivity of the atmosphere-inhomogeneous surface system: Laboratory simulation. J Atmos Sci 41:2595–2604

Mooney HA, Vitousek PM, Matson PA (1987) Exchange of materials between terrestrial ecosystems and the atmosphere. Science 238:926–931

Oltmans SJ, Komhyr WD (1986) Surface ozone distributions and variations from 1973–1984 measurements at the NOAA Geophysical monitoring for climatic change baseline observatories. J Geophys Res 91:5229–5236

Patterson EM, McMahon CK (1984) Absorption characteristics of forest fire particulate matter. Atmos Environ 18:2541–2551

Patterson EM, McMahon CK, Ward DE (1986) Absorption properties of graphitic carbon emission factors of forest fire aerosols. Geophys Res Lett 13:129–132

Pereira MC (1988) Detecção, monitoramento e analise de alguns efeitos ambientais de queimadas na Amazônia atraves da utilizaçã de imagens dos satelites NOAA e Landsat, e dados de aeronave. MSc dissertation: The Brazilian Space Research Institute/Ministry of Science and Technology- INPE/MCT, Publication No INPE-4503-TDL/326, May/88, 268 pp (in portugese)

Puschel RF, Livingston JM, Russell PB, Colburn DA, Ackerman TP, Allen DA, Zak BD, Einfeld W (1988) Smoke optical depths: magnitude, variability and wavelength dependence. JGR 93:8388–8402

Radke LF (1989) Airborne observations of cloud microphysics modified by anthropogenic forcing, AMS Symp Atmospheric Chemistry and Global Climate, Jan 29–Feb 3, Anaheim, CA 310–315

Radke LF, Hegg DA, Lyons JH, Brock CA, Hobbs PV, Weiss R, Rassmussen RA (1988) Airborne measurements of smokes from biomass burning. In: Hobbs PV, McCormick MP (eds) Aerosols and climate. Deepak, Hampton, VA

Ramanathan V, Cicerone RJ, Singh HB, Kiehl JT (1985) Trace gas trends and their potential role in climate change. J Geophys Res 90:5547–5566

Santos JR (1987) Analise preliminar da relaçã entre as repostas espectrais de dands TM/Landsat e a fitomassa dos cerrados brasileiros. In: II Symp Latino-Americano de Sensoriamento Remoto, Bogota, Colombia, May 16–20 (in press)

Seiler W, Crutzen PJ (1980) Estimates of gross and net fluxes of carbon between the biosphere and the atmosphere from biomass burning. Climate Change 2:207–247

Setzer AW, Pereira MC (1989) Amazon biomass burning in 1987 and their tropospheric emissions. (submitted to Ambio)

Setzer AW, Pereira MC, Pereira Jr AC, Almeida SAO (1988) Relatorio de atividades do projecto UBDF-INPE SEQE – ano de 1987. The Brazilian Space Research Institute/Ministry of Science and Technology-INPE/MCT, Publication No INPE-4534-RPE/565, May/88, 101 pp (in portugese)

Seinfeld JH (1986) Atmospheric chemistry and physics of air pollution. Wiley, New York, 738 pp

Stauffer R, Fisher G, Neftel A, Oeschger H (1985) Increase in atmospheric methane recorded in Antarctic ice core. Science 229:1386–1388

Tangren CD (1982) Scattering coefficient and particulate matter concentration in forest fire smoke. J Air Pollut Control Assoc 32:729–732

Tucker CJ, Holben BN, Goff TE (1984) Intensive forest clearing in Rondonia, Brazil, as detected by satellite remote sensing. Rem Sens Environ 15:255

Turco RP (1985) The photochemistry of the stratosphere. In: Levine JS (ed) The photochemistry of the atmosphere: Earth, the planets, and comets. Academic Press, Orlando, Florida, pp 77–128

Twomey SA (1977) The influence of pollution on the short wave albedo of clouds. J Atmos Sci 34:1149–1152

Twomey SA, Piepgrass M, Wolfe TL (1984) An assessment of the impact of pollution on the global albedo. Tellus 36b:356–366

Ward DE (1986) Field scale measurements of emission form open fires. Technical paper presented at the Defense Nuclear Agency Global effects review, Defense Nuclear Agency, Washington DC 20305–1000

Ward DE, Hardy CC (1984) Advances in the characterization and control of emissions from prescribed fires. 77th annual meeting of the Air Pollut Cont Ass, San Francisco, CA

Ward DE, Nelson RM, Adams DF (1979) Forest fire smoke plume documentation. 72nd annual meeting of the Air Pollut Cont Ass Cincinati, Ohio

Wayne RP (1985) Chemistry of the atmospheres. Clarendin, Oxford, 361 pp

17 NOAA-AVHRR and GIS-Based Monitoring of Fire Activity in Senegal – a Provisional Methodology and Potential Applications

P. FREDERIKSEN[1], S. LANGAAS[2], and M. MBAYE[3]

17.1 Introduction

Bushfires have been defined as uncontrolled fire occurring in the rural landscape (Cheney 1981). They occur in the three major ecological zones of Senegal; the Sahelian, Sudanian, and Guinean zones of northern, central and southern Senegal respectively, and mainly in pastoral ecosystems. Almost all are set intentionally or accidentally by the local population during the dry season (Fall 1986; UNSO 1982) as a result of badly quenched campfires, cigarettes or matches, or due to fires lit to enrichen and renew grazing lands, clear paths, hunt ground squirrels by smoking them out of their holes, kill/drive away insects and pests such as blister and millet beetles and grasshoppers, exploit the forest, or to clear the land prior to ploughing.

Their effect on the environment is multiple and is generally considered to be more negative than positive (Gillon 1983). Loss of human lives and animals, and burning of houses and crops are frequently recorded. Based on results from long-term burning experiments from similar savanna environments in Ghana (Brookman-Amissah et al. 1980), Côte d'Ivoire (Dereix and N'Guessan 1976) and Nigeria (Rose-Innes 1972), Senegalese foresters fear that repeated, late dry-season fires prevent regrowth of new tree seedlings and reduce species diversity through elimination of less fire-resistant species. Rangeland managers are primarily concerned with the fire-induced reduction of available fodder for livestock as a constraint for pastoralism. Recent work also suggests that savanna fires are much larger sources of atmospheric emissions of CO_2 and other trace gases than deforestation in the tropics (Hao et al., this Vol.). The economic resources dedicated to bushfire prevention and combating is considerable (Fall 1986). The Ministry of Nature Protection, Directorate for Water, Forests, and Hunting is responsible for limiting and combating bushfires. Four bushfire prevention and combating projects funded bi- or multilaterally complement the work of the Directorate.

No statistically reliable method for collection of data on the extent of burned areas on a national level is currently available. The official statistics published include only two categories of fires: (1) early fires set by foresters at the onset of the dry season in some of the major, protected forests (fôrets classés),

[1]Roskilde University Center, Postbox 260, DK-4000 Roskilde, Denmark
[2]Department of Surveying, Agricultural University of Norway, 1432, Aas-AUN, Norway
[3]Centre de Suivi Ecologique, United Nations Development Program, Dakar, Senegal

and (2) those late dry-season fires fought by the forestry staff. It is generally accepted that more accurate statistics are needed. A recent Senegalese workshop on the impact and extent of bushfires with participants from the above-mentioned institutions and projects concluded that it is essential for proper management to improve the statistics, and that remote sensing techniques should be used.

Satellite remote sensing techniques for monitoring active bushfires of burned areas have been used in several parts of the world. Hick et al. (1986) documented the potential of the Advanced Very High Resolution Radiometer (AVHRR) sensor on the National Oceanic and Atmospheric Administration (NOAA) series of satellites for mapping burned areas in Australia. Langaas and Muirhead (1989), Grégoire et al. (1988), and Malingreau et al. (1989) focused on NOAA AVHRR channel 3 data for detection and monitoring of actively burning fires in West Africa. For a more detailed description of the NOAA AVHRR satellite sensor system, see Justice (1986) and Kaufman et al. (this Vol.). In Burkina Faso, The Centre Regional de Télédétection de Ouagadougou (CRTO) has long worked on methodologies for visual interpretation of LANDSAT MSS hardcopies for mapping burned areas (Yergeau 1983; Parnot 1987, 1989). Systematic reconnaissance flights have been used in Senegal (Bélanger 1987; Sharman 1987), but the technique is costly, and it is difficult to obtain statistically reliable data.

The lack of reliable bushfire statistics, the documented potential of NOAA AVHRR data for mapping burned areas in other parts of the world, and the Centre de Suivi Ecologique's (CSE) long experience with NOAA AVHRR based monitoring of aboveground herbaceous biomass production, has led it to develop a semi-operational methodology for multi-temporal mapping of burned areas in Senegal. The objective is thus to provide more reliable bushfire statistics. Centre de Suivi Ecologique is a United Nations Development Program funded ecological monitoring project [UNSO/SEN/(84/X09)] placed within the Ministry of Nature Protection.

17.2 Methodology

Based on exploratory field work and image processing of imagery from the 1987/88 dry season, it had been noted that burned and unburned areas in general gave different spectral responses, that the spectral response from burned areas varied with time, and that the fires burned varying proportions of the vegetation. The first step was therefore to develop a scene model of the landscape, the second to investigate the spectral-temporal dynamics of a burned surface, the third to find out how much of an area classified as burned actually had burned, the fourth to classify the images based on previous steps, and the fifth to develop the bushfire statistics in a GIS (Geographical Information System) context.

17.2.1 Definition of a Scene Model

A simple scene model was formulated in order to direct the research and the classification of the satellite imagery. The model supposes that the Senegalese land surface consists of two classes; burned and unburned drygrass areas. It was furthermore assumed that burned drygrass areas have (1) a lower optical brightness (are "darker"), (2) a higher surface brightness temperature (are "warmer") and (3) less vegetation (activity) than unburned drygrass areas. The model and its properties are summarized in Table 1.

Table 1. A simple scene model with (bio)physical properties which can be applied to discriminate between burned and nonburned areas using AVHRR data

	Physical properties	Optical brightness	Surface brightness temperature	Vegetation amount (activity)
	AVHRR "channel"	1 and 2	4 and 5	NDVI
Class	Burned dry-grass area	Lower	Higher	Lower
	Unburned dry-grass area	Higher	Lower	Higher

The model has some limitations. Firstly, not all burned areas are completely burned. Patches of grass are left unburned due to abrupt wind changes during fire, natural borders such as tracks and roads, too high fuel load moisture, and areas of low fuel load. Secondly, woody vegetation is not included in the model. The tree cover is insignificant in northern, Sahelian Senegal, but considerable within the Guinean zone in southern Senegal. The contribution of green trees to the spectral response varies with tree coverage, phenological state of the trees, and the area on the ground shaded by the trees. It may therefore be significant in the south. This effect has previously been discussed in relation to monitoring of herbaceous biomass using the AVHRR channels 1 and 2 derived Normalized Difference Vegetation Index (NDVI) by, for example, Prince and Astle (1986) and Lamprey and De Leeuw (1988). Thirdly, land cover types other than grassland areas exist in Senegal, such as agricultural, urban, mangrove forests, bare soils, and water surfaces. Confusion caused by other land cover classes spectrally quite similar to burned areas has also been mentioned in a LANDSAT TM (Thematic Mapper) based study from Spain recently (Chuvieco and Congalton 1988). Fourthly, the spectral differences between burned and unburned drygrass areas may not always be particularly high. Sharman (1987) in northern Senegal, and Yergeau (1983) in Burkina Faso have described how rapidly the ash created by a bushfire can be removed by wind after fire in savannas, thus approaching the spectral reflectance of unburned drygrass areas. Justice et al. (1985) showed in a study of global phenology that the NDVI on the border between Mali and Guinea dropped significantly from early to late in the dry season. The potential of the NDVI for separating burned and unburned areas may therefore decrease into the dry season. These limitations mean that the model cannot be applied sensu stricto, but only as a guideline for designing the classification procedures.

17.2.2 Field Radiometric Measurements

The objectives were to evaluate the spectral-temporal dynamics of a burned surface within wavebands similar to AVHRR channels 1 and 2 and their derived NDVI, in order to determine a desired frequency with which AVHRR imagery should be acquired, and to gain some insight into the nature of the changes. This was done in the field by measuring the radiances in the red and near infrared wavelengths of a burned area every day or second day after a fire occurrence, using a NASA Mark II radiometer, described by Tucker et al. (1981). It is a three bandpass, nonscanning broadband radiometer designed for vegetation studies, and particularly for supporting LANDSAT 4 TM bands 3, 4, and 5. The correspondence in band location of the radiometer's channels 1 and 2 with channels 1 and 2 of the AVHRR has also facilitated comparison with data from this sensor. The third band in Mark II, corresponding to channel 5 of Landsat TM was not used during the field campaign. The band location in Mark II and their correspondence with related channels in LANDSAT TM and NOAA AVHRR are summarized in Table 2.

The field radiometric measurements were carried out 35 km south of Tambacounda (see Fig. 1). A site of approximately 200 × 100 m was used. The site is located in savanna woodland with a continuous stratum of Graminaceae (before burning) and woody shrubs of *Combretum* spp. and *Strychnosa spinosa*. All the grass within the site was completely burned during a bushfire occurring in the morning of December 27, 1988. The measurements started immediately after the fire on a "random" basis within the site, with approximately 150 daily measurements. Tree-shaded spots were avoided. To avoid too much variation in sun-target-sensor geometry, measurements were taken between 10.00 and

Table 2. Mark II radiometer band location

Band	Wavelength (μm)	Corresponds to:		Relates to:
1	0.63–0.69	TM ch. 3	RED	AVHRR ch. 1 (0.58–0.68 μm)
2	0.76–0.90	TM ch. 4	NIR	AVHRR ch. 2 (0.72–0.95 μm)
3	1.55–1.75	TM ch. 5	MIR	

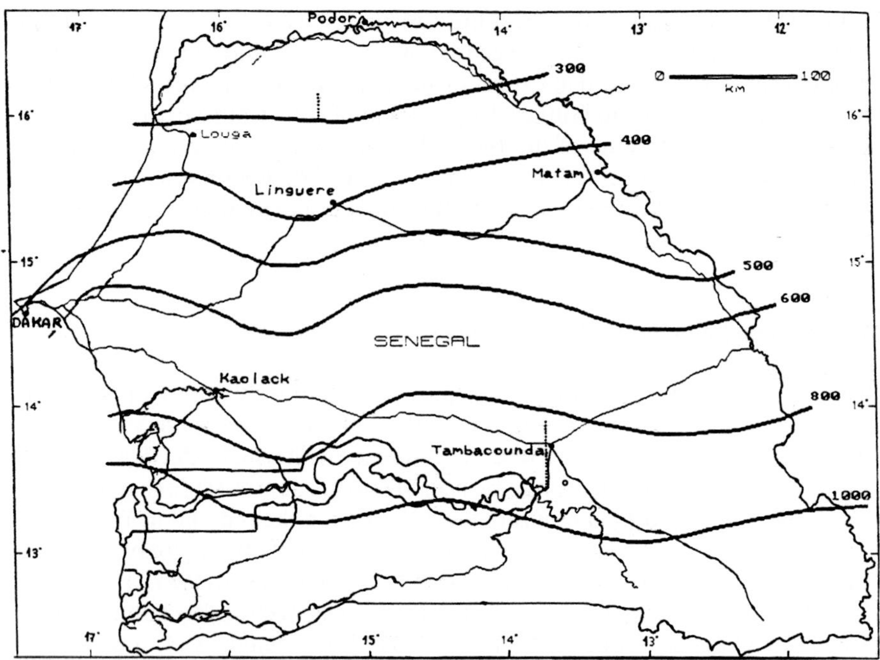

Fig. 1. Map of Senegal with isohyets superimposed and major towns indicated. Transects in Tambacounda- and Ferlo-region (*dotted lines*) and field study site near Tambacounda (*small circle*) are shown

14.00 h local time (= GMT). Between each measurement a calibration card (Kodak Gray Card) was measured in order to calculate the bidirectional reflectance. Variable atmospheric conditions may reduce the accuracy of bidirectional reflectance estimations using this method (Duggin 1980). Days with highly varying cloud conditions were consequently omitted. Measurements were also carried out over a limited number of spots characterized by a dense cover of dry leaves, over unburned drygrass close to the test site, over bare soil, over completely ash-covered surfaces and over partially burned surfaces.

17.2.3 Integrated Camera and Radiometer Measurements

The scene model assumes that the burned areas are completely burned. This is not generally true. Hopkins (1965) found that late-season fires burned an average of 84% of the herbaceous biomass, ranging from 64 to 96% in 13 experimental plots in Nigerian Guinea savanna. Hick et al. (1986) found for bushfire burned areas in schlerophyll open woodland in Australia that AVHRR channel 2 data and standard colour aerial photographs gave around 20% higher figures than LANDSAT MSS (0.8–1.1 μm). This was explained by remnants of unburned patches within the outer fire border, detected by MSS due to sufficient spatial resolution compared to AVHRR, and not detected during visual interpretation of aerial photographs.

Consequently, in order to translate AVHRR pixels classified as burned into percentage actually burned, it was decided to collect data which could give representative average values. An airborne, integrated camera and radiometer (ICAR) flown at low altitude was employed for this purpose. Mounted on an aircraft it takes coincident aerial photographs and radiometer readings over a series of sample areas along flown transects. The aerial photographs used were diapositives (Ektachrome 200), and the radiometer readings correspond to channels 1 and 2 of the AVHRR, and includes the NDVI value. A detailed description of the ICAR instrument is given by Prince et al. (1988). The ICAR was used on board a mono-engine aircraft which flew transects at 300 feet mainly over burned areas in central and northern Senegal. At this height, each sample area covered approximately 150×220 m. For the purpose of estimation of percentage actually burned only the aerial photographs were used. The diapositives were interpreted using a diaprojector and point sampling. One hundred and fifty six points located on a stratified random basis were interpreted, thus giving an estimate of the fractions burned and unburned within each area represented by a photograph. Diapositives from two transects, one from central and one from northern Senegal, considered to be representative for each region, were interpreted. These transects constituted 166 and 87 diapositives, respectively.

17.2.4 AVHRR Image Processing and Field Verification

NOAA AVHRR images were acquired on Computer Compatible Tapes from European Space Agency/Earthnet's receiving station on Maspalomas, the Canary Islands. The preprocessing of the images comprised calculation of the NDVI values for display purposes scaled using the formula NDVI*199.2–49.8 and rectification and resampling to an UTM zone 28 projection grid with 1 km^2 resolution.

Nine images, essentially cloud-free, covering Senegal from October 1988 to January 1989 were classified using threshold values on single channels or two channels in combination. The CHIPS software, developed by the Department of Geography, University of Copenhagen, was used for the image processing (Rasmussen 1989).

Classification phase 1 was done without prior field knowledge. It tested (a) channel 1 (b) channel 2 (c) channel 4 (d) NDVI and (e) channels 2 and 4 together.

The classifications were evaluated during a field campaign in northern and central Senegal in December 1988. By advancing at around 40 km h^{-1} using a vehicle, the average herbaceous cover as observed along the road, was estimated over a 10-s period. Four categories ranging from unburned to severely burned were used. In this way 200 km was covered.

Classification phase 2 was based on descriptive statistical information extracted from subregions within AVHRR images identified and mapped as burned and unburned, respectively, during the field campaign. Phase 2 only used (a) NDVI alone and (b) channels 2 and 4 in combination.

17.2.5 GIS Manipulation

By using the Geographical Information System's (GIS) functions of the CHIPS, the maps on burned areas, administrative boundaries, and aboveground herbaceous biomass production were combined to calculate the area burned per month, per administrative unit, and to give an estimate of the amount of dry matter burned by the bushfires.

17.3 Results and Discussion

17.3.1 The Spectral Evolution of a Burned Area

The results of the field radiometer campaign are shown in Figs. 2–5. Figure 2 presents the temporal dynamics of bidirectional reflectances within the red and near-infrared bands, and Fig. 3 the temporal dynamics of the NDVI derived from these two bands. Bidirectional reflectances from a nearby unburned drygrass covered surface is given for comparison on day −1.

The rapid increase in bidirectional reflectance and the NDVI after the fire can be explained by a change in fractional cover of the three elements comprising the burned site; ash, bare soil, and dry leaves. The bidirectional reflectances and NDVI is much lower for ash than for bare soil and dry leaves (see Fig. 4). During the measurement period daily maximum wind speed varied between 7 m s^{-1} and 15 m s^{-1}, and considerable amounts of ash were removed from the surface. This increased the (bare soil + dry leaves)/ash ratio of the site, which explains the increase in channels 1, 2, and the NDVI with time.

In order to compare the performance of channels 1, 2, and the NDVI to discriminate burned surfaces from unburned drygrass covered surfaces at varying post-fire dates, a post-/pre-fire index (PPFI) was defined:

$$PPFI_{1,n} = \frac{(S_{1,n} - S_{1,0})}{(S_{1,-1} - S_{1,0})},$$

where $S_{1,n}$ = spectral values (channel 1, 2, or NDVI) at day n relative to fire occurrence.

The evolution of PPFI for the three spectral measures is shown in Fig. 5. The channel 1 reflectance approaches approximately 75%, channel 2 55%, and the NDVI 45% of pre-fire levels of drygrass after 11 days. This indicates that the duration of discrimination by the NDVI is greater than for either of the individual channels, although marginally so when compared to the near-in-

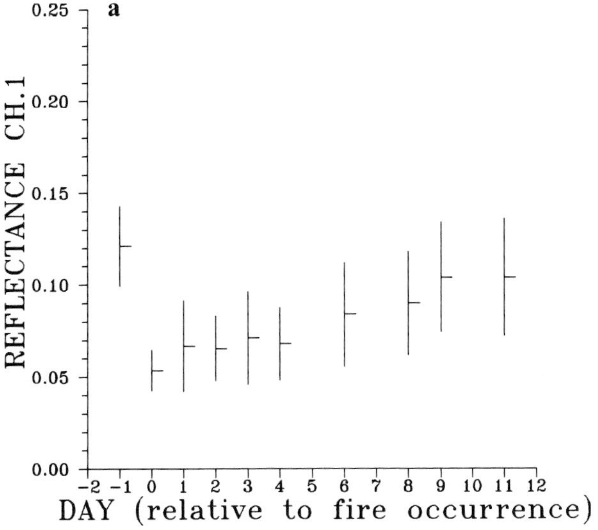

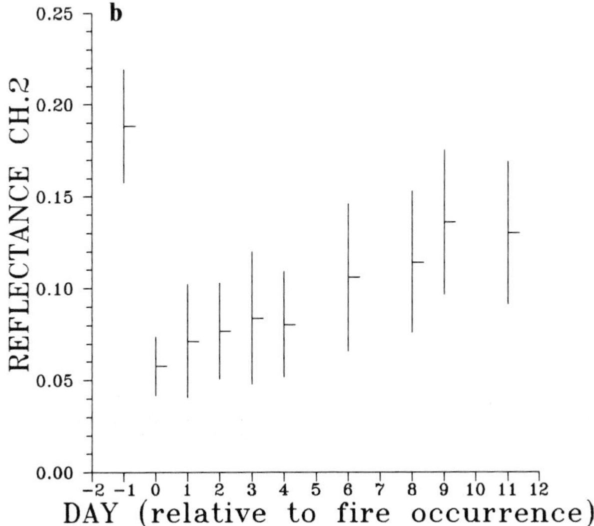

Fig. 2a,b. Temporal dynamics of reflectance in red (**a**) and near-IR (**b**) wavebands for a burned surface after fire occurrence. Reflectance from nonburned grassland is given for comparison as day −1. Values given as daily mean and one standard deviation

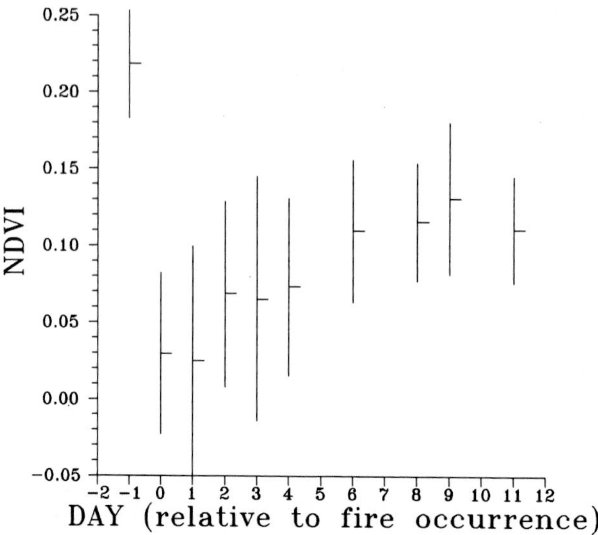

Fig. 3. Temporal dynamics of NDVI for a burned surface after fire occurrence. NDVI for nonburnt grassland is given for comparison as day −1. Values given as daily mean and one standard deviation

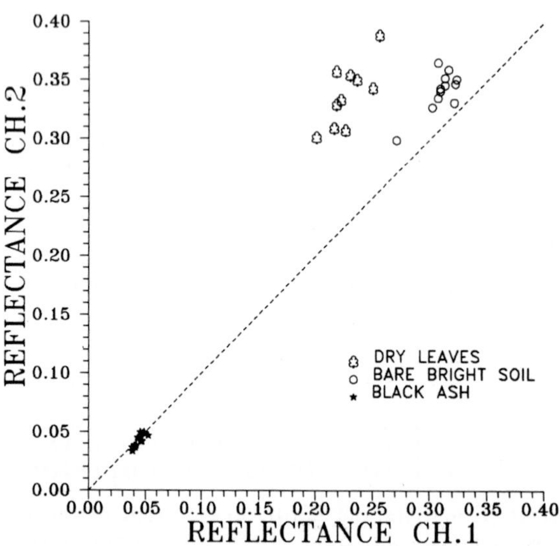

Fig. 4. Reflectances in red and near-IR wavebands derived from Mark II hand-held radiometer for the three "basic" elements comprising burned surface at field site near Tambacounda

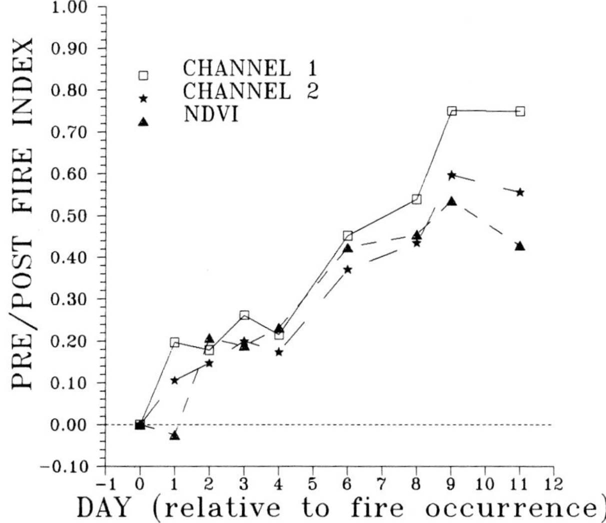

Fig. 5. Temporal dynamics of post/pre fire index for Mark II channels 1 and 2 and NDVI relative to fire occurrence

frared. It would have been valuable to continue the measurements beyond 11 days to see if the values came closer to pre-fire levels, but this was not possible within the study. It is reasonable to anticipate that the evolution of the red and near-IR reflectances (and derived NDVI and PPFI's) of a burned surface primarily is related to post-fire wind activity and microtopography, which may vary considerable both temporally and spatially. The evolution represented by Figs. 2, 3, and 5 therefore cannot be considered representative for the whole of Senegal.

The results suggest that in order to retain highest possible spectral difference within the channels considered, images on a frequent repetitive basis should be employed due to increased brightness associated with wind removal of ashes.

17.3.2 Fractional Cover Burned

Figure 6 gives the results from the interpretation of the diapositives taken by the ICAR. To compare the results from northern and central Senegal, class intervals of 10% for fractional cover burned were defined, and the relative frequency of each class calculated. The distribution patterns are right-skew for both the Sahelian and Sudanian zones, with an average of around 80% actually burned. Areas with less than 50% burned constituted only 4.5% of the photographs from the Sahelian zone and 12% in the Sudanian zone. For the 50–90% range the figures were 56.6% and 47.1%, while 39.0% and 40.9% contained areas in which more than 90–100% had burned. The results are in accordance with those of

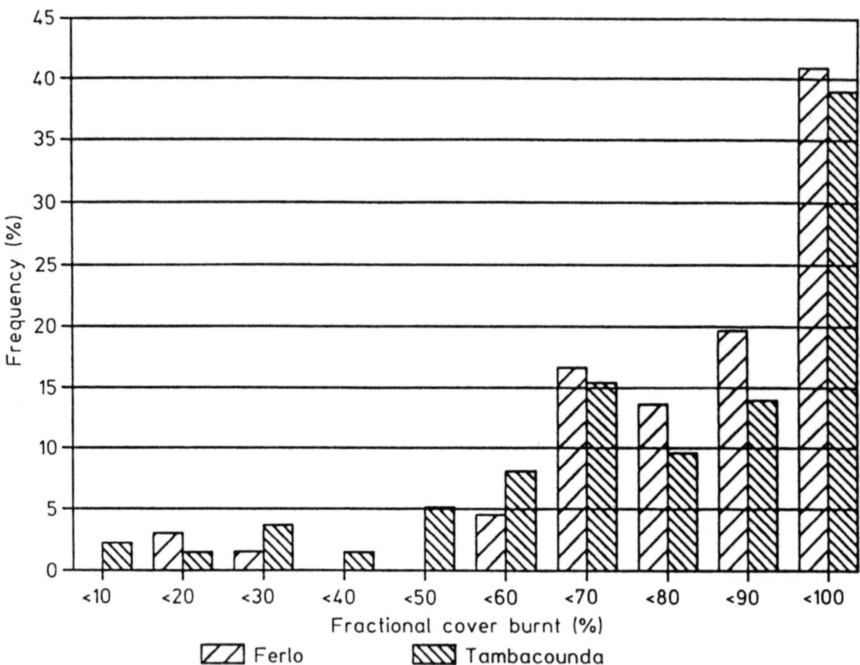

Fig. 6. Relative frequency fractional cover burnt of ICAR photos affected by fire in transects from Sénégalese Sahel (Ferlo) and Sudanian zone (Tambacounda). 10% class intervals

Hopkins (1965) and Hick et al. (1986) and imply that drygrass areas generally considered burned, often contain unburned areas. The scene model previously formulated can therefore not be used sensu stricto to estimate areas actually burned, since also remnants of unburned drygrass areas are included. A reduction factor has to be employed if the objective is to estimate areas actually burned. The results from this dataset indicate 0.8 as a reasonable figure.

17.3.3 Preliminary Bushfire Statistics

During the field campaign fire-affected and unburned areas were identified. In order to test whether the results obtained during field radiometric measurements also hold for raw AVHRR data or not, statistical image analysis was carried out for two areas in northern Senegal and two areas in central Senegal on afternoon images from 8 October (NOAA-9) and 26 October (NOAA-9). Channel 4 data was included, since it was observed from classification phase 1 that this channel could give valuable spectral discriminatory information. It has previously been shown that burning of drygrass in West-African savanna environments significantly changes the surface radiation balance for burned surfaces relative to drygrass covered (Gillon 1983). Within each region one of the areas used was burned by the 26 October, while the other was unburned on the same date (see Table 3).Each of all these six areas covered 121 km^2.

In northern Senegal the mean values of areas 1, and 2 were nearly equivalent on the 8 October when none of them were burned. Area 1 burned between the 8 October and 26 October, while area 2 did not. On the 26th, the mean values of channels 2, 4, and the NDVI had both dropped, but more for the burned area 1 than for the unburned area 2 for area 2, indicating that a bushfire produces a much larger drop in spectral valves than does natural senescence. The decrease in raw spectral values for channel 4 is associated with an increase in brightness temperature.

In central Senegal the spectral response of both the burned and unburned areas dropped, with the drop being higher for the burned area as in northern Senegal. The difference in spectral valves between burned and unburned areas is, however, lower in central than in northern Senegal. This is most probably caused by a higher tree canopy in central Senegal. The conclusion is that a drop in channel 2 and the NDVI can be observed in the AVHRR datasets, particularly when the tree canopy is insignificant. While typical tree canopy cover figures in northern Senegal are around 5%, these figures are around 30% in central Senegal.

The periods between the 2 images are 16 days. The rate of removal of ash cover found during the field radiometric measurements indicated that the temporal frequency ideally should be higher than this. Spectral overlap between fire-affected and unburned areas, as indicated by the minimum and maximum values in Table 3, were thus existent. The continuum between entirely burned and unburned areas, as shown by the aerial photographs, was further assumed to contribute to this overlap.

Table 3. Statistical AVHRR raw (8 bit) spectral response data for channels 2 and 4 and scaled NDVI from fire-affected and unburned areas

a b	CH2 mean	CH4 mean	NDVI mean	CH2 s.d.	CH4 s.d.	NDVI s.d.	CH2 min	CH2 max	CH4 min	CH4 max	NDVI min	NDVI max
October 08, 1988 (NOAA-9)												
1 U	34.00	70.33	66.48	0.70	1.40	1.34	32	35	66	73	64	70
2 U	34.04	68.23	66.29	0.66	0.95	1.12	33	35	66	71	64	70
3 U	29.56	71.75	71.31	0.52	1.32	1.68	28	30	69	74	68	76
4 U	29.40	82.42	82.36	0.56	0.63	1.43	28	31	81	83	79	86
October 26, 1988 (NOAA-9)												
1 F	19.81	48.78	52.05	2.01	3.22	3.22	17	29	44	60	50	65
2 U	34.20	78.95	67.19	1.48	7.42	1.70	31	38	57	98	64	71
3 F	18.11	55.56	50.00	0.63	1.05	0.00	17	20	53	59	50	50
4 U	22.00	68.24	61.38	0.94	3.19	4.03	19	24	60	75	50	69

[a] Area 1–2; northern (Sahelian) Senegal, area 3–4; central (Sudanian) Senegal.
[b] F – fire affected. U – unburned. Each area: n = 121 Km^2

In general the overlaps were small or nonexistent. However, for classification purposes the overlaps made it necessary to define threshold values other than those given by the minimum and maximum spectral values for fire-affected areas. These new values were defined through interactive manipulation of the images in front of the screen using digital contrast enhancement techniques, coupled with visual interpretation based on general extensive field knowledge on burn scar patterns and locations, and finally decision on representative separation pixel values using line transects and individual pixel value extractions.

The field verification after the phase 1 classification and the general field knowledge concerning the location of different land cover classes revealed that most of the classification approaches, in addition to discriminating fire-affected areas from those not affected at all, misclassified some areas which are not fire-susceptible. For instance, channel 1 included vigorously growing vegetation in Southern Senegal, mangrove forests and vegetationless laterite plateaus were included by channel 1 and channel 2, and sandy and lateritic, vegetationless surfaces by channel 4. Channels 2 and 4 combined in a parallelepiped classifier, and the NDVI alone using threshold values were considered most successful and therefore selected for classification phase 2.

The lack of extensive verification data, imply that the accuracy of the results has not yet been quantified. The results are given as a map (Fig. 7) and statistics given in Table 4. The unknown locational and statistical accuracy of the results further implies that they should be considered as preliminary. The results indicate that the officially published bushfire statistics underestimate the area burned by a factor of 10. Most of the bushfires in the first half of the 1988/89 dry season were occurring in November and December and can generally be considered destructive late dry-season fires. Most fires occurred in the less populated eastern parts of central and southern Senegal. In Table 4 are shown the temporal and spatial distribution of the fire activity as derived from the classified images coupled with a overlay of administrative units (region boundaries) using GIS-overlay techniques of CHIPS.

CSE each year uses AVHRR imagery to produce maps on a km^2 basis of total, aboveground, green biomass production based on the methodology developed by NASA/GSFC (Goddard Space Flight Center) using integrated NDVI values together with extensive field data (see Tucker et al. 1983, 1985). By overlaying the areas classified as fire-affected onto the biomass map for 1988, the total amount of dry biomass burned in Senegal in the period October 1988-January 1989 can be estimated using the formula:

$$M = A_{fa} \times B_{fa} \times c,$$

where

M = amount dry, aboveground herbaceous biomass burned,
A_{fa} = areas affected by bushfires,
B_{fa} = average values of herbaceous biomass per unit area, and
c = correction factor.

Within the areas classified as fire-affected the average herbaceous biomass production during the 1988 growth season was calculated at approximately 4000 kg ha^{-1}. However, due to different influencing factors, this figure should not be applied without consideration. The conversion factor of 0.8 for estimation areas actually burned from fire-affected areas cannot be considered as a correct conversion factor to estimate the actual amount of herbaceous biomass burned within fire-affected areas. The areas left unburned with fire-affected areas are quite often unvegetated or less vegetated (and therefore left unburned). Considering only this factor, an adjusted, higher conversion factor would consequently be more reasonable. On the other hand, loss of biomass may have occurred between the end of the growth season and the time of fire occurrence, particularly caused by grazing. Furthermore, the extensive field data collection to estimate and map the distribution of amount aboveground herbaceous biomass is only carried out in northern and central Senegal. No quantitative information on the effects of any of these factors is available for the whole of Senegal. The correction factor, c, therefore was defined based on field expe-

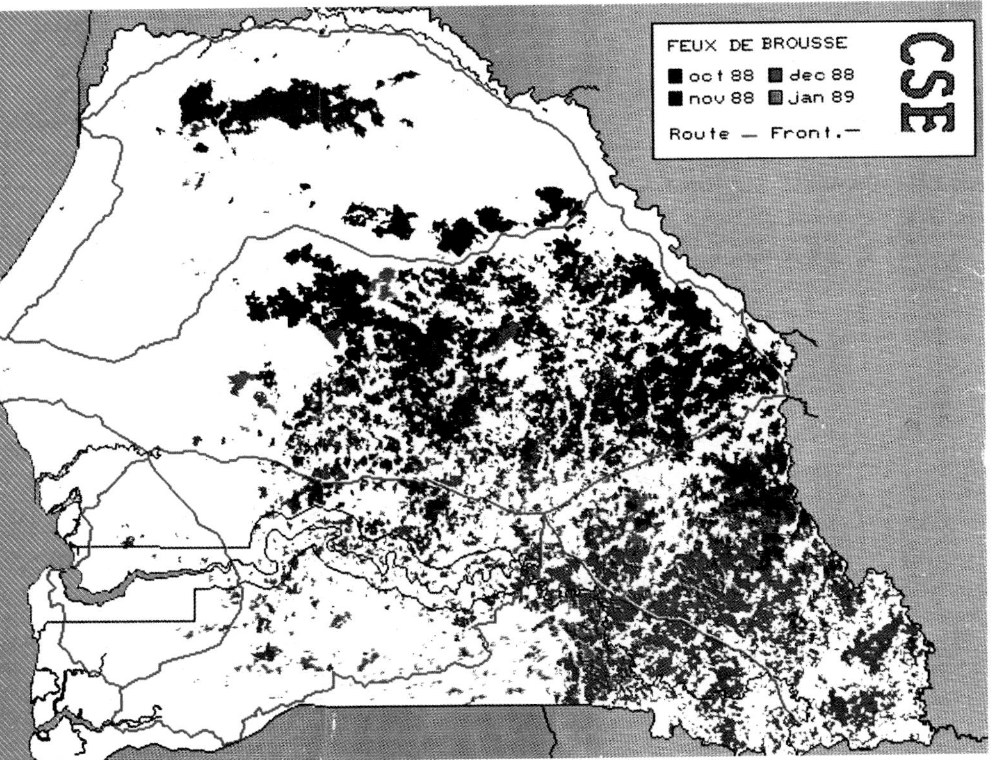

Fig. 7. NOAA-11 AVHRR November 15, 1988 afternoon contrast enhanced false colour composite image (channels 2:2:1; colours R:G:B, respectively). Image shows fire scar from fire affecting 2800 km^2 of Sahelian Acacia wooded grasslands in the Ferlo region of Senegal. UTM grid (10 km between each point), Lac de Guier, Senegal river, and major roads superimposed using GIS overlays

Table 4. Temporal and spatial distribution of bushfire activity in Senegal during the period of October 1988-January 1989

Month (total Senegal)	Areas classified as fire-affected (in km^2)	Areas supposed to be burned using conversion factor 0.8 (in km^2)
October 1988	10,128	8,102
November 1988	17,489	13,991
December 1988	21,275	17,020
January 1989	4,662	3,730
Region (total period)		
Dakar	0	0
Ziguinchor	0	0
Diourbel	0	0
Saint Louis	10,959	8,767
Tambacounda	30,788	24,630
Kaolack	3,121	2,497
Thies	0	0
Louga	5,299	4,239
Fatick	88	70
Kolda	2,347	1,878
Total Senegal	53,554	42,843

rience. A range from 0.5 to 0.9 was considered reasonable, giving figures of M ranging from 10.7 mill. tons to 19.3 mill. tons for the whole of Senegal during the period October 1988-January 1989.

The emitted amount of trace gases like CO_2, CO, and CH_4 and particulate matters released through Senegalese bushfires has not been estimated. Information on the fuel chemistry of biomass and fire behavior interactions are required to estimate such figures (Ward this Vol.). For Senegalese herbaceous biomass and fires, such information is currently not available.

17.4 Conclusions and Further Work

These results indicate that the scene model, despite its simplicity and limitations, can be applied for multi-temporal mapping of grassland/savanna areas recently affected by fires using NOAA AVHRR data. Field radiometric measurements suggest that the higher the frequency by which images are acquired and classified, the better. Interpretation of aerial photographs taken by a combined camera and radiometer in northern and central Senegal shows that areas within fire-affected areas are left unburned. A conversion factor of 0.8 to estimate actually burned areas within the fire-affected area was derived from these photographs. Two classification approaches using (1) channels 2 and 4 in a parallelepiped classifier, and (2) NDVI alone with threshold values were found best to discriminate fire-affected from unburned areas. The spectral difference between recently fire-affected and unburned areas was lower in central than in northern Senegal, assumed to be caused primarily by increased tree canopy. Nine classified AVHRR images suggest that approximately 20% or 42.8 thousand km^2 of Senegal was burned by the end of the period October 1988-January 1989. This estimated figure was derived from the figure of areas affected by bushfires (53.000 km^2), multiplied by the conversion factor of 0.8. Assuming this to be correct, the official bushfire statistics underestimate the burned areas with a factor of 10. Most areas affected were burned in November and December and located in the less populated eastern parts of central and southern Senegal. The amount of aboveground, end-of-season dry biomass burned is estimated to be within the range of 10.7–19.3 mill. tons, thus removing a considerable amount of fodder available for livestock, and causing a substantial release of trace gases and particulate matters to the atmosphere.

Future improvements of methodology are intended to be carried out within the following fields:

1. Generation of fire-nonsusceptible raster GIS overlay. This overlay may be created from a digital land cover map of Senegal, areas already classified as fire-affected, and grassland areas of low biomass. Krul and Breman (1982) and Sharman (1987) stated that 700 kg ha^{-1} dry herbaceous biomass was necessary to support fires. Georeferenced information on herbaceous biomass can be extracted from the biomass maps made annually. This GIS overlay should be used during classification.
2. Testing more classification approaches and combinations of channels. Data will be acquired on a more frequent basis, as suggested by the results from these study.
3. Accuracy evaluations using high resolution imagery (SPOT and LANDSAT) and use of a Dozier-algorithm (Dozier 1981) on AVHRR channels 3 and 4 data to discriminate pixels representing areas containing actively burning fires as an aid in verification.

Acknowledgments. The authors would like to thank Condeye Sylla, Moussa Dramé, and Abdoullayé Wellé of the CSE for the phytosociological description of the test site and for assistance during the radiometer campaign, and Niall Hanan, a consultant to the CSE, for assistance during the ICAR campaign.

References

Bélanger J (1987) Rapport concernant les patrouilles aériennes. Consultancy report/PPFS, Blais, McNeil, 14 pp

Brookman-Amissah JB Hall, Swaine MD, Attakorah JY (1980) A re-assessment of a fire protection experiment in northeastern Ghana savanna. J Appl Ecol 17:85–100

Cheney NP (1981) Fire behaviour, Chap 7. In: Gill AM, Groves RH, Noble IR (eds) Fire and the Australian biota. Aust Acad Sci, Canberra, pp 151–175

Chuvieco E, Congalton RG (1988) Mapping and inventory of forest fires from digital processing of TM data. Geocarto Int 3(4):41–53

Dereix C, N'Guessan A (1976) Etude de l'action des feux de brousse sur la vegetation. Les parcelles feux de Kokondekro. Resultats apres quarante ans de traitement. Cent Tech For Tropic Côte d'Ivoire station de Bouake (mimeo), 33 pp

Dozier J (1981) A method for satellite identification of surface temperature fields of subpixel resolution. Rem Sens Environ 11:221–229

Duggin MJ (1980) Field measurements of reflectance factors. Photogram Eng Rem Sens 46(5):221–229

Fall AO (1986) Bushfires in the Sahelian and forest countries. Paper presented at Experts meeting on interactions between forest and Sahelian ecosystems, Yamoussokro, Côte d'Ivoire, May 5–7, UNSO, 61 pp

Gillon D (1983) The fire problem in tropical savannas. Chap 30. In: Bourlière F (ed) Ecosystems of the World 13 – Tropical savannas. Elsevier, Amsterdam, pp 617–641

Grégoire J-M, Flasse S, Malingreau J-P (1988) Evaluation de l'action des feux de brousse, de novembre 1987 a février 1988, dans la région frontalière Guinée – Sierra-Léone. Report Projet Régional FED-CILLS-CCR Surveillance des Ressources Naturelles Renouveables au Sahél-Volet Guinée, 23 pp

Hick PT, Prata AJ, Spencer G, Campbell N (1986) NOAA-AVHRR satellite data evaluations for areal extent of bushfire damage in Western Australia. In: Proc 1st Aust AVHRR Conf, Perth, Western Australia, 22–24 Oct 1986

Hopkins B (1965) Observations on savanna burning in the Olokemeji Forest Reserve. J Appl Ecol 2:367–381

Justice C (1986) Editorial. Int J Rem Sens, 7(11):1385–1390

Justice CO, Townshend JRG, Holben BN, Tucker CJ (1985) Analysis of the phenology of global vegetation using meteorological satellite data. Int J Rem Sens 6(8):1271–1318

Krul JM, Breman H (1982) L'influence du feu. Chap. 6.5. In: Penning de Vris FWT, Djitèye MA (eds) La productivité des pâturages sahéliens. PUDOC, Wageningen, The Netherlands, pp 346–352

Lamprey RH, De Leeuw PN (1988) Monitoring East African Vegetation with NOAA-AVHRR Imagery. Final report, UNEP/ILCA NOAA-AVHRR Satellite Calibration Project, ILCA, Nairobi, 123 pp

Langaas S, Muirhead K (1989) 'Monitoring' bushfires in West-Africa by weather satellites. In: Proc Symp Remote Sensing of Environment 22, 20–26 Oct 1988, Abidjan, Côte d'Ivoire, ERIM, Michigan, pp 253–256

Malingreau JP, Tucker CJ, Laporte N (1989) AVHRR for monitoring global tropical deforestation. Int J Rem Sens 10(4;5):855–868

Parnot J (1987) Suivi diachronique des feux de brousse a l'aide des epreuves minute Landsat. Africa Pixel 1:81–89

Parnot J (1989) Inventaire des feux de brousse au Burkina Faso saison seche 1986–87. In: Proc Symp on Remote Sensing of Environment 22, 20–26 Oct 1988, Abidjan, Côte d'Ivoire, ERIM, Michigan, pp 563–573

Prince SD, Astle WL (1986) Satellite remote sensing of rangelands in Botswana. I. Landsat MSS and herbaceous vegetation. Int J Rem Sens 7(11):1533–1553

Prince SD, Willson PJ, Hunt DM, Halstead P (1988) An integrated camera and radiometer for aerial monitoring of vegetation. Int J Rem Sens 9(2):303–318

Rasmussen K (1989) The CHIPS Software Package. CHIPS Newslett 1, Inst Geogr, Univ Copenhagen

Rose-Innes R (1972) Fire in West-African vegetation. In: Proc Tall Timbers Fire Ecology Conference No 11, Tallahassee, Florida, pp 147–173

Sharman M (1987) Utilisation d'un avion léger dans l'inventaire et la surveillance continue des écosystèmes pastoraux saheliens. Report GEMS Serie Sahel No 3, FAO/UNEP-GEMS, Nairobi, 117 pp

Tucker CJ, Jones WH, Kley WA, Sundstrom GJ (1981) A Three-band Hand-held radiometer for field use. Science 211(4479):281–283

Tucker CJ, Vanpraet CL, Boerwinkle E, Easton A (1983) Satellite remote sensing of total dry matter accumulation in the Senegalese Sahel. Rem Sens Environ 13:461

Tucker CJ, Vanpraet CL, Sharman MJ, Van Ittersum G (1985) Satellite remote sensing of total dry matter accumulation in the Senegalese Sahel: 1980–1984. Rem Sens Environ 17:233

UNSO (1982) Republic of The Gambia – Pilot Project for bushfire control. Consultancy report prepared by Blais, McNeil, Lussier, 89 pp

Yergeau M (1983) Teledetection, feu de brousse et dynamique de la vegetation apres feu, Volta Rouge-Haute-Volta. CRTO, Burkina Faso (mimeo), 18 pp

18 Factors Influencing the Emissions of Gases and Particulate Matter from Biomass Burning

D.E. WARD[1]

18.1 Introduction

Forest fires, whether prescribed or wild, emit a complex mixture of particles and gases into the atmosphere. The diversity of composition of combustion products results from wide ranges in fuel types, fuel chemistry, and fire behavior for fires burning in the natural environment. This chapter discusses the state of knowledge concerning the effect of these factors on emissions of different smoke components into the atmosphere from the burning of biomass.

Advances in knowledge concerning the characteristics of smoke emissions in the United States have been made in response to needs for smoke management systems (Southern Forest Fire Laboratory 1976), primarily to protect visibility in National Parks and Wilderness areas and to protect air quality near population centers. More recently, concern has been growing regarding health effects on firefighters exposed to smoke and air toxics from wildland fires (Ward in press).

Although the effects are of real concern, relatively little biomass is burned in the United States in comparison to some other regions of the world. The total biomass consumed globally was estimated to be 10^4 Tg per year (Radke 1989). Seiler and Crutzen (1980) estimated global biomass burning to contribute 2 to 3.3 Tkg of carbon dioxide to the atmosphere per year. Assuming the conversion of fuel carbon to carbon dioxide to be 80% efficient, the biomass consumed would be similar (5 to 8.25 Tkg per year) to that estimated by Radke (1989). During 1988 in the United States, one of the most extreme "fire years" in recent history, 2 million ha of land were burned by wildfires including the Yellowstone fires. If we consider the average fuel consumption per ha to be 45 t, the total fuel consumed by wildfires in the United States in 1988 is 90 Tg. In addition, Chi et al. (1979) estimated that prescribed fires burn an average of 36.6 Tg of biomass per year. Thus the total fuel consumed by planned fires and wildfires in the United States represents about 2% of the estimated global biomass consumption.

Emissions from biomass burning are injected into strata of the atmosphere dependent on the rate of heat release, size of the fire, and the stability and wind velocity profile of the atmosphere. Examples of severe smoke obscuration and reduction of radiation transfer through the upper atmosphere have been recorded in the literature (Shostakovitch 1925; Plummer 1912). Radiation

[1]Intermountain Research Station, U.S. Department of Agriculture, Missoula, Montana 59807, USA

transfer properties of the atmosphere are affected by the presence of smoke particles. For a period of 4 weeks during the fire episodes of northern California in 1987, surface temperatures were noted to be 5 to 15°C lower than normal (Robock 1988). Other smoke components contribute to the "greenhouse gases" in the atmosphere and are abundantly produced from biomass burning (Crutzen et al. 1985).

Knowledge now available is not adequate to describe the range of important emissions characteristics needed to evaluate impacts on radiation transfer in the atmosphere. However, new information concerning the variance of emission rates for the various important trace materials suggests that a better job can be done in predicting the production of the trace materials if the chemistry of the biomass burned is known. New advances in describing the quantity and characteristics of emissions from biomass burning will most likely result from models that consider the relation of fuel chemistry along with the rate of heat release.

Variations in vegetative types, climatic zones, deposition of materials, and other causes affect the fuel chemistry and fire behavior. Fuel types within the United States exhibit small differences in the content of trace metals, carbon content, oxygen content, and deposited materials. The chemical diversity of biomass associated with savannah and tropical ecosystems may be large, and there is a need to characterize this diversity to effectively apply emissions characteristics data developed for fires in North America to the rest of the world.

My discussion starts with some of the chemical aspects of forest fuels important from an emissions production standpoint; then combustion processes are discussed; finally, the emissions are described according to particulate matter and gaseous fractions.

18.2 Forest Fuels Chemistry

In forest ecosystems, most plant material consists of polymeric organic compounds, generally described by the chemical formula $C_6H_9O_4$ (Byram 1959). Plant tissue is approximately 50% carbon, 44% oxygen, and 5% hydrogen by weight. The content of most wood varies between 41% and 53% cellulose, between 15% and 25% hemicellulose, and 16% and 33% lignin (Browning 1963). Lignin content is much higher (up to 65%) in decaying (punky) wood, in which the cell wall polysaccharides are partially removed by biological degradation. These generalized representations of biomass (or fuel) do not, however, explain the diversity of compounds produced as a result of burning the material.

Carbon monoxide and carbon dioxide are the major carbonaceous gases produced during the combustion of biomass fuels (Fig. 1). Other minor constituents, such as nitrogen, phosphorous, and sulfur, affect the mix of pollutants generated by burning plant tissues. Other factors are important as well, and contribute to the diversity of combustion products. For example, most vegetative material contains classes of compounds known as extractables, consisting

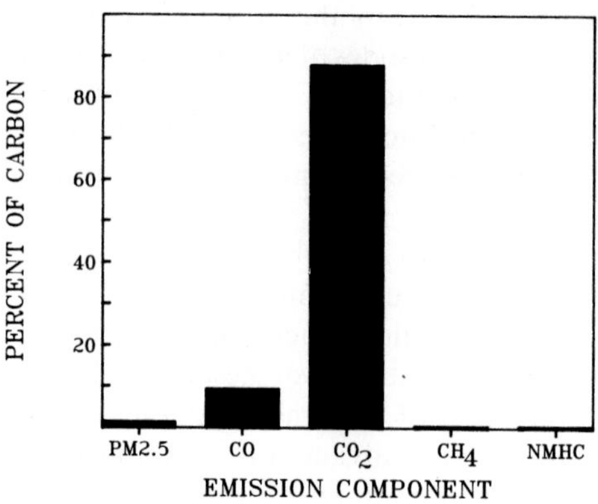

Fig. 1. The average percentage of total carbon released by biomass burning in the United States in the form of carbon monoxide, carbon dioxide, and hydrocarbons. PM2.5 is particulate matter less than 2.5 μm diameter

of aliphatic and aromatic hydrocarbons, alcohols, aldehydes, gums, and sugars. These extractables, as a group, have a higher heat value than those of the cellulose, lignin, and hemicellulosic substances. Other extractables from forest fuels contain a complex mixture of terpenes, fats, waxes, and oils. Not all of the extractable substances have low boiling points. For example, some of the extractives were shown by Shafizadeh et al. (1977) to evaporate or pyrolyze at temperatures above 300°C.

Thermal gravimetric analysis methods have been used by Susott (1980) and Susott et al. (1979) to evaluate the evolution of pyrolysis gases from solid fuels as a function of temperature. The peaks, using this technique, exhibit a spectrum reflecting the thermal stability of the fuel components as shown in Fig. 2. Each component released can have a different molecular weight and chemical form which can have significant implications regarding the formation of emissions. These materials pass through the flame structure or are released directly into the

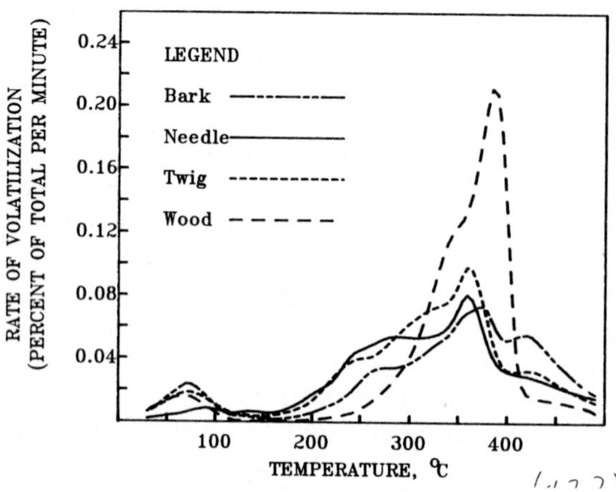

Fig. 2. Examples of different rates of fuel volatilization as a function of temperature

atmosphere. Oxidation may or may not occur at the solid fuel interface. For the woody biomass, most of the research has studied the decomposition of the cellulose and hemicellulose. Little is known about the decomposition of components of bark of which suberin – a polymer of long chain hydroxy fatty acids esterified to phenolic acids – is one substance. The degradation products in this 425 °C peak form a sizable fraction, but the chemical content may be different than for the woody carbohydrate products (Fig. 2).

Dead plant material accumulates on the forest floor or in lakes and bogs. This material is rich in nutrients and is transformed by microorganisms into humus; individual plant parts are then no longer identifiable. In the process, sulfur and nitrogen accumulate in the organic matter because many of the microorganisms require these elements for growth and reproduction. In addition, tree needles and leaves generally have a higher composition of nitrogen and sulfur than those of woody stems and limbs. Allen (1974) reported sulfur content values between 0.08% and 0.5% for plant material and 0.03% and 0.4% for organic soils. Nitrogen is one of the most dominant of the macronutrients and is of primary concern because of the large number of nitrogen-based compounds produced when biomass is burned. Nitrogen makes up as much as 0.2% of the older wood of some species, 1% of needles of some pines, and up to 2.7% of fallen hardwood leaves (Clements and McMahon 1980).

Obviously, there are large differences in the quantity, arrangement, and chemical composition of fuels found in the forest environment. Research now in progress at the Intermountain Research Station's Intermountain Fire Sciences Laboratory is examining the effect of fuel moisture content, mineral content, chemical content, and physical arrangement on emissions production.

18.3 Combustion Processes

Flaming, smoldering, and glowing combustion processes compete for available fuel and are markedly different phenomena that contribute to the diversity of combustion products. The fuel characteristics (including arrangement, distribution by size classes, moisture, and chemistry) dominate in affecting the duration of the flaming and smoldering combustion phases.

Flaming and smoldering are reasonably distinct combustion processes that involve different chemical reactions and are quite different in appearance. Flaming combustion dominates during the startup phase, with the fine fuels and surface materials supplying the volatile fuel required for the rapid oxidation reactions to be sustained in a flaming environment. The heat from the flame structure and the diffusion and turbulent mixing of oxygen at the surface of the solid fuel contribute to the heat required to sustain the pyrolysis processes. Early in the flaming phase, the volatile hydrocarbons are vaporized from the fuels. Later the cellulosic and lignin-containing cellular materials decompose through pyrolysis. These processes produce the fuel gases that sustain the visible flaming processes.

Once carbon begins to build up on the solid fuel surfaces, the pyrolytic reactions no longer produce sufficient fuel gases to maintain the flame envelope. For combustion to continue, oxygen must diffuse to the surface of the fuel. Diffusion of oxygen and the availability of oxygen at the fuel surface is enhanced through turbulence in the combustion zone and through premixing by introducing the oxygen at ground level. This allows oxidation to take place at the solid fuel surface and provides for heat evolution and heat feedback to accelerate the pyrolytic reactions and volatilization of the fuel gases from the solid fuel. The process ultimately leads to the production of charcoal, where the only combustion occurring is of the glowing type – a surface reaction of oxygen with carbon.

18.4 Smoke Production

The smoldering combustion phase produces high emissions of particulate matter and carbon monoxide. Fires of very low intensity (those in which the flaming combustion phase is barely sustained) produce high emissions of particulate matter. Heading fires generally have emission factor values two to three times greater than those for backing fires. The formation of particulate matter results primarily from two processes: (1) the agglomeration of condensed hydrocarbon and tar materials, and (2) mechanical processes which entrain fragments of vegetation and ash.

18.4.1 Release of Carbon

When forest fuels are burned, carbon is released in the form of carbon dioxide, carbon monoxide, hydrocarbons, particulate matter, and other substances in decreasing abundance (Fig. 1). A carbon mass-balance procedure is often used to characterize the fuel consumed in producing the emissions measured (Radke et al. in press; Ward and Hardy 1984). Carbon dioxide and carbon monoxide generally contain more than 95% of the carbon released during the combustion of biomass. A measurement of combustion efficiency is used that depends on the ratio of the actual carbon contained in the emissions of carbon dioxide compared to that theoretically possible if all of the carbon were released as carbon dioxide. The combustion efficiency is never 100% for biomass burned in the open environment and generally ranges from 50% to 95%. Generally, the combustion efficiency is lowest for the smoldering combustion phase and highest for those fires with good ventilation and vigorous flame action. Combustion efficiency is illustrated in Fig. 3 for a number of fuel types tested in the United States.

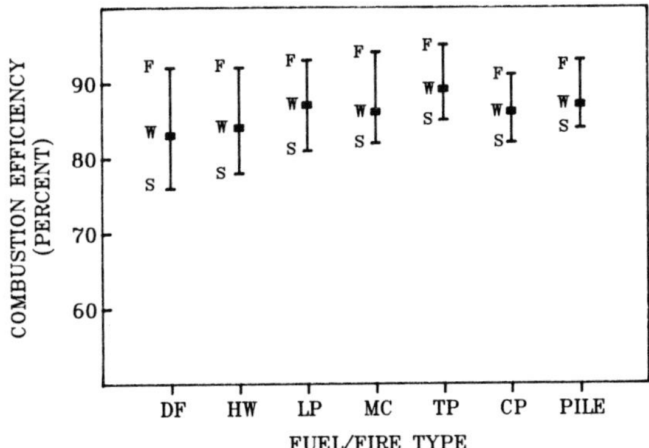

Fig. 3. Combustion efficiency for seven different fuel types for flaming, smoldering, and weighted average. *DF* Douglas fir; *HW* Hardwood; *LP* Ponderosa pine and lodgepole pine (long-needled conifers); *MC* Mixed conifer; *TP* Tractor piled; *CP* Crane piles; *Pile* Combined TP and CP

18.4.2 Formation of Particles

Forest fires are a complex form of the diffusion flame process where pyrolysis of solid fuels produces fuel gases that interdiffuse with oxygen from the atmosphere. As the interdiffusion of fuel and oxygen develops and intensifies, the flame characteristics and the chemical processes occurring in the flame zone change. After a level of fire intensity is surpassed that optimizes combustion efficiency, some of the pyrolyzed fuel may no longer pass through an active oxidation zone. At times, even in lower intensity fires, pockets of unburned, partially oxidized gaseous fuels escape the combustion zone or undergo delayed ignition. The influence of flame turbulence on combustion efficiency is not fully understood; however, as the intensity of the fire increases and the zone of complete mixing of gaseous fuel and oxygen moves farther from the solid fuel, combustion efficiency is believed to decrease and production of pollutants to increase.

Because of the increased depth and height of the flame zone, heading fires, and area fires create an extended reducing environment in which continued pyrolysis and synthesis of hydrocarbon gases and fragmented particles can occur under conditions of reduced oxygen content. In addition, heat is reradiated from the particles to the atmosphere, which can slow down the reactions as the unburned gases and particles are convected away from the active combustion zone. If the temperature in the interior of the flame zone is appropriate ($< 800°C$), rapid formation of particles and accretion of carbonaceous organic particles will occur. Consumption of the particles requires prolonged exposure at high temperatures ($> 800°C$) in a zone with near-ambient (21%) concentration of oxygen (Glassman 1977). Mass-fire experiments performed in Canada during 1988 by our Fire Chemistry Research Work Unit demonstrated the important effect of oxygen deficiency on flame structure and on emissions production. The pulsation phenomenon often observed for large fires is thought to be closely coupled to oxygen deficiency. Oxidation of the particles depends partly on the degree of premixing of pyrolyzed fuel and oxygen that takes place in the zone of active solid fuel pyrolysis. Greater premixing results in production of less particulate matter.

18.4.3 Fuel Chemistry Effects on Particle Formation

Ward (1979, 1980) used a flow-reactor combustion chamber system to model combustion environments and to test the premise that production of particulate matter is a function of flame volume. This work was for cylindrical laminar diffusion flames using a model fuel, alpha-pinene. He found emission factors for particulate matter to be positively correlated with flame volume, fuel supply rate, and air supply rate. Other fuels of varying molecular oxygen content were tested. Generally, emission factors decreased as oxygen content of the fuel increased. The results seemed to agree with those obtained from specimens of wood with a high resin content burned in the apparatus. The tests were for a nonturbulent environment, however, which makes the results difficult to apply to fires burning in forest environments where turbulence in the combustion zone is generated by mechanical and buoyancy-induced processes.

Fuel properties, such as oleoresin content or mineral content, influence flame characteristics that depend on the rate of volatilization and pyrolysis of the fuels. Heat feedback to the solid fuel, oxygen entrainment through the fuel bed, the degree of endothermicity of the pyrolysis process, and other factors control the rate of combustion and the flame characteristics (including the oxidation-reduction reactions occurring inside the flame envelope).

Other fuels, such as benzene, cyclohexane, hexane, propane, and high-resin content wood, were found to produce emission factors for particulate matter that decreased in magnitude from a high of 172 for benzene to a low of 30 g per kg for wood. These results agree with observations made for the smoke-point of diffusion flame combustion of different classes of compounds. For example, Glassman (1977) reported decreasing carbon formation for different classes of compounds in this order: aromatics $>$ alkynes $>$ alkenes $>$ alkanes. At least for the aromatic compound benzene and for alkane compounds, the smoke-point results compared favorably with the emission factors measured for particulate matter. Alpha-pinene seemed to compare more closely to benzene from the standpoint of a propensity to produce particulate matter. Ethylene glycol and ethyl alcohol produced negligible quantities of particulate matter when burned in the closed combustion chamber system. The conclusion reached was that emission factors for particulate matter are inversely related to the oxygen content of the fuel molecules.

18.4.3.1 Particle Number and Volume Distribution

Smoke particles have been measured using sophisticated instruments aboard aircraft to cover the broad distribution of particle sizes from 0.01 to 43 μm (Radke et al. 1990). The results suggest a very pronounced number concentration peak at a diameter of 0.15 μm. The volume distribution which for a first approximation represents the mass distribution shows a bimodal distribution with peaks at 0.5 μm and greater than 43 micrometers (Fig. 4).

Ward and Hardy (1984) measured a large difference in emission factors for particles less than 2.5 μm mean mass cutpoint diameter and particles without regard to size. This difference increased proportional to an increase in the rate of heat release on a square meter basis (Fig. 5). They noted a slight decrease in emission factors for particles less than 2.5 μm mean mass cutpoint diameter with an increase in total particulate matter emission factors over the range of rates of heat release tested. Radke et al. (1988) noted a similar increase in total particulate matter emission factors and concurred that this increase probably results from an increased level of turbulence in the combustion zone.

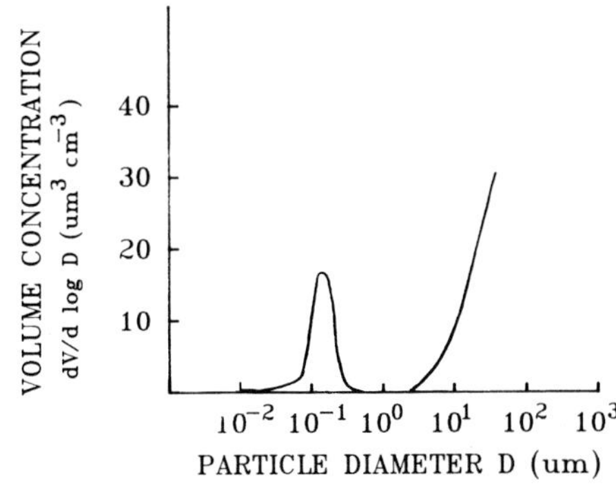

Fig. 4. Volume by size of particle fractions measured for prescribed fires of logging slash in the western United States from an airborne sampling platform. (Radke et al. 1990)

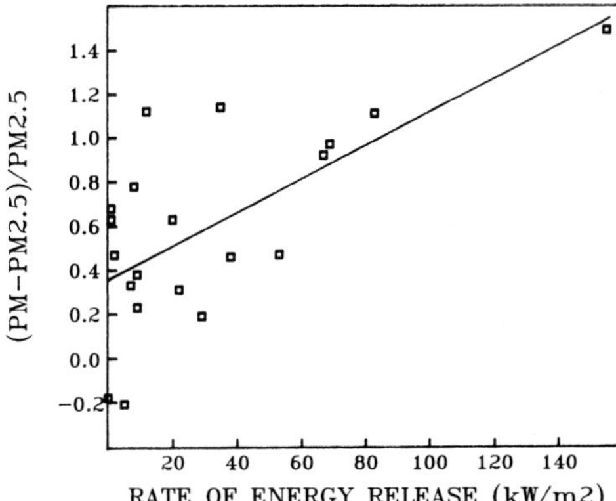

Fig. 5. The ratio of differences between emission factors for total particulate matter and particulate matter less than 2.5 μm as a function of the rate of energy release

18.4.3.2 Emission Factors for Particulate Matter

Ward et al. (1988) summarized those emission factor data available for different fuel types of the United States (Table 1). A considerable wealth of empirical data is available. Of particular concern, however, is the lack of information regarding fuel and fire conditions associated with the production of the measured emissions. Airborne sampling of "targets of opportunity" often is poorly supported by ground truth observations or measurements. Generally, investigators recognize the combustion efficiency differences between flaming and smoldering combustion phases. But the composite samples taken using airborne

Table 1. Emission factors for each fire/fuel configuration by phase of combustion. (Ward et al. 1988)

Fire/fuel configuration	Phase of[a] combustion	Emission factor (g kg^{-1}) Particulate matter PM2.5	PM10	Total
Pacific Northwest broadcast logging slash				
Hardwood	F	6	7	13
	S	13	14	20
	Fire	11	12	18
Short-needled Conifer	F	7	8	12
	S	14	15	19
	Fire	12	13	17
Long-needled Conifer	F	6	6	9
	S	16	17	25
	Fire	13	14	20
Piled logging slash				
Dozer-piled Conifer				
No mineral soil	F	4	4	5
	S	6	7	14
	Fire	4	4	6
10–30% Mineral soil	S			25
25% Organic soil	S			35
Crane-piled Conifer (No twigs or needles).				
Southwest				
Chaparral		8	9	15
Southeast line fires				
Long-needled Conifer	S/Heading		40	50
	F/Backing		20	20
Palmetto/Gallberry	Heading		15	17
	Backing		15	15
Grasslands			10	10
Sawgrass			10	10

[a]F = flaming, S = smoldering, Fire = fire-weighted average of F and S.

systems have seldom been effective differentiating either combustion phase. For many fuel types, emissions from the smoldering phase overwhelm emissions produced through flaming combustion processes – typical of measurements of smoke from wildfires and during the later stages of prescribed fires.

Ward and Hardy (1984) circumvented many of these problems by establishing experimental plots of near full-scale with sampling systems suspended from cable and tower supports over the experimental fires (Fig. 6). The system measured carbon dioxide and carbon monoxide concentrations, vertical velocity of the convection gases, and temperature of the convection gases in real time. In addition, discrete grab samples of the particulate matter and gases were collected on filters and in bags separately for the flaming and smoldering

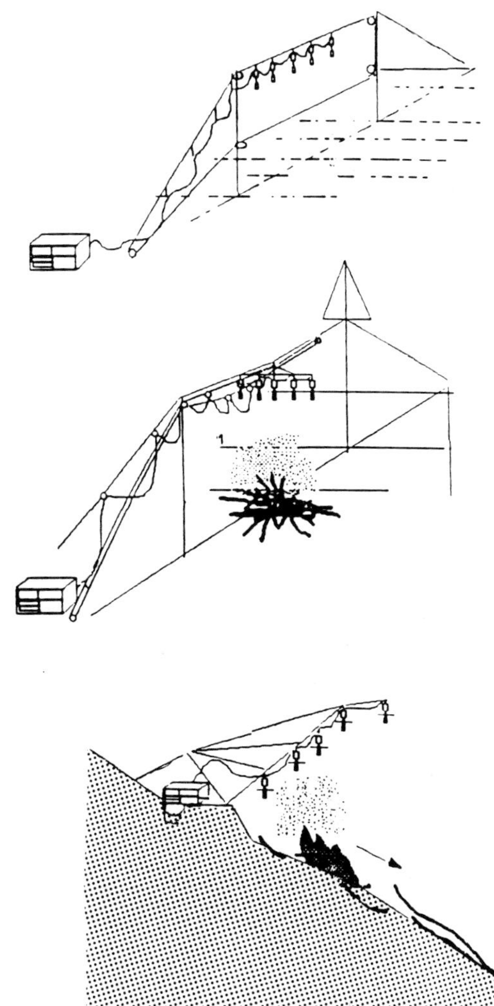

Fig. 6. Various configurations of sampling apparatus used to measure the flux of gases, particles, and heat from near full-scale test fires

combustion phases. This work demonstrated for a number of different fuel types that (1) emissions of particulate matter range over a factor of 10 depending on fire and fuel conditions that affect combustion efficiency; (2) brushy areas produce the most smoke per ton of fuel consumed and have higher rates of production of benzo[a]pyrene than nonbrushy areas; (3) fires of higher intensity (long flame lengths) produce proportionately larger particles than are found in low-intensity and smoldering combustion fires; (4) carbon monoxide is abundantly produced from open fires and, generally, on a mass basis exceeds the production of particles by a factor of 10; and (5) hydrocarbon gases are a small part of the total amount of carbon released from the combustion of forest fuels.

Emission factors have been measured for a number of different combinations of forest fuel and weather influences. Empirical data suggest that emission factors for particulate matter range from 4 to 40 g per kg for the flaming combustion phase for particles less than 10 μm in diameter and from 5 to 50 g per kg for particles measured without regard to size. The particle size distribution suggests that 40% to 95% of the mass of particulate matter consists of particles less than 2.5 μm. Particles larger than 2.5 but less than 10 μm generally make up less than 10% of the total mass. The rate of heat release has a pronounced effect on the size and mass of particles produced. Generalized models are needed, based on factors affecting fire spread and fuel consumption, for predicting the production of smoke.

Emissions of Trace Elements

The trace elements for samples of particles less than 2.5 μm in diameter (PM2.5) are shown in Fig. 7 as a percentage of the PM2.5 by combustion phase and weighted for the entire fire. All the samples of the trace elements are from broadcast burns of logging slash from coniferous species. The sodium component is especially high for the flaming phase; nearly 2.8% of the PM2.5 is sodium.

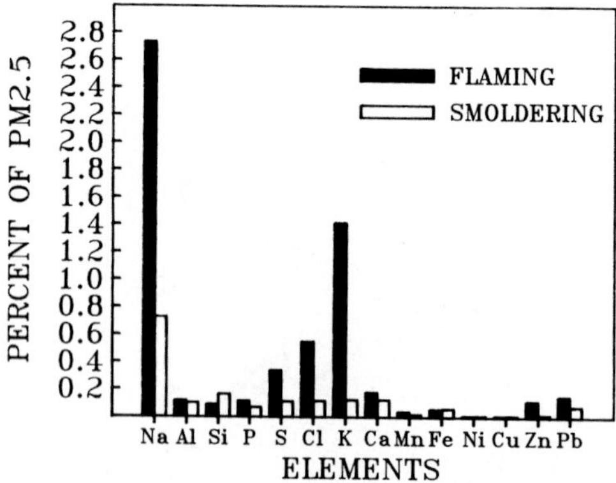

Fig. 7. Percentage composition of particulate matter less than 2.5 μm diameter in smoke from logging slash fires in the western United States

The sulfur, chlorine, and potassium contents of PM2.5 are high during the higher temperature flaming phase of the fire. Generally, as the combustion efficiency increases, more of the carbon is consumed, thus increasing the percentage of mass reported as trace elements. Ward and Hardy (1984) found that the sum of these components (S + Cl + K) is correlated with the rate of heat release ($r = 0.92$). Iron released with the PM2.5 is slightly greater during the latter periods of the smoldering phase than during the flaming phase or first part of the smoldering phase. The large difference in potassium content by combustion phase leads to accentuating the potassium-to-carbon ratio differences by combustion phase. Potassium was released proportional to the rate of heat release.

Differences in emissions of trace elements were noted as a function of fuel type by Ward and Hardy (1988, 1989). The average values for trace materials produced with the PM2.5 during the flaming phase were generally higher for the chaparral fires of California than for the logging slash broadcast fires of the Washington and Oregon areas (Fig. 8). The production of sulfur, chlorine, and potassium for the chaparral fires was an order of magnitude larger than for the slash fires. Fire intensity was much higher for some of the chaparral test fires than for the logging slash fires. The fire intensity ranged up to a maximum rate of energy release on a square meter basis of nearly 3 MW – or nearly an order of magnitude larger than for the logging slash fires. Emissions of lead from fires in southern California were very high relative to fires in the Pacific Northwest. We currently do not have adequate information to separate the effects of rate of heat release and fuel chemistry in the prediction of the content of particles. It is generally accepted that the trace elements are released in the highest proportion to the carbon contained with the particles for the highest intensity fires. However, the lead content may be higher for the California fires because of a higher deposition rate in the California area from sources outside the forest environment.

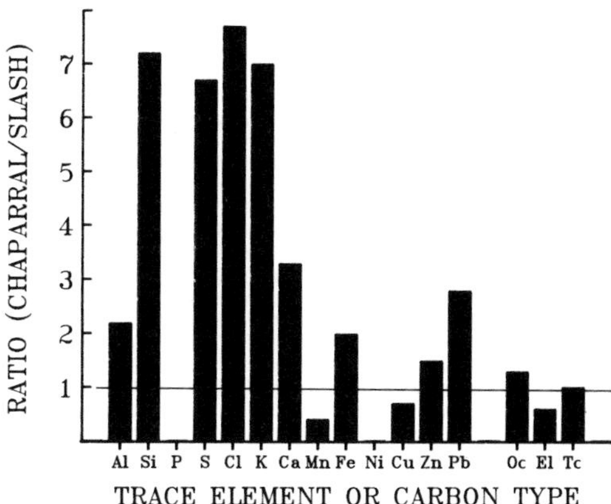

Fig. 8. Ratio of trace element content for logging slash and chaparral fires from prescribed fires in the western United States

Emissions of Graphitic and Organic Carbon

Emissions of graphitic carbon are especially important because of the contribution to the absorption of light. Since the absorption of the smoke emissions is due primarily to graphitic carbon, the specific absorption coefficient correlates well with the graphitic carbon content of the aerosol (Patterson et al. 1986). The early work for the logging slash fires of the Pacific Northwest showed emission factors for graphitic carbon ranging from 0.46 to 1.18 g per kg of fuel consumed. In tests of pine needle (slash pine) fires in a controlled environment combustion laboratory, emission factors ranged to 5.40 g per kg of fuel consumed. The results suggest an inverse correlation between specific absorption and emission factors. These results are in agreement with the inverse correlation of rate of heat release with percent graphitic carbon content reported by Ward and Hardy (1986). Generally, emission factors for PM2.5 have been found to be lower for higher intensity fires.

Organic carbon content of particulate matter is especially important because of the types of organic compounds associated with the particles. The polynuclear organic material is contained as a fraction of the organic carbon content of the particles and contains the important class of compounds known as polynuclear aromatic hydrocarbons – some of which are known to have carcinogenic properties. The carbon fraction of the organic content of particulate matter ranges between 30 and 60%. Benzo[a]pyrene is the most studied of the compounds contained in this fraction and emission ratios ranged from 2 to 274 μg per gram of particulate matter for heading and backing fires, respectively (McMahon and Tsoukalas 1978). Measured ratios of benzo[a]pyrene to particulate matter were reported in the range of 0.4 to 222 μg per g of particulate matter for fires in coniferous species logging slash in the western United States (Ward in press). The highest values occur for the smoldering combustion phase and lowest for the flaming combustion phase for the highest intensity fires.

18.4.4 Emissions of Gases

In this section, fuel chemistry and combustion efficiency effects are discussed as they affect the production of trace gases from combustion sources. In particular, studies are reviewed that have examined the production of gases containing nitrogen, sulfur, and chlorine. Other hydrocarbon gases produced, such as methane also are discussed.

18.4.4.1 Nitrogen Gases

Generally, temperatures within flame structures of forest fires do not exceed 1000°C, which suggests that molecular nitrogen gas (N_2) from the atmosphere is not disassociated to combine with free radicals within the combustion zone.

Several studies have reported the production of NO_x from burning of forest fuels. The research involved combustion laboratory fires, field fires where sampling apparatus was suspended from towers, and field fires with airborne sampling (Gerstile and Kemnitz 1967; Radke et al. 1978; Cofer et al. 1988a, 1989; Ward et al. 1982b).

Evans et al. (1977) measured NO_x in forest fire plumes in Australia. Their findings showed NO_x concentrations as high as 0.024 ppm. The conclusion was that the hotter the fire the more NO_x was produced by the fire. This has not been found to be the case for fires in the United States.

Results of Clements and McMahon (1980) suggest that the production of oxides of nitrogen increases and is highly dependent on the nitrogen content of the fuel burned. Conversion of fuel-bound nitrogen to NO_x can occur readily in oxygen-depleted air. Simple linear regression equations were developed with coefficients of determination greater than 0.70 for predicting the production of NO, NO_x, NO_2, as functions of the nitrogen content of the fuel. The quantity of fuel nitrogen converted to NO_x was found to range from 6.1 to 41.7% for wood and organic soil, respectively.

The emissions of N_2O from combustion of forest fuels showed a small variance for two extreme fuel types – the boreal forest of Canada and the chaparral fuels of California. Proximate analysis for the nitrogen content might suggest that the differences should not be real. However, for some measurements of NO_x, the difference between boreal and chaparral oxides of nitrogen emissions was quite large (Hegg et al. 1989).

Hegg et al. (1989) have measured emission factors for NH_3 from 0.1 to 2.0; for NO_x of 0.81 to 8.9; for N_2O of 0.16 to 0.41 g per kg of fuel consumed. Nitrogen loss from the biological site is significant. In addition, a fraction of the nitrogen is contained with the particulate matter. NH_4^+ is contained with the particulate matter and may contribute to as much as 0.05 g per kg of fuel consumed. Volatilization and release of nitrogen to the atmosphere can be significant on a total mass basis per unit area burned.

Crutzen et al. (1985) found an average ratio of N_2O to CO_2 of 1.5 to 3×10^{-4} for tropical biomass burning. By summing the carbon released as CO, CO_2, CH_4, NMHC, and assuming an emission rate of 1% for particulate matter (Ward and Hardy 1988), the conversion of biomass to CO_2 is 77.8% efficient. This compares favorably with average rates of combustion efficiency measured for logging slash fires of 82.8% in the western United States (Ward and Hardy 1984). The emission factor for N_2O was computed from Crutzen et al. (1985) and the result is 0.22 to 0.44 g N_2O per kg of fuel consumed. The emission factor for burning of tropical biomass is quite similar to that for boreal, chaparral, and coniferous biomass in North America as measured by Hegg et al. (1989). Ozone is not a byproduct of combustion of biomass, but forms as a product of secondary chemical reactions once the combustion products enter the atmosphere.

18.4.4.2 Sulfur Emissions (Carbonyl Sulfide)

Along with nitrogen, sulfur is one of the essential nutrients required in the synthesis of plant amino acids and other physiologically important substances (Tiedemann 1987; Grier 1975). Hence, the concern over the volatilization and loss of these important nutrients is of extreme interest in sustaining the productivity of forest lands. Nitrogen can be replaced through symbiotic N-fixation; however, sulfur is replenished mainly through atmospheric deposition. Very little work has been done in identifying the form of the sulfur- or nitrogen-containing emissions released during the combustion of biomass fuels.

Crutzen et al. (1985) measured carbonyl sulfide (COS) emission ratios to CO_2 of approximately 4 to 8×10^{-6} (5.71 to 11.42 mg per kg of fuel consumed). Ward et al. (1982a) measured emission factors for COS of 0.18 to 2.36 mg per kg of fuel consumed from controlled experiments in a combustion laboratory-burning hood facility. The experiments were for fuels of varying amounts of sulfur content ranging from 0.55% (by weight) for organic soil to a low of 0.065% for pine needles. Ward et al. (1982a) calculated emission factors from Crutzen et al. (1979) of 16.2 and 35.3 mg per kg of fuel consumed for fires in the Rocky Mountain area of the United States. The results of Ward et al. (1982a) can be used to predict emission factors for COS as a function of the rate of heat release per unit area (kW per m^2) as follows: $EF_{COS} = 0.732 - 0.0065 I_R$ where the coefficient of determination is 0.99. This would suggest that the production of COS is independent of the sulfur content of the fuel which probably is not valid. However, it may be concluded that COS production is extremely sensitive to the thermal environment reflected through the apparent dependency on I_R. Other sulfur containing compounds were measured as well including H_2S, $(CH_3)_2S$, CS_2, CH_3SSCH_3, and other unknown mercaptan compounds. The sulfur quantified made up less than 0.25 percent of the total sulfur released during the combustion experiments.

18.4.4.3 Methyl Chloride

Methyl chloride has been suggested as a natural tracer unique to the combustion of biomass fuels (Khalil et al. 1985). Ward (1986) found that methyl chloride is produced in much greater quantities in the smoldering combustion phase than in the flaming phase. Thus, to use methyl chloride as a constant ratio to the particulate matter produced could cause an error in determining the impact of smoke at receptor sites unless the mix of flaming and smoldering combustion products is known. Ward (1986) found an inverse relation between the chlorine content of the fine particles and the production of methyl chloride. Emission factors for methyl chloride are inversely proportional to the rate of heat release. For chaparral fuels, methyl chloride emission factors ranged from 16 to 47 mg per kg of fuel consumed (Ward and Hardy 1989). Reinhardt and Ward (unpublished) found that most, if not all, of the methyl chloride is produced from the smoldering combustion process.

18.4.4.4 Carbon Monoxide

Carbon monoxide is the second most abundant carbon-containing gas produced during the combustion of biomass (Fig. 1). Combustion efficiency is nearly perfectly correlated with the ratio of the production of carbon monoxide relative to carbon dioxide (Fig. 9). Other correlations have been found with carbon monoxide. Ward (in press) found particulate matter concentration to be correlated with carbon monoxide concentration (r = 0.89). Reinhardt (1989) found the concentration of formaldehyde to be correlated with the concentration of carbon monoxide (r = 0.93). Generally, emission factors for carbon monoxide on a mass basis are 10 times greater than for the fine particle fraction. Emission factors for carbon monoxide range from 60 to over 300 g per kg of fuel consumed.

The calculated range of emission factors for the measurements made by Crutzen et al. (1985) in Brazil of 167 to 209 g per kg is generally higher than the fire-weighted average emission factors for logging slash fires in the western United States (average of 171 g per kg). This difference may be a result of either vegetation or moisture content differences, or both.

18.4.4.5 Methane and Nonmethane Hydrocarbons

Methane is produced in much larger quantities during the smoldering combustion phase than in the flaming phase (Ward and Hardy 1984). Emission factors are about 2 to 3 times greater for the smoldering phase, ranging from 5.7 to 19.4 g per kg of fuel consumed. The flaming phase emission factors range from 1 to 4.2 g per kg. These values are consistent with emission factors reported by Crutzen et al. (1985) in Brazil of 11 to 22 g per kg of fuel consumed.

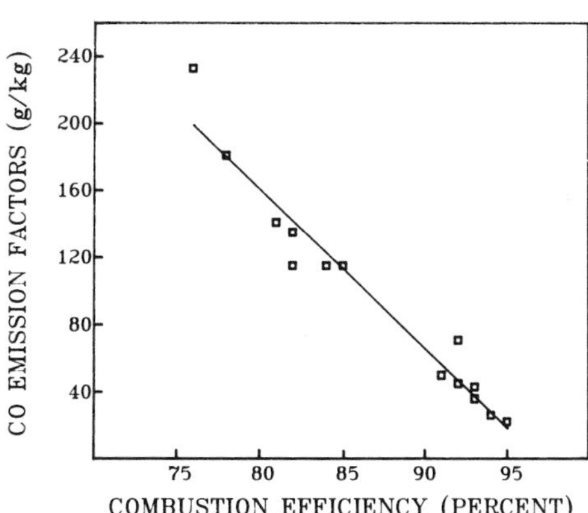

Fig. 9. Functional relation for carbon monoxide and combustion efficiency

18.5 Summary

Global biomass burning is approximately 10^4 Tg per year, with biomass burning in the United States accounting for less than 2% of the total. New information is needed regarding the combined effects of fuel chemistry, physical arrangement, and resulting fire behavior to satisfy the need for predicting emissions produced from the large number of different sites where biomass burning occurs.

Flaming and smoldering combustion processes affect the production of emissions. Carbon monoxide and carbon dioxide combined account for 90 to 95% of the carbon released during biomass burning. Combustion efficiency ranges from 50 to 80% for smoldering combustion and from 80 to 95% for flaming combustion. Many of the compounds released during biomass burning are correlated with combustion efficiency and with carbon monoxide concentration.

The size distribution for particles produced from biomass burning is bimodal, with particle mass peaks occurring near 0.5 μm and greater than 43 μm. The mass of particulate matter between 1 and 10 μm makes up less than 10% of the total mass. Models are needed, based on factors affecting fire spread and fuel consumption, for predicting the production of smoke emissions. Deposition of materials from sources other than forest environments can affect the types of materials resuspended in the atmosphere through biomass burning. An example is lead in the Los Angeles, California, area.

Graphitic and organic carbon content of the particles are important because the graphitic carbon affects the absorption of light by the particles. The organic carbon fraction contains the polynuclear aromatic hydrocarbon fraction in which some of the compounds are carcinogenic.

Nitrogen, sulfur, and chlorine content of biomass have been studied, as have a few of the compounds released during the combustion of biomass that contain these elements. A systematic study has not been performed to examine factors leading to the production of many of the compounds known to be released. Some of the nitrogen compounds are released in proportion to the nitrogen content of the fuel. Sulfur compounds studied are released as a function of the rate of heat production on a unit area basis. Methyl chloride seems to be produced primarily during the low-temperature combustion of biomass.

Similarity seems to exist for many of the compounds between measurements taken in Brazil and measurements taken in North America. Further study is needed to confirm the similarity based on a study of fuel chemical properties and the resulting emissions production when these fuels are burned.

References

Allen SE (1974) Chemical analysis of ecological materials. Wiley, New York, 565 pp

Browning BL (1963) Methods of wood chemistry, Vol I. Wiley, New York, 259 pp

Byram GM (1959) Combustion of forest fuels. In: Davis KP (ed) Forest Fire Control and Use. McGraw-Hill, New York, 565 pp

Chi CT, Horn DA, Reznik RB et al. (1979) Source assessment: prescribed burning, state of the art. Research Triangle Park, NC: November, EPA-600/2-79-019h

Clements HB, McMahon CK (1980) Nitrogen oxides from burning forest fuels examined by thermogravimetry and evolved gas analysis. Thermochim Acta 35:133-139

Cofer III WR, Levine JS, Riggan PJ et al. (1988a) Trace gas emissions from a mid-latitude prescribed chaparral fire. J Geophys Res 93:1653-1658

Cofer III WR, Levine JS, Riggan PJ et al. (1988b) Particulate emissions from a mid-latitude prescribed chaparral fire. J Geophys Res 93:5207-5212

Cofer III WR, Levine JS, Riggan PJ et al. (1989) Trace gas emissions from chaparral and boreal forest fires. J Geophys Res 94:2255-2259

Crutzen PJ, Heidt LE, Krasnec JP, Pollock WH, Seiler W (1979) Biomass burning as a source of the atmospheric gases CO, H_2, N_2O, NO, CH_3Cl, and COS. Nature (Lond) 282:253-256

Crutzen PJ, Delany AC, Greenberg J et al. (1985) Tropospheric chemical composition measurements in Brazil during the dry season. J Atmos Chem 2:233-256

Evans LF, Weeks IA, Eccleston AJ, Packham DR (1977) Photochemical ozone in smoke from prescribed burning of forests. Environ Sci Technol 11(9):896-900

Gerstile RS, Kemnitz DA (1967) Atmospheric emissions from open burning. J Air Pollut Control Assoc 17:324-327

Glassman I (1977) Combustion. Academic Press, New York, 275 pp

Grier C (1975) Wildfire effects on nutrient distribution and leaching in a coniferous ecosystem. Can J Soil Sci 5:599-607

Hegg DA, Radke LF, Hobbs PV et al. (1989) Emissions of some biomass fires. In: Proc 1989 National Air and Waste Management Association meeting. 1989 June 25-30, Anaheim, CA, No 089-025 003

Khalil MAK, Edgerton SA, Rasmussen RA (1985) Gaseous tracers for sources of regional scale pollution. J Air Pollut Control Assoc 35(8):838-840

McMahon CK, Tsoukalas SN (1978) Polynuclear aromatic hydrocarbons in forest fire smoke. In: Jones PW, Freudenthal RI (eds) Carcinogenesis, Vol 3: Polynuclear Aromatic hydrocarbons. Raven, New York, pp 61-73

Patterson EM, McMahon CK, Ward DE (1986) Absorption properties and graphitic carbon emission factors of forest fire aerosols. Geophys Res Lett 13(1):129-132

Plummer FG (1912) Forest fires: Their causes, extent and effects, with a summary of recorded destruction and loss. Tech Bul 117 Washington, DC: US Dep Agric, 39 pp

Radke LF (1989) Airborne observations of cloud microphysics modified by anthropogenic forcing. In: Proc Symp on the role of clouds in atmospheric chemistry and global climate. Am Meteorol Soc 1989 Jan 29-Feb 3, Anaheim, CA, pp 310-315

Radke LF, Stith JL, Hegg DA, Hobbs PV (1978) Airborne studies of particulates and gases from forest fires. J Air Pollut Control Assoc 28:30-33

Radke LF, Hegg DA, Lyons JH, Brock CA, Hobbs PV, Weiss R, Rasmussen R (1988) Airborne measurements on smokes from biomass burning. J Geophys Res

Radke LF, Lyongs JH, Hegg DA et al. (1990) Airborne monitoring and smoke characterization of prescribed fires on forest lands in western Washington and Oregon. Gen Tech Rep PNW-251. Portland, OR: US Dep Agric For Service, Pacific Northwest Research Station, 81 pp

Reinhardt TE (1989) Monitoring firefighter exposure to air toxics at prescribed burns of forest and range biomass: Proc 1989 March 8-10; Seattle, WA; Pacific Northwest International Section of the Air and Waste Management Association

Reinhardt TE, Ward DE (unpublished) Factors affecting methyl chloride emissions from burning forest biomass. Unpublished manuscript on file at Seattle, WA: US Dep Agric For Service, Pacific Northwest Research Station

Robock Alan (1988) Enhancement of surface cooling due to forest fire smoke. Science 242:911–913

Seiler W, Crutzen PJ (1980) Estimates of gross and net fluxes of carbon between the biosphere and the atmosphere from biomass burning. Climatic Change 2:207–248

Shafizadeh F, Chin PS, DeGroot WF (1977) Effective heat content of green forest fuels. For Sci 23:81–89

Shostakovitch VB (1925) Forest conflagrations in Siberia with special reference to the fires of 1915. J For 23:365–371

Southern Forest Fire Laboratory (1976) Southern forestry smoke management guidebook. Gen Tech Rep SE-10. Asheville, NC: US Dep Agric For Service, Southeastern Forest Experiment Station, 140 pp

Susott RA (1980) Thermal behavior of conifer needle extractives. For Sci 26:347–360

Susott RA, Shafizahdeh F, Aanerud TW (1979) A quantitative thermal analysis technique for combustible gas detection. J Fire Flammability 10:94–104

Tiedemann AR (1987) Combustion losses of sulfur from forest foliage and litter. For Sci 33:(1)216–223

Ward DE (1979) Particulate matter and aromatic hydrocarbon emissions from the controlled combustion of alpha-pinene. Dissertation, Seattle: Univ Washington, 244 pp

Ward DE (1980) Particulate matter production from cylindrical laminar diffusion flames. In: Proc Fall meeting of the Western States Section of the Combustion Institute. Los Angeles, CA 18 pp

Ward DE (1986) Characteristic emissions of smoke from prescribed fires for source apportionment. In: Proc Annual Meeting of the Pacific Northwest International Section, Air Pollution Control Association, Eugene, Oregon (Nov. 19–21, 1986)

Ward DE (in press) Air toxics and fireline exposure. In: Proc Tenth Conf on Fire and Forest Meteorology. 1989 April 17–21, Ottawa, Canada

Ward DE, Hardy CC (1984) Advances in the characterization and control of emissions from prescribed fires. In: Proc 78th annual meeting of the Air Pollution Control Association. 1984 June 24–29, San Francisco, CA, Paper No 84-363

Ward DE, Hardy CC (1986) Advances in the characterization and control of emissions from prescribed broadcast fires of coniferous species logging slash on clearcut units. Final report, Project EPA DW12930110-01-3/DOE DE-A179-83BP12869, on file at For Sci Lab, Seattle, WA, 125 pp

Ward DE, Hardy CC (1988) Organic and elemental profiles for smoke from prescribed fires. In: Watson JG (ed) Receptor models in air resources management. Proc Int Spec Conf 1988 February, San Francisco, CA, pp 299–321, Air and Waste Management Association, Pittsburgh, PA

Ward DE, Hardy CC (1989) Emissions from burning of chaparral. In: Proc of the 1989 National Air and Waste Management Association Meeting. 1989 June 25–30, Anaheim CA

Ward DE, McMahon CK, Adams DF (1982a) Laboratory measurements of carbonyl sulfide and total sulfur emissions from open burning of forest biomass. In: Proc of the 75th Annual Meeting of the Air Pollution Control Association. 1982 June 20–25, New Orleans, LA

Ward DE, Sandberg DV, Ottmar RD et al. (1982b) Measurement of smoke from two prescribed fires in the Pacific Northwest. In: Proc of the 75th Annual Meeting of the Air Pollution Control Association. 1982 June 20–25, New Orleans, LA

Ward DE, Hardy CC, Sandberg DV (1988) Emission factors for particles from prescribed fires by region in the United States. In: Mathai CV, Stonefield CH (eds) Pm-10: Implementation of Standards. Proc of APCA/EPA Int Spec Conf 1988 February, San Francisco, CA, pp 372–386

19 Ozone Production from Biomass Burning in Tropical Africa. Results from DECAFE-88

M.O. ANDREAE[1], G. HELAS[1], J. RUDOLPH[2], B. CROS[3], R. DELMAS[3], D. NGANGA[3], and J. FONTAN[4]

The DECAFE (Dynamique Et Chimie Atmosphérique en Forêt Equatoriale) project was set up to study meteorological and chemical processes in and above the rain forest of equatorial Africa. This project makes use of a combination of aircraft, balloon, and ground measurements of chemical and physical parameters at several sites in equatorial Africa. The DECAFE-88 experiment was carried out near Impfondo in the rain forest region of the northern Congo during February 1988.

The distribution of ozone was measured from the surface up to 4 km altitude, using tethered balloons and instrumented aircraft. Simultaneous determinations of meteorological parameters and various trace components, including Aitken nuclei, CO, CO_2, organic acids, sulfur gases, and NO_x, were carried out. A pronounced ozone maximum with concentrations up to 70 ppbv was found at altitudes between 1 and 3 km. This maximum coincided with high levels of CO, CO_2, organic acids, and Aitken nuclei. A typical vertical profile of ozone and Aitken nuclei concentrations is shown in Fig. 1. While these high levels aloft persisted during the entire measurement campaign, a distinct diel variation of ozone was observed at lower altitudes. Regional survey flights showed that elevated ozone levels were present throughout the region, with no significant change evident over distances of over 300 km. Above an altitude of about 3–4 km, ozone decreased sharply to values typical of the remote mid-troposphere, typically about 30 ppb.

Measurements of the meteorological parameters revealed a complex pattern, indicating the presence of several layers of differing origins. Air-mass trajectories and synoptic charts suggest that the ozone-enriched layer formed in air masses which originate in northern Africa and subsequently advect over dry tropical regions where large amounts of aerosols, CO, NO, and hydrocarbons from biomass burning are introduced (Andreae 1990). Satellite imagery from the NOAA AVHRR instrument shows the presence of a large number of fires in the regions over which the air masses had traveled some 2–4 days before arriving in the northern Congo (Y.J. Kaufman pers. commun., 1989). These air masses then became trapped in the equatorial region between the near-surface monsoon flow from the southeast and the easterly flow above 3–4 km. Visual

[1]Max Planck Institute for Chemistry, Biogeochemistry Department, 6500 Mainz, FRG
[2]Nuclear Research Center Jülich, 5170 Jülich, FRG
[3]University Marien Ngouabi, DGRST, Brazzaville, P.R. Congo
[4]University Paul Sabatier, 31062 Toulouse, France

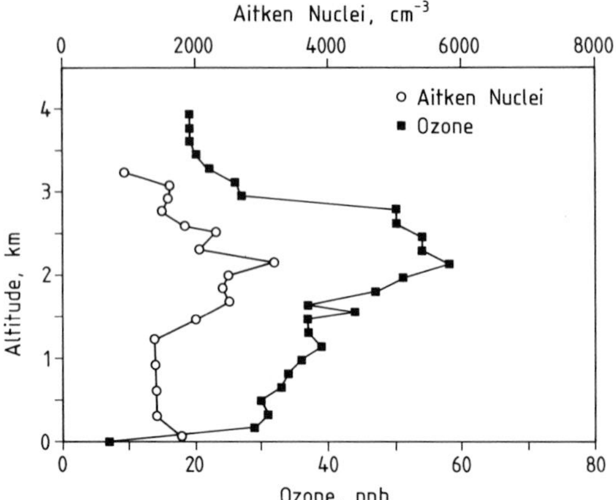

Fig. 1. Vertical profile of ozone and Aitken nuclei over the equatorial rain forest of the northern Congo (23 Feb. 1988, 07.40 h local time)

observations from the research aircraft showed the presence of numerous fires also within the area through which the monsoon airmasses had traveled, contributing further amounts of pyrogenic emissions. Photochemical reactions involving the oxidation of CO and hydrocarbons in the presence of oxides of nitrogen lead to the production of ozone in this layer. The ozone-enriched layer observed during DECAFE is sufficient to explain the tropospheric ozone anomaly observed by remote sensing in the tropics. This anomaly reaches from Africa across the Atlantic to South America (Fishman and Larson 1987). Similar ozone-enriched layers resulting from biomass burning in Brazil had been observed previously over that country (Delany et al. 1985; Andreae et al. 1988).

Diel variability in the vertical distribution of ozone is driven by removal of O_3 by surface uptake and reactions with NO and hydrocarbons, leading to surface O_3 mixing ratios near zero at night and a steep O_3 gradient through the subcloud layer. During the day, this gradient is reduced by convective mixing and by photochemical production of ozone. The ozone concentrations observed just above the forest during some of our flights, and previously in the southern part of the Congo Republic by Cros et al. (1987, 1988) are high enough to cause concern about chronical damage of the vegetation due to a combination of ozone damage and acid precipitation. Acid rain, predominantly due to a combination of nitric acid and organic acids, has been observed in this region by Lacaux et al. (1987, 1988).

A detailed evaluation of the results from DECAFE-88 is now underway. The results will be published in a special issue of the Journal of Geophysical Research which is expected to be available in 1990.

References

Andreae MO (1990) Biomass burning in the tropics: Impact on environmental quality and global climate. Popul Develop Rev (in press)

Andreae MO, Browell EV, Garstang M, Gregory GL, Harriss RC, Hill GF, Jacob DJ, Pereira MC, Sachse GW, Setzer AW, Silva Dias PL, Talbot RW, Torres AL, Wofsy SC (1988) Biomass-burning and associated haze layers over Amazonia. J Geophys Res 93:1509–1527

Cros B, Delmas R, Clairac B, Loemba-Ndembi J, Fontan J (1987) Survey of ozone concentrations in an equatorial region during the rainy season. J Geophys Res 92:9772–9778

Cros B, Delmas R, Nganga D, Clairac B, Fontan J (1988) Seasonal trends of ozone in equatorial Africa: Experimental evidence of photochemical formation. J Geophys Res 93:8355–8366

Delany AC, Haagensen P, Walters S, Wartburg AF, Crutzen PJ (1985) Photochemically produced ozone in the emission from large-scale tropical vegetation fires. J Geophys Res 90:2425–2429

Fishman J, Larsen JC (1987) The distribution of total ozone and stratospheric ozone in the tropics: Implications for the distribution of tropospheric ozone. J Geophys Res 92:6627–6634

Lacaux JP, Servant J, Baudet JGR (1987) Acid rain in the tropical forests of the Ivory Coast. Atmos Environ 21:2643–2647

Lacaux JP, Servant J, Huertas ML, Cros B, Delmas R, Loemba-Ndembi J, Andreae MO (1988) Precipitation chemistry from remote sites in the African equatorial forest. Eos Trans AGU 69:1069

20 Estimates of Annual and Regional Releases of CO_2 and Other Trace Gases to the Atmosphere from Fires in the Tropics, Based on the FAO Statistics for the Period 1975–1980

WEI MIN HAO, MEI-HUEY LIU, and P.J. CRUTZEN[1]

20.1 Introduction

Deforestation in the tropics plays an important role in determining the atmospheric CO_2 content and therefore the earth's climate (e.g., Myers 1980; Seiler and Crutzen 1980; Lanly 1982; Brown and Lugo 1982, 1984; Houghton et al. 1985; Melillo et al. 1985; Detwiler et al. 1985; Detwiler and Hall 1988). The burning of biomass is also a significant source of chemically reactive trace gases, such as NO_x, N_2O, CO, CH_4, and other hydrocarbons (Crutzen et al. 1979; Logan 1983; Greenberg et al. 1984; Crutzen et al. 1985; Andreae et al. 1988). Among these, CO_2, CH_4, and N_2O are also important "greenhouse" gases, which have an impact on the global climate (e.g., WMO 1985). The oxidation of CO, CH_4, and other hydrocarbons in the presence of NO_x leads to more ozone in the troposphere and thus influences its oxidation efficiency. In the stratosphere, increases in N_2O lead to additional NO_x, a loss of ozone above about 25 km and an increase below 25 km. Worldwide observations of CO_2, CH_4, and N_2O have shown that the concentrations of these gases are currently increasing annually by about 0.4, 1, and 0.25% (WMO 1985).

Seiler and Crutzen (1980) estimated the amount of CO_2 emitted from burning biomass based on the then available statistics in the changes of land use and rural population, land requirements per capita in permanent agriculture and shifting (or fallow) cultivation, biomass densities of various kinds of vegetation, and burning efficiencies of trees and grass. There were large uncertainties in the data on which the study was based. Considerable efforts were made in the early 1980's to improve the statistics which are necessary for understanding the global carbon cycle. Matthews (1983, 1985) compiled the data on global vegetation and land use in digital form at 1° latitude × 1° longitude resolution. Two recent surveys on the changes of land use in tropical countries were conducted by Myers (1980) and the Food and Agricultural Organization (FAO) of the United Nations (Lanly 1982). The annual deforestation of closed broadleaved forests in the two reports differs by only about 20% (Melillo et al. 1985; Molofsky et al. 1986).

The FAO survey for the period 1975–1980 is the most extensive and systematic study on tropical deforestation and agricultural land use changes, covering 76 countries in tropical America, Africa, and Asia (Lanly 1982). The classification of various kinds of vegetation used in the FAO report (Table 1) is

[1]Max-Planck-Institute for Chemistry, Atmospheric Chemistry Department, 6500 Mainz, FRG

Table 1. Classifications and definitions of various vegetation used by the FAO report (Lanly 1982) and the corresponding symbols used in the Matthews maps. (Matthews 1985)

Classification	Definition	Symbols (in Matthews maps)
Closed nonfallow forests	Forests without a continuous layer of grass and not cleared recently by shifting cultivation	1,2,9
–Undisturbed productive forests	Primary or old secondary forests which have not been logged in the last 60 to 80 years	1,2,9
–Previously logged forests	Forests which have been logged at least once in the last 60 to 80 years	1,2,9
–Unproductive forests	Forests which are not productive because of low quality of wood, difficult terrain conditions or legal protection	1,2,9
Closed fallow forests	Complex woody vegetation without a continuous layer of grass and cleared recently by shifting cultivation	1,2,9
Open nonfallow forests	Mixed vegetation of broadleaved forests and grassland trees with a continuous layer of grass, not cleared recently by shifting cultivation	N,O,P,Q,R
–Productive forests	Mixed vegetation which has been cleared recently	N,O,P,Q,R
–Unproductive forests	Mixed vegetation which is not productive because of low quality of wood, difficult terrain conditions or legal protection	N,O,P,Q,R
Open fallow forests	Mixed vegetation of broadleaved forests and grassland trees with a continuous layer of grass, recently cleared by shifting cultivation	N,O,P,Q,R
Grass layer of open forests (humid savannas)	Continuous layer of dense grass in open forests	N,O,P,Q,R

partially based on their economic uses and thus different from that used in other papers (Whittaker and Likens 1975; Seiler and Crutzen 1980). Forests are categorized as closed nonfallow and fallow forests, and open nonfallow and fallow forests in the FAO report (Lanly 1982). Nonfallow forests are primary forests which have not been cleared recently for shifting cultivation. Fallow forests are secondary forests which are used for shifting cultivation. Deforestation is defined as the transformation of forests either to land used for shifting cultivation or to land cleared permanently. Closed forests are defined as having

100% canopy cover without a continuous layer of herbaceous grass on the ground. Closed nonfallow and fallow forests are equivalent to virgin forests and secondary forests, respectively, used by Seiler and Crutzen (1980). Open forests in the FAO report, corresponding to the category of humid savannas in the paper of Seiler and Crutzen (1980), are broadleaved forests and woodlands mixed with a continuous and extensive layer of grass, which provides fuel for widespread fires during the dry season.

In this chapter we present new estimates of the amounts of CO_2 and other trace gases released from deforestation and savanna fires based on the data provided by the FAO report (Lanly 1982). The amounts of biomass burned from various types of vegetation in tropical America, Africa, Asia, and Australia are tabulated. The regional monthly emissions of CO_2 from biomass burning are subdivided into grid cells of 5° latitude by 5° longitude. The values can also be converted into emissions of other trace gases by applying measured emission factors, and the results are thus available for use in the global photochemical models of the atmosphere.

20.2 Computational Approach

To determine the amount of biomass burned from various types of vegetation in tropical America, Africa, Asia, and Australia, we follow the method of Seiler and Crutzen (1980). The following categories of ecosystems are considered for this study: closed nonfallow forests, closed fallow forests, open nonfallow forests, open fallow forests and the grass layer of open forests.

The amount of biomass burned per year (M) in each of these ecosystems is described by the equation

$$M = A \times B \times \alpha \times \beta, \tag{1}$$

where A denotes the area of land cleared annually (ha a^{-1}; 1 ha = 10^4 m^2), B the biomass density (t ha^{-1}; 1 t = 10^6 g), α the fraction of biomass in the ecosystem which is aboveground, and β the fraction of aboveground biomass that is burned. Representative values for α of 81 $\pm$ 5% for closed forests and 71 $\pm$ 5% for open forests have been derived from field measurements (Rodin and Bazilevich 1966; Bazilevich et al. 1971; Whittaker and Likens 1975; Rodin et al. 1975), as summarized by Seiler and Crutzen (1980). The uncertainty of α for these categories may be assumed to be less than 10%. The fraction of aboveground biomass that is burned (β) in tropical forests depends on agricultural practices. Large areas of forests have been cleared traditionally for shifting cultivation or slash-and-burn agriculture. Increasingly, forests are also cleared for permanent agriculture, coffee plantations, and cattle ranches. After cutting and felling, fires may consume only a relatively small fraction ($\approx$ 10–20%) of the biomass. However, unburned wood may be stacked up and burned again especially if permanent deforestation for agricultural development takes place. Fearnside (1985) estimated that about 30% of the aboveground biomass was burned at an experimental site in the Amazonian primary forests. The estimate is close to the value 25% used by Seiler and Crutzen (1980). The parameter β of secondary (fallow) forests may be larger than that of primary (nonfallow) forests due to the smaller sizes of the vegetation. Hence, we adopt a value of β equal to 0.3 for primary and 0.4 for secondary forests, realizing that the uncertainty of these values could be about 30%. The uncertainty may not be very critical for our analysis, as we will show that most of the biomass burning takes place in the savanna regions. The uncertainty may affect the timing of slow release of CO_2 from decomposition of unburned materials by a few years, which is not a significant factor in estimating the net release of CO_2 to the atmosphere.

In contrast to forest fires, most of the material burned in savanna fires consists of grass. The fraction of grass burned in tropical savanna fires was estimated to be 84% in Nigeria (Hopkins 1965), and 67 and 98% on the Ivory Coast (Delmas 1982; Lamotte 1985). In our calculation, the average β value of 0.83, with an uncertainty of $\pm$ 20%, is taken as the fraction of aboveground biomass which is burned in tropical savannas. The amount of litter burned is not separately calculated, as it constitutes only a few percent of the aboveground biomass in tropical savannas (Lacey et al. 1982; Huntley and Morris 1982).

The information on annually deforested areas (A) of closed and open nonfallow forests in tropical America, Africa, and Asia is summarized by Lanly (1982). This report does not include the areas cleared annually for shifting cultivation. Values of A for closed fallow forests have been estimated to be 13.5 million ha per year by Houghton et al. (1985) and 18 million ha per year by Detwiler et al. (1985) with an additional 18.6 million ha per year for open fallow forests (Detwiler et al. 1985). Seiler and Crutzen (1980) estimated the ranges of A of about 12–37 million ha and 8–24 million ha per year for closed and open fallow forests respectively, indicating that the estimates of Houghton et al. (1985) and Detwiler et al. (1985), which are adopted in this study, may be low. The areas of grass layer cleared per year in open forests (humid savannas) are calculated by dividing the savanna area by the average time between fires. The uncertainty of A may be estimated to be about 20% for closed broadleaved forests based on the discrepancy between the FAO report (Lanly 1982) and the report of Myers (1980). The uncertainty of land cleared annually in fallow forests is calculated to be about 30%, based on the different estimates of Seiler and Crutzen (1980), Detwiler et al. (1985), and Houghton et al. (1985).

There are considerable differences in the estimates of biomass densities in various ecosystems. The estimates of Brown and Lugo (1984), 89–238 t ha^{-1} for closed broadleaved forests and 21–79 t ha^{-1} for open nonfallow forests, are substantially lower than those which have been previously reported (Whittaker and Likens 1975; Detwiler et al. 1985). Both the low and high values of biomass densities are used in our calculation, since it is not clear at present which values are better.

For future use in chemical models of the atmosphere, the data on biomass burned due to deforestation and savanna fires in tropical America, Africa, and Asia are resolved into 5° latitude $\times$ 5° longitude grid cells. The compilations are based on the FAO report (Lanly 1982) and the high resolution digital data base of global vegetation (Matthews 1985). For each grid cell, the areas affected by deforestation and agricultural conversions are estimated from the FAO report, which provides the information collected from individual countries, and from the distributions of various vegetation categories in the Matthews report, which are digitized with a $1^\circ \times 1^\circ$ grid resolution. It is therefore necessary to develop a method linking the two data sets. The relevant types of vegetation used by Matthews can be assigned to the FAO vegetation types adopted in the present study (Table 1), since both reports basically follow the UNESCO (1973) classification system. However, nonfallow and fallow forests were not distinguished in the Matthews report. Consequently, closed broadleaved nonfallow and fallow forests are identified in the Matthews maps by the symbols 1, 2, or 9, closed coniferous forests by the symbol 7, open nonfallow and fallow forests by N, O, P, Q, or R. The areas of annual deforestation of closed broadleaved and other nonfallow forests for a particular $5^\circ \times 5^\circ$ grid cell (Ag) are calculated by using the statistics compiled for individual countries by the FAO (Lanly 1982) and summing overall countries in the grid cell. If the size of the country is larger than the grid cell, we use the formula:

$$\mathrm{Ag} = \mathrm{Ac} \times \frac{\mathrm{Ng}}{\mathrm{Nc}}, \tag{2}$$

where

Ag = forest area cleared annually in the grid cell,
Ac = forest area cleared annually in the country,
Ng = number of symbols 1, 2 or 9 in the grid cell,
Nc = total number of symbols 1, 2, or 9 in the country.

The areas of closed or open fallow forests cleared per year in each country (Ac) were not given by the FAO report, but are estimated by the equation:

$$\mathrm{Ac} = \mathrm{At} \times \frac{\mathrm{Wc}}{\mathrm{Wt}}, \tag{3}$$

where

At = area of closed or open fallow forests cleared annually in the tropics, estimated by Houghton et al. (1985) and Detwiler et al. (1985),
Wc = area of closed or open fallow forests in the country,
Wt = total area of closed or open fallow forests in the tropics.

The areas of annual deforestation in Brazil are calculated by using higher resolution information than those used in other parts of the tropics. The Amazon forest is the largest tropical forest in the world, covering about twenty two $5^\circ \times 5^\circ$ grid cells. The degree of deforestation varies greatly among and within the grid cells. However, because the areas deforested per year were available at a resolution of $1^\circ \times 1^\circ$ for the entire Amazon forest (Fearnside 1986), these data could be averaged to the $5^\circ \times 5^\circ$ grid cells.

20.3 Results

From the information given above, the amount of biomass that is burned annually in each category of vegetation was calculated by using Eq. (1). The results are compiled in Tables 2–5 for each of the tropical continents. Biomass burning is assumed to occur in the last 5 months of the dry season, which are determined from the monthly rainfall statistics (Jaeger 1976). We assume the distributions of burning to be 12.5% for the first and the last month, and 25% for each of the middle 3 months. The overall CO_2 emitted monthly from biomass burning in each $5° \times 5°$ grid cell is calculated by the summation of CO_2 emissions from the burning of various types of vegetation. These data are presented in Fig. 1a,b, and c.

We will briefly discuss the results obtained for the various ecosystems.

20.3.1 Closed Nonfallow Forests (Table 2)

The FAO statistics for 1980 showed that about 0.6% of the remaining area of closed nonfallow forests (1201 million ha) was cleared annually in each of the tropical continents (Lanly 1982). The amounts of biomass cleared annually in each category of Table 2 were calculated from the rates of deforestation in the FAO report (Lanly 1982) multiplied by the biomass density estimates of Brown and Lugo (1984) or Detwiler et al. (1985). The results indicate that about $2.3–4.9 \times 10^8$ t of biomass are burned per year due to deforestation of closed nonfallow forests. More than 80% of the biomass burned in closed nonfallow forests is from undisturbed productive forests and previously logged forests. Tropical America contains the largest area of closed nonfallow forests (679 million ha), compared to 217 million ha in tropical Africa and 306 million ha in tropical Asia (Lanly 1982). Hence, almost half of the biomass burned in closed nonfallow forests is from tropical America, and the remaining half from tropical Africa and Asia.

The cleared land of closed nonfallow forests was converted to shifting cultivation (3.4 million ha per year) and to permanent agriculture and other uses (4.1 million ha per year), according to Lanly (1982). Different estimates, based on the analysis of Myers (1980) data, were derived by Houghton et al. (1985), who suggested that 1.9 million ha of closed broadleaved forests were cleared annually for shifting cultivation and 5.1 million ha for permanent uses. Of the permanently cleared land areas annually, 3.9 million ha were estimated to be converted to agricultural use, 0.7 million ha to pasture and 0.5 million ha were deforested for fuel wood (Houghton et al. 1985).

20.3.2 Closed Fallow Forests (Table 3)

Lanly (1982) estimated the area of closed fallow forests (or secondary forests) to be 239 million ha in 1980, but did not give any information on the conversion and

Table 2. Biomass cleared and burned in closed nonfallow forests

Region	Undisturbed Productive			Previously Logged			Unproductive			Total	Total[d]
	Deforestation[a] rate (10^6 ha a^{-1})	Biomass[b] density (t ha^{-1})	Biomass[c] cleared (10^6 t a^{-1})	Deforestation[a] rate (10^6 ha a^{-1})	Biomass[b] density (t ha^{-1})	Biomass[c] cleared (10^6 t a^{-1})	Deforestation[a] rate (10^6 ha/a)	Biomass[b] density (t ha^{-1})	Biomass[c] cleared (10^6 t a^{-1})	Biomass cleared (10^6 t a^{-1})	Biomass burned (10^6 t a^{-1})
Broadleaved forests											
Tropical America	1.135	155–328	176–372	1.684	118–238	199–401	0.988	89–328	88–324	463–1097	112–267
Tropical Africa	0.220	238–328	52–72	1.036	179–238	185–247	0.063	130–328	8–21	246–339	60–82
Tropical Asia	0.483	196–328	95–158	1.174	93–238	109–279	0.110	132–328	15–36	218–474	53–115
Subtotal	1.838		323–603	3.894		493–927	1.161		110–380	927–1910	225–464
Coniferous and bamboo forests											
Tropical America	0.102	136–328	14–33	0.128	50–238	7–30	0.082	49–328	4–27	24–91	6–22
Tropical Africa	0.002	119–328	0–1	0.004	52–238	0–1	0.008	89–328	1–3	1–4	0–1
Tropical Asia	0.020	145–328	3–7	0.020	113–238	2–5	0.008	104–328	1–3	6–14	2–3
Subtotal	0.124		17–41	0.152		9–36	0.098		6–32	31–109	8–26
Total tropics	1.962		340–644	4.046		502–963	1.259		116–413	958–2020	233–491

[a] Lanly (1982).
[b] The low values from Brown and Lugo (1984) and the high values from Detwiler et al. (1985).
[c] Biomass cleared = deforestation rate × biomass density.
[d] Total biomass burned = Σ biomass cleared × α (81%) × β (30%).

Table 3. Biomass cleared and burned in closed fallow forests

Region	Area[a] (10^6 ha)	Land cleared (10^6 ha a^{-1})			Biomass[b] density (t ha^{-1})	Total[c] biomass cleared (10^6 t a^{-1})	Total[c,d] biomass burned (10^6 t a^{-1})
		Houghton et al. (1985)		Detwiler et al. (1985)			
		Permanent	Shifting cultivation	Shifting cultivation			
Tropical America	109	4.6	1.5	8.2	126	771–1029	250–333
Tropical Africa	62	2.6	0.9	4.6	126	438–584	142–189
Tropical Asia	69	2.9	1.0	5.2	126	492–656	159–213
Total tropics	239	10.1	3.4	18.0		1701–2268	551–735

[a] Lanly (1982).
[b] Detwiler et al. (1985).
[c] The low values are derived from Houghton et al. (1985), and the high values from Detwiler et al. (1985).
[d] Total biomass burned = Σ biomass cleared $\times \alpha$ (81%) $\times \beta$ (40%).

clearing rates. There are considerable discrepancies between the analyses by Houghton et al. (1985) and Detwiler et al. (1985) regarding the fates of closed fallow forests, whether cleared for permanent agriculture or for shifting cultivation. In shifting cultivation, fallow forests are generally cleared and cultivated for about 2 years, after which the land is abandoned for about 10 to 20 years to allow regrowth of secondary forests (fallow period). The secondary forests are cleared and cultivated again at the end of the fallow period. Detwiler et al. (1985) suggested that 18 million ha of closed fallow forests were cleared per year for shifting cultivation. Houghton et al. (1985) calculated that 13.5 million ha were cleared annually based on the data tabulated by Myers (1980), of which 10.1 million ha were cleared permanently and only 3.4 million ha remained in shifting cultivation. Of the permanently cleared land, 6.8 million ha were used for agriculture, 2.0 million ha for fuel wood, and 1.3 million ha for pasture (Hougton et al. 1985). Since we cannot decide between the estimates of Detwiler et al. (1985) and Houghton et al. (1985) without further detailed information, both estimates are used in the present calculation (Table 3). The areas cleared annually on each continent are weighted by the distribution of areas of closed fallow forests in the three regions.

Biomass densities of fallow forests were not determined by Brown and Lugo (1984). The average biomass density of closed fallow forests was estimated to be 126 t ha^{-1} by Detwiler et al. (1985), which is low compared to the estimates 150–200 t ha^{-1} used by Seiler and Crutzen (1980). Based on the above information, it is estimated that $5.5–7.4 \times 10^8$ t of biomass are burned each year in closed fallow forests (Table 3), which is larger than the high estimate for closed nonfallow forests. Although the biomass densities in fallow forests are smaller than those in nonfallow forests, the larger area cleared in fallow forests more than compensates. Hence, land clearance of closed fallow forests for shifting cultivation and permanent uses could be as important as deforestation of closed nonfallow forests as a source of atmospheric trace gases.

20.3.3 Open Nonfallow Forests (Table 4)

According to Lanly (1982), 3.8 million ha or about 0.5% of the area of open nonfallow forests (or primary open forests) was cleared each year and converted either to shifting cultivation (1.7 million ha) or to permanent agriculture (2.1 million ha). As a result, $4.3–6.5 \times 10^7$ t of biomass are burned each year in open nonfallow forests (Table 4), with about equal amounts from tropical America and Africa. The low values are derived from the low biomass densities, which are in the range of 46–79 t ha^{-1} for productive forests and 21–33 t ha^{-1} for unproductive forests (Brown and Lugo 1984). The high values are derived from the high biomass density (80 t ha^{-1}) estimated by Detwiler et al. (1985). Since both the biomass densities and the rates of deforestation are low, the amount of biomass burned in open nonfallow forests is only 10–20% of that burned in closed nonfallow forests.

Table 4. Biomass cleared and burned in open nonfallow forests

Region	Productive			Unproductive			Total biomass cleared (10^6 t a^{-1})	Total[c] biomass burned (10^6 t a^{-1})
	Deforestation[a] rate (10^6 ha a^{-1})	Biomass[b] density (t ha^{-1})	Biomass cleared (10^6 t a^{-1})	Deforestation[a] rate (10^6 ha a^{-1})	Biomass[b] density (t ha^{-1})	Biomass cleared (10^6 t a^{-1})		
Tropical America	1.22	77–80	94–98	0.05	33–80	2–4	96–102	20–22
Tropical Africa	1.93	46–80	90–155	0.41	21–80	9–33	98–188	21–40
Tropical Asia	0.09	79–80	7–7	0.10	26–80	3–8	10–15	2–3
Total tropics	3.24		191–260	0.56		13–45	204–305	43–65

[a] Lanly (1982).
[b] The low values from Brown and Lugo (1984), the high values from Detwiler et al. (1985).
[c] Total biomass burned = Σ biomass cleared $\times$ α (71%) $\times$ β (30%).

20.3.4 Open Fallow Forests (Table 5)

The fallow cycle of open fallow forests is shorter than that of closed fallow forests, with a fallow period of about 9 years. Taking the area of open fallow forests of 170 million ha (Lanly 1982), Detwiler et al. (1985) calculated that 18.6 million ha (or 11% of the area) of open fallow forests were cleared for shifting cultivation in 1980. The area cleared in open fallow forests is as large as the area cleared in closed fallow forests. However, since the biomass densities in open fallow forests are relatively low (38 t ha^{-1}), the amount of biomass burned each year in open fallow forests (2×10^8 t per year) is only about one-third of the biomass burned in closed fallow forests (see Table 3).

Table 5. Biomass cleared and burned in open fallow forests

Region	Area[a] (10^6 ha)	Land cleared for shifting cultivation (10^6 ha a^{-1})	Biomass[b] density (t ha^{-1})	Total biomass cleared (10^6 t a^{-1})	Total[c] biomass burned (10^6 t a^{-1})
Tropical America	62	6.8	38	256	73
Tropical Africa	104	11.4	38	434	123
Tropical Asia	4	0.4	38	17	5
Total tropics	170	18.6[b]		707	201

[a] Lanly (1982).
[b] Detwiler et al. (1985).
[c] Total biomass burned = Σ biomass cleared $\times$ α (71%) $\times$ β (40%).

20.3.5 Grass Layer of Open Forests (Humid Savannas) (Table 7)

The area of tropical savannas (1530 million ha) is larger than the area of closed forests (1440 million ha) (Lanly 1982). About 60% of the total savanna area is humid savanna (annual rainfall $\geq$700 mm) and 40% is arid savanna (annual rainfall $<$700 mm) (Lanly 1982). The humid savannas are defined as open forests and the arid savannas as shrub formations in the FAO report. The grass layer of humid savannas (or open forests) is burned regularly by man during the dry season, although fires may be caused occasionally by lightning. Fire is mostly used for agricultural purposes, e.g., to kill insects and pests and to reduce the quantities of dead grass, thus promoting regrowth of young grass for animal grazing during the rainy season. The arid savannas are rarely burned because of a lack of a continuous layer of grass to sustain the fire (Harris 1980; Bucher 1982; Huntley 1982). Savanna trees generally are resistant to fires and are not significantly burned (Coutinho 1982).

Table 6. The highest observed biomass densities of grass in humid savannas

Country	Biomass density ($t\ ha^{-1}$)	Annual rainfall (mm)	Reference
Tropical America			
Venezuela	5.7, 6.53	≈ 1500	San José and Medina (1976)
Venezuela	3–4	1334	San José and Medina (1975)
Venezuela	7, 8, 9.16		González-Jiménez (1979)
Brazil	6–7		Coutinho (1982)
Tropical Africa			
Nigeria	5.2		Hopkins (1965)
Nigeria	3.8	1000	Haggar (1970)
Ivory Coast	6.2, 6.0, 8.7, 7.5, 7.0, 8.9, 6.7	1300	Menaut and César (1982)
S. Africa	5.7	695	Huntley and Morris (1982)
Tropical Asia			
India	4.94		Singh and Misra (1978)

The amount of biomass in the grass layer of savannas depends on annual rainfall. The highest seasonal biomass densities of aboveground grass in humid savannas near the end of growing season are listed in Table 6. The average biomass densities are 6.6 $t\ ha^{-1}$ in tropical America (San José and Medina 1975, 1976; González Jiménez 1979; Coutinho 1982), 6.6 $t\ ha^{-1}$ in tropical Africa (Hopkins 1965; Haggar 1970; Menaut and César 1982; Huntley and Morris 1982), and 4.9 $t\ ha^{-1}$ in tropical Asia (Singh and Misra 1978), with an uncertainty of 30% based on the available field measurements. The time between fires varies from region to region (Table 7). The entire humid savanna is burned once every 1 or 2 years in the Brazilian cerrado (Eiten 1972; Sarmiento and Monasterio 1975), and about 75% of the humid savannas is burned every year in Africa (Menaut and Cesar 1982; Menaut 1983; Menaut pers. commun.). Although there may be considerable uncertainty about the frequency of fires, it may not influence the quantities of biomass burned proportionally, as a plot which has not been burned in one year may contain more biomass during the following dry season. We estimate the uncertainty in the areas burned to be about 30%.

The amount of biomass burned in Australian savannas (4.3×10^8 t per year) is calculated based on the detailed information of time periods between fires and the corresponding biomass densities in various regions provided by Lacey et al. (1982). Seventy-five percent of the biomass burned in Australia is over the northern coastal area, which is associated with the short period between burning (1.5 years) and the relatively high biomass densities.

The amounts of grass burned in the humid savannas of tropical America, Africa, Asia, and Australia are summarized in Table 7. Overall, about 3.7×10^9 t

Table 7. Biomass burned in the grass layer of open forests

Region	Area[a] (10⁶ ha)	Aboveground biomass (t ha⁻¹)	Average period between fires (years)	Biomass exposed to fire (10⁶ t a⁻¹)	Total[c] biomass burned (10⁶ t a⁻¹)
Tropical America	279	6.6 ± 1.8 (n = 7)	2	924	767
Tropical Africa	591	6.6 ± 1.6 (n = 10)	1.33[d]	2925	2428
Tropical Asia	35	4.9 (n = 1)	2	86	71
Australia	402	2.1–6	1.5–30	512[b]	425
Total				4447	3691

[a] The same as the areas of open forests in Lanly (1982).
[b] Calculated from the information of biomass densities and corresponding average periods between fires in various regions of Australia (Lacey et al. 1982).
[c] Total biomass burned = $\frac{\text{biomass exposed to fire}}{\text{average period between fires}} \times \beta$ (83%).
[d] Assumed 75% of the area burned each year (Menaut pers. commun.).

of grass are burned annually in the tropical savannas. About two-thirds of the biomass are burned in tropical Africa, which contains the largest area of savannas, about 75% of which is burned every year (Menaut and Cesar 1982; Menaut 1983; Menaut pers. commun.). The amount of savanna grass burned in tropical America is only about one third of the amount burned in tropical Africa. It is important to note that the total amount of biomass burned in humid savannas is about ten times larger than that burned in closed nonfallow forests, even though the biomass densities in the savannas are much lower than those in tropical forests. The main reasons are (1) the savanna area cleared each year (748 million ha) is about 100 times larger than the deforested area (7.2 million ha), and (2) the fraction of biomass burned in savanna grass (83%) is much higher than that in forest trees (30%).

20.3.6 Distribution of CO_2 Emissions in 5° × 5° Grid Cells

The distributions of monthly gross CO_2 emissions from all tropical biomass burning in 5° × 5° grid cells are summarized in Fig. 1. The period of the most intensive burning usually occurs from December to March in the northern hemisphere and from June to September in the southern hemisphere. The results in Fig. 1a indicate that high emissions of CO_2 ($\approx 1.0 \times 10^{12}$ g CO_2-C/month) caused by forest burning occur in Brazil in the region of 5–20°S and 40–65°W, which covers the states of Maranhao, Goias, Mato Grosso and Rondonia. Most of the savanna fires in tropical America are concentrated in the cerrado region of 5–15°S and 45–65°W.

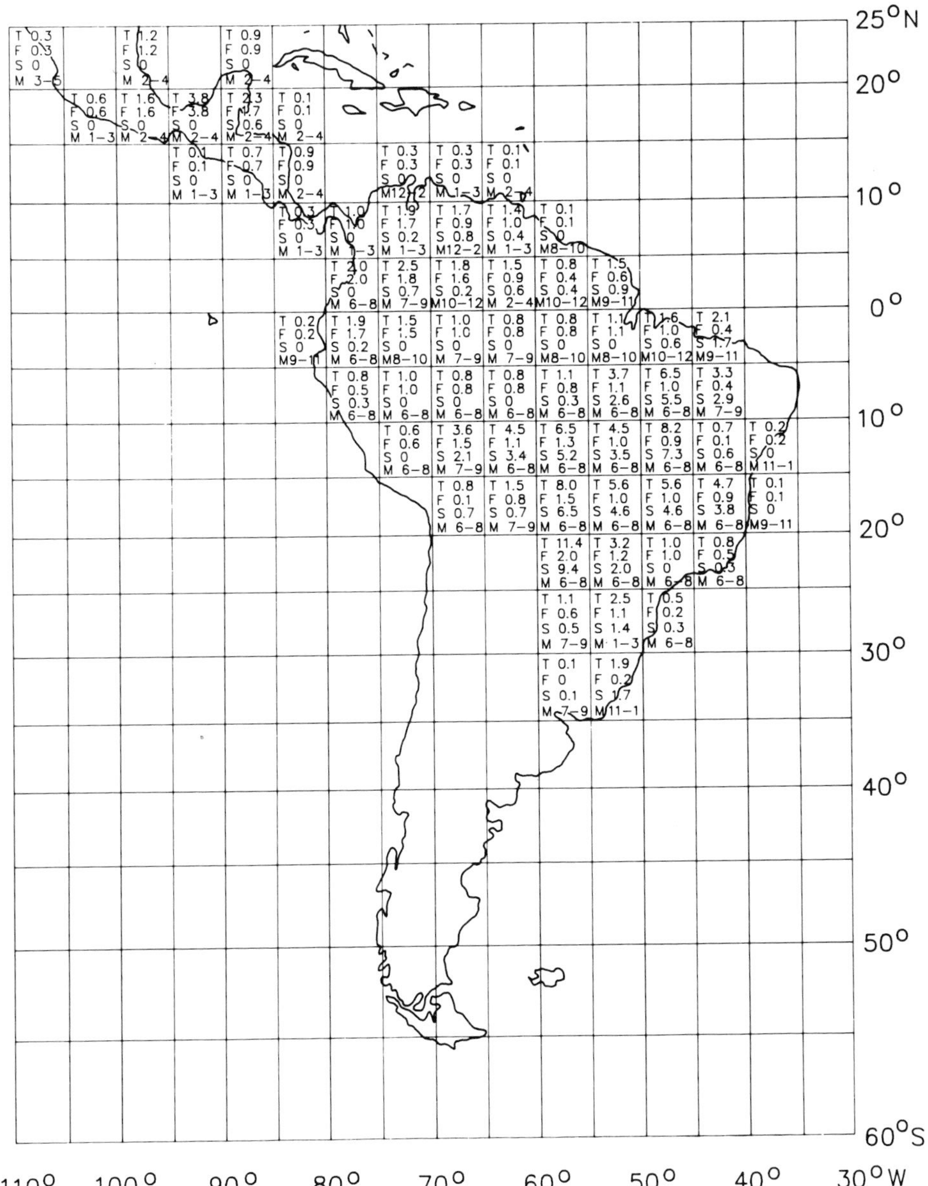

Fig. 1a. Distributions of CO_2 emissions from biomass burning in tropical America during the dry season. *T* total; *F* forest; *S* savanna; *M* months of intensive burning. (unit 10^{12} g CO_2-C month^{-1})

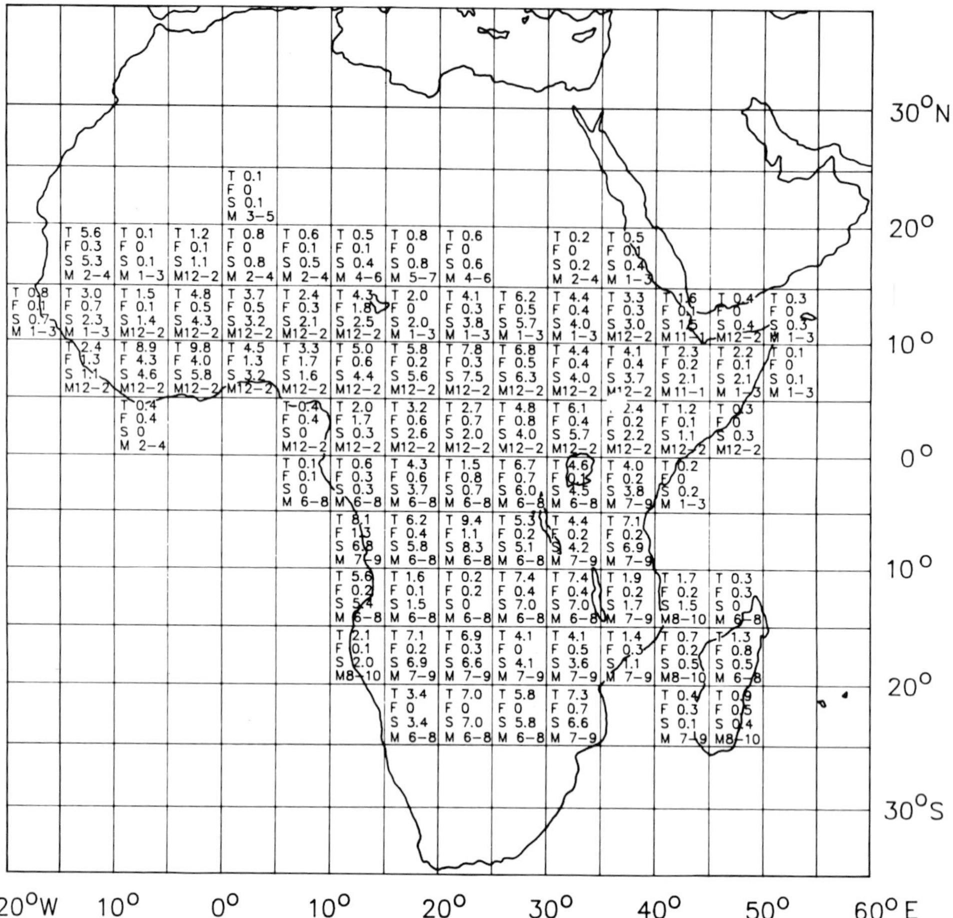

Fig. 1b. Distributions of CO_2 emissions from biomass burning in tropical Africa during the dry season. *T* total; *F* forest; *S* savanna; *M* months of intensive burning. (unit 10^{12} g CO_2-C month^{-1})

In tropical Africa, about 85% of the CO_2 (1.2×10^{15} g C a^{-1}) emitted by fires is from the savannas and only 15% from the forests. A large fraction of the savanna fires occurs between 15°N and 15°S (Fig. 1b). Ivory Coast and Nigeria are the two countries having the largest CO_2 emissions due to forest fires.

The amount of CO_2 emitted from burning biomass in tropical Asia is small (1.4×10^{14} g C a^{-1}) compared to the amount of CO_2 emitted from tropical America and Africa. Most of the burning occurs in Indonesia, Thailand, Malaysia, and India, where deforestation is being actively pursued (Fig. 1c).

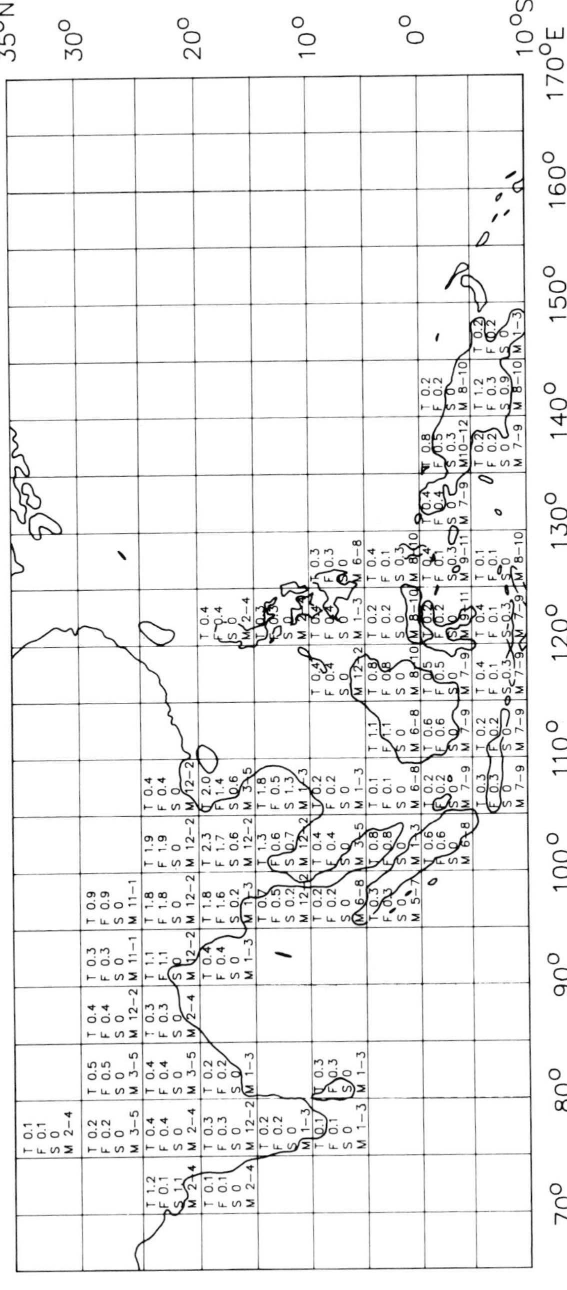

Fig. 1c. Distributions of CO_2 emissions from biomass burning in tropical Asia during the dry season. *T* total; *F* forest; *S* savanna; *M* months of intensive burning. (unit 10^{12} g CO_2-C $month^{-1}$)

20.4 Discussion

The amounts of biomass burned due to deforestation and savanna fires in the tropics are summarized in Table 8. About 4.7–5.2 × 10^9 t of biomass are burned and 2.1–2.3 × 10^{15} g CO_2-C emitted to the atmosphere each year, depending on the values of biomass densities; about 75% is due to fires in the savannas and only about 25% due to deforestation. Seiler and Crutzen (1980) estimated that 1.4–3.4 × 10^9 t of biomass from the forests and 1.4–2.3 × 10^9 t of biomass from the savannas were burned annually in the tropics. The total amount of biomass burned calculated from this work is close to the high estimate by Seiler and Crutzen. However, there is a significant difference in the source distribution of CO_2 between this study and the study by Seiler and Crutzen (1980), in which the amount of biomass burned in the forests is comparable to that in the savannas. This work, however, indicates that savanna fires are far more important than forest fires as a source of atmospheric CO_2. The biomass densities of nonfallow forests used in the Seiler and Crutzen calculation are about one-third higher than the values used in this study. They also may have underestimated the biomass densities of savanna grass (3.6 t ha^{-1}) by about 80%.

It is difficult at present to determine the uncertainties of parameters used for calculating CO_2 emissions from biomass burning. Based on the limited information of estimated uncertainties of A, B, α, and β, the overall uncertainty is calculated from the equation: $E = \sum_i f_i \sqrt{\sum_j E_{ij}^2}$, where E_j is the uncertainty of parameters A, B, α, and β for a given ecosystem i, and the weighting factor f_i is equal to the amount of biomass burned from ecosystem i divided by that of total

Table 8. Estimates of total biomass burned in the tropics (10^6 t a^{-1})

Region	Forests		Savanna grass		Total biomass burned	Total[c] CO_2-C produced
	Biomass[a,b] exposed to fire	Biomass[b] burned	Biomass exposed to fire	Biomass burned		
Tropical America	1659 (1269–2049)	589 (461–717)	924	767	1356	555
Tropical Africa	1063 (933–1192)	390 (346–435)	2925	2428	2818	1153
Tropical Asia	774 (599–949)	280 (221–339)	86	71	351	144
Australia			512	425	425	174
Total	3496 (2801–4191)	1259 (1028–1491)	4447	3691	4950	2026

[a] Biomass exposed to fire = biomass cleared × fraction of biomass aboveground (α).
[b] The low (high) values are derived from the low (high) biomass densities.
[c] Carbon content = 0.45 × biomass, the average molar ratio of CO to CO_2 emissions is taken to be 0.1.

biomass burned. The overall uncertainty is then estimated to be about ± 50%. Improved data are needed particularly for the fire frequencies, the amounts of biomass which are aboveground in tropical forests, and the burning efficiencies of tropical forests and savannas. Remote sensing from space seems to be the most promising way to obtain the improved information on the critical factors determining the global extents and spatial and temporal distributions of biomass burning and resulting trace gas emissions.

The burning of agricultural wastes, such as straw, leaves, and stubble, is another important source of atmospheric CO_2 and other trace gases. Seiler and Crutzen (1980) estimated that about 1.3×10^9 t of agricultural wastes were produced in developing countries. Assuming that 80% of the harvested agricultural wastes are burned with an efficiency of 90% and that 70% of the agricultural wastes in developing countries is in the tropics, then 6.6×10^8 t of agricultural wastes are burned each year in the tropics, adding about 15% of the total amount of biomass burned from deforestation and savanna fires.

The use of fuel wood as a source of energy in tropical countries could lead to significant production of CO_2 and other trace gases. According to the FAO statistics of forest products in 1980 (FAO 1987), 8.8×10^8 m^3 of fuel wood were produced from nonconiferous forests and 3.2×10^7 m^3 from coniferous forests. About half of the fuel wood is burned in tropical Asia and 30% in tropical Africa and 20% in tropical America. Taking the average wood density of 625 kg m^{-3} for coniferous forests and 750 kg m^{-3} for nonconiferous forests, and assuming that 90% of the fuel wood is burned, we estimate that 2.8×10^{14} g CO_2-C are emitted annually from burning fuel wood in tropical countries, contributing about additional 14% to the total CO_2 produced from forest and savanna fires. Overall, 2.6×10^{15} g CO_2-C are emitted annually from deforestation, savanna fires, and the burning of agricultural wastes and fuel wood.

It is interesting to compare the gross amount of CO_2 emitted from biomass burning with the net amount of CO_2 released from tropical forests. The net release of CO_2 from tropical forests is mainly caused by the changes of land use from forests to shifting cultivation, pasture, agricultural lands, and roads. The net fluxes of CO_2 from tropical forests into the atmosphere were estimated to be $0.4–1.6 \times 10^{15}$ g C per year (Detwiler and Hall 1988) or $0.9–2.5 \times 10^{15}$ g C per year (Houghton et al. 1985) based on the land surveys of Myers (1980) and Lanly (1982) and the data on biomass densities (Whittaker and Likens 1975; Brown and Lugo 1984). By using a similar method, the data of Lanly (1982) and Myers (1980) and the estimated biomass densities of Brown and Lugo (1984) and Detwiler et al. (1985), we calculate the net amount of CO_2 released to the atmosphere from tropical forests to be about $0.7–2.0 \times 10^{15}$ g C per year (Table 9). Intensive cultivation of cleared forest soils could also enhance the oxidation of organic matter in the soils, resulting in an additional net release of CO_2 of about 25% (Detwiler and Hall 1988), yielding a range of $0.9–2.5 \times 10^{15}$ g C a^{-1}. Almost half of the net CO_2 released to the atmosphere is caused by the conversions of forests to permanent agriculture and cattle holdings. The other major causes are the conversion of forests to shifting cultivation and pasture, the

Table 9. Net CO_2 released to the atmosphere due to changes of land use

Land use change	Area of annual deforestation (10^6 ha a^{-1})			CO_2 released[b] (10^{12} g C a^{-1})		
	Closed forests		Open forests	Closed forests		Open forests
	Lanly (1982)	Myers (1980)	Lanly (1982)	Lanly (1982)	Myers (1980)	Lanly (1982)
Primary forest to permanent clearance	4.1			304–652		
– to pasture		0.35[a]	1.3		26–56	29–46
– to fuel wood		0.25[a]			19–40	
– to agriculture			0.8			18–28
Primary to logged forest	3.7	4.5		94–167	114–203	
Logged forest to permanent clearance						
– to pasture		0.35[a]			17–40	
– to fuel wood		0.25[a]			12–29	
– to agriculture and others		3.9			190–445	
Secondary (fallow) forest to permanent clearance						
– to pasture		1.3			67–91	
– to fuelwood		2.0			103–140	
– to agriculture and others		6.8			352–476	
Primary forest to shifting cultivation	3.4	1.9	1.7	252–541	141–302	38–60
Total				650–1360	1041–1822	85–134

[a] Assumed to be converted evenly from primary and logged forests.
[b] Calculated by multiplying the area of annual deforestation by the biomass densities. Low values derived from low biomass density estimates of Brown and Lugo (1984). High values derived from high biomass density estimates of Detwiler et al. (1985). Carbon content = biomass × 0.45.

demand for fuel wood, and the conversion of primary to logged forests. Of the total net release of CO_2 to the atmosphere, only 0.1–0.3 × 10^{15} g CO_2-C per year are promptly emitted to the atmosphere from the burning of nonfallow forests. An additional 0.6–1.7 × 10^{15} g CO_2-C per year are released within a decade from decomposition of unburned material.

During the process of burning biomass, organic carbon and nitrogen in the vegetation are volatilized and oxidized, producing trace gases, e.g., CO, CH_4, nonmethane hydrocarbons, NO_x, N_2O, in addition to CO_2 (Crutzen et al. 1979; Greenberg et al. 1984; Crutzen et al. 1985; Andreae et al. 1988). Average mole ratios of emitted CO to CO_2 were observed to be 0.12 (Greenberg et al. 1984; Crutzen et al. 1985) and 0.085 (Andreae et al. 1988) in the plumes of Brazilian fires during two extensive field measurement programs. Taking the average emission ratio of CO to CO_2 to be 0.10 ± 0.02, 1.2–4.0 × 10^{14} g C of CO are estimated to be emitted to the atmosphere each year during the burning season in the tropics. By using average emission ratios of 0.8–1.6% for CH_4, 1–1.6% for nonmethane hydrocarbons (NMHC) which are mostly alkenes and alkanes, 0.2% for NO_x and 1.5–3 × 10^{-4} for N_2O (Crutzen et al. 1985; Andreae et al. 1988), we estimate that 1.2–5.0 × 10^{13} g CH_4-C, 1.5–5.2 × 10^{13} g NMHC, 3.0–9.1 × 10^{12} g NO_x-N and 0.5–2.2 × 10^{12} g N_2O-N were emitted to the atmosphere in 1980 from burning biomass in the tropics. The annual production of global trace gases was estimated to be 1.1 × 10^{15} g CO-C, 3.8 × 10^{14} g CH_4-C (WMO 1985), 1 × 10^{14} g NMHC (Ehhalt and Rudolph 1984), 2.4–6.4 × 10^{13} g NO_x-N (Logan 1983), and 1.5 × 10^{13} g N_2O-N (WMO 1985). Therefore, the contribution of fires in tropical regions to the global sources of atmospheric trace gases is about 11–36% for CO, 3–13% for CH_4, 15–52% for NMHC, 7–20% for NO_x, and 4–15% for N_2O. Biomass burning is therefore a significant source for these trace gases. It has been shown that burning causes a substantial effect on the photochemistry of the tropics during the dry season, e.g., through the formation of ozone (Crutzen et al. 1985; Andreae et al. 1988).

The amounts of various atmospheric trace gases emitted from deforestation and savanna fires in each 5° × 5° grid cell can be readily estimated by multiplying the amount of CO_2 emitted to the atmosphere in the grid cell (Fig. 1) by the emission ratios of various trace gases to CO_2. Such information is necessary for modeling the effects of biomass burning on global atmospheric chemistry.

20.5 Conclusion

Based on the FAO survey for the period 1975–1980 (Lanly 1982), we presented estimates of emissions of CO_2 and other trace gases from deforestation and savanna fires in 5° latitude × 5° longitude grid cells in the tropical region. About 2–2.5 × 10^{15} g CO_2-C per year are emitted to the atmosphere, but the uncertainty, although difficult to determine, may be at least a factor of 2. The amounts of CO_2 and other trace gases emitted by savanna fires are about three times larger than those emitted from deforestation. About two-thirds of the savanna fires occur in tropical Africa. Improved statistics on land use changes and biomass densities in tropical forest regions, the fraction of aboveground forest which is burned and the frequencies of savanna fires are especially needed in order to calculate more reliably the amounts of CO_2 and other trace gases emitted from burning biomass. The net release of CO_2 to the atmosphere in tropical forests is estimated to be $0.9–2.5 \times 10^{15}$ g C per year, caused mainly by the conversion of primary forests to permanent agriculture and shifting cultivation and the demand for fuel wood. The source of atmospheric trace gases from burning biomass may, however, have increased in the present decade because of the increasing demand for land and fuel wood by the rapidly growing populations in the tropical world.

References

Andreae MO, Browell EV, Garstang M, Gregory GL, Harriss RC, Hill GF, Jacob DJ, Pereira MC, Sachse GW, Setzer AW, Silva Dias PL, Talbot RW, Torres AL, Wofsy SC (1988) Biomass-burning emissions and associated haze layers over Amazonia. J Geophys Res 93:1509–1527

Bazilevich NI, Rodin L, Rozov NM (1971) Geographical aspect of biological productivity. Sov Geogr Rev Transl 12:293–317

Brown S, Lugo AE (1982) The storage and production of organic matter in tropical forests and their role in the global carbon cycle. Biotropica 14:161–187

Brown S, Lugo AE (1984) Biomass of tropical forests: A new estimate based on forest volumes. Science 223:1290–1293

Bucher EH (1982) Chaco and Caatinga- South American arid savannas, woodlands and thickets. In: Huntley BJ, Walker BH (eds) Ecology of Tropical Savannas. (Ecological Studies 42) Springer, Berlin Heidelberg New York Tokyo, pp 48–79

Coutinho LM (1982) Ecological effects of fire in Brazilian cerrado. In: Huntley BJ, Walker BH (eds) Ecology of Tropical Savannas. (Ecological Studies 42) Springer, Berlin Heidelberg New York Tokyo, pp 273–291

Crutzen PJ, Heidt LE, Krasnec JP, Pollock WH, Seiler W (1979) Biomass burning as a source of atmospheric gases CO, H_2, N_2O, NO, CH_3Cl and COS. Nature (Lond) 282:253–256

Crutzen PJ, Delany AC, Greenberg J, Haagenson P, Heidt L, Lueb R, Pollock W, Seiler W, Wartburg A, Zimmerman P (1985) Tropospheric chemical composition measurements in Brazil during the dry season. J Atmos Chem 2:233–256

Delmas R (1982) On the emission of carbon, nitrogen and sulfur in the atmosphere during bushfires in intertropical savanna zones. Geophys Res Lett 9:761–764

Detwiler RP, Hall CAS (1988) Tropical forests and the global carbon cycle. Science 239:42–47

Detwiler RP, Hall CAS, Bogdonoff P (1985) Land use change and carbon exchange in the tropics: II. Estimates for the entire region. Environ Manage 9:335–344

Eiten G (1972) The cerrado vegetation of Brazil. Bot Rev 38:201–341

Ehhalt DH, Rudolph J (1984) On the importance of light hydrocarbons in multiple atmospheric systems. Ber Kernforschunganlage, Jülich GmbH, July

FAO (1987) Yearbook of Forest Products 1976–1987. Food and Agriculture Organization, Rome

Fearnside PM (1985) Burn quality prediction for simulation of the agricultural system of Brazil's Transamazon Highway colonists for estimating human carrying capacity. In: Govil GV (ed) Ecology and Resource Management in the Tropics. Int Soc Tropic Ecol, Varanasi, India

Fearnside PM (1986) Spatial concentration of deforestation in the Brazilian Amazon. Ambio 15:74–81

Golley FZ (1975) Productivity and mineral cycling in tropical forests. NSF report ISBN-0-309-02317-3, pp 106–116
González-Jiménez E (1979) Primary and secondary productivity in flooded savannas. In: UNESCO/UNEP/FAO (ed) Tropical grazing land ecosystems of Venezuela. Nat Resour Res 16:620–625
Greenberg JP, Zimmerman PR, Heidt L, Pollock W (1984) Hydrocarbon and carbon monoxide emissions from biomass burning in Brazil. J Geophys Res 89:1350–1354
Haggar RJ (1970) Seasonal production of *Andropogon gayanus*, 1. Seasonal changes in field components and chemical composition. J Agric Sci 74:487–494
Harris DR (1980) Tropical savanna environments: Definition, distribution, diversity and development. In: Harris DR (ed) Human ecology in savanna environments. Academic Press, New York, pp 3–27
Hopkins B (1965) Observations on savanna burning in the Olokemeji Forest Reserve, Nigeria. J Appl Ecol 2:367–381
Houghton RA, Boone RD, Melillo JM, Palm CA, Woodwell GM, Myers N, Moore III B, Skole DL (1985) Net flux of carbon dioxide from tropical forests in 1980. Nature (Lond) 316:617–620
Huntley BJ (1982) South African savannas. In: Huntley BJ, Walker BH (eds) Ecology of tropical savannas. (Ecological Studies 42) Springer, Berlin Heidelberg New York Tokyo, pp 101–119
Huntley BJ, Morris JW (1982) Structure of Nylsvley savanna. In: Huntley BJ, Walker BH (eds) Ecology of tropical savannas. (Ecological Studies 42) Springer, Berlin Heidelberg New York Tokyo, pp 433–455
Jaeger L (1976) Monatskarten des Niederschlags für die ganze Erde. Ber Dtsch Wetterdienst 139
Lacey CJ, Walker J, Noble IR (1982) Fire in Australian tropical savannas. In: Huntley BJ, Walker BH (eds) Ecology of tropical savannas. (Ecological Studies 42) Springer, Berlin Heidelberg New York Tokyo, pp 246–272
Lamotte M (1985) Some aspects of studies on savanna ecosystems. Trop Ecol 26:89–98
Lanly JP (1982) Tropical forest resources. FAO For Pap 30, FAO, Rome
Logan JA (1983) Nitrogen oxides in the troposphere: Global and regional budgets. J Geophys Res 88:10785–10807
Lugo AE, Brown S (1986) Brazil's Amazon forest and the global carbon cycle. Interciencia 11:57–58
Matthews E (1983) Global vegetation and land use: New high-resolution data bases for climate studies. J Climate Appl Meteorol 22:474–487
Matthews E (1985) Atlas of archived vegetation, land-use and seasonal albedo data sets. NASA Tech Memorandum 86199
Melillo JM, Palm CA, Houghton RA, Woodwell GM, Myers N (1985) A comparison of two recent estimates of disturbance in tropical forests. Environ Conserv 12:37–40
Menaut JC (1983) The vegetation of African savannas. In: Bourlière F (ed) Tropical savannas. (Ecosystems of the World 13) Elsevier, Amsterdam, pp 109–149
Menaut JC, César J (1982) The structure and dynamics of a west African savanna. In: Huntley BJ, Walker BH (eds) Ecology of tropical savannas. (Ecological Studies 42) Springer, Berlin Heidelberg New York Tokyo, pp 80–100
Molofsky J, Hall CAS, Myers N (1986) A comparison of tropical forest surveys. DOE/NBB-0078, US Dep Energy
Myers N (1980) Conversion of tropical moist forests. Nat Acad Sci, Washington DC, USA
Rodin L Ye, Bazilevich NI (1966) Production and mineral cycling in terrestrial vegetation. Oliver and Boyd, Edinburgh, 288 pp
Rodin L Ye, Bazilevich NI, Rozov NN (1975) Productivity of the world's main ecosystems. In: Productivity of World Ecosystems. Nat Acad Sci, Washington DC, USA
San José JJ, Medina E (1975) Effect of fire on organic matter production and water balance in a tropical savanna. In: Golley FB, Medina E (eds) Tropical ecological systems. (Ecological Studies 11) Springer, Berlin Heidelberg New York, pp 251–264
San José JJ, Medina E (1976) Organic matter production in the Trachypogon savannas in Venezuela. Trop Ecol 17:113–124
Sarmiento G, Monasterio M (1975) A critical consideration of the environmental conditions associated with the occurrence of savanna ecosystems in tropical America. In: Golley FB,

Medina E (eds) Tropical ecological systems. (Ecological Studies II) Springer, Berlin Heidelberg New York, pp 223–250

Seiler W, Crutzen PJ (1980) Estimates of gross and net fluxes of carbon between the biosphere and the atmosphere from biomass burning. Climatic Change 2:207–247

Singh KP, Misra R (1978) Structure and functioning of natural, modified and silvicultural ecosystems of Eastern Uttar Pradesh. Tech Rep MAB Research Project, Banaras Hindu Univ, Varanasi, 161 pp (mimeograph)

UNESCO (1973) International classification and mapping of vegetation. UNESO, Paris, 93 pp

Whittaker RH, Likens GE (1975) The biosphere and man. In: Lieth H, Whittaker R (eds) Primary productivity of the biosphere. Springer, Berlin Heidelberg New York, pp 305–328

World Meteorological Organization (1985) Atmospheric Ozone. Global Ozone Research and Monitoring Project – Report No 16, Geneva, Switzerland

21 Global Change: Effects on Forest Ecosystems and Wildfire Severity

M.A. Fosberg[1], J.G. Goldammer[2], D. Rind[3] and C. Price[3]

21.1 Introduction

Climate change, as a result of the greenhouse effect, is expected to take place within the next 100 years – a time span comparable to the planting-to-harvest interval of many commercial tree species. The predicted increases of temperature are expected to be comparable to those that have taken place since the end of the last ice age 15,000 years ago. The 5 °C warming that occurred between 15,000 and 7000 years ago resulted in major changes in the location and abundance of North America's tree species (Bernabo and Webb 1977). The rate of temperature change predicted from the increase in greenhouse gases, that is 5 °C in 100 years as compared to the rate experienced in the early Holocene, is unprecedented in history. There is thus a great need to determine the impact of this predicted change on North America's ecosystems and, in particular, on our forest resources.

Our concern is not only with global warming from the greenhouse effect, but also with the various stresses these ecosystems will experience as a result of changed precipitation patterns, trauma such as fire, insects, disease, and air pollution, ultraviolet radiation as a result of stratospheric ozone depletion, and from changes in the ability of all species, plants, animals, and microorganisms to compete for limited energy, water, and nutrients.

21.2 Scientific Bases for the Greenhouse Effect

The greenhouse theory is based on the energy balance between incoming solar energy and the energy radiated to space from the earth. If there is not a balance between the incoming and outgoing energy, then the earth would either warm or cool. The energy from the sun reaches the earth primarily as visible light. Some of this incoming energy is reflected back to space from clouds, a small portion is absorbed by the atmosphere, approximately 43% reaches the earth's surface, where it is absorbed. This absorbed energy warms the earth. A portion

[1]U.S. Department of Agriculture, Forest Service, Washington, DC 20090, USA
[2]Department of Forestry, University of Freiburg, 7800 Freiburg, FRG
[3]NASA Goddard Institut for Space Studies, New York, NY 10025, also at Lamont-Doherty Geological Observatory of Columbia University, Palisades, NY 10964, USA

of the received energy warms the atmosphere directly through heat transfer. A portion is also radiated towards space as long wave infrared radiation. Certain trace gases such as carbon dioxide, methane, and water vapor absorb a small percentage of the outgoing long wave radiation and warm the atmosphere further. If the total incoming solar energy is balanced by the total energy returned to space, the temperature of the earth remains constant. The equilibrium temperature for the earth is currently 15°C. If the atmosphere were not able to absorb the infrared radiation, the equilibrium temperature would be -7°C (Haltiner and Martin 1957). Essentially, the greenhouse effect is based on sound physics. The greenhouse effect is also beneficial, raising the earth's temperature above the freezing point of water.

Current interest in the greenhouse effect is not focused on the theory, but on whether the equilibrium temperature has been disturbed through increase of the radiatively active gases such as carbon dioxide, methane, and others (Hansen et al. 1987). Precise monitoring of the amount of carbon dioxide in the atmosphere (Keeling 1984) has shown a steady increase since 1958, the beginning of the record (Fig. 1). Extrapolating this record back in time, it is estimated that before the industrial revolution concentration of carbon dioxide was 270 ppm, as compared to the current level of 350 ppm. Similar increases in other radiative active gases such as methane have also been observed (Hansen 1987). All the increases are attributable to human activity, burning of fossil fuels, forest burning, agricultural production, such as me-

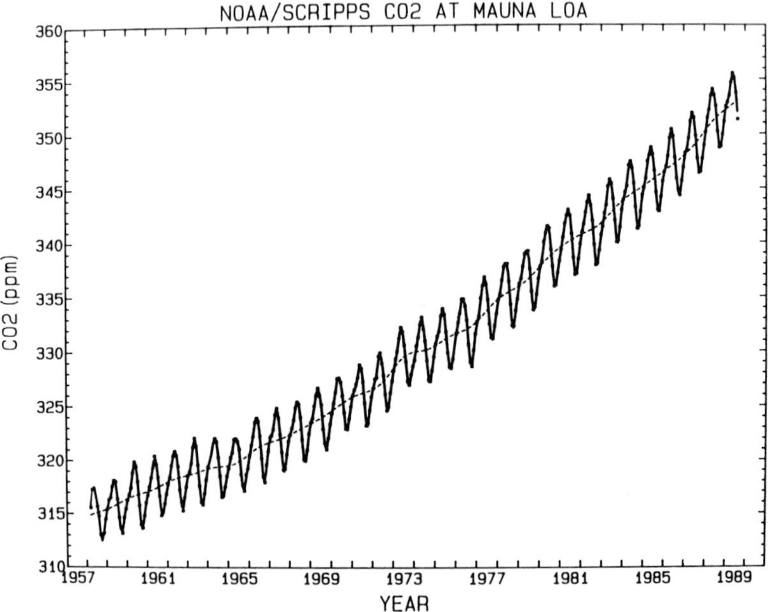

Fig. 1. Mean monthly concentrations of atmospheric carbon dioxide at Mauna Loa. (Keeling 1984; Thoning et al. 1989)

thane from cattle and rice production, and from manufactured chemicals such as chlorofluorocarbons.

That temperature and atmospheric concentrations of carbon dioxide are strongly correlated is not in doubt. Ice core data extending back 160,000 years (Fig. 2) clearly demonstrate this correlation (Barnola et al. 1987).

Projections of future concentrations of these greenhouse gases are based on forecasts of energy consumption, energy efficiency, and population growth. Current projections indicate that with present technology and population growth, the concentrations of the radiative active gases will double by 2030 A.D. and that even with highest levels of energy conservation and efficiency concentrations, should double by 2075 A.D. (Mintzer 1987). There is, then, no doubt that the earth will warm in the future.

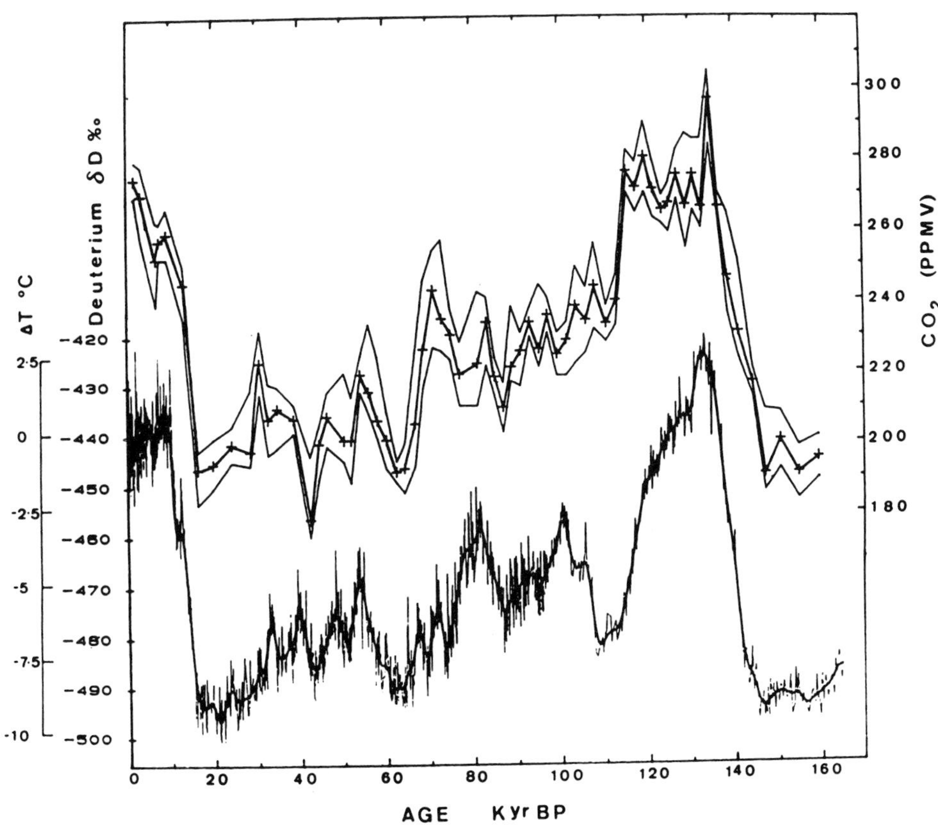

Fig. 2. Ice core concentrations of carbon dioxide and temperatures for the past 160,000 years (After Barnola et al. 1987). Reprinted by permission from Nature (London) Vol. 329 pp. 410. Copyright (c) 1987 Macmillan Magazines Ltd

21.3 Modeling the Atmospheric Response to the Greenhouse Effect

The fact that greenhouse warming will take place is not sufficient information to determine the impact on forests and related ecosystems. Heterogeneity of land and water distribution on the surface of the earth renders any simple mean value calculation useless. Current science and technology permit only an approximation of impacts through use of mathematical models of the general circulation of the atmosphere and oceans. These general circulation models (GCM) are the equations representing the physical concepts of conservation of mass, energy, and momentum. Such models describe the atmosphere and oceans with a large number of discrete points for which forecasts of temperature, pressure, water (for the atmosphere), and salinity (for the oceans) are made. These forecasts permit calculation of clouds, wind, precipitation, and exchange of energy between the biosphere, hydrosphere, and the geosphere (Schneider 1988; Dickenson 1982). Spatial resolution of these models is very coarse, a few degrees of latitude and longitude, when compared to ecosystem dimensions. Physical processes associated with small physical dimension phenomenon such as individual clouds cannot be described in these models. Instead, such processes are represented by an expected mean effect on the energy, mass, and momentum budgets. This shortcoming is particularly acute in mountainous regions where ecosystem dimensions are very small, and where local climate variations are large (Schlesinger 1988).

Representation of clouds in these GCM's is particularly important because of their influence on both incoming solar radiation and outgoing infrared radiation. Intercomparison of the different GCM predictions has attempted to reduce the uncertainty resulting from how clouds influence the model results (Schlesinger 1988), but there is still need for improvement before this problem can be considered resolved. Effects of clouds on the local energy exchange is three times that predicted from the greenhouse gases (Ramanathan et al. 1989).

A second major deficiency in these GCM's is in how the oceans are depicted. In two of the models, oceans are represented by a shallow ocean in which pre-determined sea surface temperatures are used to regulate the atmospheric circulation. In the other two models, a deep ocean with ocean currents, to redistribute the heat, is used, but these models do not give the same results (Schlesinger 1988; Byran 1988).

The Oregon State University (OSU), National Center for Atmospheric Research (NCAR), NASA Goddard Institute for Space Studies (GISS), and the Geophysical Fluid Dynamics Laboratory (GFDL) models show some consistency in predicting the future temperature rise and the regional distribution of temperatures. Also, all four models predict that the global precipitation will increase, primarily because a warmer atmosphere has a higher saturation vapor pressure (Mitchell 1988).

Intercomparison of model results for regional precipitation distribution shows far less consistency (Kellogg and Zhao 1988). This lack of consistency is particularly troublesome because water distribution more than temperature determines the distribution and composition of ecosystems.

Major weaknesses and sources of uncertainty in applying these models to predict future ecosystems are then (1) coarse spatial resolution for ecosystem studies, (2) inadequate representation of the role of clouds in the energy balance, and (3) an inconsistent prediction of the hydrologic cycle.

The rate at which climate will change is also important. If the climate evolves slowly, the biosphere may be able to adapt. Three of these models attempt only to predict the equilibrium climate of the future. Only the GISS model allows the greenhouse gases to increase with time and give an estimate of the rate of climate change.

21.4 Coupling the Biosphere to the Geosphere

Direct, interactive coupling of the biosphere to the atmosphere so that there is a direct exchange of mass and energy (Abramopoulos et al. 1988) will need to be improved dramatically before results useful to resource managers will be available. Current approaches only attempt to describe the heat and water vapor exchange and make no reference to the structure and composition of the ecosystem or to the abundance of individual species.

The current level of science uses predicted atmospheric change to drive the biosphere changes with no mediating effects of the biosphere in the atmosphere. There are two main approaches currently in use. The Holdridge life zone concept (Holdridge 1964) is based on correlating temperature and precipitation to major ecosystem structure, e.g., southern pines, spruce-fir ecosystems. In the Holdridge approach, if the climate changed, the forest would change to the optimal ecosystem for the new climate. If the climate were displaced geographically, the optimal forest would migrate with the climate. That is, the forest would look just like it does now, but just be somewhere else. Such migration may be possible if the climate were to change slowly, but if the climate were to change rapidly, maintaining an intact forest would be difficult or even impossible, because each species in the forest will migrate at a different rate (Davis 1981). This disassociation of species in migrating forests was clearly observed during the early Holocene (Bernabo and Webb 1977). Some caution must be exercised in using this approach in predicting future forests resulting from climate change. A more realistic method of predicting future forests associated with climate change is with the gap phase models (Botkin et al. 1972). The gap phase models predict the germination, growth, and death of individual trees. These models account for competition for light, water, and nutrients between species. Such models allow individual species to die in the ecosystem and to be replaced by new species which are better adapted to the environment, or that are more competitive for light, water, and nutrients. Shortcomings of these models are not conceptual, but only that they are relatively new and therefore have yet to reach their full potential. Such shortcomings are that trauma (fire, insects, disease, and pollutants) are not yet incorporated, nor are microorganisms or animals, particularly herbiverous. Despite these current shortcomings, the gap phase models are the best available approach to assessing the impacts of climate change.

21.5 Sensitivity of the Ecosystem Forecasts to Uncertainties in the GCM's

The four GCM's predict a range of temperatures for global warming. These temperature predictions range from a low of 2.8°C to a high of 4.2°C (Schlesinger 1988) for mean surface temperature. North America regional and seasonal distribution of these temperature increases differs by as much as 8°C for summer and by 4°C for winter (Schlesinger 1988). Spatial and seasonal distribution of precipitation is expressed as soil moisture. During the winter, the southwest, south, and southeastern states are expected to be drier. During the summer, however, the entire United States is expected to be drier (Kellogg and Zhao 1988). There are marked differences between each of the model predictions in both winter (Fig. 3) and in summer (Fig. 4). Areas of greatest discrepancy, and therefore uncertainty, are in winter precipitation for the west coast, the Great Basin, Rocky Mountains, the mid-west, and the northeast. There is therefore greater consistency between the predictions and confidence during the summer.

Natural variations in annual precipitation and mean temperatures have always existed. During the past 100 years, the long-term temperature record for the United States has not shown any systematic change, but has ranged from 10.6 to 12.8°C (Karl and Jones 1989). Over the last 2700 years, which includes the little ice age of the 17th century, North America has experienced a natural variability of 1.5°C (Bernabo 1981). Similarly, precipitation has shown large year-to-year variability during the last 2000 years (Stahle et al. 1988). The Palmer Drought Index (Fig. 5) for which a value of –2 represents extreme drought, shows that abnormally wet or abnormally dry periods are more common than normal precipitation.

Given the uncertainty in the predictions, and the natural variability, which climate change must exceed before we can detect effects?; what can we say about impacts on the ecosystem? Two independent analyses of climate change impact have been completed for the Lake States. Both of these studies used a gap phase ecosystem model, and both were based on GCM predictions of climate for a doubled carbon dioxide concentration in the atmosphere. The difference between these two predictions is only that two different GCM's were used. Solomon and West (1987) used the NCAR model, while Botkin et al. (1988) used the GISS model. These two analyses provide us with some measure of the sensitivities of the ecosystem model to climate change prediction. Differences between the two simulations are that in one, conifers will be totally replaced by hardwoods, and in the other, conifers will be retained. Also, the two differ in the

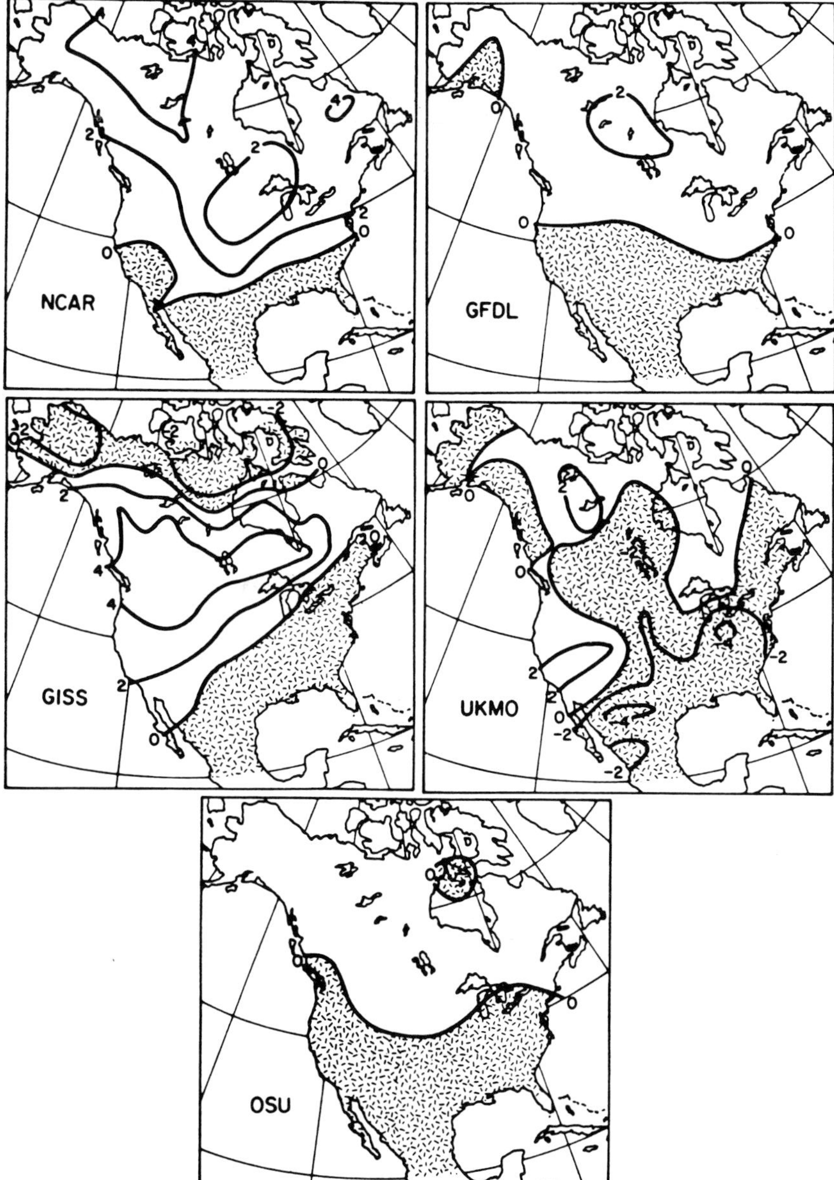

Fig. 3. Predicted soil moisture changes as a result of doubled carbon dioxide for winter from five GCM's. *Shaded areas* are for decreases. (Kellogg and Zhao 1988)

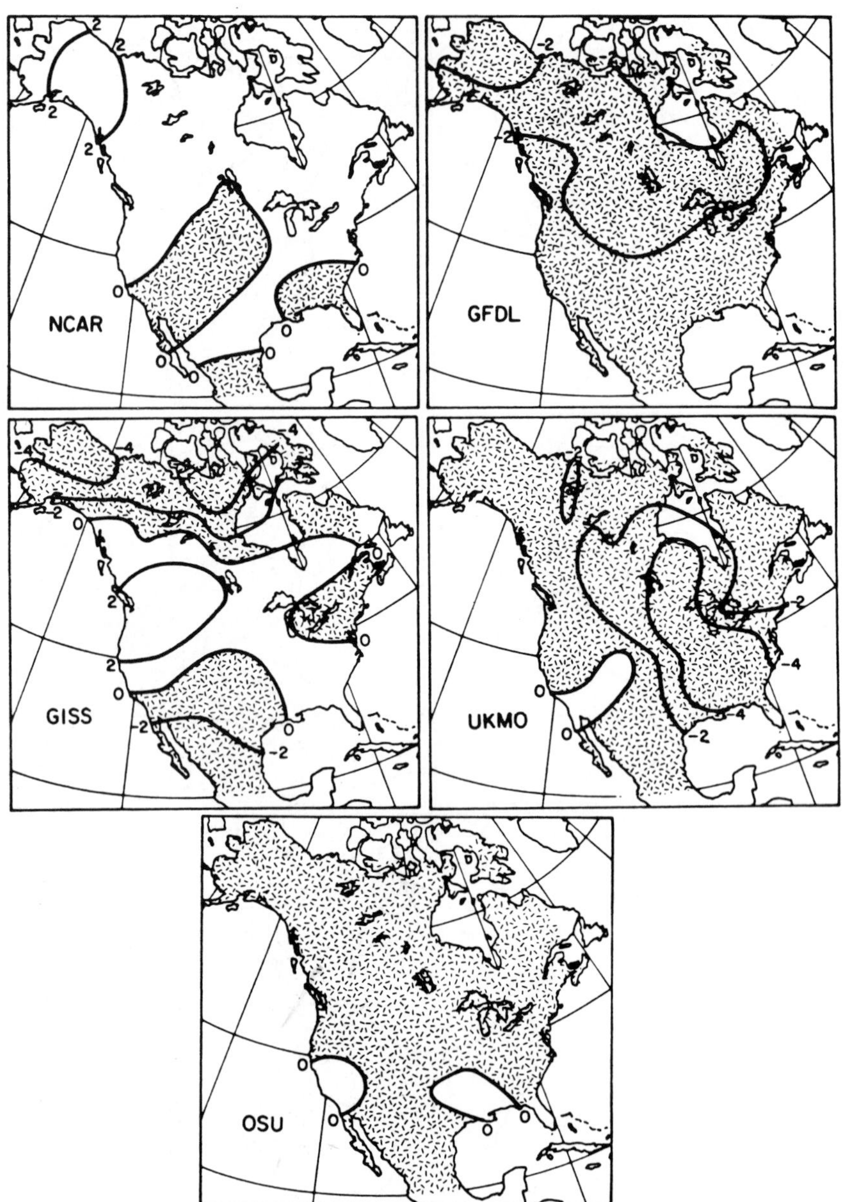

Fig. 4. Predicted soil moisture changes as a result of doubled carbon dioxide for summer from five GCM's. *Shaded areas* are decreases. (Kellogg and Zhao 1988)

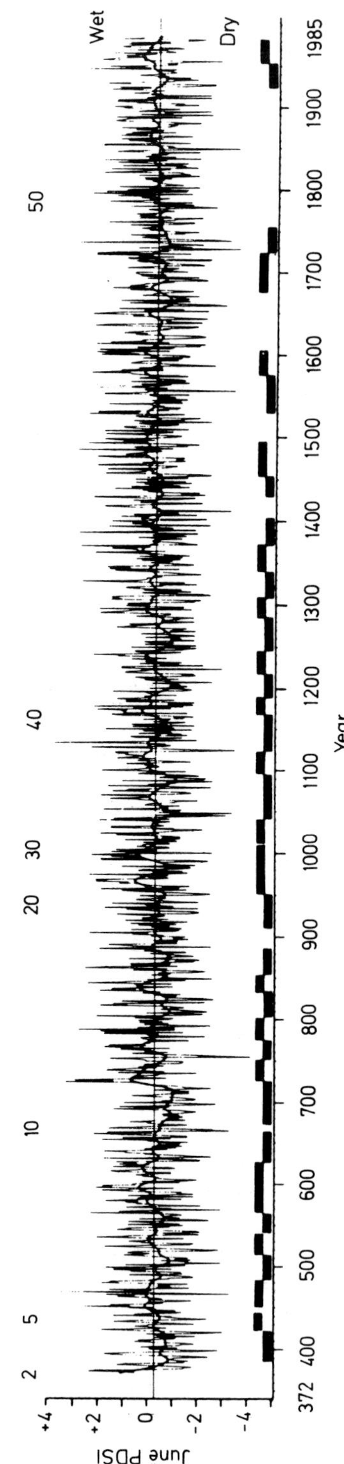

Fig. 5. Palmer Drought Index as inferred from tree rings. Record is for the period 372 up to 1985 A.D. for a site in the southern United States. (Stahle et al. 1988)

number of trees per hectare, one showing an increase and the other a decrease. Similarities are that both simulations show a decrease in total biomass. The sensitivity analysis suggests that some confidence may be placed in prediction of total biomass, but less confidence in the structure and abundance of individual species.

21.6 Ecosystem Stresses Which Have Not Been Included

Impact of climate change of forest ecosystems has focused on primary production. While these analyses include temperature and water stress, they fail to recognize the importance of other stresses such as insects, disease, fire, or air pollution. Also, these analyses do not address the impact of climate change on the frequency and severity of the traumatic events and how these in turn will impact primary productivity. For example, if hardwoods are to become the dominant forest where conifers now exist, gypsy moth defoliation will certainly influence primary productivity unless mitigating action is taken (Winget 1988). Also, if cottonwood becomes a more important species for pulp and paper, impact of climate change on melampsora leaf rust (Hepting 1971; McCracken et al. 1984) must be taken into account (Fosberg 1988).

Increased insect- and disease-caused losses in the world's forests will become one of the first observed effects of climate change. Evidence of this can easily be found in the pest-caused epidemics which now occur as a result of periodic droughts or rainy periods. Changes in climate through effects either on the pest or on the host may increase or decrease pest-caused losses. High temperatures and reduced precipitation cause insect epidemics when these climatic factors stress the tree (host) to the point that they lose their inherent resistance to native pests. Increased moisture may favor disease which historically was limited in distribution and infection success because of unfavorable low moisture conditions. As the climate changes we can expect frequency, period, and geographic extent of new epidemics, while currently important pest problems may all but disappear. These pest attacks will often determine the new geographic distribution of tree species in the new climatic conditions.

Fire frequency and severity are also missing in assessing the impact of climate change on forests. As the structure, composition, and total biomass of the forest change, so will the behavior of fire. Much of the structure and composition of a forest will remain long after climate change-induced stress has prevented regeneration of seedlings. Also, new species will take hold, so that during the transition from one equilibrium ecosystem to another, a transitional forest containing elements of both will exist.

Paleo-analysis of charcoal in sea sediments (Herring 1985) has shown a weak, but definite trend of charcoal deposition over the last 50 million years. Combining these data with temperature relations (Fifield 1988), we see a weak but positive correlation between temperature and charcoal (Fig. 6). Because this correlation is post K-T boundary (massive extinction period for a number of

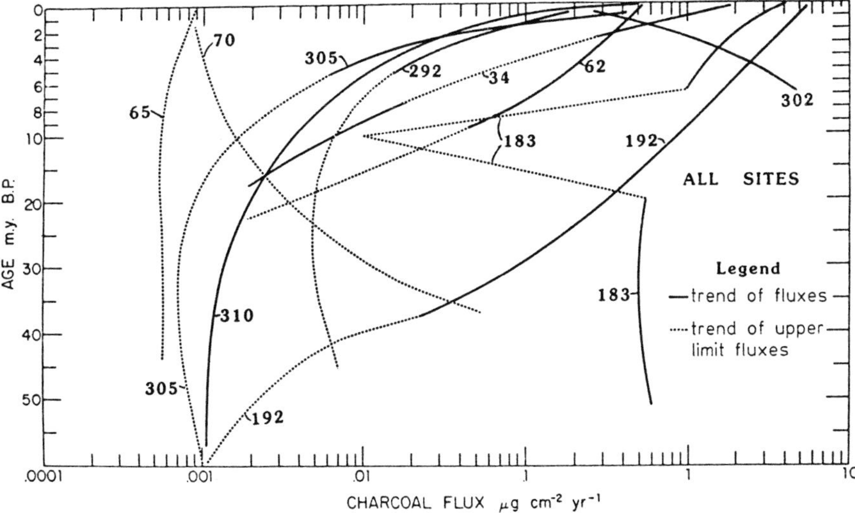

Fig. 6. Charcoal concentrations in deep sea sediments from the Pacific Ocean. (Herring 1985)

species) and before human discovery of fire, this record should reflect total biomass burned and the fire relationship to climate.

To understand the relationship between fire and climate change, we need to understand fire behavior. Our knowledge of fire behavior integrates the amount and structure of the vegetation (the fuels) and weather. Two measures of fire behavior are of interest here, rate of propagation and flame length or fire line intensity. These two measures tell us area burned, damage, and effort required to suppress a fire.

Forests are composed of fuel elements of varying size, from conifer needles, small twigs and leaves, to the large boles of trees. Each of these elements responds differently to weather. Fine dead fuels (twigs, etc.) respond rapidly to diurnal variations of relative humidity while large dead fuels may take several months to dry following winter rains. Living vegetation is also consumed in a fire. Vegetation under drought stress has lowered moisture content, frequently contributing to fire propagation and intensity. The fine fuels have high surface area to volume ratios and determine the rate at which fires propagate. Large fuels contain the energy that will be released during fires and determine the intensity of a fire. The mixture of fuel sizes and the amount of fuel of each size determine the potential fire behavior of a given ecosystem. Grasslands contain no large fuels. In grasslands, fire can spread rapidly and typically results in less significant damage. Heavy timber ecosystems which have been cut and allowed to cure have large fuel elements. In the cured state, much of the fine fuels, the foliage, has fallen to the ground and become somewhat compacted. Fire spreads slowly through this complex, but can be very intense, and substantial damage may result (Fosberg 1989).

Rothermel (1972) developed a model which provides a crude method of translating the structure, composition, and total biomass of the ecosystem into a method of predicting fire behavior. For fire spread, the biomass is classified by a dicotomous living or dead classification, and by discrete groupings of surface area to volume ratios. For fire intensity, weight of contribution is by biomass in each surface area to volume ratio group. This discretization classification is based on how dead fuels respond to environmental change (Fosberg 1971). This physical discretion of the ecosystem is an approximation and is consistent with the Holdridge life zone concept (Fosberg and Furman 1973). Static descriptions (models) have yet to be linked to dynamic ecosystems in which change is either normal because of life cycle and natural variability, or because climate-induced changes in the composition and structure of the ecosystem occur.

21.7 Ecosystems and Potential Fire Behavior

In this chapter, we use a simplified version of the U.S. National Fire Danger Rating System (Deeming et al. 1972). This simplification results in potential fire behavior for a number of North American ecosystems and is based on measured structure and composition of those ecosystems as expressed by the size and amount of fuels. The classification has a general relevance for various vegetation types of the world because similar forest fuels and other wildland fuels are found outside North America. For both rate of fire spread and flame length, we assign a value of 100 to the maximum in each category. For example, for grass fuels, fuel A in Table 1, the rate of spread is 100, while the flamelength has a value of 13. Chaparral brush fields, fuel B, have a rate of spread of 52 on this scale and a flamelength of 100. The remaining fuel types are:

C. Open overstory forest with grasses or other herbaceous plants as a ground fuel. Young conifer plantations, open Ponderosa, sugar, longleaf, slash, and sand pines, as well as pinyon juniper stands characterize this class.

Table 1. Relation between ecosystem description and fire behavior

Fuel type	Relative rate of spread	Relative flame length
A	100	13
B	52	100
C	54	41
D	23	45
E	11	19
F	8	7
G	8	55
H	5	21
I	19	85

D. Ecosystems in which there is heavy loading of fuels, 2 cm or less in diameter, and in which living fuels burn readily. The low pocosins of the Atlantic states and black spruce stands at high latitudes are presented by this model.
E. Hardwood and mixed conifer-hardwood stands during the dormant season, before leaf fall has been compacted.
F. Young brush fields which contain little or no dead materials. Laurel, mountain mahogany, and young stands of chamise and manzanita are represented here.
G. Dense conifer stands where heavy build-up of downed timber has accumulated.
H. Closed short-needle conifer stands, hardwoods, and mixed hardwood-conifer stands after leaf fall has been compacted.
I. Clearcut timber where little material has been removed.

For determination of effects of climate change on forest fires, only the flame length will be used because flame length is directly correlated with required suppression effort.

On the scale described in Table 1, 20 represents the limits of personnel working a fire line, 50 represents the limits of direct attack, and 60 represents crowning fires where suppression forces are withdrawn for safety reasons. The Yellowstone fires of 1988 (fuel type G) ranked about 100 to 150 on this scale. The expected potential fire behavior for fuel type G was 55. The most intense fire in recent years, the Sundance fire of 1967 in Montana, also fuel type G, had a suppression effort of 200. This fire totally reduced the forest to black ash.

There are few definitive studies of direct effects of climate change on fire frequency and severity. Direct effects are used here to define the changes in drought frequency, humidity, precipitation, and other weather elements that determine day-to-day variation and interannual variability in fire behavior. Fried and Torn (1988) compared the changes in area burned under the current and a double CO_2 climate. They found that there would be a twofold increase in modest sized fires (a few hundred hectares) and a threefold increase in fires greater than 1000 hectares. Fried and Thorn based their studies on an area of the Californian Sierra Nevada in which the ecosystem is expected to remain unchanged in a future climate.

An analysis of drought from A.D. 372 to 1985 (Stahle et al. 1988) showed increased interannual variability of drought when decadal variability was high. Stahle correlated tree rings with the Palmer Drought Index for a site in the southeastern United States. Because the Palmer Drought Index is used to calculate the moisture content of live vegetation, it is an ideal analog for the moisture content of live vegetation, and therefore for fire danger variability.

21.8 The Effect of Climate Change on Lightning

The major causes of present and future wildland fires, especially in the tropical world, are anthropogenic. However, lightning is the most important cause of natural fires which have shaped and will further influence the dynamics of vegetation development within the tropics and the continental regions of the temperate and boreal zone. Therefore, in order to study how future climate change may affect wildfires, one needs not only to consider changes in temperatures and precipitation, but also the possible changes in lightning frequency.

Satellites have been used to observe lightning on the global scale (Turman and Edgar 1982; Orville and Henderson 1986), and all these studies show that spatial and temporal distributions of global lightning agree remarkably well with the general circulation of the atmosphere.

For this reason it is possible to use GCM's to simulate lightning, and so predict changes in lightning frequencies in a warmer climate. Lightning frequency has been shown to be closely related to the height of convective cloud tops (Williams 1985). The relationship shows that the flash frequency is proportional to the fifth power of the cloud height.

Using the GISS GCM to calculate changes in penetrating convection, it is possible to determine changes in lightning frequency for the $2 \times CO_2$ climate. In our model this corresponds to a global warming of approximately 4°C. Figure 7a shows the zonally averaged lightning frequencies as observed by satellite at dusk for the period August 1977 to June 1978. It is clear that the highest concentration of lightning occurs in the tropics along the Intertropical Convergence Zone (ITCZ). Figure 7b shows the zonally averaged lightning frequency as calculated using the GISS GCM control run, for today's climatology. There is a fair agreement between the observations and the model predictions. Both measurements and observations show more lightning in the Northern Hemisphere. The reason for this is that lightning is closely related to the distribution of landmasses in each hemisphere. Figure 7c gives the percentage and absolute change in lightning frequency as a function of latitude for a $2 \times CO_2$ atmosphere as calculated from the model, after correcting for model deficiencies (e.g. Figure 7a,b).

The model shows a significant increase in lightning frequency at all latitudes. When the percentage change is integrated over all latitudes, a mean global increase of approximately 26% is obtained. Considering that around 100 lightning flashes occur each second around the globe (Turman and Edgar 1982), this amounts to nearly 2.23×10^6 extra lightning flashes per day.

Fig. 7. **a** Zonally averaged lightning frequency as observed by satellite at dusk for August 1977-June 1978. (After Turman and Edgar 1982). **b** Zonally averaged lightning frequency as calculated using the GISS GCM. **c** The calculated percentage and absolute change in lightning frequency for the $2 \times CO_2$ climate

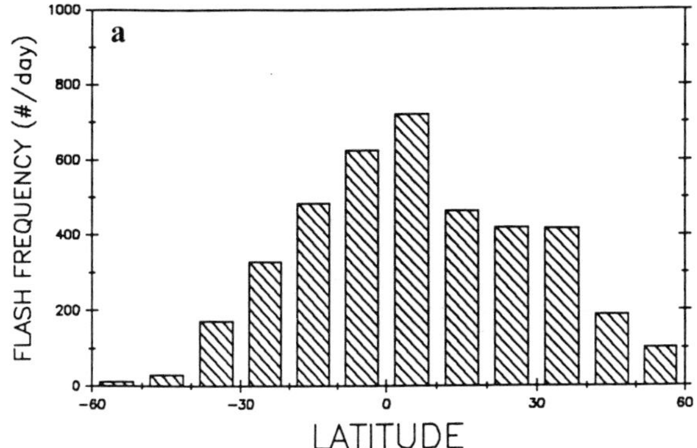
Zonally Averaged Lightning Frequency
for Dusk August 1977–June 1978
ANNUAL MEAN
a
FLASH FREQUENCY (#/day)
1000
800
600
400
200
0
-60
-30
0
30
60
LATITUDE

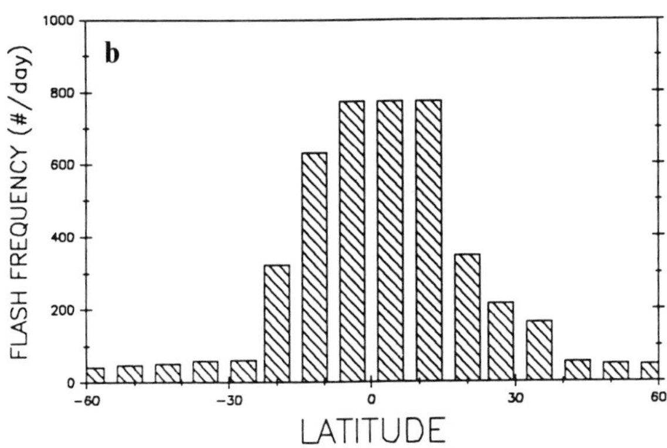
Zonally Averaged Lightning Frequency
for the Control Run of the GISS GCM
ANNUAL MEAN
b
FLASH FREQUENCY (#/day)
1000
800
600
400
200
0
-60
-30
0
30
60
LATITUDE

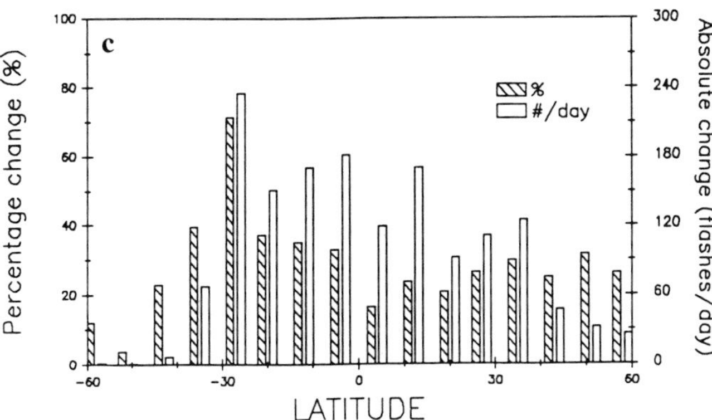
Changes in Lightning Frequency for 2XCO2
as calculated from the GISS GCM
ANNUAL MEAN
c
%
#/day
Percentage change (%)
Absolute change (flashes/day)
100
80
60
40
20
0
300
240
180
120
60
0
-60
-30
0
30
60
LATITUDE

The above changes are for all lightning strikes, both intracloud and ground flashes. Since the fraction of lightning flashes that reach the ground is a function of latitude, further studies are needed to determine how ground flashes may be influenced by climate change.

However, the implication for fires is obvious, and with the expected increase in temperature and evaporation, the increase in lightning may have a dramatic effect on the frequency of wildfires in the future.

21.9 Regional Predictions: Impact of Climate Change on Distribution of Forest Biomes and Other Effects

21.9.1 North America

Assessment of forest resources 50 to 100 years from now as a result of the greenhouse effect has not been done in any systematic fashion. Coverage of the United States is not uniform, with several estimates made for some regions and only one estimate for others. Also, different methods, namely the Holdridge life zone technique (Holdridge 1964), or the gap phase model (Botkin et al. 1972) have been used for different regions. Fortunately, there are two regions, the south-southeast region and the Lake States, for which estimates have been made using both techniques. Intercomparison of the estimates by more than one method and by more than one future climate scenario will give us some measure of our confidence in these estimates. In the context of this book the findings on future distribution of *Pinus* spp. seem to be of particular relevance because of the role of this genus in tropical fire climax communities and afforestation activities.

The U.S. Environmental Protection Agency assessment for the Great Lakes area of North America (Smith 1988) predicts that conifers will remain in the northern portion of the Great Lakes Region and potentially migrate to James Bay, in Canada, with a Holdridge-GISS scenario, and that with a Holdridge-GFDL scenario conifers would be totally lost from the Great Lakes area. Sugar maple would show similar migration patterns under these two scenarios. The gap phase model simulations show less dramatic changes. Using the GISS output, Botkin et al. (1988) predicted that conifers would disappear by 2040 A.D. and that sugar maple and other hardwoods would dominate the ecosystem, with a decreased biomass. Solomon and West (1987), using the NCAR GCM, also show hardwoods preferred over conifers, but conifers retained in the ecosystem. This simulation predicted a decrease in biomass.

The EPA assessment (Titus 1988) for the south and southeast of the United States shows that southern pines would be greatly reduced or eliminated in Mississippi, South Carolina, and Georgia and would experience a 30% reduction of biomass in Tennessee. This prediction was based on a gap phase model. Miller et al. (1987), using the Holdridge method, confirm this prediction of marked reduction in Mississippi, Georgia, and South Carolina, and show

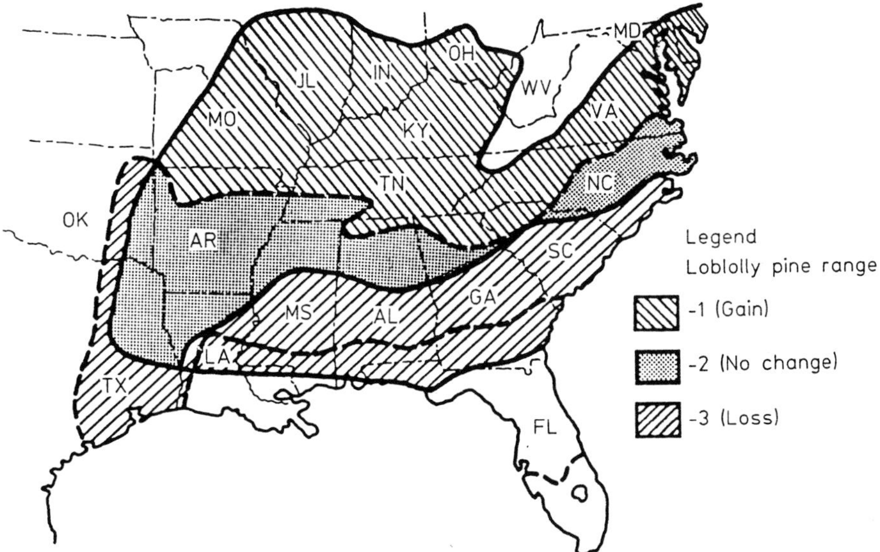

Fig. 8. Possible redistribution of loblolly pine in the eastern United States as a result of a double carbon dioxide atmosphere. Current distribution is shown with the *cross-hatching* sloping *upward to the right* and *in the stippled area*. Future distribution is shown with the *stippled area* and the *cross-hatching* sloping *downward towards the right*. (Miller et al. 1987)

loblolly pine invading into Tennessee (Fig. 8). The southern pines extended northward but did not move out of the south and southeast under the GFDL-Holdridge and OSU-Holdridge scenarios (Winjum and Neilson 1988).

Conifers would retreat into Maine under the more severe GFDL-Holdridge simulation and there would not be appreciable change under the milder GISS Holdridge scenario (Smith 1988). All of the simulation models for the Central United States show a marked decline or elimination of hardwoods by 2050 A.D. (Smith 1988).

Major species in the Pacific Northwest of the United States are Douglas-fir (*Pseudotsuga menziesii*) and Ponderosa pine (*Pinus ponderosa*). Leverenz and Lev (1987), using the Holdridge life zone method and an unspecified GCM, predicted that Douglas-fir would retreat to higher elevations. Ponderosa pine (Fig. 9) would show a reduction in area in interior Washington. Little change would be expected for lodgepole pine (*Pinus contorta*) (Fig. 10). In the extreme south of the Californian Sierra Nevada, Ponderosa pine would be eliminated in the Southern Rocky Mountains (Fig. 9).

There is no complete agreement on the impacts of climate change on forest resources, but there is some consistency in that some species will not be severely affected, others will experience reduced ranges or total replacement, and that biomass will decrease. The only moderately positive prediction is that the loblolly pine (*Pinus taeda*) may migrate northward.

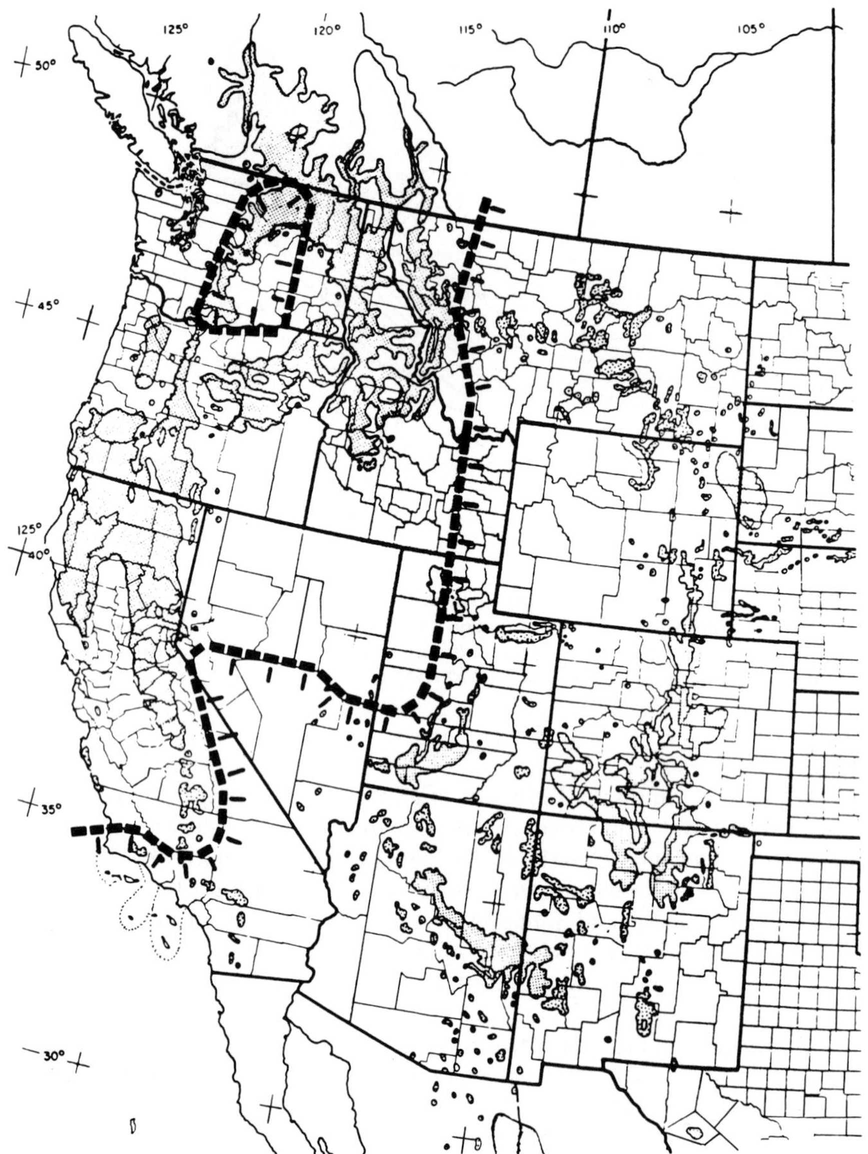

Fig. 9. Current and future distribution of Ponderosa pine (*Pinus ponderosa*). *Hatching* is directed towards zones of decreasing acreage as projected from a doubled carbon dioxide atmosphere. (Leverenz and Lev 1987)

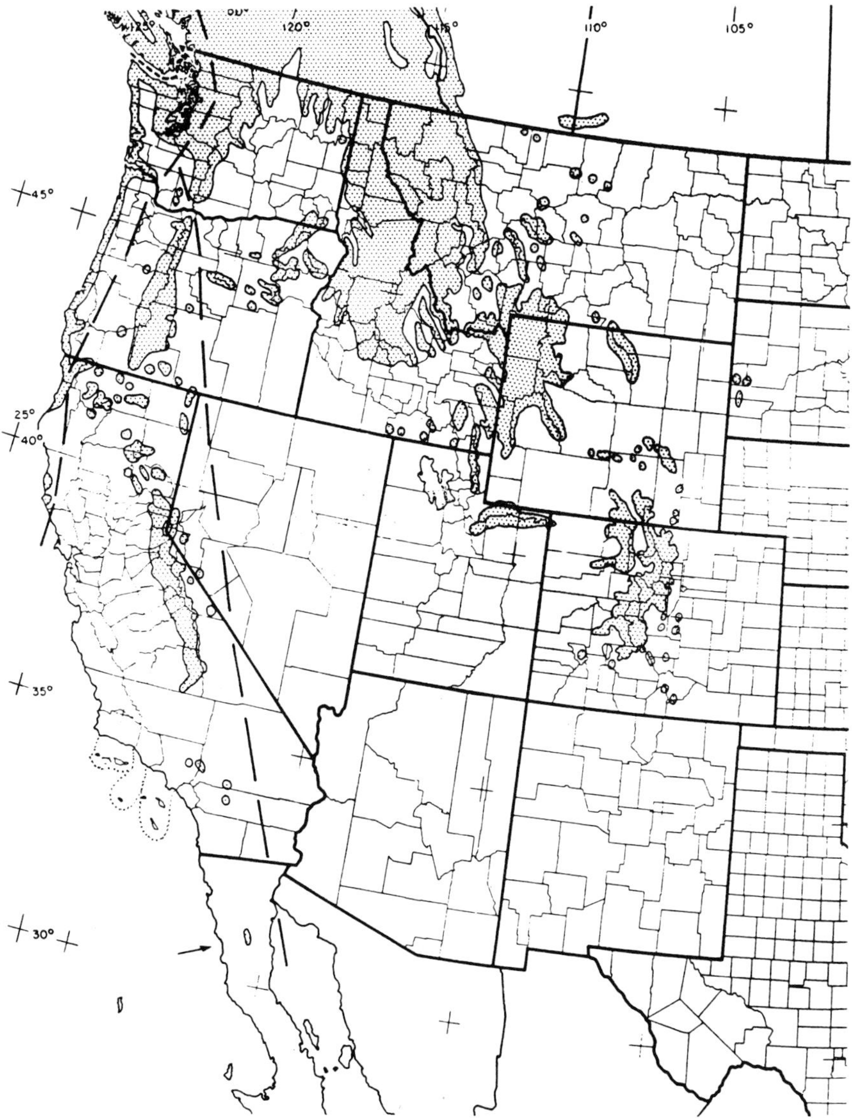

Fig. 10. Current and future distribution of lodgepole pine (*Pinus contorta*). *Hatching* is directed towards zones of decreasing acreage as projected from a doubled carbon dioxide atmosphere. (Leverenz and Lev 1987)

21.9.2 The Tropics

Regional predictions on the impact of climate change on tropical forest species and other vegetation are not yet available. However, paleoecological evidence on the biogeography of tropical vegetation may be helpful to predict basic patterns of change. Various findings on the distribution of tropical vegetation between the last glacial maximum and the present climate (palynological evidence) indicate altitudinal shifts of vegetational zones. During the late Quaternary, the maximum depression of the forest limit in the New Guinea highlands amounted to ca. 1000 to 1500 m below today's levels (Hope 1976; Flenley 1979). Tentative reconstructions of similar upward and downward migration of vegetational zones have been established for the Andes of South America (van der Hammen 1974; Flenley 1979).

It is expected that a generally warmer climate in the tropics will lead to a further upward movement of vegetational zones in which two main processes will be involved. In those mountainous lands that are below the potential present forest limit, an upward zonal migration will result in a total loss of species confined to the upper vegetation zones (Goldammer 1991). In those mountains that are higher than the present forest limit, the vegetation will be affected less severely. However, snow fall in higher elevations and perennial snow packs will be reduced.

The effects of an altered hydrological regime (loss of snow pack and possible change of upland precipitation patterns) may have severe downstream effects. One of the potential effects would be the reduction of continuity of perennial water supply to lowland aquifers and water bodies. This would have considerable impact on lowland water tables, fish life, riparian forests, and probably even on the coastal mangrove swamps.

The feed-back mechanisms of anthropogenic forest degradation and climate change may also lead to changing precipitation regimes. These effects will probably be most dramatic in today's rain forest biomes, which will develop toward pyrophytic and xerophytic communities characterized by short-return interval fires and further contribute to the change of the atmosphere (Fig. 11; see also Mueller-Dombois and Goldammer this Vol.). The vast amount of tropical rain forests in the lowlands, even if untouched by human disturbances, would probably not survive in a generally warmer and drier climate. Thus, the equatorial rain forest would migrate or be confined to small refugia in higher elevations.

21.10 The Future

While we have yet to detect the first signals of greenhouse warming, either through direct measurements of temperature or through impacts on forest ecosystems, we need to begin preparing for the inevitable changes.

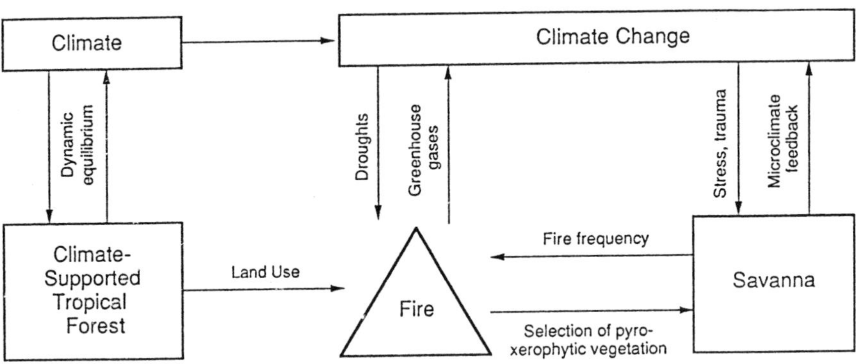

Fig. 11. The role of fire in the degradation of the tropical rain forest and climate-feedback mechanisms. (After Goldammer 1990)

Our policy options are to conserve what we currently have in forest resources, to adapt to change, to develop strategies to mitigate the effects of climate changed, or some combination of those three. The conservation option is undoubtedly the most difficult to achieve simply because the external force, the changed climate, will ultimately prevail and we will have a different ecosystem. We can conserve some elements, albeit at a high cost. Such conservation actions might include installation of irrigation systems in plantations or use of fertilizers to compensate for reductions in growth rates. The second option, that of mitigating the effects of climate change, involves the global community. Energy conservation or use of nonfossil fuel energy will slow global warming. Such actions require a global policy rather than a local land management policy, and therefore are also difficult to achieve. Energy conservation or use of alternate energy sources cannot reverse the build-up of greenhouse gases in the atmosphere that have already occurred. The process of photosynthesis removes carbon dioxide (one of the greenhouse gases) from the atmosphere and turns it into wood; energy conservation will control the rate of greenhouse gases build-up but cannot reverse the build-up. Through aggressive reforestation and afforestation we can offset some of the anthropogenic emissions of trace gases. If this aggressive tree planting action is accompanied by high levels of utilization of the wood, in the form of long-term durable forest product, we can sequester carbon dioxide, and continue to remove additional carbon dioxide with young, vigorously growing trees. Only when forests become mature do they become carbon neutral, in that growth rates are offset by decay rates. The third option, that of adaptation, offers the greatest flexibility in managing forest in a changing climate. Adaptive strategies involve developing new technologies to utilize the resources of the future forest, importing new industries or businesses which are compatible with the resources of the future forests, or relocating existing activities in anticipation of a changed climate. Adaptive strategies also include developing or introducing species which are compatible with the changed climate. Because forests are complex ecosystems, and because uses of the forests are so varied, there is no set formula which can be prescribed for all forests. Future forest management will undoubtedly contain elements of all three options to address the problems arising from climate change.

Because of the uncertainties in the current prediction of impact of climate change on America's forests, we will need to continue careful monitoring and surveillance of our forest ecosystems, particularly those components which are highly sensitive to greenhouse effect in order to refine management strategies.

Also, because our current capability to predict impacts is imprecise, we must continue to carry out research on the effects of multiple stresses on our forests in order to assure the health and productivity in a changing atmospheric environment.

References

Abramopoulos F, Rosenzweig C, Choudhury B (1988) Improved ground hydrology calculations for global climate Models (GCMs): Soil water movement and evapotranspiration. Climate 1:921–941

Barnola J-M, Raynaud D, Korotkevich YS, Lorius C (1987) Vostok ice core provides 160,000 year record of atmospheric CO_2. Nature (Lond) 329:408–414

Becker M (1987) Quality and utilization of wood from damaged trees. In: Impact of air pollution damage to forests for round supply and forest products market. United Nations, New York, pp 1–8

Bernabo JC (1981) Quantitative estimates of temperature changes over the last 2700 years in Michigan based on pollen data. Quat Res 15:143–159

Bernabo JC, Webb T III (1977) Changing patterns in the holocene pollen record of Northeastern North America. A mapped summary. Quat Res 8:64–96

Botkin DB, Janak JF, Wallis JR (1972) Some ecological consequences of a computer model of forest growth. J Ecol 60:849–872

Botkin DB, Nisbet RA, Keynales TE (1988) Effects of climate change on forest of the Great Lakes. Final report to EPA Univ California, Santa Barbara, 40 pp

Byran K (1988) Climate response to greenhouse warming: the role of the oceans. In: Berger A, Schneider S, Duplessy CL (eds) Climate and geo-sciences: a challenge for science and society in the 21st century. NATO ASI Series C. Mathematical and Physical Sciences, vol 285. Kluwer, Dordrecht, pp 435–446

Davis MB (1981) Quaternary history and the stability of forest communities. In: West DC, Shugart HH, Botkin DB (eds) Forest succession: Concepts and application. Springer, Berlin Heidelberg New York Tokyo, pp 132–153

Deeming JE, Lancaster JW, Fosberg MA, Furman RW, Schroeder MJ (1972) National fire danger rating system. USDA For Serv Pap RM-84, 165 pp

Dickenson RG (1982) Modeling climate changes due to carbon dioxide increases. In: Clark WC (ed) Carbon dioxide review. Clarendon, Oxford, pp 100–135

Fifield R (1988) Frozen assets of the ice cores. New Sci 118:28–29

Flenley JR (1979) The equatorial rain forest: a geological history. Butterworths, London

Fosberg MA, Furmann III RW (1973) Fire climates in the southwest. Agric Meteor 12:27–34

Fosberg MA (1971) Climatological influences on moisture characteristics of dead fuels: Theoretical analysis. For Sci 17:64–72

Fosberg MA (1988) Forest health and productivity in a changing atmospheric environment. In: Berger A, Schneider S, Duplessy CL (eds) Climate and geo-sciences: a challenge for science and society in the 21st century. NATO ASI Series C. Mathematical and Physical Sciences, vol 285. Kluwer, Dordrecht, pp 681–688

Fosberg MA (1989) Climate change and forest fires. In: Topping JC (ed) Coping with climate change. Proc Second North American Conference on Preparing for Climate Change. Climate Institute, Washington, pp 292–296

Fried JS, Torn MS (1988) The altered climate fire model simulating the effects of climate change on the effectiveness of a wildland fire initial attack program. (in preparation, Univ California, Berkeley)

Friedli J, Lotscher H, Deschaer H, Siegenthaler U, Stauffer B (1986) Ice record of the $^{13}C/^{14}C$ ratio of atmospherice CO_2 in the past two centuries. Nature (Lond) 324:237–238

Goldammer JG (1990) Waldumwandlung und Waldverbrennung in den Tiefland-Regenwäldern des Amazonasbeckens: Ursachen und ökologische Implikationen. In: Hoppe A (ed) Der tropische Regenwald Südamerikas. Versuch einer interdisziplinären Annäherung. Ber Naturf Ges 79, Freiburg, pp 119–142

Goldammer JG (1991) Feuer in Waldökosystemen der Tropen und Subtropen. Birkhäuser Basel-Boston (in preparation)

Haltiner GJ, Martin FL (1957) Dynamical and physical meteorology. McGraw-Hill, New York

Hammen T van der (1974) The Pleistocene changes of vegetation and climate in tropical South America. J Biogeogr 1:3–26

Hansen J, Fung I, Lacis A, Lebedoff S, Rind D, Reudy R, Russell G, Stone P (1987) Prediction of near term climate evolution: What can we tell decision makers now? In: Proc Preparing for climate change. Gov Inst Inc, Washington, pp 35–47

Hepting GH (1971) Diseases of forest and shade trees of the United States. USDA For Serv Handbook 386

Herring JR (1985) Charcoal fluxes into sediments of the North Pacific Ocean: The Cenozoic record of burning. In: The carbon cycle and atmospheric CO_2: Natural variations Archean to Present. Geophys Monogr 32:419–442 Am Geophys Union, Washington

Holdridge LR (1964) Life zone ecology. Tropic Sci Cent, San José, Costa Rica

Hope GS (1976) The vegetational history of Mt Wilhelm, Papua New Guinea. J Ecol 64:627–663

Keeling C (1984) Atmospheric CO_2 concentration at Manua Loa Observatory Hawaii, 1958–1983. US DOE report NDP-011, Washington DC

Kellogg WW, Zong-ci Zhao (1988) Sensitivity of soil moisture to doubling of carbon dioxide in climate model experiment of climate. Part 1: North America. J Climate 1:348–366

Karl TR, Jones PD (1989) Urban bias in area averaged surface temperature trends. Bull Am Meteor Soc 70:265–270

King GA, DeVelice RL, Neilson RP, Worrest RC (1988) California. In: Smith JB, Tirpak DA (eds) The potential effects of global climate change on the United States, Vol 1, Chapter 4. US Environ Protect Agency, Washington

Leverenz JW, Lev DJ (1987) Effects of carbon dioxide-induced climate changes on the natural ranges of six major commercial tree species in the Western United States. In: Shands WE, Hoffmann JS (eds) The greenhouse effect, climate change, and U.S. forests. The Conservation Foundation, Washington, pp 123–155

McCracken F, Schipper AL, Widin KD (1984) Observation on occurrence of cottonwood leaf rust in the Central United States. Eur J For Path 14:226–233

Miller WF, Dougherty PM, Switzer GL (1987) Effects of rising carbon dioxide and potential climate change on Loblolly pine distribution, growth, survival and productivity. In: Shands WE, Hoffmann JS (eds) The greenhouse effect, climate change, and US forests. The Conservation Foundation, Washington, pp 157–189

Mintzer L (1987) Energy policy and the greenhouse problem: A challenge to sustainable development. In: Proc First North American Conf on preparing for climate change. Govt Inst Inc, Washington, pp 18–34

Mitchell JFB (1988) Climate sensitivity: Model dependence of results. In: Berger A, Schneider S, Duplessy CL (eds) Climate and geo-sciences: a challenge for science and society in the 21st century. NATO ASI Series C. Mathematical and Physical Sciences, vol 285. Kluwer, Dordrecht, pp 417–434

Orville RE, Henderson RW (1986) Global distribution of midnight lightning: September 1977 to August 1978. Mon Weather Rev 114:2640–2653

Peters RL (1987) Effects of global warming on biodiversity: An overview. In: Proc First North American Conf on Preparing for Climate Change. Govt Inst Inc, Washington, pp 169–185

Ramanathan V, Cess RD, Harrison EF, Minnis P, Barkstrom BR, Ahmad E, Hartmann D (1989) Cloud-radiative forcing and climate: Results from Earth Radiation Budget Experiment. Science 243:57–63

Rothermel RC (1972) A mathematical model for predicting fire spread in wildland fuels. USDA For Serv Res Pap INT-115, Ogden, Utah

Schlesinger MG (1988) Model projections of the climate changes induced by increased atmospheric CO_2. In: Berger A, Schneider S, Duplessy CL (eds) Climate and geo-sciences: a challenge for science and society in the 21st century. NATO ASI Series C. Mathematical and Physical Sciences, vol 285. Kluwer, Dordrecht, pp 375–416

Schneider SH (1988) The greenhouse effect: What we can or should do about it. In: Proc First North American Conf on Preparing for Climate Change. Govt Inst Inc, Washington, pp 18–34

Smith JB (1988) Great Lakes. In: Smith JB, Tirpek DA (eds) The potential effects of global climate change in the United States. Draft EPA report to Congress, Chapter 5

Solomon AM, West DC (1987) Simulating forest ecosystem Responses to expected climate change in Eastern North America: Applications to decision making in the forest industry. In: Shands WE, Hoffmann JS (eds) The greenhouse effect, climate change, and US forests. The Conservation Foundation, Washington, pp 189–218

Stahle DW, Cleveland MK, Hehr JG (1988) North Carolina climate changes reconstructed from tree rings: AD 372 to 1985. Science 240:1517–1519

Titus JG (1988) Southeast. In: Smith JB, Tirpek DA (eds) The potential effects of global climate change in the United States. Draft EPA report to Congress Chapter 6

Thoning KW, Tans PP, Komhyr WD (1989) Atmospheric carbon dioxide at Mauna Loa Observatory. 2. Analysis of the NOAA GMCC Data, 1974–1985. J Geophys Res 94:8549–8565

Turman BN, Edgar BC (1982) Global lightning distributions at dawn and dusk. J Geophys Res 87:1191–1206

Williams ER (1985) Large-scale charge separation in thunderclouds. J Geophys Res 90:6013–6025

Winget CH (1988) Forest management strategies to address climate change. In: Proc First North American Conf on Preparing for Climate Change. Govt Inst Inc, Washington, pp 328–333

Winjum JK, Neilson RP (1988) In: Smith JB, Tirpek DA (eds) The potential effects of global climate change in the United States, vol 2. Draft EPA report to Congress, Chapter 8

Appendix: The Freiburg Declaration on Tropical Fires

The Freiburg Declaration on Tropical Fires was released by the participants of the 3rd International Symposium on Fire Ecology held at Freiburg University, Federal Republic of Germany, 16–20 May 1989.

The Role of Fires in Tropical Ecosystems

Fires in the forest and other vegetation of the tropics and subtropics and the changing tropical land use have increasing regional and global impact on the environment. The smoke plumes from tropical biomass fires carry vast amounts of atmospheric pollutants, including CO_2, CO, NO_x, N_2O, CH_4, nonmethane hydrocarbons, and aerosols. Smog-like photochemistry produces ozone concentrations comparable to those found in the industrialized regions. These perturbations of the tropical atmosphere are on such a scale that they can be easily detected by remote sensing from space. Alterations of the hydrological regimes can have severe environmental and human consequences for the regions being burned and in neighboring regions. The consequences of biomass burning, such as the aggravation of the greenhouse effect, affect nontropical regions most strongly. The catastrophic fires on the island of Borneo in 1982/83 indicate the danger that possible climatic changes pose to the survival of the tropical forests themselves.

On the other hand, fires play a central role in the maintenance of many natural ecosystems and in the practice of agriculture and pastoralism. The various types of savannas are burned frequently both by human- and non-human-caused fires. Burning is used as a tool in maintaining tree plantations and natural forests, especially in the subtropics. Forests in the moist tropics have long been used in shifting cultivation to support low population densities of traditional agriculturalists without degrading either the forest or the productive potential of the soil. This situation has changed radically by accelerating shifting cultivation cycles under the influence of market economies and because of increasing population pressure, both from demographic growth and from reduced access to land. Nonsustainable slash-and-burn pioneer agriculture, without the long fallows of traditional systems, is practised by populations that are either attracted to or forced to migrate to tropical forest areas, or that are transported to these regions under government colonization or transmigration programs. Both shifting cultivation and pioneer farmers depend on burning to produce crops at acceptable labor input intensities. Burning is also the key process in maintaining the cattle pastures that are replacing tropical forest in vast areas of tropical Latin America.

In the enormous areas of savannas – especially in Africa – where burning is a part of the natural cycle, the frequency of fires has greatly increased, and with this the impact of uncontrolled fires is more and more detrimental. The dual role of fire must be recognized, being both a natural agent of ecosystem maintenance and a potentially disastrous cause of ecosystem destruction.

Where Do We Stand?

Fire control has been the traditional fire policy in many parts of the world. An increasing number of countries have adopted fire management policies instead, in order to maintain the function of fire in removing the accumulation of fuel loads that would otherwise lead to damaging wildfires, and in order to arrest succession at stages that are more productive to humans than are forests that would predominate in the absence of fire. Frequently, inappropriate choices are made – often because decisions are influenced by other regions where conditions differ. Such influence may come through misguided international aid programs, through visiting consultants and researchers, or through the temperate-zone bias of local technical staff trained abroad. Researchers and policy-makers must be sensitive to the different functions of fire in each ecosystem and to the needs of the people who depend on it.

When current burning practices are correctly identified as damaging, as in the case of the recent explosion of deforestation and burning in lowland Amazonia, the measures taken are often ineffective. Prohibiting burning, and attempting to enforce this through inspection and punishment, is bound to fail. The motives for burning must be removed, such as land speculation, tax or other incentives, and land documentation criteria that reward deforestation. Migration of farmers to infertile rain forest areas must not be facilitated by highway construction and must not be augmented by policies that expel populations from other regions through land tenure concentration and through lack of employment alternatives. Sound policies to bring the use of fire under rational control must be based on an accurate understanding of why burning is done, what its costs and benefits are, and who enjoys the benefits and suffers the impacts of present and alternative practices.

An Action Plan

Both more research and immediate action are needed. Education must begin now to bring about long-term changes in attitudes towards fire and nature. Global monitoring systems must be expanded and coordinated. For example, the rain forests of the Congo Basin have so far been almost untouched by fire, but must be watched because the situation could change rapidly, as it has in other tropical areas. Temperate zone countries can contribute greatly to research efforts through financial contributions and by participating in intellectual exchange with tropical countries. The International Geosphere-Biosphere Programme (IGBP) offers a promising channel for international cooperation in fire research and the Intergovernmental Panel on Climate Change (IPCC), under the auspices of the United Nations Environmental Program (UNEP), will provide response strategies to these environmental threats. It is essential, however, that the IGBP focus its resources on the large ecosystems that play major roles in global geochemical processes. Tropical rain forest, for example, must be understood in the Brazilian Amazon and in Equatorial Africa rather than being studied primarily in isolated remnants of forest in Puerto Rico, Panama, Costa Rica, or Hawaii.

Without waiting for further results, much could be done to translate what we already know into action. These actions include reforming the policies of international lending institutions and development assistance programs to give greater consideration to the environmental impacts of policies that either provoke or eliminate fires. Recent increase in the environmental review capabilities of the World Bank is a hopeful sign, but it is only a tiny beginning.

Institutional mechanisms must be developed to distribute fairly – both within and between nations – the costs and benefits of changes in fire policy. The questions of "fire for whom?" and "fire control for whom?" must be answered clearly if sound and fair policies are to be formulated. Policies must respect national sovereignties. Fortunately, the interests of different nations almost always point in the same direction: limiting deforestation is not only in the long-term interest of the people of the tropical countries where forests are being cleared, but is also beneficial to other nations concerned by the loss of biodiversity and by the danger of atmospheric impacts in temperate latitudes.

Subject Index

abiotic effects 88
Aboriginal burning 159, 163
Aboriginal fire regimes 163
Acacia saligna 185
Acre 391
adaptations 132
adaptive traits 46
aerosol 382
Africa 358
  Central Africa 362
  Congo Basin 362, 365
  Ivory Coast 360
  West Africa 359
  West Africa fire belt 362
agricultural residues 68
air temperature 88
Albizzia procera 54
alcohols 420
aldehydes 420
alien shrubs 185
alien woody weeds 190
allelopathy 7
alpha-pinene 58, 59, 424
Altamira (Pará) 110
Amazon Basin 106, 117, 351, 371
Amazonas 106
ancient fires 106
Andropogon leucostachyrus 154
Andropogon virginicus 8
anthropogenic activities 117
anthropogenic fire 291
anthropogenic fire climax forest 6
anthropogenic savannas 6
Antidesma frutescens 54
Araguia river 354
Araucaria angustifolia 69
Araucaria araucana 75
arid savannas 198
Aripo Savannas 154
Arundinaria sp. 34, 35
ash content 92
atmospheric response 466
Atta spp. 93
Australia 159, 296, 330
autotoxicity 7
AVHRR 107, 339, 400, 437

Babaçu Palm Forest 66
backfires 36
backing fire 238, 422
Bahamas 217
baked mudstone 14
Balikpapan 19
bark properties 126
bark thickness 126
Belém 391
Belize 135, 139, 217
benzene 424
biodiversity 2, 4, 9
biomass 120
  burning 371, 418
  consumption 125
  load 6
biotic effects 96
Bobok 48
Borassodendron borneensis 26
Bornean ironwood 22
Borneo 12, 13, 356
Brachystegia-Julbernardia woodlands 34
Brazil 217
  Acre 391
  Altamira (Pará) 110
  Amazon Basin 106, 117, 351, 371
  Amazonas 106
  Belém 391
  Goiás 107
  Legal Amazon 107, 352
  Maranhão 107, 391
  Mato Grosso 88, 354
  Pará 106, 354
  Paraná 69
  Porto Velho 391
  Rondônia 106, 352, 371, 391
  Transamazon Highway 111
  Xingu river 354
British India 323
Bukit Soeharto forest reserve 15
Burma 32, 51, 358

burning coal seams 14
burning pile 243
burning windrow 243
burning-off 332
burnt clay 14
Byrsonima crassifolia 152

Caatinga 67
Callitris forest 161
callus tissue 47
cambium 47
Cambodia 45
Campo Cerrado 83
Campo Limpo 83
Campo Sujo 83
Cape Province 202
Cappilipodium parviflorum 54
carbon 422
  dioxide 373, 419, 422, 440
  dioxide sinks 9
  flux 274
  monoxide 419, 422
Cathedral Peak State Forest 194
cattle trampling 48
Celebes 296
cellulose 420
center firing 243
Central America 135
  Bahamas 217
  Belize 135, 139, 217
  Costa Rica 135, 140, 217
  Guatemala 139
  Honduras 135, 139, 217
  Nicaragua 135, 140, 217
  Panama 135, 141, 217
Central Kalimantan 347
Cerradão 83
Cerrado 82
Chaco 74
Charcoal 12, 86, 106, 117, 473
  deposition 472
  flux 473
Chiang Mai 34, 358
China 45
chlorine 429
Chrysopogon accilculatus 54
climate change 1, 142, 463
climax 83
  climatic climax 83, 85
  fire climax 180, 218
  pedoclimax 83
  peinoclimax 83
  stage 63
cloud condensation nuclei 374
coal fires 15

combustion processes 421
Congo Basin 362, 365
controlled burning 329
coppice 50, 127
cortex 50
  formation 48
Costa Rica 135, 140, 217
crown scorch 236
Curatella americana 152
cyclohexane 424

Dacrydium elatum 21
damage
  cambium 252
  crown 254
  root 252
  secondary 256
debris burning 69, 224
DECAFE 437
decomposer activity 6
deforestation 2, 107, 115
degraded savannas 6
demography 165
Dendroctonus frontalis 257
depth of burn 235
Dicranopteris pectinata 148
Diospyros melanoxylon 35
dipterocarp forest 56
dopterocarp rain forest 11
Dipterocarpus spp.
  intricatus 39
  obtusifolius 39
  tuberculatus 35, 37, 39
dispersal mechanismus 130
diversfication 28
dormant buds 127
Drakensberg 194
dry forest 150

early burning 325
early fires 359
early seasonal burn 148
East Kalimantan 11, 12
ecological plasticity 60
ecological stress 33
ecosystem stresses 472
edge burning 242
El Niño drought 347
El Niño-Southern Oscillation 11, 13
El Salvador 135, 140
emission factor 426
emission rate 384
emissions 175, 418
  carbon dioxide 373, 419, 422, 440, 464
  carbon monoxide 419, 422, 433

hydrocarbons 420
methane 373, 433
methyl chloride 432
nonmethane hydrocarbons 433
sulfur 432

Enterolbium cyclocarpum 153
epicormic sprouting 126, 128
erosion 6, 43, 56, 59
Etosha National Park 210
Eucalypt forests 161, 168
Eucalypt woodlands 168
Eucalyptus spp. 77
E. tetrodonta 168
Eugenia cumini 8
Eulalia quadrinervis 53
Eulalia trispicata 53
Eusideroxylon zwageri 22, 25, 27
evolution 278
exotic plantations 217
experimental fires 125
expert systems 173

fauna 100
Fiji 217
fire
ancient fires 106
backfires 36
backing fire 238, 422
behavior 125, 236, 474
behavior prediction 233
early fires 359
fire climate 162
fire climax 46, 52, 77, 180, 218
fire conservancy 319
frequency 6, 147, 162, 185, 193, 483
hazard reduction 185
history 118, 320
intensity 54, 200
management 41, 101, 159, 179
management systems 41
monitoring 366
patterns 34
regimes 6, 192
resistance 96
responses 125
return interval 54
season 189, 195
selection 46
stand replacement fires 6

fireline intensity 234
flaming fires 379
flowering 99, 131
forest
closed fallow forests 445
closed nonfallow forests 445
coastal monsoon forests 170
forest demography 7
lowland tropical rain forest 11
monsoon forest 32, 166
monsoon vine forests 170
open fallow forests 450
open nonfallow forests 448
regeneration 23
subtropical forest 68
transition forest 363

forest fire danger index 162
forest floor dynamics 258
fragility 6
Freiburg Declaration 487
fuel 119
accumulation 50
appraisal 221
arrangement 119
available 223
biomass 119
chemistry 419
load 119
moisture 225

fuel stick moisture 124
Fynbos 179, 182

gases 418
GCM's 466
Geographical Information System 173, 400
Ghana 360
GIS 173, 400
global change 1, 337, 463
global monitoring 337
Global Vegetation Index 346
global warming 9
Goiás 107
grass layer 450
grass stage 38, 39, 48, 50
grassland 180
succession 152

grazing pressure 6, 57
greenhouse effect 114, 373, 463
growth 255
Guatemala 139
Guazuma ulmifolia 153
GVI 346

Hakea sericea 185, 190
Haldwani 42
headfires 36
heading fires 422
heat per unit area 235
heating process 58
hemicellulose 421
herb stage 4

herbivores 274
Heterobasidion annosum 256
hexane 424
historical fire regimes 47
Hluhluwe/Umfolozi Game Reserves 203
Holdrige Life Zone Concept 467
Holocene 12, 117, 463
hominid use of fire 288
Hominoidae 281
Honduras 135, 139, 217
human disturbances 63
human population density 6
humid savannas 450
hydrocarbons 420
Hyparrhenia rufa 4, 102, 151

IGBP 2, 489
ignition sources 162
Imperata brasiliensis 4, 113
Imperata cylindrica 4, 26, 53, 54, 55
industrial pine plantations 216
Intergovernmental Panel on Climate Change 489
International Geosphere-Biosphere Program 2, 489
Intertropical Convergence Zone 476
invasive plants 174
IPCC 489
Ips calligraphus 58
Ips interstitialis 58, 256
ipsdienol 59
Isostigma peucedanifolium 92
ITCZ 476
ITTO 26
Ivory Coast 360

Java 14

kaingin fires 52
Kalimantan 356
Kampuchea 33
Kedang Pahu river 19
Korat 34
Kota Bangun 16
Kruger National Park 207
Kutai district 19
Kutai National Park 12

LANDSAT 109, 164, 378
landscape mosaic 6
landscape stability 59
Lantana spp. 174
Lantana montevidensis 92
Laos 33, 358
large-scale fires 121
Legal Amazon 107, 362
Leptocorphium lanatum 154
let burn 59
light burning 329
lightning 476
  fires 50, 199
  frequency 476
  scars 35
  strikes 106
lignin 420
litter fires 36
litter loading 258
logged-over forest 24
Long Sungai Barang, Bulungan 16
lowland deciduous forest 32
Luzón 46, 52

Macaranga spp. 25, 26
Madagascar 301
Mahakam 19, 356
Makassar Strait 12
Malesia 12
mammals 280
Manipur State 33, 358
Maranhão 107, 391
Mato Grosso 88, 354
Mauna Loa 464
megaherbivores 287
Melaleuca spp. 160
Melientha suavis 35
Melinis minutiflora 4, 102
meristematic tissues 126
methane 373, 433
microclimate 121
microclimate feedback 483
milpa 143
Mimosa priga 153
Mimosa scabrella 72
Miscanthus sinensis 53
modeling 466
moisture content 124
moisture of extinction 124
monoterpenes 58
monsoon forest 32, 166
mortality 255
mossy forest 52
Muara Kaman 19
Muara Lawa 19

Natal Drakensberg 192
natural fire climax forest 6
natural fires 275
NDVI 346
needle litter fall 50
New Guinea 12, 296

New Zealand 303
Nicaragua 135, 140, 217
NOAA 339, 375, 400, 437
non-wood forests products 6
nonforest 4
Normalized Difference Vegetation Index 346
normalized emission ratio 375
Nothofagus spp. 75
nutrients 143

Orbignya martiana 66
Orinoco basin 65
overgrazing 208
ozone 380, 437

Palmer Drought Index 471
palynological information 12
Panama 135, 141, 217
Pará 106, 354
Paramo 154
Paraná 69
particulate matter 418
particulates 383, 418, 423
Paspalum pulchellum 154
pasture burns 112
pasture improvement 68
pastures 107
patch dynamics 3
peat swamp forest 24
pedoclimax 85
Pennisetum polystachyon 5
pest management 219
Philippines 46, 296
phyto-production 275
phytomass production 90
Pilanesberg National Park 205
Piliostigma malabricum 54
pine-dipterocarp savannas 38
pine-grassland biomes 45
pine savannas 36
Pinus spp. 45, 478
  caribaea 5, 146, 216
  contorta 479
  echinata 50
  elliottii 216
  kesiya 38, 42, 47
  merkusii 35, 38, 47
  oocarpa 50, 146
  patula 194, 216
  pinaster 216
  ponderosa 479
  pseudostrobus 150
  radiata 77, 216
  rigida 50
  roxburghii 47
  taeda 50, 216, 479
  wallichiana 47
Pithecellobium saman 153
Pleistocene 1, 12, 106
Plio-Pleistocene 287
point source (grid) ignition 240
population pressure 1
Porto Velho 391
post-fire landscapes 350
potassium 429
precipitation 232
  regimes 482
prehistory 273
prescribed burning 42, 184, 196, 216
  effects 252
  objectives 218
  plans 244
  techniques 237
prescribed fire 216
prescribed grazing 42
primary forest 23
propagules 130
propane 424
protection 319
Pseudotsuga menziesii 479
Pteridium aquilinum 53, 148
Pterocarpus macrocarpus 40
pulsation phenomenon 424
pyroclimax 85
pyrolysis 421
pyrolyzed fuel 424
pyrophytic 96
  communities 482
  grass life form 2
  life forms 11

Quebracho 74

radiometer 403
radiometric information 12
reaction intensity 234
refuge theory 28
relative humidity 230
remote sensing 337, 371
residence time 234
Rhizina undulata 256
ring firing 243
Rondônia 106, 352, 371, 391
Roraima 113
runoff 59

(S)-cis-verbenol 59
Sabah 12, 16, 19
SAGE 380
Samarinda 16

Sangkulirang 12
Sarawak 16
satellite monitoring 338
savannas 135, 198
  arid 198
  Aripo 154
  degraded 6
savannization 5
scleromorphism 90
sclerophyllous heathlands 179
Sebangau river 12
secondary forest 24
sediment yield 56
seed banks 129
seeds 100, 145
  dispersion 100
  germination 100
  longivity 129
Senegal 400
Sesbiana emerus 153
shifting agriculture 139
shifting cultivation fallows 112
Shorea spp. 22
  lamellata 27
  obtusa 39
  robusta 33, 40, 41, 47
  siamensis 40
shrub stage 4
Sierra Leone 360
silvopastoral land use 219
Sisaket 38
slash-and-burn 51
slope 233
smoke
  analysis 387
  particles 425
  production 422
  tracking 394
smoldering 379
soil 7, 261
  chemical effects 262
  denudation 59
  fauna 260
  moisture 232
  physical effects 261
  sterilization 152
  temperature 88
  water-repellent 262
Sook Plain 19
Sorghum spp. 159
  intrans 168
South Asia 13
  Haldwani 42
  Manipur State 33, 358
  Uttar Pradesh 42
South East Asia 356
  Balikpapan 19
  Borneo 12, 13, 356
  Bukit Soeharto forest reserve 15
  Cambodia 45
  Celebes 296
  Central Kalimantan 347
  Chiang Mai 34, 358
  East Kalimantan 11, 12
  Java 14
  Kalimantan 356
  Kampuchea 33
  Kedang Pahu river 19
  Korat 34
  Kota Bangun 16
  Kutai district 19
  Kutai National Park 12
  Laos 33, 358
  Long Sungai Barang, Bulungan 16
  Luzón 46, 52
  Mahakam 19, 356
  Makassar Strait 12
  Malesia 12
  Muara Kaman 19
  Muara Lawa 19
  Philippines 46, 296
  Sabah 12, 16, 19
  Samarinda 16
  Sangkulirang 12
  Sarawak 16
  Sebangau river 12
  Sisaket 38
  Sook Plain 19
  South Kalimantan 12
  Sulawesi 296
  Sumatra 14, 45
  Sunda shelf 12, 18
  Ubon Ratchathani 38
  Viet Nam 33, 45
South Kalimantan 12
Southern Africa 179
  Cathedral Peak State Forest 194
  Cape Province 202
  Drakensberg 194
  Etosha National Park 210
  Hluhluwe/Umfolozi Game Reserves 203
  Kruger National Park 207
  Natal Drakensberg 192
  New Guinea 12, 296
  Pilanesberg National Park 205
Spain 217
species diversity 6
Spondia mombin 153
SPOT 339, 378
spotting 36

sprouting 50
stability 6, 98
stand replacement fires 6
stem tissues 127
steppe 67
steppic savanna 74
Stratospheric Aerosol and Gas Experiment 380
strip-heading fire 240
subsaharian savanna 359
subterranean organs 128
succession 147
  pattern 3
Sulawesi 296
sulfur 429
Sumatra 14, 45
Sunda shelf 12, 28
surface runoff 55
swamp 24
swidden agriculture 52
Sylia kerrii 39
Syzygium cumini 54

Tasmania 296
Tectona grandis 33, 40, 43
temperature 231
Terai forest 40
Terminalia alata 40
Thailand 32, 36, 45, 51
Themeda australis 167
Themeda triandra 53, 54, 55, 194
thermal gravimetric analysis 420
thermoluminescence analysis 14
Total Ozone Mapping Spectrometer (TOMS) 380
trace gases 373, 440
trade wind forest 65
Transamazon Highway 111
trauma 463, 467
troposhere 380

Ubon Ratchathani 38
underburning 224
understory vegetation 257
United Nations Environmental Program (UNEP) 489
Uttar Pradesh 42

vegetation adaptations 125
vegetative sprouting 128
Venezuela 217
Viet Nam 33, 45

wet-dry tropics 159
wildfire hazard reduction 218
wildfire severity 463
wildland fire 319
wildland/residential interface 51
wildlife 100, 273
wind 228

xerophytic communities 482
Xingu river 354
xylem tissues 128

Yellowstone fires 418